Lecture Notes in Computer Science   14555

The series Lecture Notes in Computer Science (LNCS), including its subseries Lecture Notes in Artificial Intelligence (LNAI) and Lecture Notes in Bioinformatics (LNBI), has established itself as a medium for the publication of new developments in computer science and information technology research, teaching, and education.

LNCS enjoys close cooperation with the computer science R & D community, the series counts many renowned academics among its volume editors and paper authors, and collaborates with prestigious societies. Its mission is to serve this international community by providing an invaluable service, mainly focused on the publication of conference and workshop proceedings and postproceedings. LNCS commenced publication in 1973.

Stevan Rudinac · Alan Hanjalic · Cynthia Liem ·
Marcel Worring · Björn Þór Jónsson · Bei Liu ·
Yoko Yamakata
Editors

MultiMedia Modeling

30th International Conference, MMM 2024
Amsterdam, The Netherlands, January 29 – February 2, 2024
Proceedings, Part II

 Springer

Editors

Stevan Rudinac 🆔
University of Amsterdam
Amsterdam, The Netherlands

Alan Hanjalic 🆔
Delft University of Technology
Delft, The Netherlands

Cynthia Liem 🆔
Delft University of Technology
Delft, The Netherlands

Marcel Worring 🆔
University of Amsterdam
Amsterdam, The Netherlands

Björn Þór Jónsson 🆔
Reykjavik University
Reykjavik, Iceland

Bei Liu 🆔
Microsoft Research Lab – Asia
Beijing, China

Yoko Yamakata 🆔
The University of Tokyo
Tokyo, Japan

ISSN 0302-9743   ISSN 1611-3349 (electronic)
Lecture Notes in Computer Science
ISBN 978-3-031-53307-5   ISBN 978-3-031-53308-2 (eBook)
https://doi.org/10.1007/978-3-031-53308-2

Preface

These four proceedings volumes contain the papers presented at MMM 2024, the International Conference on Multimedia Modeling. This 30th anniversary edition of the conference was held in Amsterdam, The Netherlands, from 29 January to 2 February 2024. The event showcased recent research developments in a broad spectrum of topics related to multimedia modelling, particularly: audio, image, video processing, coding and compression, multimodal analysis for retrieval applications, and multimedia fusion methods.

We received 297 regular, special session, Brave New Ideas, demonstration and Video Browser Showdown paper submissions. Out of 238 submitted regular papers, 27 were selected for oral and 86 for poster presentation through a double-blind review process in which, on average, each paper was judged by at least three program committee members and reviewers. In addition, the conference featured 23 special session papers, 2 Brave New Ideas and 8 demonstrations. The following four special sessions were part of the MMM 2024 program:

- FMM: Special Session on Foundation Models for Multimedia
- MDRE: Special Session on Multimedia Datasets for Repeatable Experimentation
- ICDAR: Special Session on Intelligent Cross-Data Analysis and Retrieval
- XR-MACCI: Special Session on eXtended Reality and Multimedia - Advancing Content Creation and Interaction

The program further included four inspiring keynote talks by Anna Vilanova from the Eindhoven University of Technology, Cees Snoek from the University of Amsterdam, Fleur Zeldenrust from the Radboud University and Ioannis Kompatsiaris from CERTH-ITI.

In addition, the annual MediaEval workshop was organised in conjunction with the conference. The attractive and high-quality program was completed by the Video Browser Showdown, an annual live video retrieval competition, in which 13 teams participated.

We would like to thank the members of the organizing committee, special session and VBS organisers, steering and technical program committee members, reviewers, keynote speakers and authors for making MMM 2024 a success.

December 2023

Stevan Rudinac
Alan Hanjalic
Cynthia Liem
Marcel Worring
Björn Þór Jónsson
Bei Liu
Yoko Yamakata

Organization

General Chairs

Stevan Rudinac   University of Amsterdam, The Netherlands
Alan Hanjalic   Delft University of Technology, The Netherlands
Cynthia Liem   Delft University of Technology, The Netherlands
Marcel Worring   University of Amsterdam, The Netherlands

Technical Program Chairs

Björn Þór Jónsson   Reykjavik University, Iceland
Bei Liu   Microsoft Research, China
Yoko Yamakata   University of Tokyo, Japan

Community Direction Chairs

Lucia Vadicamo   ISTI-CNR, Italy
Ichiro Ide   Nagoya University, Japan
Vasileios Mezaris   Information Technologies Institute, Greece

Demo Chairs

Liting Zhou   Dublin City University, Ireland
Binh Nguyen   University of Science, Vietnam National
     University Ho Chi Minh City, Vietnam

Web Chairs

Nanne van Noord   University of Amsterdam, The Netherlands
Yen-Chia Hsu   University of Amsterdam, The Netherlands

Video Browser Showdown Organization Committee

Klaus Schoeffmann	Klagenfurt University, Austria
Werner Bailer	Joanneum Research, Austria
Jakub Lokoc	Charles University in Prague, Czech Republic
Cathal Gurrin	Dublin City University, Ireland
Luca Rossetto	University of Zurich, Switzerland

MediaEval Liaison

Martha Larson	Radboud University, The Netherlands

MMM Conference Liaison

Cathal Gurrin	Dublin City University, Ireland

Local Arrangements

Emily Gale	University of Amsterdam, The Netherlands

Steering Committee

Phoebe Chen	La Trobe University, Australia
Tat-Seng Chua	National University of Singapore, Singapore
Kiyoharu Aizawa	University of Tokyo, Japan
Cathal Gurrin	Dublin City University, Ireland
Benoit Huet	Eurecom, France
Klaus Schoeffmann	Klagenfurt University, Austria
Richang Hong	Hefei University of Technology, China
Björn Þór Jónsson	Reykjavik University, Iceland
Guo-Jun Qi	University of Central Florida, USA
Wen-Huang Cheng	National Chiao Tung University, Taiwan
Peng Cui	Tsinghua University, China
Duc-Tien Dang-Nguyen	University of Bergen, Norway

Special Session Organizers

FMM: Special Session on Foundation Models for Multimedia

Xirong Li	Renmin University of China, China
Zhineng Chen	Fudan University, China
Xing Xu	University of Electronic Science and Technology of China, China
Symeon (Akis) Papadopoulos	Centre for Research and Technology Hellas, Greece
Jing Liu	Chinese Academy of Sciences, China

MDRE: Special Session on Multimedia Datasets for Repeatable Experimentation

Klaus Schöffmann	Klagenfurt University, Austria
Björn Þór Jónsson	Reykjavik University, Iceland
Cathal Gurrin	Dublin City University, Ireland
Duc-Tien Dang-Nguyen	University of Bergen, Norway
Liting Zhou	Dublin City University, Ireland

ICDAR: Special Session on Intelligent Cross-Data Analysis and Retrieval

Minh-Son Dao	National Institute of Information and Communications Technology, Japan
Michael Alexander Riegler	Simula Metropolitan Center for Digital Engineering, Norway
Duc-Tien Dang-Nguyen	University of Bergen, Norway
Binh Nguyen	University of Science, Vietnam National University Ho Chi Minh City, Vietnam

XR-MACCI: Special Session on eXtended Reality and Multimedia - Advancing Content Creation and Interaction

Claudio Gennaro	Information Science and Technologies Institute, National Research Council, Italy
Sotiris Diplaris	Information Technologies Institute, Centre for Research and Technology Hellas, Greece
Stefanos Vrochidis	Information Technologies Institute, Centre for Research and Technology Hellas, Greece
Heiko Schuldt	University of Basel, Switzerland
Werner Bailer	Joanneum Research, Austria

Program Committee

Alan Smeaton	Dublin City University, Ireland
Anh-Khoa Tran	National Institute of Information and Communications Technology, Japan
Chih-Wei Lin	Fujian Agriculture and Forestry University, China
Chutisant Kerdvibulvech	National Institute of Development Administration, Thailand
Cong-Thang Truong	Aizu University, Japan
Fan Zhang	Macau University of Science and Technology/Communication University of Zhejiang, China
Hilmil Pradana	Sepuluh Nopember Institute of Technology, Indonesia
Huy Quang Ung	KDDI Research, Inc., Japan
Jakub Lokoc	Charles University, Czech Republic
Jiyi Li	University of Yamanashi, Japan
Koichi Shinoda	Tokyo Institute of Technology, Japan
Konstantinos Ioannidis	Centre for Research & Technology Hellas/Information Technologies Institute, Greece
Kyoung-Sook Kim	National Institute of Advanced Industrial Science and Technology, Japan
Ladislav Peska	Charles University, Czech Republic
Li Yu	Huazhong University of Science and Technology, China
Linlin Shen	Shenzhen University, China
Luca Rossetto	University of Zurich, Switzerland
Maarten Michiel Sukel	University of Amsterdam, The Netherlands
Martin Winter	Joanneum Research, Austria
Naoko Nitta	Mukogawa Women's University, Japan
Naye Ji	Communication University of Zhejiang, China
Nhat-Minh Pham-Quang	Aimesoft JSC, Vietnam
Pierre-Etienne Martin	Max Planck Institute for Evolutionary Anthropology, Germany
Shaodong Li	Guangxi University, China
Sheng Li	National Institute of Information and Communications Technology, Japan
Stefanie Onsori-Wechtitsch	Joanneum Research, Austria
Takayuki Nakatsuka	National Institute of Advanced Industrial Science and Technology, Japan
Tao Peng	UT Southwestern Medical Center, USA

Thitirat Siriborvornratanakul	National Institute of Development Administration, Thailand
Vajira Thambawita	SimulaMet, Norway
Wei-Ta Chu	National Cheng Kung University, Taiwan
Wenbin Gan	National Institute of Information and Communications Technology, Japan
Xiangling Ding	Hunan University of Science and Technology, China
Xiao Luo	University of California, Los Angeles, USA
Xiaoshan Yang	Institute of Automation, Chinese Academy of Sciences, China
Xiaozhou Ye	AsiaInfo, China
Xu Wang	Shanghai Institute of Microsystem and Information Technology, China
Yasutomo Kawanishi	RIKEN, Japan
Yijia Zhang	Dalian Maritime University, China
Yuan Lin	Kristiania University College, Norway
Zhenhua Yu	Ningxia University, China
Weifeng Liu	China University of Petroleum, China

Additional Reviewers

Alberto Valese
Alexander Shvets
Ali Abdari
Bei Liu
Ben Liang
Benno Weck
Bo Wang
Bowen Wang
Carlo Bretti
Carlos Cancino-Chacón
Chen-Hsiu Huang
Chengjie Bai
Chenlin Zhao
Chenyang Lyu
Chi-Yu Chen
Chinmaya Laxmikant Kaundanya
Christos Koutlis
Chunyin Sheng
Dennis Hoppe
Dexu Yao
Die Yu

Dimitris Karageorgiou
Dong Zhang
Duy Dong Le
Evlampios Apostolidis
Fahong Wang
Fang Yang
Fanran Sun
Fazhi He
Feng Chen
Fengfa Li
Florian Spiess
Fuyang Yu
Gang Yang
Gopi Krishna Erabati
Graham Healy
Guangjie Yang
Guangrui Liu
Guangyu Gao
Guanming Liu
Guohua Lv
Guowei Wang

Gylfi Þór Guðmundsson
Hai Yang Zhang
Hannes Fassold
Hao Li
Hao-Yuan Ma
Haochen He
Haotian Wu
Haoyang Ma
Haozheng Zhang
Herng-Hua Chang
Honglei Zhang
Honglei Zheng
Hu Lu
Hua Chen
Hua Li Du
Huang Lipeng
Huanyu Mei
Huishan Yang
Ilias Koulalis
Ioannis Paraskevopoulos
Ioannis Sarridis
Javier Huertas-Tato
Jiacheng Zhang
Jiahuan Wang
Jianbo Xiong
Jiancheng Huang
Jiang Deng
Jiaqi Qiu
Jiashuang Zhou
Jiaxin Bai
Jiaxin Li
Jiayu Bao
Jie Lei
Jing Zhang
Jingjing Xie
Jixuan Hong
Jun Li
Jun Sang
Jun Wu
Jun-Cheng Chen
Juntao Huang
Junzhou Chen
Kai Wang
Kai-Uwe Barthel
Kang Yi

Kangkang Feng
Katashi Nagao
Kedi Qiu
Kha-Luan Pham
Khawla Ben Salah
Konstantin Schall
Konstantinos Apostolidis
Konstantinos Triaridis
Kun Zhang
Lantao Wang
Lei Wang
Li Yan
Liang Zhu
Ling Shengrong
Ling Xiao
Linyi Qian
Linzi Xing
Liting Zhou
Liu Junpeng
Liyun Xu
Loris Sauter
Lu Zhang
Luca Ciampi
Luca Rossetto
Luotao Zhang
Ly-Duyen Tran
Mario Taschwer
Marta Micheli
Masatoshi Hamanaka
Meiling Ning
Meng Jie Zhang
Meng Lin
Mengying Xu
Minh-Van Nguyen
Muyuan Liu
Naomi Ubina
Naushad Alam
Nicola Messina
Nima Yazdani
Omar Shahbaz Khan
Panagiotis Kasnesis
Pantid Chantangphol
Peide Zhu
Pingping Cai
Qian Cao

Qian Qiao
Qiang Chen
Qiulin Li
Qiuxian Li
Quoc-Huy Trinh
Rahel Arnold
Ralph Gasser
Ricardo Rios M. Do Carmo
Rim Afdhal
Ruichen Li
Ruilin Yao
Sahar Nasirihaghighi
Sanyi Zhang
Shahram Ghandeharizadeh
Shan Cao
Shaomin Xie
Shengbin Meng
Shengjia Zhang
Shihichi Ka
Shilong Yu
Shize Wang
Shuai Wang
Shuaiwei Wang
Shukai Liu
Shuo Wang
Shuxiang Song
Sizheng Guo
Song-Lu Chen
Songkang Dai
Songwei Pei
Stefanos Iordanis Papadopoulos
Stuart James
Su Chang Quan
Sze An Peter Tan
Takafumi Nakanishi
Tanya Koohpayeh Araghi
Tao Zhang
Theodor Clemens Wulff
Thu Nguyen
Tianxiang Zhao
Tianyou Chang
Tiaobo Ji
Ting Liu
Ting Peng
Tongwei Ma

Trung-Nghia Le
Ujjwal Sharma
Van-Tien Nguyen
Van-Tu Ninh
Vasilis Sitokonstantinou
Viet-Tham Huynh
Wang Sicheng
Wang Zhou
Wei Liu
Weilong Zhang
Wenjie Deng
Wenjie Wu
Wenjie Xing
Wenjun Gan
Wenlong Lu
Wenzhu Yang
Xi Xiao
Xiang Li
Xiangzheng Li
Xiaochen Yuan
Xiaohai Zhang
Xiaohui Liang
Xiaoming Mao
Xiaopei Hu
Xiaopeng Hu
Xiaoting Li
Xiaotong Bu
Xin Chen
Xin Dong
Xin Zhi
Xinyu Li
Xiran Zhang
Xitie Zhang
Xu Chen
Xuan-Nam Cao
Xueyang Qin
Xutong Cui
Xuyang Luo
Yan Gao
Yan Ke
Yanyan Jiao
Yao Zhang
Yaoqin Luo
Yehong Pan
Yi Jiang

Yi Rong
Yi Zhang
Yihang Zhou
Yinqing Cheng
Yinzhou Zhang
Yiru Zhang
Yizhi Luo
Yonghao Wan
Yongkang Ding
Yongliang Xu
Yosuke Tsuchiya
Youkai Wang
Yu Boan
Yuan Zhou
Yuanjian He
Yuanyuan Liu
Yuanyuan Xu
Yufeng Chen
Yuhang Yang
Yulong Wang

Yunzhou Jiang
Yuqi Li
Yuxuan Zhang
Zebin Li
Zhangziyi Zhou
Zhanjie Jin
Zhao Liu
Zhe Kong
Zhen Wang
Zheng Zhong
Zhengye Shen
Zhenlei Cui
Zhibin Zhang
Zhongjie Hu
Zhongliang Wang
Zijian Lin
Zimi Lv
Zituo Li
Zixuan Hong

Contents – Part II

Self-distillation Enhanced Vertical Wavelet Spatial Attention for Person Re-identification

Yuxuan Zhang[1], Huibin Tan[1], Long Lan[1], Xiao Teng[1], Jing Ren[1(✉)], and Yongjun Zhang[2]

[1] Institute for Quantum Information & State Key Laboratory of High Performance Computing, College of Computer Science and Technology, National University of Defense Technology, Changsha 410073, China
{yuxuanzhang21,tanhb,long.lan,tengxiao14,renjing}@nudt.edu.cn
[2] Artificial Intelligence Research Center (AIRC), National Innovation Institute of Defense Technology (NIIDT), Beijing 100071, China
jyzhang@nudt.edu.cn

Abstract. Person re-identification is a challenging problem in computer vision, aiming to accurately match and recognize the same individual across different viewpoints and cameras. Due to significant variations in appearance under different scenes, person re-identification requires highly discriminative features. Wavelet features contain richer phase and amplitude information as well as rotational invariance, demonstrating good performance in various visual tasks. However, through our observations and validations, we have found that the vertical component within wavelet features exhibits stronger adaptability and discriminability in person re-identification. It better captures the body contour and detailed information of pedestrians, which is particularly helpful in distinguishing differences among individuals. Based on this observation, we propose a vertical wavelet spatial attention only with the vertical component in the high frequency specifically designed for feature extraction and matching in person re-identification. To enhance spatial semantic consistency and facilitate the transfer of knowledge between different layers of wavelet attention in the neural network, we introduce a self-distillation enhancement method to constrain shallow and deep spatial attention. Experimental results on Market-1501 and DukeMTMC-reID datasets validate the effectiveness of our model.

Keywords: wavelet spatial attention · person Re-ID · self-distillation

1 Introduction

Person re-identification, an important task within computer vision, aims to identify a specific individual in a video sequence or set of images captured by non-overlapping cameras. It plays a crucial role in scenarios such as tracking fugitives, locating missing persons, and tracing abducted individuals, particularly women and children.

Y. Zhang and H. Tan—Co-first authors of the article.

S. Rudinac et al. (Eds.): MMM 2024, LNCS 14555, pp. 1–13, 2024.
https://doi.org/10.1007/978-3-031-53308-2_1

Typically, person re-identification models utilize various features, including appearance, body shape, and clothing, for recognizing individuals. However, real-world scenarios introduce challenges such as blurred images, variations in shooting distance and angle, occlusions, and changes in pedestrian posture. Moreover, the inherent similarity of features among different individuals further complicates the task, making it more challenging than traditional target recognition. Consequently, achieving high discriminability in features becomes imperative for effective person re-identification.

The accurate extraction of valid pedestrian features becomes a key issue in improving the precision of person re-identification. Current mainstream research approaches have explored pedestrian features from several perspectives, including pixel features [1], local and global feature [2], and cross-domain invariant features [3], which have achieved good results. In a recent study, frequency features of images were introduced to the vision tasks [4]. The authors demonstrated that global average pooling is a special case of frequency domain compression and introduced discrete cosine transform (DCT) to extend signal compression towards richer multi-spectrum information, shifting our attention to the spatial frequencies of images. Meanwhile, [5] shown that low-frequency information, which represents semantic and labeling information, is a primary feature commonly used in computer vision tasks. On the other hand, high-frequency information contains valuable image texture and boundary details but also introduces noise that hampers model training. Traditional neural network models primarily utilize the low-frequency semantic information and are trained without explicitly considering the high-frequency information. However, as model accuracy improves, overfitting tendencies due to the dominance of high-frequency noise emerge. In summary, the frequency domain signal contains more information, but needs to be analyzed and selected.

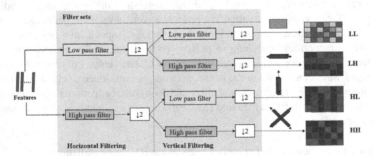

Fig. 1. Decomposition of images by 2D DWT. Low pass filter represents scale filter, high pass filter represents wavelet filter, and "↓2" means downsampling. They form the discrete wavelet transform.

The wavelet transform is a method of transforming a signal between the time and frequency domains. Wavelet functions of different scales and frequencies can be used to capture changes in the different frequency components of a signal. Large wavelet coefficients correspond to the more significant frequency components of the signal, while smaller wavelet coefficients correspond to the less significant frequency components. By selectively processing these wavelet coefficients, we can extract the key frequency domain features in the signal and perform signal analysis, processing and feature extraction. Figure 1 shows the Haar wavelet as an example of how the decomposition of image spatial frequencies in different directions can be achieved using the two-dimensional discrete wavelet transform. Wavelet features have shown good performance in various vision tasks such as image classification [6], target detection [7] and image segmentation [8], and have also been used for tasks such as blur detection [9], denoising [10] and improving the mathematical interpretability of neural networks [11].

In person re-identification, we observe that pedestrians usually stand vertically, while roads and occlusions in the background are generally oriented horizontally. Therefore, the spatial features of pedestrians should be continuous in the vertical direction, and the segmentation line with the background is also essentially vertical. Based on the above assumptions, we extracted the vertical component of the wavelet features on the dataset. Experiments verified that the component can better capture the body contour and detail information of pedestrians, and has stronger adaptability and is more discriminative in the person re-identification task. Due to the low resolution of the person re-identification data, we select a pedestrian image from an open-source library for better displaying the effect of the wavelet transform. As shown in Fig. 2(a), the upper image is the horizontal high-frequency component contains in the original image, and the bottom image shows the vertical high-frequency component. The red box is enlarged to clearly show that the pedestrian profile information is disambiguated in the horizontal direction, while it is displayed more clearly in the vertical component.

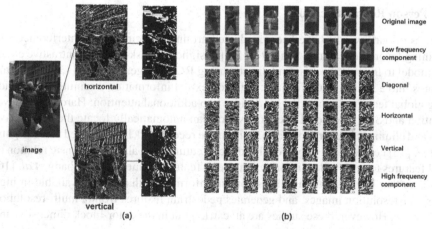

Fig. 2. Spatial structure preservation effect of images under wavelet transform.

Based on the aforementioned observations, this paper presents a vertical wavelet spatial attention module for feature extraction and matching in person re-identification, which uses the Haar wavelet transform to extract the vertical components of an image and constructs a vertical wavelet attention to obtain a useful high frequency. Experiments show that the model is able to better capture body contour and detail information of pedestrians, which helps to distinguish differences between different pedestrians. Furthermore, to enhance the consistency of spatial semantics and facilitate knowledge transfer across different layers of the neural network, this paper introduces a self-distillation enhancement mechanism. This mechanism constrains the spatial attention in both shallow and deep layers by inserting a vertical wavelet attention module into each layer. These two modules generate separate spatial attention maps after extracting different network features, which are then interconnected through distillation loss functions. This enables bidirectional knowledge transfer within the model, facilitating improved prediction performance through self-learning.

Main contributions of this paper:

1. A vertical wavelet spatial attention module based on the Haar wavelet transform is proposed to selectively extract the vertical component from the high-frequency information of the image, so as to better capture the body contour and detail information of pedestrians and improve the discriminability of features.
2. A self-distillation enhancement mechanism is introduced to enhance the coherence of spatial attention in the shallow and deep layers and to induce the transfer of spatial attention knowledge in the neural network between different layers.
3. A self-distillation enhanced vertical wavelet spatial attention is applied to the person re-identification task and the effectiveness of the model is verified.

2 Related Work

2.1 Person Re-identification

A series of early works have been conducted around reducing noise interference and obtaining favorable features. Song et al. [12] designed a mask-guided contrastive attention model to filter the background directly using RGB image. Yu et al. [13] divides the features into a series of sub-features, with contextual information retaining. To considering global features, Jiao et al. [14] designs two additional attention: Hard region-level attention and Soft pixel-level attention, which can automatically locate the most active parts and eliminate background noise. Some more recent work has provided new insights. Li et al. [15] consider not only global and local features, but also the intrinsic relation of local features to retrieve lost information due to feature segmentation. Zhang et al. [16] considers that low-resolution images contain information that is not available in high and super-resolution images, and generates pedestrian features from a multi-resolution perspective. However, these studies are all carried out in the color-block dimension and have not been translated to the frequency domain.

2.2 Wavelet Transforms in CNN

The wavelet transform is an advancement of the Fourier transform and is widely used in the field of image processing. After the rise of neural networks, researchers found that large-scale models went under better than using wavelet algorithms alone, and began trying to combine wavelets and neural networks [17]. In general, wavelet transforms are more often used in the low-level tasks such as deraining [18] and deblurring [19].

Wave-CNet [7] replaces the maxpool with a wavelet transform and retains only the low-frequency part when downsampling to achieve Noise-Robust. Wave-kernel [8] replaces the first convolution layer directly with the wavelet transform. DWAN [20] points out that global average pooling has the disadvantage of insufficient channel information, so it switches to wavelet transform and introduces the Haar wavelet [21],which has the advantages of simple formulas, low computational effort, and easy implementation.

2.3 Knowledge Distillation Mechanism

Knowledge distillation is a method of model compression proposed by Hinton [22], there are teacher network and student network in the training stage. And the self-distillation is a deformation of knowledge distillation to train without teacher networks. Zhang *et al.* [23] add bottleneck and FC layers after each network block as a separate classifier, and calculate loss with the final layer of classification results. Xu *et al.* [24] pass data separately into the feature extraction network, and then use the output to calculate KL scatter, making the original and distorted images guide each other. Li *et al.* [25] split the batch into two networks, and get the target network by sharing the weights and averaging. Depending on the task, the exact structure of the self-distillation varies.

3 Method

In this section, we propose a novel approach for person re-identification called Self-distillation Enhanced Vertical Wavelet Spatial Attention. Our approach enhances the ResNet50 network by adding a vertical wavelet spatial attention module to transform the feature representation and filter noise in high-frequency components, which strengthens pedestrian features. Additionally, we employ a self-distillation enhancement mechanism to improve the network's performance. We present the method's architecture in detail, organized in a bottom-up order of network construction, which includes the framework in Sect. 3.1, vertical wavelet spatial attention module in Sect. 3.2, and the self-distillation enhancement mechanism in Sect. 3.3.

3.1 Overall Model Framework

Baseline Model Settings for Person Re-identification. In this study, we select article [26] as the baseline for our experiments and largely follow the proposed tricks in our replication experiments. We decide to use triplet loss instead of center loss since it is a more classical method, and based on the reported results in the article, center loss results in only 0.2 mAP increase in the model's performance.

Our model employs ResNet50 as the backbone network. The final downsampling parameter is changed to 1. This increases the spatial resolution of the output feature maps, helping to capture more detail of the pedestrian images. To further improve the performance of the model, we randomly erase parts of the input images. This encourages it to learn better representations of pedestrian images with occlusions and increases the robustness. We use a learning rate warm-up strategy during training to ensure that the learning rate is always within an appropriate range. When calculating the loss, we use the sum of the triplet loss (Eq. (1)) and the cross-entropy loss of label smoothing (Eq. (2)) to prevent overfitting, where a and p are from the same category, a and n are from the different category. And d is calculating distance. The ε is a constant. The y indicates the true label and p_i is prediction logits. As the cross-entropy loss mainly optimizes the cosine distance, while the triplet loss concentrates on the Euclidean distance, we introduce a BN layer after the backbone network to explore the feature space more exhaustively.

$$Loss1 = \max(d(a,p) - d(a,n) + \varepsilon, 0) \tag{1}$$

$$Loss2 = \begin{cases} \sum_{i=1}^{N} -\left(1 - \frac{N-1}{N}\varepsilon\right)\log p_i, y = i \\ \sum_{i=1}^{N} -\frac{\varepsilon}{N}\log p_i, y \neq i \end{cases} \tag{2}$$

The Architecture of Self-distillation Enhanced Vertical Wavelet Spatial Attention for Person Re-identification. We present a novel approach for person re-identification using a 4-layers pipelined backbone network, as depicted in Fig. 3. Our proposed framework comprises of the original resnet50 network represented by the blue cube, a feedforward network containing the yellow and grey cubes, the vertical wavelet spatial attention module represented by the green parts and the self-distillation mechanism represented by the faint yellow part. Features from different stages are extracted to calculate different losses.

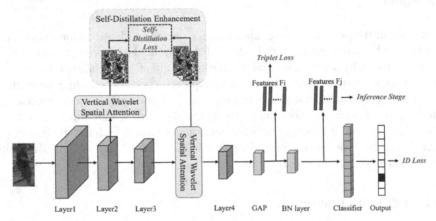

Fig. 3. The overall architecture of the method (Color figure online).

Our vertical wavelet spatial attention module can be encoded into any layer of the network based on the specific task requirements and experimental data. In this work, we put it between the third and fourth layers. By doing so, we can significantly enhance the model's performance in the person re-identification task.

To avoid increasing the parameter space of the backbone network, we employ an independent auxiliary network for self-distillation. Since the Haar wavelets are fully reconstructed, our wavelet spatial attention module does not change the feature size before and after decomposition and reconstruction, thus being plug and play and requiring only minor code changes for the module to shift between different layers.

3.2 Vertical Wavelet Spatial Attention Module

Inspired by wavelet spatial attention (WSA) used in image processing tasks, we introduce vertical wavelet spatial attention (VWSA) into the person re-identification model using the Haar wavelet transform.

Wavelet Spatial Attention in Person Re-identification. The WSA block comprises three main components, namely the wavelet decomposition, aggregation, and attention generation modules. Since Haar wavelet transform can be seen as the permutation and combination of the low-pass filter and high-pass filter, the two-dimensional discrete wavelet transform (2D DWT) is able to decompose features extracted from the network into four components: low-frequency approximation features (LL), horizontal features (LH), vertical features (HL), and diagonal features (HH) as illustrated in Fig. 1. Through summing up the high-frequency components and concatenating them with the low-frequency components, WSA produces aggregated features Fw. These aggregated features are then passed through the attention transformation network to determine the attention coefficients, forming the output wavelet attention, ATTw. Multiplying the attention and original feature results in the final output. The process is depicted in Fig. 4(a).

The WSA block functions as a structure-independent add-on module integrated into existing network architectures due to its modular design. It can be added to each layer of the network and ensure constant processing of the feature size. The effects of this module vary due to the differences in the sizes of the feature space in each layer.

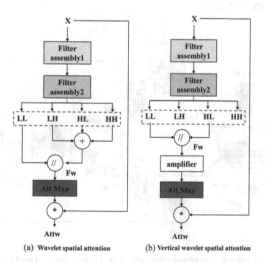

(a) Wavelet spatial attention　　　(b) Vertical wavelet spatial attention

Fig. 4. Vertical Wavelet Spatial Attention Module. Fig(a) is the original wavelet spatial attention module, and Fig(b) is a vertical wavelet attention module. In this figure, "+" means add, "//" means splice and "*" means multiply.

Vertical High Frequency Extraction and Noise Filtering. For person re-identification tasks, we analyze the characteristics of the task. We believe that there are more continuous contours in the vertical direction than the horizontal direction in pedestrian, and propose a method to extract contours by filtering high-frequency information from the vertical direction.

We utilize the wavelet transform to segment low frequency and different components of high-frequency information. We introduce the vertical wavelet spatial attention (VWSA) module, which integrates component features without affecting the feature size by discarding some high-frequency information. The preservation of the low-frequency component is crucial due to the significant amount of image approximation information it retains. Moreover, we directly splice the low-frequency component with the vertical component to maximize the retention of feature information, which enhances the pedestrian profile.

As shown in Fig. 4(b), the VWSA module extracts smaller wavelet feature values after discarding some components. Hence, we add signal amplifiers to obtain a more distinguishable attention map to avoid entering the inert region of the activation function.

3.3 Self-distillation Enhancement

We propose a self-distillation enhancement mechanism to enhance consistency in the ResNet50 network. Specifically, we aim to add the VWSA module after layer 2 and layer 3, and enable them to learn from each other. To achieve this goal, we extract the wavelet spatial attention after layer 2 and layer 3, and employ them to calculate an additional distillation loss function L3, such that the features converge and both can better focus on pedestrian features (Fig. 5).

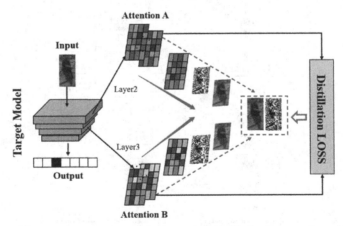

Fig. 5. Self-distillation enhancement mechanism. The Distillation Loss push two features from middle layers of the target network to achieve complementarity. The red arrow parts indicate that the attention A and attention B may focus on boundary at different locations and gradually complement each other under the constraint of the loss function.

To avoid increasing the size of the backbone network while improving its performance, a VWSA block is added as the additional auxiliary network after layer 2. The added network structure is illustrated in the faint yellow area of Fig. 3. We measure the distillation loss using MSE Loss as shown in Eq. (3). The overall loss function is shown in Eq. (4), where att_i represents the spatial attention of different layers, and the α, β, and γ are hyperparameters.

$$Loss3 = \frac{1}{N} \sum_{k=1}^{N} \left(att_k^i - att_k^j \right)^2 \tag{3}$$

$$Loss = \alpha * Loss1 + \beta * Loss2 + \gamma * Loss3 \tag{4}$$

4 Experience

In this section, we present the experimental setup, datasets, and results of our proposed approach. We conducted experiments on the Market1501 and DukeMTMC-reID datasets, and evaluated the performance of our model using the Rank-1 and mAP metrics. Further, we analyzed the effectiveness of each modification by evaluating the performance of different stages of model adjustment and ablation study.

4.1 Datasets and Settings

Market1501 [27] and DukeMTMC-reID [28] are most widely used datasets for person re-identification. These two datasets have the similar organization structure of train, query and gallery sets. In contrast to the Market1501 dataset, the DukeMTMC-reID dataset has pedestrians that only appear in one camera as interference terms.

To ensure the preciseness of the experiment, we basically followed the same experimental setup as in the baseline [26]. For the loss function, we proposed Eq. (4). We keep the same part unchanged as the baseline, i.e. $\alpha = 1$ and $\beta = 1$.

4.2 Experience Results

To observe the effect of the discrete wavelet transform based on the Haar wavelet on the person re-identification dataset more intuitively, we conducted experiments on some images. As shown in Fig. 2(b), we selected several photos with different resolutions for the discrete wavelet transform, and the results show that the wavelet transform is able to better maintain the spatial structure in the images. The low-frequency component maintains the general information of the image well, but blurs textures and contours. The high frequency contains information in different directions and is comprehensive but cluttered. The information in the diagonal direction gives little indication of pedestrian features, and the horizontal frequency contains information in which the pedestrian contours are cut off and the pedestrian features are connected to the horizontal background. Pedestrian contours and details, such as leg information, can be better retained in the information contained in the vertical direction.

We make incremental improvements to the baseline to verify the effect of each step on the improvement of the model. The experiments' results are shown in Table 1. It shows that the combination of the wavelet transform and the person re-identification task is able to achieve good improvements in both Rank-1 and mAP metrics. Based on it, the high-frequency component obtained by vertically high-pass filtering the wavelet attention is also able to obtain an improvement, which is more evident on mAP. The further addition of the self-distillation module is able to continue to push the mAP even higher while maintaining the Rank-1 level. The performance of the model firstly demonstrates that CNN can learn high-frequency information that is not visible to people. Secondly, it demonstrates that the person re-identification model based on wavelet attention can achieve better performance by filtering out noisy information from the high-frequency component. Finally, it also demonstrates the effectiveness of self-distillation architecture using attention between layers. As can be seen, our model is competitive among other advanced methods.

Table 1. Results of the model experiment on Market1501 and DukeMTMC-reID datasets.

Method		Datasets			
		Market1501		DukeMTMC-reID	
		Rank-1 (%)	mAP (%)	Rank-1 (%)	mAP (%)
PCB [2]	ECCV18	93.8	81.6	83.3	69.2
CAMA [29]	CVPR19	94.7	84.5	85.8	72.9
SNR [30]	CVPR20	94.4	84.7	85.9	73.0
GASM [31]	ECCV20	95.3	84.7	88.3	74.4
RANGEv2 [32]	PR22	94.7	86.8	87.0	78.2
Baseline [26]		94.1	85.7	86.2	75.9
+wavelet		94.4	86.4	86.8	76.9
+vertical wavelet		94.6	87.0	87.6	77.2
+vertical wavelet & self-distillation		95.2	87.4	87.9	77.4

4.3 Ablation Study

We propose a vertical wavelet spatial attention with a frequency selection mechanism, and in order to investigate the effectiveness of frequencies in each direction for the person re-identification, we conduct corresponding ablation experiments on Market1501. In addition, the wavelet attention module can be inserted after different layers of the model, and we also present the results of the ablation experiments for where the module is inserted here.

The Effect of Choosing Different Directions in High Frequency. Table 2 shows the effect of choosing to retain different frequencies for the person re-identification task. LL represents the low-frequency component, LH represents the horizontal high frequency,

Table 2. The effect of choosing different directions in high frequency.

Feature selection	Rank-1 (%)	mAP (%)
LL + LH	94.3	86.3
LL + HL	94.6	87.0
LL + HH	94.6	86.5
LL + LH + HL + HH	94.4	86.4

HL represents the vertical high frequency, and HH represents the diagonal high frequency. It indicates that the horizontal component cuts off the pedestrian profile feature and makes it discrete, while the diagonal direction contains compromising information and the vertical direction can best maintain the continuity of the pedestrian profile feature. This also verifies our hypothesis.

The Effect of Choosing Different Insertion Positions. We tried to encode the wavelet spatial attention module after different layers of the model and compare the final effect. The VW represents the wavelet attention module that completes the vertical frequency selection. The results are shown in Table 3. The results demonstrate that the wavelet attention module is movable and the effect varies with the size of the feature space size. This experiment result becomes the basis for constructing a self-distillation enhancement mechanism.

Table 3. The effect of choosing different insertion positions.

Insertion after	Rank-1 (%)	mAP (%)
Layer1	94.4	85.9
Layer2	94.5	85.9
Layer3	94.4	86.4
Layer4	93.9	86.0
Layer1(VW)	94.6	86.8
Layer2(VW)	94.9	86.9
Layer3(VW)	94.6	87.0
Layer4(VW)	94.3	86.2

5 Conclusion

Summarizing our work, in order to break through the limitations of the pixel level for the person re-identification, we introduce Haar wavelets into the CNN to build Haar wavelet attention in spatial; to make better use of the effective high-frequency components contained in the pictures for accurate knowledge, we select vertical components that better match the upright pedestrian from the high frequency; to enhance learning and transfer knowledge in the CNN, we add a self-distillation enhancement mechanism to enable vertical wavelet spatial attention between different layers to learn from each other. Our proposed method, so far, offers an effective way to refine the attention maps of the ResNet50 network and improve its performance.

Acknowledgment. This work was supported by: The Self networks (CNNs). Furthermore, the Transformer has also-directed Project of State Key Laboratory of High Performance Computing: 202101-18.

References

1. Matsukawa, T., Okabe, T., Suzuki, E., Sato, Y.: Hierarchical Gaussian descriptor for person re-identification. In: 2016 IEEE CVPR, pp. 1363–1372 (2016)
2. Sun, Y., Zheng, L., Yang, Y., Tian, Q., Wang, S.: Beyond part models: person retrieval with refined part pooling (and a strong convolutional baseline) (2018). http://arxiv.org/abs/1711.09349
3. Liu, J., Zha, Z.-J., Chen, D., Hong, R., Wang, M.: Adaptive transfer network for cross-domain person re-identification. In: 2019 IEEE/CVF CVPR, pp. 7195–7204 (2019)
4. Qin, Z., Zhang, P., Wu, F., Li, X.: FcaNet: frequency channel attention networks. In: 2021 IEEE/CVF International Conference on Computer Vision, pp. 763–772 (2021)
5. Wang, H., Wu, X., Huang, Z., Xing, E.P.: High frequency component helps explain the generalization of convolutional neural networks, http://arxiv.org/abs/1905.13545 (2020)
6. Li, Q., Shen, L., Guo, S., Lai, Z.: Wavelet integrated CNNs for noise-robust image classification. In: 2020 IEEE/CVF Conference on Computer Vision and Pattern Recognition (CVPR), pp. 7243–7252 (2020)
7. Alaba, S.Y., Ball, J.E.: WCNN3D: wavelet convolutional neural network-based 3D object detection for autonomous driving. Sensors **22**, 7010 (2022)
8. Li, Q., Shen, L.: WaveSNet: wavelet integrated deep networks for image segmentation. http://arxiv.org/abs/2005.14461 (2020)
9. Tong, H., Li, M., Zhang, H., Zhang, C.: Blur detection for digital images using wavelet transform. In: 2004 IEEE International Conference on Multimedia and Expo (ICME) (IEEE Cat. No.04TH8763), pp. 17–20. IEEE, Taipei, Taiwan (2004)
10. Luo, X., Zhang, J., Hong, M., Qu, Y., Xie, Y., Li, C.: Deep wavelet network with domain adaptation for single image demoireing. In: 2020 IEEE/CVF Conference on Computer Vision and Pattern Recognition Workshops (CVPRW), pp. 1687–1694 (2020)
11. Leterme, H., Polisano, K., Perrier, V., Alahari, K.: On the shift invariance of max pooling feature maps in convolutional neural networks. http://arxiv.org/abs/2209.11740 (2022)
12. Song, C., Huang, Y., Ouyang, W., Wang, L.: Mask-guided contrastive attention model for person re-identification. In: 2018 IEEE/CVF Conference on Computer Vision and Pattern Recognition, pp. 1179–1188. IEEE, Salt Lake City, UT (2018)
13. Yu, R., Zhou, Z., Bai, S., Bai, X.: Divide and fuse: a re-ranking approach for person re-identification. In: Proceedings of the British Machine Vision Conference 2017, p. 135. British Machine Vision Association, London, UK (2017). https://doi.org/10.5244/C.31.135
14. Jiao, S., Wang, J., Hu, G., Pan, Z., Du, L., Zhang, J.: Joint attention mechanism for person re-identification. IEEE Access **7**, 90497–90506 (2019)
15. Li, W., et al.: Collaborative attention network for person re-identification. J. Phys. Conf. Ser. **1848**, 012074 (2021). https://doi.org/10.1088/1742-6596/1848/1/012074

16. Zhang, G., Chen, Y., Lin, W., Chandran, A., Jing, X.: Low resolution information also matters: learning multi-resolution representations for person re-identification. In: Proceedings of the Thirtieth International Joint Conference on Artificial Intelligence, pp. 1295–1301, (2021)
17. Fujieda, S., Takayama, K., Hachisuka, T.: Wavelet convolutional neural networks, http://arxiv.org/abs/1805.08620 (2018)
18. Huang, H., Yu, A., Chai, Z., He, R., Tan, T.: Selective wavelet attention learning for single image deraining. Int. J. Comput. Vis. **129**, 1282–1300 (2021)
19. Zou, W., Jiang, M., Zhang, Y., Chen, L., Lu, Z., Wu, Y.: SDWNet: a straight dilated network with wavelet transformation for image deblurring. In: 2021 IEEE/CVF International Conference on Computer Vision Workshops (ICCVW), pp. 1895–1904 (2021)
20. Yang, Y., et al.: Dual wavelet attention networks for image classification. IEEE Trans. Circuits Syst. Video Technol. 1 (2022)
21. Mallat, S.G.: A Wavelet Tour of Signal Processing: The Sparse Way. Elsevier/Academic Press, Amsterdam, Boston (2009)
22. Hinton, G., Vinyals, O., Dean, J.: Distilling the knowledge in a neural network, http://arxiv.org/abs/1503.02531 (2015)
23. Zhang, L., Song, J., Gao, A., Chen, J., Bao, C., Ma, K.: Be your own teacher: improve the performance of convolutional neural networks via self distillation, http://arxiv.org/abs/1905.08094 (2019)
24. Xu, T.-B., Liu, C.-L.: Data-distortion guided self-distillation for deep neural networks. AAAI **33**, 5565–5572 (2019). https://doi.org/10.1609/aaai.v33i01.33015565
25. Li, G., Togo, R., Ogawa, T., Haseyama, M.: Self-knowledge distillation based self-supervised learning for Covid-19 detection from chest X-ray images. In: ICASSP 2022, pp. 1371–1375 (2022)
26. Luo, H., Gu, Y., Liao, X., Lai, S., Jiang, W.: Bag of tricks and a strong baseline for deep person re-identification. In: 2019 IEEE/CVF Conference on Computer Vision and Pattern Recognition Workshops (CVPRW), pp. 1487–1495. IEEE, Long Beach, CA, USA (2019)
27. Zheng, L., Shen, L., Tian, L., Wang, S., Wang, J., Tian, Q.: Scalable Person re-identification: a benchmark. In: 2015 IEEE International Conference on Computer Vision (ICCV), pp. 1116–1124. IEEE, Santiago, Chile (2015)
28. Zheng, Z., Zheng, L., Yang, Y.: Unlabeled samples generated by GAN improve the person re-identification baseline in vitro, http://arxiv.org/abs/1701.07717 (2017)
29. Yang, W., Huang, H., Zhang, Z., Chen, X., Huang, K., Zhang, S.: Towards rich feature discovery with class activation maps augmentation for person re-identification. In: 2019 IEEE/CVF Conference on Computer Vision and Pattern Recognition (CVPR), pp. 1389–1398. IEEE, Long Beach, CA, USA (2019)
30. Jin, X., Lan, C., Zeng, W., Chen, Z., Zhang, L.: Style normalization and restitution for generalizable person re-identification. In: 2020 IEEE/CVF Conference on Computer Vision and Pattern Recognition (CVPR), pp. 3140–3149. IEEE, Seattle, WA, USA (2020). https://doi.org/10.1109/CVPR42600.2020.00321
31. He, L., Liu, W.: Guided saliency feature learning for person re-identification in crowded scenes. In: Vedaldi, A., Bischof, H., Brox, T., Frahm, J.-M. (eds.) ECCV 2020. LNCS, vol. 12373, pp. 357–373. Springer, Cham (2020). https://doi.org/10.1007/978-3-030-58604-1_22
32. Wu, G., Zhu, X., Gong, S.: Learning hybrid ranking representation for person re-identification. Pattern Recogn. **121**, 108239 (2022)

High Capacity Reversible Data Hiding in Encrypted Images Based on Pixel Value Preprocessing and Block Classification

Tao Zhang, Ju Zhang, Yicheng Zou, and Yu Zhang(✉)

College of Computer and Information Science, Southwest University,
Chongqing 400715, China
zhangyu@swu.edu.cn

Abstract. Reversible data hiding in encrypted images (RDHEI) can simultaneously achieve secure transmission of images and secret storage of embedded additional data, which can be used for cloud storage and privacy protection. In this paper, an RDHEI scheme based on pixel value preprocessing and block classification (PVPBC-RDHEI) is proposed. The content owner first preprocesses the pixel values of the original image and then encrypts the image by combining bitwise exclusive-or operation and block permutation, so that the correlation between neighboring pixels in the encrypted image block is preserved. Upon receiving the encrypted image, the data hider classifies all blocks into six types based on the number of continuously consistent value bit planes counted from top-down in the pixel block, and generates corresponding indicators for six block types using Huffman coding. Therefore, each pixel block in the image can vacate room for data embedding. The receiver can separately extract the additional data or recover the original image according to the different keys. The experimental results show that the embedding rate of the proposed scheme is significantly superior to the state-of-the-art schemes, and the security and reversibility are guaranteed.

Keywords: Reversible data hiding · Encrypted images · Pixel value preprocessing · Block classification

1 Introduction

With the rapid development of information technology, people's demand for data security and privacy protection is becoming increasingly urgent. To protect sensitive information from unauthorized access and tampering, data hiding technology has been extensively studied. Traditional data hiding methods can cause permanent distortion of the image by embedding additional data into the carrier image [1]. To solve this problem, reversible data hiding (RDH) technology has emerged. Reversible data hiding can recover the original image losslessly after extracting the embedded data. Due to the reversibility, RDH can be applied to some special scenarios, e.g., medical and military imagery, law forensics and

S. Rudinac et al. (Eds.): MMM 2024, LNCS 14555, pp. 14–27, 2024.
https://doi.org/10.1007/978-3-031-53308-2_2

precision machining. To date, numerous RDH methods have been proposed and they can be roughly classified into five categories: differential expansion (DE) [2–4], histogram shifting (HS) [5–7], lossless compression [8–10], prediction-error expansion (PEE) [11,12], and pixel value ordering (PVO) [13,14].

In recent years, with the popularity of cloud storage, more and more users tend to upload data to the cloud for storage or sharing. However, due to certain security risks in cloud storage, the problem of data privacy leakage has become increasingly prominent. In response to this challenge, the combination of reversible data hiding and image encryption, i.e., reversible data hiding in encrypted images (RDHEI) has become a new research hotspot. This approach enables both secure transmission of images and secret storage of embedded additional data. Up to now, many RDHEI methods have been proposed. According to the point of view of vacating room, the existing methods can mainly be divided into two categories: reserving room before encryption (RRBE) [15–18] and vacating room after encryption (VRAE) [19–25].

In RRBE schemes, the content owner preprocesses the original image using its spatial correlation to reserve room before image encryption. The data hider uses the spare room to embed additional data after receiving the encrypted image. Ma et al. [15] first proposed the RRBE method by using the traditional RDH method to embed the least significant bit (LSB) of some pixels into other pixels to reserve room. Chen and Chang [16] combined extended run-length encoding and a block-based most significant bit (MSB) plane rearrangement mechanism to compress the MSB planes of the image, thus generating room for high-capacity embedding. Puteaux and Puech [17] recursively processed each bit-plane of an image from MSB to LSB by combining error prediction, reversible adaptation, encryption and embedding. Gao et al. [18] located prediction errors by generating the error location binary map, and marked the location of prediction errors with error blocks and embedded additional data with message blocks.

Unlike RRBE schemes, in VRAE schemes, the content owner only needs to choose a suitable encryption algorithm to encrypt the original image. Then, the data hider reserve room in the encrypted image to embed additional data. Zhang [19] used stream cipher technology to encrypt the original image, and then the data hider divides the encrypted image into non-overlapping blocks and embeds a secret bit by flipping the last three LSBs of half pixels in each block. Qian and Zhang [20] proposed a separable method for encrypted images using Slepian-Wolf sources coding which achieved a high embedding payload and good image reconstruction quality. Later, block-based image encryption schemes have been proposed and applied to VRAE. Qin et al. [21] employed an efficient encryption method to encrypt each original block while transferring redundancy from MSB to LSB. The additional data is then embedded into the LSB of the encrypted block by the sparse matrix encoding method. Fu et al. [22] divided the encrypted blocks into embeddable and non-embeddable blocks by analyzing the distribution of MSB layers and then used Huffman coding to adaptively compress embeddable blocks according to the occurrence frequency of the MSB, and the vacated room can be used for additional data embedding. Wang et al. [23] classified all blocks into usable blocks (UBs) and unusable blocks (NUBs) while preserving intra-block pixel correlation. Then, since the pixels in the block share the same MSBs, the UB is reconstructed to vacate room for data embedding. Wang et al. [24] classified and encoded all encrypted blocks based on the number of MSB bit planes where all values are '1' or '0'. After embedding the indicators that are generated by Huffman coding into the first MSB bit plane of each block, the remaining bits of these MSB bit planes can be vacated for additional data embedding. Wang and He [25] proposed a novel RDHEI method based on adaptive MSB prediction. The encrypted image is first divided into 2×2 sized non-overlapping blocks and then the other three pixels are predicted with the upper-left pixel within the encrypted block such that the embedding room is vacated. Although the embedding rate of these block-based encryption schemes has increased, there is still some room for improvement.

In this paper, we propose an efficient reversible data hiding scheme for encrypted images based on pixel value preprocessing and block classification (PVPBC-RDHEI) to further improve the embedding rate. In our scheme, the content owner first preprocesses the pixel values of the original image and then encrypts the original image block by block. The data hider classifies all blocks into six types based on the number of continuously consistent value bit planes counted from top-down in the pixel block, and uses Huffman coding to generate corresponding indicators for six block types. Therefore, each pixel block in the image can embed data. On the receiving side, data extraction and image recovery can be separated according to the image encryption key and the data hiding key. The embedding rate of our scheme is superior to some existing RDHEI schemes and the main contributions of our scheme are summarized as follows:

1. By preprocessing the pixel values of the original image, each pixel block in the image can embed data.
2. Block-level encryption of the image combined with bitwise XOR operation and block permutation preserves the correlation of pixels within the block.
3. Our scheme achieves a higher embedding rate than state-of-the-art schemes, and the data extraction and image recovery are also separable and error-free.

The rest of this paper is organized as follows. Sect. 2 describes the detailed procedures of the proposed PVPBC-RDHEI scheme. In Sect. 3, the experimental results and analysis are provided. Finally, conclusions are presented in Sect. 4.

2   The Proposed PVPBC-RDHEI Scheme

This section describes our proposed PVPBC-RDHEI scheme. Figure 1 shows the framework of the proposed PVPBC-RDHEI scheme. There are three types of users: content owner, data hider and receiver. First, the content owner preprocesses the pixel values of the original image, then encrypts the image in combination with bitwise XOR operation and block permutation, and embeds the location map generated by preprocessing the pixel values into the encrypted image. On the data hider side, the data hider first classifies all blocks of the encrypted image, then encrypts the additional data using the data hiding key to enhance security and embeds the encrypted additional data into the received encrypted image. Finally, the receiver can extract additional data or recover the original image by using different keys.

2.1 Image Encryption

In our scheme, encryption of the image includes three steps: pixel value preprocessing, block-level stream encryption and block permutation, and location map embedding.

Pixel Value Preprocessing. For the original image I of size $M \times N$, the pixel value $x(i, j)$ of each pixel is preprocessed to a value less than or equal to 31, in which $1 \leq i \leq M$, $1 \leq j \leq N$.

Step 1: Make each pixel value $x(i, j)$ less than or equal to 127 by Eq. (1),

$$x(i, j) = \begin{cases} 255 - x(i, j), x(i, j) > 127 \\ x(i, j), otherwise \end{cases} \tag{1}$$

Step 2: Use Eq. (2) to make each pixel value $x(i, j)$ less than or equal to 63,

$$x(i, j) = \begin{cases} 127 - x(i, j), x(i, j) > 63 \\ x(i, j), otherwise \end{cases} \tag{2}$$

Step 3: Obtain the final pixel value for each pixel by Eq. (3).

$$x(i, j) = \begin{cases} 63 - x(i, j), x(i, j) > 31 \\ x(i, j), otherwise \end{cases} \tag{3}$$

Note that to correctly determine whether the pixels in the original image have changed in three steps, three location maps sized $M \times N$ are required for marking, with '1' indicating no change and '0' indicating a change.

Block-Level Stream Encryption and Block Permutation. After preprocessing the pixel values of the original image I, the image is encrypted using block-level stream encryption and block permutation. First, the $M \times N$ sized image I is divided into non-overlapping blocks sized 2×2 given by

$$I = \left\{ P_k(i, j) \middle| 1 \leq i \leq \lfloor \frac{M}{2} \rfloor, 1 \leq j \leq \lfloor \frac{N}{2} \rfloor, 1 \leq k \leq 4 \right\} \tag{4}$$

where $P_k(i, j)$ denote the k-th pixel of the block located at (i, j).

Next, a $\lfloor \frac{M}{2} \rfloor \times \lfloor \frac{N}{2} \rfloor$ sized pseudo-random matrix H is generated using the image encryption key K_e. Then, each block is stream ciphered by

$$P_k^{en}(i, j) = P_k(i, j) \oplus h(i, j), 1 \leq k \leq 4 \tag{5}$$

where $h(i, j)$ denote the element located at (i, j) in the pseudo-random matrix H and $P_k^{en}(i, j)$ denote the k-th encrypted pixel of the block located at (i, j) and $\oplus$ denotes the bitwise exclusive-or operation.

After the stream encryption, the cover image is divided into 2×2 sized non-overlapping blocks once again. Then, all blocks are permutated using the same encryption key K_e to generate the encrypted image I_e.

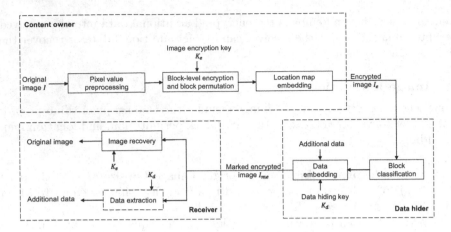

Fig. 1. The framework of the proposed scheme.

Location Map Embedding. Embed the three location maps into the encrypted image I_e as three bit planes, as shown in Fig. 2. Specifically, all pixels of the encrypted image I_e are converted into the 8-bit binary sequence, and 8-bits of all encrypted pixels are extracted sequentially to generate 8 encrypted bit planes. Then, the last three LSB bit planes are moved to the front as the first three MSB bit planes of the encrypted image, while keeping the position of the two bit planes in the middle unchanged. Next, the three location maps are embedded in the encrypted image as the last three LSB bit planes in turn. Finally, the 8-bit binary sequence of all pixels is converted to decimal system to regain the encrypted image I_e.

2.2 Block Classification

After receiving the encrypted image I_e, the embedding room for data hiding is vacated by the data hider. First, all pixels of the encrypted image are converted into the 8-bit binary system, and 8-bits of all encrypted pixels are extracted sequentially to generate 8 encrypted bit planes. Then, the last three LSB bit planes are extracted to obtain the previously embedded three location maps. Next, the first three MSB bit planes are extracted and recovered to the original LSB position of the encrypted image. Finally, the values of the first three MSB bit planes of the encrypted image are set to '0' so that all pixel values of the encrypted image are less than or equal to 31. Note that to ensure reversibility, the three location maps need to be compressed and saved as auxiliary information. This paper uses the bit-plane rearrangement and the bit-stream compression scheme [16] to compress the location map.

Next, the encrypted image I_e is divided into 2×2 sized non-overlapping pixel blocks. The structure of pixel block is shown in Fig. 3 (a). Then, for each block, all pixels are converted into the 8-bit binary system. For one block, 8-bits of all encrypted pixels are extracted sequentially to generate 8 encrypted bit planes

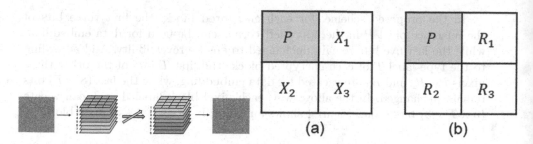

Fig. 2. Location map embedding process.   **Fig. 3.** The structure of a block.

E_λ ($\lambda = 1, 2, \cdots, 8$), and the first MSB encrypted bit plane is denoted as E_1. As shown in Fig. 3 (a), each pixel block consists of four pixels P, X_1, X_2 and X_3. The pixel P marked by yellow is selected as the reference pixel. First, three values R_1, R_2 and R_3 are obtained by

$$R_i = P \oplus X_i, i = 1, 2, 3 \tag{6}$$

Then, the three pixels X_1, X_2 and X_3 in the pixel block are replaced with R_1, R_2 and R_3 respectively. The reconstructed pixel block is shown in Fig. 3 (b). With R_1, R_2 and R_3, the block type-label T for each block can be calculated as

$$T = 8 - lenbin\left(\max\left(R_1, R_2, R_3\right)\right) \tag{7}$$

where $lenbin\,(d)$ returns the number of bits in the binary sequence of the decimal number d. For example, we have $lenbin\,(\max\,(30, 16, 9)) = 5$ since $\max\,(30, 16, 9) = 30 = (11110)_2$. The block type-label T records the number of consecutive uniformly valued bit-planes, i.e., totally '0' or '1', in the block counted from top-down. Therefore, according to how many bit planes are all '0' or all '1', all blocks can be scanned and classified into six different types as listed in Table 1. Note that the value of the block type-label T ranges from '3' to '8'.

Finally, Huffman coding is applied to generate corresponding indicators for six block types. Six Huffman codes are predefined to represent the six labels, namely $\{0, 10, 110, 1110, 11110, 11111\}$. Sort the six block type-labels by the number of blocks and use the shorter code to represent the label with the larger number of blocks. Taking the Lena image as an example, the distribution of each block type and its corresponding Huffman coding are shown in Table 2.

After the Huffman coding is determined, all block indicators are linked into a sequence of block type indicators and its total length LH can be calculated by

$$LH = \sum_{t=3}^{8} (n_t \times l_t), t = 3, 4, \cdots, 8 \tag{8}$$

where n_t is the number of pixel blocks with block type-label $T = t$, and l_t is the length of the corresponding Huffman code.

In the proposed scheme, for each encrypted block, the first three bits of the reference pixel P in the upper-left corner can be used for data embedding, while the last five bits remain unchanged to ensure reversibility. And according to the type-label T of each encrypted block, the first T bits of the other three pixels R_1, R_2 and R_3 are vacated for data embedding, while the last $(8 - T)$ bits remain unchanged. In the above way, each pixel block labeled as T can vacate $(3 \times T + 3)$ bits room. Figure 4 shows an example of six types of pixel blocks.

Table 1. Six block types and corresponding descriptions.

Block types	Descriptions
Type 3	E_1 to E_3 are all '0'
Type 4	E_1 to E_4 are all '0' or all '1'
Type 5	E_1 to E_5 are all '0' or all '1'
Type 6	E_1 to E_6 are all '0' or all '1'
Type 7	E_1 to E_7 are all '0' or all '1'
Type 8	E_1 to E_8 are all '0' or all '1'

Table 2. Lena's block type distribution and Huffman coding.

Block types	Amount	Indicator
Type 3	21125	0
Type 4	19860	10
Type 5	16192	110
Type 6	6685	1110
Type 7	1349	11110
Type 8	325	11111

2.3 Data Embedding

In this phase, two sets of data, auxiliary information and additional data, are embedded into the encrypted image I_e after block classification to obtain the marked encrypted image I_{me}. Detailed descriptions are as follows.

To reversibly recover the image, auxiliary information includes the total length of the block type indicator sequence, the block type indicator sequence, the length of the compressed location map, the compressed location map, and a 20-bits Huffman coding list. The Huffman coding list, the total length of the block type indicator sequence, the block type indicator sequence, the length of the three compressed location maps, and their corresponding compressed location maps are linked in turn to form complete auxiliary information. To ensure that the Huffman coding list and block type indicator sequence can be completely extracted in later phases, auxiliary information is first embedded into the first three bits of the upper-left pixel of each block. Then, the data hider uses the data hiding key K_d to encrypt the additional data and embeds the encrypted additional data into the remaining vacated room.

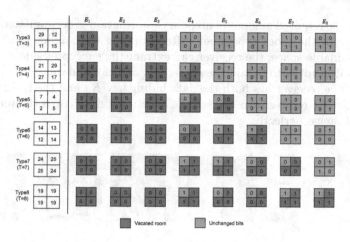

Fig. 4. Examples of block classification.

2.4 Data Extraction and Image Recovery

For the proposed scheme, data extraction and image recovery can be performed separately. The permission for data extraction and image recovery depends on the number of keys for the receiver. There are three cases:

Case 1: The receiver only has the image encryption key K_e, the original image can be recovered losslessly. The process is as follows.

1. Divide the marked encrypted image I_{me} into 2×2 sized non-overlapping blocks and extract the first three bits of the upper-left reference pixel in each block. Next, the Huffman coding list and the block type indicator sequence are extracted from the extracted data to determine the block type-label T for each block.
2. Extract three compressed location maps from the extracted data, and then decompress them to obtain the original three location maps.
3. For each pixel block, firstly, fill the first three bits of the upper-left reference pixel in the block with '0' to obtain P. Then, according to the block type-label T of each block, fill the first T bits of the other three pixels in the block with '0' to obtain R_1, R_2 and R_3. Next, X_1, X_2 and X_3 are recovered as

$$X_i = P \oplus R_i, i = 1, 2, 3 \tag{9}$$

4. After the encrypted pixel values of all blocks are recovered, the original image with the pixel values preprocessed is obtained by decryption using the key K_e.
5. Finally, apply three location maps in reverse order to recover each pixel value to obtain the original image I.

Case 2: The receiver only has the data hiding key K_d. The receiver first determines the block type-label T for each block as described in the previous case. Then, the required amount of auxiliary information is accumulated through the guidance of the block type-label. After skipping the auxiliary information in the embeddable space, the receiver can extract the encrypted additional data stream and decrypt it to obtain the embedded additional data.

Case 3: The receiver has both the data hiding key K_d and the image encryption key K_e. The receiver can extract the additional data correctly and recover the original image perfectly.

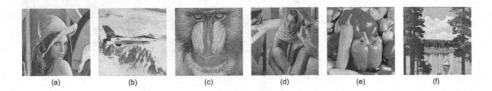

Fig. 5. Six test images. (a) Lena; (b) Airplane; (c) Baboon; (d) Barbara; (e) Peppers; (f) Lake.

3   Experimental Results and Analysis

Experiments are conducted to evaluate the performance of the proposed scheme. First, in Sect. 3.1, the embedding rate of the proposed scheme is calculated and the reversibility is analyzed. The security of the proposed scheme is then analyzed in Sect. 3.2. In Sect. 3.3, three image datasets including BOSSBase [26], BOWS-2 [27] and UCID [28] are used to illustrate the universality of the proposed scheme. Finally, the proposed scheme is compared with some state-of-the-art RDHEI schemes in Sect. 3.4. Six commonly used 512×512 sized grayscale images: Lena, Airplane, Baboon, Barbara, Peppers, and Lake are selected to perform the experiments, as shown in Fig. 5. The embedding rate (ER) presented by bpp (bits per pixel) is the key indicator. In addition, PSNR (Peak signal-to-noise ratio) and SSIM (structural similarity) are also used to verify reversibility.

3.1   Embedding Rate and Reversibility

The embedding rate of the proposed scheme on six test images is calculated. Table 3 presents the distribution of pixel blocks of different block types, the total length of auxiliary information and the ER on six test images, where NT_k denote the number of pixel blocks labeled with block type-label $T = k \, (k = 3, 4, \cdots, 8)$, and LAU denote the total length of auxiliary information. The ER of the $M \times N$ sized image is calculated as

$$ER = \frac{1}{M \times N} \times \left(\sum_{k=3}^{8} (3 \times k + 3) \times NT_k - LAU \right). \tag{10}$$

As shown in Table 3, for the image 'Baboon', the LAU is 590314 bits and the ER is 1.0941 bpp. For the image 'Airplane', the LAU is 364376 bits and the ER is

2.7356 bpp. 'Airplane' is known to be much smoother than 'Baboon'. Therefore, it can be concluded that the ER generally depends on image smoothness.

Table 3. The distribution of pixel blocks of different block types on six test images.

Image	NT_3	NT_4	NT_5	NT_6	NT_7	NT_8	LAU (bits)	ER (bpp)
Lena	21125	19860	16192	6685	1349	325	348861	2.5769
Airplane	18600	16195	16170	9585	3865	1121	364376	2.7356
Baboon	42424	17100	5045	836	118	13	590314	1.0941
Barbara	28651	16999	12883	5684	1136	183	454114	2.0147
Peppers	20972	23744	15387	4658	682	93	361635	2.4408
Lake	29857	21792	9033	3492	1143	219	432787	1.9899

After theoretically calculating the ER of the image, the reversibility of the proposed scheme can be analyzed experimentally. Taking the Lena image as an example, Fig. 6 is the experimental results of each phase obtained by our scheme. The results show that the ER of the proposed scheme on the Lena image reaches 2.5769 bpp. Figure 6 (d1) and (a1) are identical, confirming that the proposed scheme can completely recover the original image. In addition, the quality of the recovered image is evaluated by PSNR and SSIM in the simulation results, where PSNR is close to $+\infty$ and SSIM is equal to 1. This indicates that the recovered image is the same as the original image. The theoretical analysis and simulation results verify the reversibility of the proposed scheme.

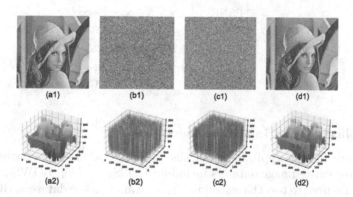

Fig. 6. Results in different phases for Lena. (a1) Original image; (b1) Encrypted image; (c1) Marked encrypted image with the ER = 2.5769 bpp; (d1) Recovered image (PSNR $\rightarrow +\infty$, SSIM = 1); (a2-d2) Pixel value distributions of the images a1-d1.

3.2 Security Analysis

To evaluate the security level of the proposed scheme, the statistical features of the images obtained at different stages are analyzed first. As can be seen from

Fig. 6 (a2-c2), the pixel value distribution of the original image has significant feature information, while the pixel value distribution of both the encrypted image and the marked encrypted image are uniform. It is difficult to find any content of the original image from the encrypted image or the marked encrypted image by using statistical analysis, which reinforces the security.

To further illustrate the security of this scheme, we analyze the key space of the encryption method. Our scheme is based on block-level stream encryption and block permutation to encrypt images. When dividing the $M \times N$ sized image into $s \times s$ non-overlapping blocks, the number of blocks is $n = \frac{M \times N}{s \times s}$. Therefore, the number of possible combinations of block-level stream encryption is $\theta_1 = 256^n$ and the number of possible combinations of block permutation is $\theta_2 = n!$. Overall, the entire key space of the encryption method is $\theta = \theta_1 \times \theta_2 = 256^n \times n!$. Assume that the 512×512 sized image is divided into 2×2 sized non-overlapping pixel blocks, the total number of image blocks is $n = 65536$. Therefore, the entire key space is $\theta = 256^{65536} \times 65536!$, which is a large number and it is almost impossible to decrypt the image correctly without the key. This can guarantee the security of the encrypted image.

Table 4. Experimental results on three entire image datasets.

Dataset	metric	Best	Worst	Average
BOSSBase	ER(bpp)	6.3852	0.4781	3.0207
	PSNR	$+\infty$	$+\infty$	$+\infty$
	SSIM	1	1	1
BOWS-2	ER(bpp)	5.8511	0.2652	2.8910
	PSNR	$+\infty$	$+\infty$	$+\infty$
	SSIM	1	1	1
UCID	ER(bpp)	4.7000	0.0824	2.3669
	PSNR	$+\infty$	$+\infty$	$+\infty$
	SSIM	1	1	1

3.3   Applicability to Image Datasets

To further verify the applicability of the proposed scheme to various feature images, three entire image datasets including BOSSBase [26], BOWS-2 [27] and UCID [28] are used to test the embedding rate. Among these datasets, BOSSBase and BOWS-2 both have 10000 grayscale images sized 512×512 while UCID has 1338 color images sized 384×512 or 384×512. For UCID, we first used the Image module in Python's Pillow library to convert 1338 color images into grayscale images. Table 4 lists the best, worst, and average cases on the three image datasets. It can be seen that the PNSRs of all images are close to $+\infty$ and the SSIMs of all images are equal to 1. This indicates that all images can be recovered losslessly. Consequently, the proposed scheme has good embedding performance in terms of ER.

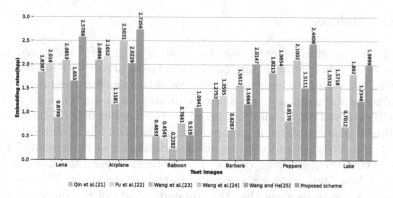

Fig. 7. Comparison of ER (bpp) on six test images.

3.4 Comparisons with Other State-of-the-art Schemes

In order to better illustrate the superiority of our scheme, the proposed scheme is compared with five state-of-the-art schemes [21–25]. All five schemes are VRAE-based scheme. As shown in Fig. 7, we first compare the ER on the six test images as shown in Fig. 5. It can be observed that the ER of the six test images obtained by our scheme are higher than these five schemes. In addition, our scheme can achieve good performance for rough images such as 'Baboon'.

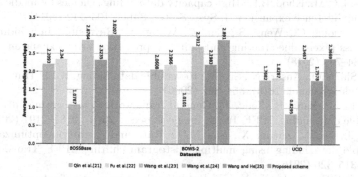

Fig. 8. Average ER (bpp) comparison on three image datasets.

Moreover, the experimental results on the three datasets of BOSSBase [26], BOWS-2 [27] and UCID [28] further confirm that our scheme outperforms the comparison schemes [21–25]. As shown in Fig. 8, the average ER of the proposed scheme on BOSSBase, BOWS-2 and UCID are 3.0207 bpp, 2.8910 bpp and 2.3669 bpp, respectively, which are higher than the average ER of the comparison schemes.

4 Conclusions

In this paper, an efficient reversible data hiding scheme for encrypted images based on pixel value preprocessing and block classification (PVPBC-RDHEI) is proposed. In the proposed scheme, the image is encrypted by preprocessing the pixel values of the original image and combining bitwise exclusive-or operation and block permutation. In addition, based on the number of consecutive consistent value bit planes counted from top-down in the pixel block, all blocks can be classified into six types and corresponding indicators can be generated for the six block types using Huffman coding. Therefore, each pixel block in the image can embed data, thus exploring a larger data embedding space. The experimental results and analysis show that the embedding rate of the proposed scheme is significantly better than the state-of-the-art schemes, and the security and reversibility are guaranteed. At the same time, the data extraction and image recovery are separable and error-free. In the future, we will investigate more efficient RDHEI schemes to further improve the embedding rate.

References

1. Jiang, R., Zhou, H., Zhang, W., Yu, N.: Reversible data hiding in encrypted three-dimensional mesh models. IEEE Trans. Multimedia **20**(1), 55–67 (2017)
2. Tian, J.: Reversible data embedding using a difference expansion. IEEE Trans. Circuits Syst. Video Technol. **13**(8), 890–896 (2003)
3. Al-Qershi, O.M., Khoo, B.E.: High capacity data hiding schemes for medical images based on difference expansion. J. Syst. Softw. **84**(1), 105–112 (2011)
4. He, W., Xiong, G., Weng, S., Cai, Z., Wang, Y.: Reversible data hiding using multi-pass pixel-value-ordering and pairwise prediction-error expansion. Inf. Sci. **467**, 784–799 (2018)
5. Ni, Z., Shi, Y.Q., Ansari, N., Su, W.: Reversible data hiding. IEEE Trans. Circuits Syst. Video Technol. **16**(3), 354–362 (2006)
6. Li, X., Li, B., Yang, B., Zeng, T.: General framework to histogram-shifting-based reversible data hiding. IEEE Trans. Image Process. **22**(6), 2181–2191 (2013)
7. Wang, J., Ni, J., Zhang, X., Shi, Y.Q.: Rate and distortion optimization for reversible data hiding using multiple histogram shifting. IEEE Trans. Cybern. **47**(2), 315–326 (2016)
8. Zhang, W., Hu, X., Li, X., Yu, N.: Recursive histogram modification: establishing equivalency between reversible data hiding and lossless data compression. IEEE Trans. Image Process. **22**(7), 2775–2785 (2013)
9. Lin, C.C., Liu, X.L., Tai, W.L., Yuan, S.M.: A novel reversible data hiding scheme based on AMBTC compression technique. Multimedia Tools Appl. **74**, 3823–3842 (2015)
10. Lin, C.C., Liu, X.L., Yuan, S.M.: Reversible data hiding for VQ-compressed images based on search-order coding and state-codebook mapping. Inf. Sci. **293**, 314–326 (2015)
11. Ou, B., Li, X., Zhao, Y., Ni, R., Shi, Y.Q.: Pairwise prediction-error expansion for efficient reversible data hiding. IEEE Trans. Image Process. **22**(12), 5010–5021 (2013)

12. Kumar, R., Jung, K.H.: Enhanced pairwise IPVO-based reversible data hiding scheme using rhombus context. Inf. Sci. **536**, 101–119 (2020)
13. Li, X., Li, J., Li, B., Yang, B.: High-fidelity reversible data hiding scheme based on pixel-value-ordering and prediction-error expansion. Signal Process. **93**(1), 198–205 (2013)
14. Qu, X., Kim, H.J.: Pixel-based pixel value ordering predictor for high-fidelity reversible data hiding. Signal Process. **111**, 249–260 (2015)
15. Ma, K., Zhang, W., Zhao, X., Yu, N., Li, F.: Reversible data hiding in encrypted images by reserving room before encryption. IEEE Trans. Inf. Forensics Secur. **8**(3), 553–562 (2013)
16. Chen, K., Chang, C.C.: High-capacity reversible data hiding in encrypted images based on extended run-length coding and block-based MSB plane rearrangement. J. Vis. Commun. Image Represent. **58**, 334–344 (2019)
17. Puteaux, P., Puech, W.: A recursive reversible data hiding in encrypted images method with a very high payload. IEEE Trans. Multimedia **23**, 636–650 (2020)
18. Gao, G., Tong, S., Xia, Z., Shi, Y.: A universal reversible data hiding method in encrypted image based on MSB prediction and error embedding. IEEE Trans. Cloud Comput. **11**(2), 1692–1706 (2022)
19. Zhang, X.: Reversible data hiding in encrypted image. IEEE Signal Process. Lett. **18**(4), 255–258 (2011)
20. Qian, Z., Zhang, X.: Reversible data hiding in encrypted images with distributed source encoding. IEEE Trans. Circuits Syst. Video Technol. **26**(4), 636–646 (2015)
21. Qin, C., Qian, X., Hong, W., Zhang, X.: An efficient coding scheme for reversible data hiding in encrypted image with redundancy transfer. Inf. Sci. **487**, 176–192 (2019)
22. Fu, Y., Kong, P., Yao, H., Tang, Z., Qin, C.: Effective reversible data hiding in encrypted image with adaptive encoding strategy. Inf. Sci. **494**, 21–36 (2019)
23. Wang, Y., Cai, Z., He, W.: High capacity reversible data hiding in encrypted image based on intra-block lossless compression. IEEE Trans. Multimedia **23**, 1466–1473 (2021)
24. Wang, X., Chang, C.C., Lin, C.C.: Reversible data hiding in encrypted images with block-based adaptive MSB encoding. Inf. Sci. **567**, 375–394 (2021)
25. Wang, Y., He, W.: High capacity reversible data hiding in encrypted image based on adaptive MSB prediction. IEEE Trans. Multimedia **24**, 1288–1298 (2022)
26. Bas, P., Filler, T., Pevný, T.: Break our steganographic system: the ins and outs of organizing BOSS. In: Filler, T., Pevný, T., Craver, S., Ker, A. (eds.) IH 2011. LNCS, vol. 6958, pp. 59–70. Springer, Heidelberg (2011). https://doi.org/10.1007/978-3-642-24178-9_5
27. Bas, P., Furon, T.: Image database of bows-2. Accessed 20 Jun 2017
28. Schaefer, G., Stich, M.: UCID: an uncompressed color image database. In: Storage and Retrieval Methods and Applications for Multimedia 2004. vol. 5307, pp. 472–480. SPIE (2003)

HPattack: An Effective Adversarial Attack for Human Parsing

Xin Dong[1,2] , Rui Wang[1,2(✉)] , Sanyi Zhang[1,2] , and Lihua Jing[1,2]

[1] Institute of Information Engineering, Chinese Academy of Sciences,
No.19 Shucun Road, Haidian District, Beijing 100085, China
[2] School of Cyber Security, University of Chinese Academy of Sciences,
Beijing, China
{dongxin,wangrui,zhangsanyi,jinglihua}@iie.ac.cn

Abstract. Adversarial attacks on human parsing models aim to mislead deep neural networks by injecting imperceptible perturbations to input images. In general, different human parts are connected in a closed region. The attacks do not work well if we directly transfer current adversarial attacks on standard semantic segmentation models to human parsers. In this paper, we propose an effective adversarial attack method called HPattack, for human parsing from two perspectives, *i.e.*, sensitive pixel mining and prediction fooling. By analyzing the characteristics of human parsing tasks, we propose exploiting the human region and contour clues to improve the attack capability. To further fool the human parsers, we introduce a novel background target attack mechanism by leading the predictions away from the correct label to obtain high-quality adversarial examples. Comparative experiments on the human parsing benchmark dataset have shown that HPattack can produce more effective adversarial examples than other methods at the same number of iterations. Furthermore, HPattack also successfully attacks the Segment Anything Model (SAM) model.

Keywords: Adversarial attack · Human parsing · Sensitive pixel mining · Prediction fooling

1 Introduction

In the past few years, many adversarial attack methods have been proposed to generate adversarial examples for image classification tasks [8,15,19–21]. The core idea is implementing very small perturbations on the clean image to mislead the deep neural networks. While for dense prediction tasks, for example, the goal of the adversarial attacks on the semantic segmentation task [3–6] dedicates to segmenting each pixel as the wrong class, they are also vulnerable to adversarial attacks similar to the classification task [2,9–13,26,27]. Human parsing [17,22] is a specific semantic segmentation task that focuses on recognizing each pixel of the human regions with the correct human part, which has great potential

S. Rudinac et al. (Eds.): MMM 2024, LNCS 14555, pp. 28–41, 2024.
https://doi.org/10.1007/978-3-031-53308-2_3

in the areas of shopping platforms, human-computer interaction, image editing, and posture analysis in medical rehabilitation domain, and so on. Like semantic segmentation models, human parsers are still vulnerable to adversarial attacks, which in turn can cause applications that use human parsers to give incorrect information or be paralyzed, creating a great potential for life. Therefore, the study of adversarial examples against human parsers is an important research step to prevent it from being attacked.

Many existing adversarial attack methods designed for semantic segmentation can not achieve good attack results on the human parsing models, so there is still lots of room for improvement since they ignore the independent characteristics of the human parsing task. The core difference between human parsing and semantic segmentation is that human parsing focuses on segmenting the human region, and the pixels outside the human region are grouped into one class, i.e., the background class. And the human parts are usually connected in a closed spatial region. An effective adversarial attack should meet the conditions of implementing imperceptible perturbations on the vulnerable pixels and misleading more pixels to be classified as the wrong classes. Motivated by these inherent properties and the attack demand, in this paper, we propose an effective adversarial attack method for human parsing called **HPattack**, it formulates two distinct mechanisms into a unified framework, i.e., sensitive pixel mining and prediction fooling. For sensitive pixel mining, we find that the loss of pixels in the background region impedes improving the attack capability, so pixel selection should ignore this part. Apart from that, the pixels in the contour region [4,22] are wrongly recognized will lead to the whole segmentation accuracy decrease. Thus, the adversarial attack method should not only focus on the pixels inside the human region but also emphatically perturb the pixels in the contour region. Except for pixel mining, misleading the parsing predictions is also essential to further improve the attack capability. The dodging attack [7,24] and impersonation attack [1,23] are designed to keep the prediction results away from the ground truth, and close to a pre-defined target, respectively. To inherit the advantages of dodging and impersonation attacks, we introduce a novel background target attack fusing two types of attacks together to guide the generation of better adversarial examples, i.e., the prediction results of each pixel are close to the background class. Extensive experiments on the human parsing benchmark dataset LIP [18] have shown the effectiveness of the proposed HPattack, it can achieve the best attack performance over state-of-the-art adversarial attack methods. In addition, we also conduct experiments on the Segment Anything model (SAM) [14], the proposed HPattack obtains better attack results than other methods. The main contributions can be summarized in the following:

– We propose an effective adversarial attack method for human parsing, HPattack, which can exploit and inject imperceptible perturbations on sensitive human body pixels and effectively fool human parsers into classifying more pixels as wrong labels.

- The human body and contour regions are introduced to exploit useful pixels for decreasing the accuracy of the whole human parsing. We find that pixels in the background region are harmful, and those in the contour region are conducive to the adversarial attack.
- A novel background target attack that combines dodging and impersonation attacks is proposed to further enhance the attack capability for generating high-quality adversarial examples.
- Extensive experimental results show that HPattack can generate effective adversarial examples. It achieves over 93.36% success rate using only three iterations. In addition, HPattack is also more capable of attacking SAM than other methods.

2   Related Work

2.1   Adversarial Attacks on Segmentation Models

Adversarial attacks aim at adding small and imperceptible perturbations to the input of a deep neural network to generate adversarial examples that interfere with the model's capabilities. In semantic segmentation [4,6,30], most of the current work is dedicated to discovering the vulnerability of semantic segmentation models using existing adversarial attack methods [9,13,26,27]. In addition, the study [2] proposes a method for benchmarking the robustness of segmentation models, showing that segmentation models have different performances under FGSM [8] and BIM [15] attacks than classification methods, and demonstrating the robustness of segmentation models under multi-scale transformations. Hendrik *et.al.* [12] used universal adversarial perturbation to generate adversarial examples against semantic segmentation. MLAttack [11] adds the loss of the output of multiple intermediate layers of the model to avoid the impact of multi-scale analysis on attack performance. SegPGD [10] generates effective and efficient adversarial examples by analyzing the relationship between correctly and incorrectly classified pixels, calculated the loss function of both separately, and tested the capability to attack against robust semantic segmentation models, and found that these adversarially trained models are not resistant to the adversarial attack of SegPGD. All these works show that existing semantic segmentation models are highly against well-designed adversarial examples, and it is essential to study more effective attack methods and apply them to defense such as adversarial training. In this paper, the proposed approach is mainly improving the attack capability of white-box attacks on human parsing.

2.2   Human Parsing

Human parsing models aims to classify each image pixel as the correct human part, i.e., a pixel point at a given location is judged to belong to which of the categories of head, arm, leg, etc. The current solutions mainly exploit valuable clues to improve the parsing capability. Researchers have developed various useful clues, such as multi-scale context information is useful to solve various scale

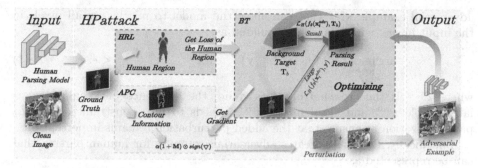

Fig. 1. The framework of HPattack. The input image is first fed into the pixel mining module, *i.e.*, the human region and the contour region (red and yellow background area), which leads the HPattack focusing on the sensitive pixels. Then the background target attack (purple background area) is further employed to optimize the model generating error prediction results. Finally, HPattack outputs a new adversarial image with imperceptible perturbation.

problems (*e.g.*, JPPNet [18] exploited the correlation between the two tasks of parsing and pose estimation, making them mutually reinforcing, CE2P [22] added edge information into the loss function to enhance the ability of human parsing, PGECNet [28] employed a Gather-Excite operation to accurately reflect relevant human parts of various scales), edge information across adjacent human classes helps to obtain better boundaries [22], fusing edge, human pose and parsing together *e.g.*, CorrPM [29], and human hierarchy structure [16,25]. In addition, some other mechanisms are also proposed, for example, self-correction mechanism SCHP [17] is proposed to utilize the pseudo label assistance and edge consistency, and so on.

3 HPattack: Adversarial Attack for Human Parsing

The proposed HPattack contains two main parts, *i.e.*, adversarial sensitive pixel mining (human body and contour regions) and prediction fooling (background target attack), the detailed HPattack is shown in Fig. 1.

3.1 PGD for Human Parsing

In this section, we first make a formulaic statement on the adversarial attack for human parsing by the baseline PGD [19]. In human parsing, given a human parsing model f_θ parameterized by θ, a clean input image $\mathbf{x}^{clean} \in \mathbb{R}^{H \times W \times 3}$ and its corresponding ground truth $\mathbf{y} \in \mathbb{R}^{H \times W}$, each value of $\mathbf{y}$ belongs to $C = \{0, 1, ..., N-1\}$, where N is the total number of categories. The model classifies each pixel of the input image $f_\theta\left(\mathbf{x}^{clean}\right) \in \mathbb{R}^{H \times W \times N}$, where (H, W) is the sizes of the input image. PGD for human parsing aims at finding an adversarial example $\mathbf{x}^{adv}$ that can allow the model to misclassify all pixels in the input image, which can be formulated as:

$$\arg\max_{\mathbf{x}^{adv}} \mathcal{L}^{CE}\left(f_\theta\left(\mathbf{x}^{adv}\right), \mathbf{y}\right), \quad ||\mathbf{x}^{adv} - \mathbf{x}^{clean}||_p < \epsilon, \tag{1}$$

where $\mathcal{L}^{CE}$ is the cross-entropy loss value, the difference between adversarial example $\mathbf{x}^{adv}$ and clean image $\mathbf{x}^{clean}$ needs to be ϵ-constraint at the L_p parametrization to ensure that the added perturbation remains imperceptible.

Specifically, PGD [19] creates adversarial examples for human parsing that can be represented as:

$$\mathbf{x}_{t+1}^{adv} = \text{Clip}_{\epsilon,\mathbf{x}^{clean}}\left(\mathbf{x}_t^{adv} + \alpha \times sign\left(\nabla_{\mathbf{x}_t^{adv}}\mathcal{L}^{CE}\left(f_\theta\left(\mathbf{x}_t^{adv}\right), \mathbf{y}\right)\right)\right), \tag{2}$$

where α and ϵ represent each iteration's step size and perturbation range. $\mathbf{x}_t^{adv}$ is the adversarial example generated at the t-th iteration, and the initial value $\mathbf{x}_0^{adv}$ is set to $\mathbf{x}_0^{adv} = \mathbf{x}^{clean} + \mathcal{U}\left(-\epsilon, +\epsilon\right)$, i.e., random initialization of the perturbation on the input image. The Clip $(\cdot)$ function constrains the perturbation to the L_p parametrization under the ϵ-constraint. $\mathcal{L}^{CE}$ is required to get gradually larger during the attacking process, i.e., leading adversarial perturbation added according to the direction of the gradient.

3.2   Adversarial Pixel Mining of HPattack

Human Region Loss of HPattack (HRL). For the adversarial attack on human parsing, the core goal is to misclassify as many pixels of the human region as possible, and misclassification of the background region is not necessary. However, the background pixels may take up a large ratio of the whole image, so the loss value of the background's pixels will affect the optimization direction of the pixel gradient in the human region and the attack capability of the adversarial example. To reduce the influence, we ignore the pixels of the background region computed in the loss function to guide the adversarial example generation. Thus, original loss $\mathcal{L}$ can be expressed as the sum of two separate components, i.e.,

$$\mathcal{L} = \mathcal{L}_{B+H} = \frac{1}{H \times W}\left(\sum_{i \in S_B} \mathcal{L}_i^{CE} + \sum_{j \in S_H} \mathcal{L}_j^{CE}\right), \tag{3}$$

where $\mathcal{L}^{CE}$ is the cross-entropy loss for each pixel, S_B is the set of pixels in the background region, and S_H is the set of pixels in the human region.

Based on the above considerations, we first use ground truth $\mathbf{y}$ to select pixels that are in the human region, then keep only the loss of these pixels to guide the generation of the adversarial example, as Eq. (4) shows.

$$\mathcal{L}_H = \frac{1}{H \times W}\sum_{j \in S_H} \mathcal{L}_j^{CE}. \tag{4}$$

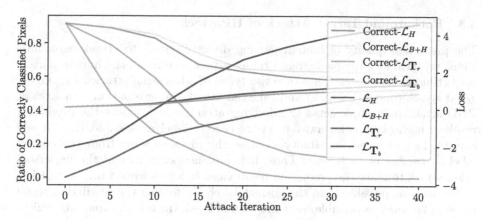

Fig. 2. The effectiveness validation of the loss design implemented on pixels only in the human region ($\mathcal{L}_H$) and background target attack ($\mathcal{L}_{T_b}$) to improve the attack capability of PGD when attacking CE2P.

To demonstrate the validity of Eq. (4), we compare the attack capability of Eq. (4) and Eq. (3) as loss functions in PGD, respectively. We calculate the ratio of pixels still correctly classified by CE2P [22] and the value of the loss function. As shown in Fig. 2, Eq. (4) has fewer correctly classified pixels and larger loss values compared to Eq. (3). This indicates that the loss of pixels in the background region affects the direction of the gradient and blocks the improvement of attack capability.

Aggravated Perturbation on Contour of HPattack (APC). The correct classification of the contour regions can sharpen and refine the segmentation results and thus further help the pixels within each class to be correctly classified, it has been widely verified for improving the performance of human parsing and semantic segmentation models [6,22]. Inspired by this, we add one extra perturbation to the pixels belonging to the contour region of the clean image. Specifically, we extract contour mask $\mathbf{M} \in \mathbb{R}^{H \times W}$ for the ground truth of the input image as shown in Fig. 1 via

$$\mathbf{M} = g(\mathbf{y}), \tag{5}$$

where $g(\cdot)$ is a contour extraction algorithm used in CE2P [22]. The value $\mathbf{M}_i \in \{0,1\}$ for each position, 1 means that the point at that position lies on the contour, and 0 means that it does not. Then we add one more perturbation value to the points on the contour via

$$\mathbf{x}_{t+1}^{adv} = \text{Clip}_{\epsilon, \mathbf{x}^{clean}} \left(\mathbf{x}_t^{adv} + \alpha \left(1 + \mathbf{M} \right) \odot sign \left(\nabla_{\mathbf{x}_t^{adv}} \mathcal{L}_H \left(f_\theta \left(\mathbf{x}_t^{adv} \right), \mathbf{y} \right) \right) \right). \tag{6}$$

3.3   Background Target Attack of HPattack

The prediction quality of human parsing directly relates to attack capability. Thus, we introduce the background target attack to mislead the human parsers predicting wrong classes. There are two types of adversarial attacks, $i.e.$, untargeted attack (dodging attack) and targeted attack (impersonation attack). The PGD attack only implements the dodging attack, which makes the prediction result of each pixel as far away from the correct label as possible. Still, it will cause a lack of attack capability because the distance away from the correct label is insufficient each time. Thus, it is also necessary to add the impersonation attack to make the prediction result close to a pre-defined target, which is as far away as possible from the correct label. By forcing the prediction result to be as far away as possible from the correct label, the attack capability will be significantly enhanced.

Since dense human parsing focuses on classifying each pixel as the correct label, we only set a target value for each pixel for convenience. We have stated that the pixels in the background region limit the attack ability. Thus, we set the target for each pixel in the human region to be 0, $i.e.$, leading the model to classify pixels of the human region as background class. We define it as background target $\mathbf{T}_b \in \mathbb{R}^{H \times W}$, where each value is 0. So the loss function $\mathcal{L}_H$ in Eq. (6) can be replaced as

$$\mathcal{L}_{\mathbf{T}_b} = \mathcal{L}_H \left(f_\theta \left(\mathbf{x}_t^{adv} \right), \mathbf{y} \right) - \mathcal{L}_H \left(f_\theta \left(\mathbf{x}_t^{adv} \right), \mathbf{T}_b \right), \tag{7}$$

where $\mathcal{L}_H(f_\theta(\mathbf{x}_t^{adv}), \mathbf{y})$ represents the loss of the dodging attack, which has to become larger to move away from the correct labels. $\mathcal{L}_H(f_\theta(\mathbf{x}_t^{adv}), \mathbf{T}_b)$ represents the loss of impersonation attack, which has to keep getting smaller as to move closer to the background target, we define this module as **BT**.

Regarding the different choices of the target label, in addition to choosing the background class 0, it is also possible to randomly select any class other than the correct label for each pixel. Specifically, we design a comparison that randomly chooses another class as the target label. For a pixel $j \in S_H$ with a correct class $\mathbf{y}_j$, then randomly choose from $C \setminus \mathbf{y}_j$. Combining the results of random selection per pixel, we denote this random target as $\mathbf{T}_r \in \mathbb{R}^{H \times W}$.

Again, we compared the two choices of the target in Fig. 2 and find that the loss function of $\mathbf{T}_b$ increases more than twice as much as that of $\mathbf{T}_r$. The ratio of pixels correctly classified by $\mathbf{T}_b$ decreases faster than that of $\mathbf{T}_r$ and is already close to 0% at the 40-th iteration, which shows that $\mathbf{T}_b$ will be more capable of attacking than $\mathbf{T}_r$. After that, a detailed comparison experiment of their attacking capability is conducted in Subsect. 4.3.

In summary, HPattack first use ground truth to select pixels belonging to the human region and get the contour mask, then calculates the loss of the pixels in the human region with background target and ground truth, finally finding the gradient of the loss function $\mathcal{L}_{\mathbf{T}_b}$ with respect to $\mathbf{x}_t^{adv}$ and adding one more perturbation at the contour region. The overall flow is shown in Alg. 1.

Algorithm 1. HPattack

Input: The human parsing model f_θ, the loss function $\mathcal{L}_{\mathbf{T}_b}$, a clean image $\mathbf{x}^{clean}$, the number of iteration T, the corresponding ground truth $\mathbf{y}$.
Output: The adversarial example $\mathbf{x}^{adv}$.
1: $\mathbf{x}_0^{adv} \leftarrow \mathbf{x}^{clean} + \mathcal{U}(-\epsilon, +\epsilon)$;
2: **for** $t = 0$ to $T - 1$ **do**
3:     Get pixels of human region and calculate the $\mathcal{L}_{\mathbf{T}_b}$ via $\mathcal{L}_{\mathbf{T}_b} = \mathcal{L}_H\left(f_\theta\left(\mathbf{x}_t^{adv}\right), \mathbf{y}\right) - \mathcal{L}_H\left(f_\theta\left(\mathbf{x}_t^{adv}\right), \mathbf{T}_b\right)$;
4:     Compute the gradient $\nabla_{\mathbf{x}_t^{adv}} \mathcal{L}_{\mathbf{T}_b}$;
5:     Get contour mask $\mathbf{M}$ and generate adversarial example $\mathbf{x}_{t+1}^{adv}$ via $\mathbf{x}_{t+1}^{adv} = \text{Clip}_{\epsilon, \mathbf{x}^{clean}}\left(\mathbf{x}_t^{adv} + \alpha\left(1 + \mathbf{M}\right) \odot sign\left(\nabla_{\mathbf{x}_t^{adv}} \mathcal{L}_{\mathbf{T}_b}\right)\right)$;
6: **end for**
7: **return** $\mathbf{x}^{adv}$.

---

4   Experiments

In this section, we first introduce the experimental setup, and then divide experiments into evaluation of attack capability, ablation study, attack segment anything model, and performance against defense methods, all of which indicate that our HPattack has high performance.

4.1   Experimental Settings

We choose the large-scale human parsing benchmark dataset LIP [18] to verify the effectiveness of the proposed HPattack. There are 50,462 images in total, 30,462/10,000/10,000 images are chosen for training, validation, and testing, respectively. All images are annotated with categories $C = \{0, 1, ..., 19\}$, *i.e.*, a background class 0, and 19 categories belonging to the human region. Specifically, we implement adversarial attacks on the LIP validation set.

To verify the effectiveness of the attack, we choose two state-of-the-art models, CE2P [22] and SCHP [17], as the two attacked models for human parsing. The evaluation metric is the standard mIoU [22], which is commonly used in human parsing, *i.e.*, IoU is first calculated for each category, followed by the average value. We use mIoU for $C \setminus 0$ to reflect the attack success rate, the smaller the value of mIoU, the higher the attack success rate. The comparison methods we chose include two traditional adversarial attack methods, BIM [15] and PGD [19], and also two adversarial attack methods, MLAttack [11] and SegPGD [10] for semantic segmentation.

For the setting of hyperparameters, we choose a maximum perturbation value of $\epsilon = \frac{8}{255}$, a step size of $\frac{\epsilon}{T}$, and a constraint of L_∞ according to the setting of SegPGD [10]. We test the attack capability from the number of iterations T starting from 3 and up to 100.

Table 1. Comparison of the attacking capability, the smaller the mIoU score means the stronger the attack capability, the best results are shown in bold.

Model		mIoU						
CE2P [22]	Clean	52.86						
	Method/iter	3	5	7	10	20	40	100
	BIM [15]	15.34	12.73	12.15	10.01	9.23	8.47	7.82
	PGD [19]	14.70	11.85	11.47	9.92	8.63	7.77	7.33
	MLAttack [11]	13.11	10.70	10.31	8.64	7.48	6.13	5.79
	SegPGD [10]	10.06	7.80	7.13	5.93	5.07	4.57	4.03
	HPattack (Ours)	**3.20**	**1.43**	**0.64**	**0.37**	**0.16**	**0.11**	**0.09**
SCHP [17]	Clean	57.06						
	Method/iter	3	5	7	10	20	40	100
	BIM [15]	19.03	17.98	16.90	14.83	13.99	12.58	12.10
	PGD [19]	18.17	15.48	14.26	13.36	12.36	11.29	11.05
	MLAttack [11]	16.74	13.14	12.79	11.46	10.93	10.57	9.91
	SegPGD [10]	10.10	8.59	8.03	7.62	6.90	6.37	5.86
	HPattack (Ours)	**4.12**	**2.90**	**2.37**	**2.05**	**1.51**	**1.22**	**1.18**

4.2 Evaluation of Attack Capability

In this section, we conduct comparative experiments to show the effectiveness of our HPattack. Specifically, we first test the models' parsing capability on the clean images and then compare our HPattack with four state-of-the-art adversarial attack methods.

As shown in Table 1, the results for the clean images of CE2P [22] and SCHP [17] are 52.86 and 57.06, respectively. It achieves the worst attack capability for the BIM [15] attack, while PGD [19] has a better attack effect than BIM. MLAttack [11] reduces to about 5.79 at the 100-th iteration tested on the CE2P model, but SegPGD [10] method can reduce to about 5.93 just with 10 iterations. If we adopt the attack mechanism of our HPattack, it just needs three iterations which attack the CE2P [22] and the SCHP [17] both drop too much less than 5. This indicates that our HPattack method can generate adversarial examples with strong attack capability in a short time. In addition, if we increase more iterations implemented with our HPattack, the attack success rate can close to 100%, *i.e.*, let mIoU be almost 0. This means that the human parsing model is almost invalid.

Besides, we also provide qualitative comparison results in Fig. 3, where segmentation results are obtained by testing adversarial examples on the CE2P and SCHP model with 3 attack iterations. In particular, we choose the SegPGD as a comparison, which performed better in quantitative experiments than other comparison methods. It can be observed that the human parsing results of the HPattack misclassified almost all pixels of the whole image region compared to the SegPGD method.

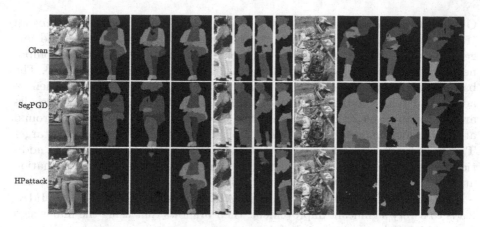

Fig. 3. The qualitative analysis of the adversarial attack. For each colorful image, the first column is itself, the second is the prediction result of CE2P on it, the third is the prediction result of SCHP on it, and the forth is the ground truth.

Furthermore, since we add the additional perturbations to the contour region in our HPattack, there is a risk of making the perturbation more prominent and thus resulting in the adversarial examples being vulnerable to human perception. Still, as shown in Fig. 3, the generated adversarial examples have no significant perturbations compared to the clean images and the adversarial examples generated by SegPGD. Overall, it can be seen that the attack capability of HPattack is significantly better than SegPGD.

4.3   Ablation Study

Table 2. Ablation study for HPattack, the smaller the mIoU, the stronger the attack capability, the best results are shown in bold.

Model	Method	mIoU						
		3	5	7	10	20	40	100
CE2P [22]	PGD [19]	14.70	11.85	11.47	9.92	8.63	7.77	7.33
	+HRL	10.49	8.11	7.05	5.09	4.29	3.37	3.05
	+HRL+APC	10.02	7.74	6.16	5.00	4.03	3.34	3.01
	+HRL+APC+RT	7.16	5.10	3.42	3.14	1.76	1.33	0.91
	+HRL+APC+IMA	4.77	2.13	1.35	1.07	0.66	0.53	0.49
	+HRL+APC+BT (HPattack)	**3.20**	**1.43**	**0.64**	**0.37**	**0.16**	**0.11**	**0.09**
SCHP [17]	PGD [19]	18.17	15.48	14.26	13.36	12.36	11.29	11.05
	+HRL	11.98	9.34	8.51	7.69	6.81	5.97	5.58
	+HRL+APC	11.23	8.66	7.85	7.03	6.13	5.23	4.87
	+HRL+APC+RT	7.58	6.83	5.00	4.08	3.82	3.47	3.18
	+HRL+APC+IMA	6.59	3.67	2.91	2.58	1.92	1.69	1.55
	+HRL+APC+BT (HPattack)	**4.12**	**2.90**	**2.37**	**2.05**	**1.51**	**1.22**	**1.18**

Our HPattack method consists of three components: the loss of human region (HRL), aggravated perturbation on contour region (APC), and background target attack (BT). Thus, we conduct ablation experiments on these three components to test the attack capability for attacking CE2P [22] and SCHP [17]. The baseline method is implemented with the standard PGD attack method. Then we add three modules to the PGD inch by inch, and the detailed comparison results are shown in Table 2. To further explain the effectiveness of the background attack mechanism, we provide a comparison that replaces the background target $\mathbf{T}_b$ with a random target $\mathbf{T}_r$, and we denote it as (+HRL+APC+RT). In addition, to test the effectiveness of using the minus sign to combine impersonation attacks, we tested the attack capability of the second part in Eq. 7 ($\mathcal{L}_{\mathbf{T}_b}$), *i.e.*, $-\mathcal{L}_H(f_\theta(\mathbf{x}_t^{adv}), \mathbf{T}_b)$, denoted as (+HRL+APC+IMA). If (+HRL+APC+IMA) leads to a very significant improvement, then it shows that using the minus sign for combining does not affect performance.

As shown in Table 2, the human region loss (HRL) module reduces the mIoU to two-thirds of PGD, and aggravated perturbation on contour (APC) only reduces the mIoU by about 0.5 compared to (+HRL) because it only adds one more perturbation to the pixels in the contour region. In contrast, adding the background target (BT) can obtain a performance drop of about 80% than the PGD result tested on the CE2P model with 3 attack iterations, which is the most significant improvement of the attack capability. From Table 2, we can observe that the random target (RT) will play some role in reducing mIoU relative to human region loss and aggravated perturbation on contour. However, the performance reduction is not significant enough relative to the background target attack, which verifies the analysis in Sect. 3.3 that the random target is not as effective as the background target in improving the attack capability. The mIoU scores of (PGD+HRL+APC+IMA) show that using the minus sign as the combination strategy can reduce mIoU to below 5 in many cases, *i.e.*, adopting the minus sign mechanism is effective.

4.4   Attack Segment Anything Model

We have shown the effectiveness of the HPattack on the standard human parsing models, an interesting discussion is whether the HPattack mechanism can be transferred to the large-scale pre-trained segmentation model, such as Segment Anything Model (SAM) [14]. SAM is a state-of-the-art image segmentation model trained with 11 million images and 1.1 billion masks, and SAM has very high accuracy in various semantic segmentation tasks. Specifically, we choose the pre-trained ViT-Base model of SAM to attack and use the SegPGD [10] as a comparison method, which performed relatively better in quantitative experiments. We should note that the current SAM can only output the segmentation result of the whole image, but the actual class of each pixel is not given.

The detailed comparison results are shown in Fig. 4. The highlighted regions with red circles show that HPattack fools the SAM model with worse segmentation results than SegPGD. For example, the first, second, and fourth columns of the image after being attacked by our HPattack compared to SegPGD, will

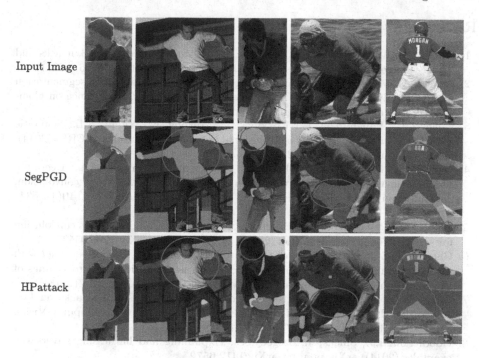

Fig. 4. Comparison of attacking SAM. HPattack performs better than SegPGD.

make SAM unable to segment the upper clothes or pants; the third and fifth columns will make the SAM fail to segment hair and shoes, respectively.

5   Conclusion

In this paper, we propose an effective adversarial attack method for human parsing, which boosts the adversarial attack capability by progressively combining the loss of human region, the aggravated perturbation on contour, and the background target attack. HPattack owns good attack ability on the human parsing models and applies to large-scale pre-trained segmentation model SAM. In conclusion, our approach can provide a meaningful reference for subsequent research on human parsing and even semantic segmentation.

Acknowledgements. This work is supported in part by the National Natural Science Foundation of China Under Grants No.62176253, No.62202461 and China Postdoctoral Science Foundation No.2022M723364.

References

1. Akhtar, N., Mian, A., Kardan, N., Shah, M.: Advances in adversarial attacks and defenses in computer vision: a survey. IEEE Access **9**, 155161–155196 (2021)
2. Arnab, A., Miksik, O., Torr, P.H.: On the robustness of semantic segmentation models to adversarial attacks. In: Proceedings of the IEEE Conference on Computer Vision and Pattern Recognition, pp. 888–897 (2018)
3. Chen, L.C., Papandreou, G., Kokkinos, I., Murphy, K., Yuille, A.L.: Semantic image segmentation with deep convolutional nets and fully connected CRFs (2014). arXiv preprint arXiv:1412.7062
4. Chen, L.C., Papandreou, G., Kokkinos, I., Murphy, K., Yuille, A.L.: DeepLab: semantic image segmentation with deep convolutional nets, atrous convolution, and fully connected CRFs. IEEE Trans. Pattern Anal. Mach. Intell. **40**(4), 834–848 (2017)
5. Chen, L.C., Papandreou, G., Schroff, F., Adam, H.: Rethinking atrous convolution for semantic image segmentation (2017). arXiv preprint arXiv:1706.05587
6. Chen, L.C., Zhu, Y., Papandreou, G., Schroff, F., Adam, H.: Encoder-decoder with atrous separable convolution for semantic image segmentation. In: Proceedings of the European Conference on Computer Vision (ECCV), pp. 801–818 (2018)
7. Dong, Y., et al.: Efficient decision-based black-box adversarial attacks on face recognition. In: Proceedings of the IEEE/CVF Conference on Computer Vision and Pattern Recognition, pp. 7714–7722 (2019)
8. Goodfellow, I.J., Shlens, J., Szegedy, C.: Explaining and harnessing adversarial examples (2014). arXiv preprint arXiv:1412.6572
9. Gu, J., Zhao, H., Tresp, V., Torr, P.: Adversarial examples on segmentation models can be easy to transfer (2021). arXiv preprint arXiv:2111.11368
10. Gu, J., Zhao, H., Tresp, V., Torr, P.H.S.: SegPGD: an effective and efficient adversarial attack for evaluating and boosting segmentation robustness. In: Avidan, S., Brostow, G., Cissé, M., Farinella, G.M., Hassner, T. (eds.) Computer Vision – ECCV 2022. ECCV 2022. LNCS, vol. 13689. Springer, Cham (2022). https://doi.org/10.1007/978-3-031-19818-2_18
11. Gupta, P., Rahtu, E.: MLAttack: fooling semantic segmentation networks by multilayer attacks. In: Fink, G.A., Frintrop, S., Jiang, X. (eds.) DAGM GCPR 2019. LNCS, vol. 11824, pp. 401–413. Springer, Cham (2019). https://doi.org/10.1007/978-3-030-33676-9_28
12. Hendrik Metzen, J., Chaithanya Kumar, M., Brox, T., Fischer, V.: Universal adversarial perturbations against semantic image segmentation. In: Proceedings of the IEEE International Conference on Computer Vision, pp. 2755–2764 (2017)
13. Kang, X., Song, B., Du, X., Guizani, M.: Adversarial attacks for image segmentation on multiple lightweight models. IEEE Access **8**, 31359–31370 (2020)
14. Kirillov, A., et al.: Segment anything (2023). arXiv preprint arXiv:2304.02643
15. Kurakin, A., Goodfellow, I., Bengio, S.: Adversarial examples in the physical world. arXiv e-prints pp. arXiv-1607 (2016)
16. Li, L., Zhou, T., Wang, W., Li, J., Yang, Y.: Deep hierarchical semantic segmentation. In: Proceedings of the IEEE/CVF Conference on Computer Vision and Pattern Recognition, pp. 1246–1257 (2022)
17. Li, P., Xu, Y., Wei, Y., Yang, Y.: Self-correction for human parsing. IEEE Trans. Pattern Anal. Mach. Intell. **44**(6), 3260–3271 (2022)
18. Liang, X., Gong, K., Shen, X., Lin, L.: Look into person: Joint body parsing & pose estimation network and a new benchmark. IEEE Trans. Pattern Anal. Mach. Intell. **41**(4), 871–885 (2018)

19. Madry, A., Makelov, A., Schmidt, L., Tsipras, D., Vladu, A.: Towards deep learning models resistant to adversarial attacks (2017). arXiv preprint arXiv:1706.06083
20. Moosavi-Dezfooli, S.M., Fawzi, A., Frossard, P.: DeepFool: a simple and accurate method to fool deep neural networks. In: Proceedings of the IEEE Conference on Computer Vision and Pattern Recognition, pp. 2574–2582 (2016)
21. Papernot, N., McDaniel, P., Jha, S., Fredrikson, M., Celik, Z.B., Swami, A.: The limitations of deep learning in adversarial settings. In: 2016 IEEE European Symposium on Security and Privacy (EuroS&P), pp. 372–387. IEEE (2016)
22. Ruan, T., Liu, T., Huang, Z., Wei, Y., Wei, S., Zhao, Y.: Devil in the details: towards accurate single and multiple human parsing. In: Proceedings of the AAAI Conference on Artificial Intelligence. vol. 33, pp. 4814–4821 (2019)
23. Sharif, M., Bhagavatula, S., Bauer, L., Reiter, M.K.: A general framework for adversarial examples with objectives. ACM Trans. Priv. Secur. (TOPS) **22**(3), 1–30 (2019)
24. Sun, L., Tan, M., Zhou, Z.: A survey of practical adversarial example attacks. Cybersecurity **1**, 1–9 (2018)
25. Wang, W., Zhou, T., Qi, S., Shen, J., Zhu, S.C.: Hierarchical human semantic parsing with comprehensive part-relation modeling. IEEE Trans. Pattern Anal. Mach. Intell. **44**(7), 3508–3522 (2021)
26. Xiao, C., Deng, R., Li, B., Yu, F., Liu, M., Song, D.: Characterizing adversarial examples based on spatial consistency information for semantic segmentation. In: Proceedings of the European Conference on Computer Vision (ECCV), pp. 217–234 (2018)
27. Xie, C., Wang, J., Zhang, Z., Zhou, Y., Xie, L., Yuille, A.: Adversarial examples for semantic segmentation and object detection. In: Proceedings of the IEEE International Conference on Computer Vision, pp. 1369–1378 (2017)
28. Zhang, S., Qi, G.J., Cao, X., Song, Z., Zhou, J.: Human parsing with pyramidical gather-excite context. IEEE Trans. Circuits Syst. Video Technol. **31**(3), 1016–1030 (2020)
29. Zhang, Z., Su, C., Zheng, L., Xie, X.: Correlating edge, pose with parsing. In: Proceedings of the IEEE/CVF Conference on Computer Vision and Pattern Recognition, pp. 8900–8909 (2020)
30. Zhao, H., Shi, J., Qi, X., Wang, X., Jia, J.: Pyramid scene parsing network. In: Proceedings of the IEEE Conference on Computer Vision and Pattern Recognition, pp. 2881–2890 (2017)

Dynamic-Static Graph Convolutional Network for Video-Based Facial Expression Recognition

Fahong Wang[1], Zhao Liu[2], Jie Lei[1(✉)], Zeyu Zou[2], Wentao Han[2], Juan Xu[2], Xuan Li[2], Zunlei Feng[3], and Ronghua Liang[1]

[1] College of Computer Science, Zhejiang University of Technology, Hangzhou, China
{fhwang,jasonlei,rhliang}@zjut.edu.cn
[2] Ping an Life Insurance of China, Ltd., Shanghai, China
{liuzhao556,zouzeyu313,xujuan635,lixuan208}@pingan.com.cn,
wthan@zjut.edu.cn
[3] College of of Computer Science, Zhejiang University, Hangzhou, China
zunleifeng@zju.edu.cn

Abstract. Most of the current methods for video-based facial expression recognition (FER) in the wild are based on deep neural networks with attention mechanism to capture the relationships between frames. However, these methods suffer from the large variations of expression patterns and data uncertainties. This paper proposes a Dynamic-Static Graph Convolutional Network (DSGCN), which mainly consists of a Static-Relational graph (SRG) and a Dynamic-Relational graph (DRG). The SRG aims to guide the network to learn the static spatial relationship of facial expressions in each video frame, strengthening the salient areas of the face through the dependencies of context nodes. The DRG learns the dynamic temporal relationship of facial expressions by aggregating video sequence features, constructing a graph with other samples within a batch to share facial expression features with different contexts, thus promoting feature diversity to improve robustness. The proposed DSGCN framework achieves state-of-the-art results on the FERV39K, DFEW and AFEW benchmarks, and ablation experiments verify the effectiveness of each module.

Keywords: video-based facial expression recognition · graph convolutional networks · dynamic-static relation

1 Introduction

Currently, automatic recognition of facial expressions, which plays a crucial role in various human-computer interaction systems [1], including medical treatment, driver assistance, and other areas, has been a popular subject for researchers. The goal of facial expression recognition (FER) is to classify the input images into seven basic expressions: neutral, happy, sad, surprised, afraid, disgusted,

S. Rudinac et al. (Eds.): MMM 2024, LNCS 14555, pp. 42–55, 2024.
https://doi.org/10.1007/978-3-031-53308-2_4

and angry. According to different types of input data, the FER systems can be divided into image-based FER and video-based FER. Early FER methods mainly focused on image-based facial expressions. However, since facial expression is a dynamic process characterized by the interplay of muscle movements in different regions of the face, understanding the temporal sequence of expressions, plays a more important role than classifying static images. As a result, video-based FER research has received increasing attention in recent years.

According to different data scenarios, video-based FER datasets can be mainly divided into lab-controlled and in-the-wild. For the lab-controlled datasets, all video sequences are collected in a controlled laboratory environment, and the videos are relatively simple and free from occlusion. Representative datasets include CK+ [2], Oulu-CASIA [3], and MMI [4]. For the in-the-wild datasets (e.g., AFEW [5], Ferv39k [6], and DFEW [7]), video sequences are collected from real-world scenes, which are closer to natural facial events. Furthermore, the in-the-wild datasets are captured from thousands of subjects in complex scenes, which greatly increases the diversity of the data. Nowadays, the focus of video-based FER research has shifted from laboratory controls to challenges under field conditions.

Early methods for solving video-based FER were primarily based on hand-crafted features, such as the LBP-TOP [8], STLMBP [9], and HOG-TOP [10]. In addition, Liu et al. [11] introduced a spatio-temporal manifold(STM) method to model the video clips, and Liu et al. [12] used different Riemannian kernels measuring the similarity distance between sequences. In recent years, deep learning based methods gradually replace traditional methods. Among these methods, the RNN-based method performed better at capturing the temporal relationship between frames, and the spatial self-attention emerged [13] was proposed as a powerful tool for guiding the extraction of image features and determining the importance of each local feature. However, these methods focused on limited attention features or relationships from a single perspective, thus neglecting large variations of different perspective expression patterns and data uncertainties. The 3D CNN-based method [14] was able to learn both spatial and temporal features in the sequence, but failed to effectively utilize long-distance attention-dependent information to extract rich emotional features. Meanwhile, CNN-based methods require stacking multiple layers of convolutional layers to enlarge the receptive field. However, this often leads to the loss of input information, increases computational load, and may even result in gradient vanishing issues.

Motivated by the above shortages of existing methods, in this paper we propose a novel dynamic-static graph convolutional network (DSGCN) for video-based FER. DSGCN consists of the Static-relational graph (SRG) and the Dynamic-relational graph (DRG). Specifically, in SRG, our method first focuses on the static spatial features extracted from the input facial expression images, and then constructs GCN for these features, the features from each frame are used as the vertex and the spatial similarity are used as the edges. Thus the constructed SRG strengthens the salient areas of the face through the dependencies of context nodes, and weakens the impact of in-the-wild factors (illumination changes, non-frontal head poses, facial occlusions) on the final recognition. In DRG our method first aggregates the features of the entire input video sequence to learn the dynamic temporal information, then constructs GCN on other video samples in the same batch. The sample nodes in the batch share features through similarity, improving the robustness of facial expressions extracted in a single situation, thus better dealing with complex and changeable real situations. At last, the video-based FER task is transferred into a node classification problem in the graphs constructed on the batches.

In summary, this paper has the following contributions:

(1) We propose DSGCN that simultaneously captures static spatial feature relationships, long-distance dynamic temporal dependencies and sample similarity relationships to gain efficient expression-related features.
(2) We present a graph-based approach for solving the task of video-based facial expression recognition by casting it as a node classification problem.
(3) Extensive experiments demonstrate DSGCN is able to outperform the baseline model significantly and achieve state-of-the-art results on three popular video-based FER datasets. Ablation studies verify the effectiveness of the composed modules (*i.e.*, SRG, DRG).

2   Related Work

2.1   Image-Based FER in the Wild

The Image-based FER mainly consists of three stages, namely face detection, feature extraction and expression recognition. In the face detection stage, methods such as MTCNN [15] and Dlib [16] are usually used to locate faces in complex situations. In the feature extraction stage, early methods mostly use hand-extracted features. Among them, texture-based features include HOG [10], Histograms of LBP [8], Gabor wavelet coefficients. At the same time, there are many methods of extracting features based on landmark points such as noses, eyes, and mouths, and using multiple feature combinations to obtain richer representations. Currently deep learning based methods are widely used. Fasel [17] found that shallow CNNs are robust to facial poses. Tang and Kahou et al. [18] used deep CNN for feature extraction and won the FER2013 and Emotiw2013 challenges respectively. Liu et al. [19] proposed a CNN architecture based on facial action units for expression recognition. The next stage after feature extraction is to feed the features into supervised classifiers such as support vector machines (SVM), softmax layers, and logistic regression to assign facial expression categories.

2.2   Video-Based FER in the Wild

In order to capture the spatio-temporal information in the video, methods based on CNN and RNN have emerged. Most of the CNN-RNN based DFER methods first use CNN to learn spatial facial features for each video frame, and then RNN processes the temporal information between video frames. Some methods use VGG or ResNet to extract spatial features, and long short-term memory (LSTM) or Gated Recurrent Unit (GRU) to extract temporal features. For example, Baddar et al. [20] proposed a pattern varied LSTM to encode spatio-temporal features that are robust to unseen changing patterns. For 3D CNN-based methods [14], spatial and temporal feature representations of video sequences are jointly extracted through 3D convolutions. Some 3D-CNN-based methods extract temporal and spatial features of video sequences through 3D convolutions. These methods [21] extract spatio-temporal facial features by directly adopting 3D-CNN, and such spatio-temporal features are usually combined with other types of facial features. Recently, Liu et al. [22] leveraged graph convolutional networks (GCNs) to learn frame-based features that focus on specific expression regions. Lee et al. [23] proposed a Multi-modal Recurrent Attention Network (MRAN) for learning spatio-temporal attention maps for robust DFER in the wild. Zhao et al. [24] first introduced the transformer to the DFER task, they designed CS-Former and T-Former for extracting spatial and temporal features.

2.3 GNN for Video Understanding

In recent years, transformer and GNN based methods have demonstrated excellent performance in the field of video understanding, especially in improving the performance of CNN/RNN-based methods. In the field of video understanding, GNNs have been applied in dialogue modeling, video retrieval, emotion recognition and action detection. There are also video representation frameworks that can be used for multiple downstream tasks. For example, Arnab et al. [25] created a fully connected graph using foreground nodes extracted from video frames in a sliding window fashion. They established connections between the foreground nodes and the context nodes of adjacent frames. Liu et al. [22] introduced the GCN layer in the general CNN-RNN based model of video-based FER, but they only focused on the relationship between frames, and did not focus on the similarity between samples. Differently, our work is dedicated to construct a graph structure that can capture more relationships.

3 Proposed Method

As shown in Fig. 1, the proposed DSGCN mainly consists of a static-relational graph (SRG) and dynamic-relational graph (DRG). The input of DSGCN are dynamically sampled fixed-length facial expression sequences from raw videos. SRG takes video series as input, dividing the video to single frames and extracting spatial facial features for each frame. Subsequently, SRG constructs a graph by using the spatial feature of each frame as nodes, the similarity between nodes as edges, thus capturing the long-distance dependencies of expressions. DRG aggregates the spatial feature sequence enhanced by SRG, and constructs GCN from other sample videos in the same batch, sharing feature information through the similarity between samples. Finally, the classification results are obtained by a full-connected (FC) layer.

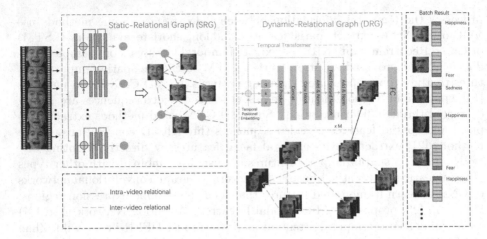

Fig. 1. The proposed model (DSGCN) architecture, which mainly consists of a Static-Relational graph (SRG) and a Dynamic-Relational graph (DRG).

3.1   Static-Relational Graph (SRG)

SRG mainly builds GCNs from frame nodes with rich spatial features. Given a facial expression video as input, a fixed-length sequence of facial expressions dynamically sampled from the raw video sequence is fed into the model. The frames in the sequence are first transformed to features carrying rich facial spatial information through the spatial network module, and then GCNs are constructed based on the features to strengthen the salient facial expression regions.

Static Spatial Feature: Fixed-length clip $X \in \mathbb{R}^{T \times 3 \times H \times W}$ are obtained as input by dynamically sampling raw video. Specifically, we split the video sequence into S segments, and randomly select V frames in each segment. We thus obtain an input clip of fixed length $T = S \times V$.

For building static-relational graph, extracting rich spatial representation from the frame, we use a Spatial Transformer [24]. The Spatial Transformer consists of five convolution blocks and N spatial encoders. The previous four convolution blocks, including conv1, conv2, conv3 and conv4, are used to extract local facial spatial features $M \in \mathbb{R}^{C \times H' \times W'}$. After this, we flatten the feature and add positional information $P_{spatial}$ to feed it into N spatial encoder. The spatial encoders consist of a multi-head self-attention and feed forward network to model global spatial relationships. The final convolution block conv5 is used to refine the final facial features. Therefore, input the Spatial Transformer of the t-length clip, and the output is $F \in \mathbb{R}^{T \times f}$ that carries sufficient spatial information.

Intra-Video Graph: In order to capture the long-range dependencies of facial regions in videos, we propose a graph-based module to capture expression changes. We construct a GCN layer by obtaining T features with spatial-temporal relations from the previous module, and model the contextual correlation by learning the dynamic adjacency matrix A. All nodes tend to be influenced by expression informative frames and update themselves as more contributing ones.

The inputs are representation maps $\hat{F} = \{\hat{f}_1, \ldots, \hat{f}_T\}$ extracted by spatial transformer from the original video. To begin with, we use cosine similarity coefficient to calculated the similarity between different representations as:

$$cossim(f_i, f_j) = \frac{f_i * f_j}{\|f_i\|\|f_j\|} \tag{1}$$

At the same time, we construct the adjacency matrix A through the cosine similarity coefficient, and $A_{i,j}$ represents the similarity between node i and node j. And in each time step, as the node features are updated, the adjacency matrix A will also update the similar state between nodes.

$$A_{i,j} = cossim(f_i, f_j) \tag{2}$$

$i, j \in \{1, 2, \ldots, T\}$. then, we employ GCN as:

$$F^{l+1} = \bar{D}^{-\frac{1}{2}} \bar{A} \bar{D}^{-\frac{1}{2}} F^l W^l \tag{3}$$

where l represents the lth time step, $\bar{A} = A + I$ is the sum of un-directed graph A and the identity matrix, $\bar{D}$ is the diagonal matrix from A, which is $\bar{D}_{i,i} = \sum_j A_{i,j}$.

F^l and F^{l+1} are the corresponding input and output representations on the l_th level, and W^l are the trainable parameters on this level. At each time step, the GCN layer shares the features of each node to neighbor nodes based on the adjacency matrix A, and accepts update messages from neighbor nodes.

3.2 Dynamic-Relational Graph (DRG)

Dynamic Temporal Feature: Using the output F' from the SRG as input, DRG aims at capturing the dynamic temporal relation for the feature nodes that have obtained spatial information, and mining the facial expression movement information between nodes. In the method, we first use the implemented Temporal Transformer [24]. The Temporal Transformer consists of M temporal encoders, each of which includes a multi-head self-attention and a feed-forward network. For T spatial features X' from the Spatial relation graph, they will be input to the temporal encoder after adding position information $P_{temporal}$. Through the multi-head self-attention and a feed-forward network in the temporal encoder, the global temporal information is modeled to output features h with rich spatial-temporal information.

Inter-Video Graph: Not limited to learning contextual relations in videos, we then extend the DRG module to learn the similarity between the input video samples. Our module accepts B video samples in the same batch, and the features of each video are transformed into a feature h carrying rich spatio-temporal information through the above steps, then we construct GCNs on these features. Our graph structure learns different scene knowledge of similar expressions for each video node in a single scene by sharing video samples of different expressions.

The inputs are representation maps $\hat{H} = \{\hat{h}_1, \ldots, \hat{h}_B\}$ extracted by temporal transformer from the sample video under the same batch. We will construct the adjacency matrix A based on Eqn. (2), and update the video nodes in the same batch based on Eqn. (3). After l rounds of updating node features, each node successfully learns different scene knowledge of similar expressions to deal with the large variations of expression patterns and data uncertainties.

Node Classification: In the previous steps, we have described our graph construction procedure that converts a batch video into a graph where each node has its own spatio-temporal feature vector. During the training process, we feed all videos in a batch simultaneously into the proposed model, and add fc layers at the end of the outputs, transferring the original video-based facial expression recognition into a seven-category node classification problem in the constructed graph.

4 Experiments

4.1 Datasets

FERV39k. [6] Currently represents the largest in-the-wild DFER dataset, containing 38,935 video clips collected from four different scenarios, which can be recursively divided into 22 fine-grained scenes, such as daily life, talk shows, business, and crime. All scene video clips are randomly shuffled and 80% of clips are allocated to the training set, while 20% of clips are reserved for the test set to avoid dataset overlaps. Therefore, in order to conduct a fair comparison, we directly use the training and testing sets divided by FERV39k.

DFEW [7] is a database that contains 16,372 video clips from more than 1,500 movies. All samples have been divided into five equally sized parts (fd1 fd5). Five-fold cross-validation is used as the evaluation scheme. In each fold, a portion of the samples is used for testing while the remaining data is reserved for training. Finally, all predicted labels are used to compute an evaluation metric by comparing them with the ground truth.

AFEW. [5] Dataset serves as the evaluation platform for the annual EmotiW challenge from 2013 to 2019. AFEW contains 1809 video clips collected from different movies and TV series. Consistent with DFEW, each video clip in AFEW is assigned to one of seven basic expressions. The test clips are not publicly available, so we train our model on train clips and test on validation clips.

Table 1. Comparison with state-of-the-art methods on FERV39k.

Methods	Accuracy of Each Emotion (%)							Metrics (%)	
	Happiness	Sadness	Neutral	Anger	Surprise	Disgust	Fear	UAR	WAR
C3D	48.20	35.53	52.71	13.72	3.45	4.93	0.23	22.68	31.69
P3D	61.85	42.21	49.80	42.57	10.50	0.86	5.57	30.48	40.81
R2Plus1D	59.33	42.43	50.82	42.57	16.30	4.50	4.87	31.55	41.28
3DR18	57.64	28.21	59.60	33.29	4.70	0.21	3.02	26.67	37.57
R18+LSTM	61.91	31.95	61.70	45.93	14.62	0.00	0.70	30.92	42.59
VGG13+LSTM	66.26	51.26	53.22	37.93	13.64	0.43	4.18	32.42	43.37
Two C3D [6]	54.85	52.91	60.67	31.34	5.96	2.36	6.96	30.72	41.77
Two R18+LSTM [6]	59.00	45.87	61.90	40.15	9.87	1.71	0.46	31.28	43.2
Two VGG13+LSTM [6]	69.65	47.31	52.55	47.88	7.68	1.93	2.55	32.79	44.54
Former-DFER [24]	65.65	51.33	56.74	43.64	21.94	8.57	12.52	37.20	46.85
STT [26]	69.77	47.81	59.14	47.41	20.22	**10.49**	9.51	37.76	48.11
NR-DFERNet [27]	69.18	54.77	51.12	49.70	13.17	0.00	0.23	33.99	45.97
DSGCN	**86.90**	**61.95**	**72.32**	**55.68**	**31.19**	9.21	**16.24**	**47.64**	**59.88**

4.2 Implementation Details

Training Setting: For all the three datasets, we train our model from scratch with a batch size of 32, initialize the learning rate to 0.01, and divide it by 5 every 50 epochs. Due to the small number of data samples in AFEW dataset, in order to make a fair comparison, we first pre-train our model and other models on DFEW (fd1), and then fine-tune on AFEW with the same setting.

Evaluation Metrics: Without loss of generality, We choose Unweighted Average Recall (UAR, i.e. the accuracy of each category divided by the number of

Table 2. Comparison with state-of-the-art methods on DFEW.

Methods	Metrics(%)	
	UAR	WAR
3DR18	46.52	58.27
R18+LSTM	51.32	63.85
R18+GRU	51.68	64.02
EC-STFL [7]	45.35	56.51
Former-DFER [24]	53.69	65.70
EST [28]	53.43	65.85
STT [26]	54.58	66.65
NR-DFERNet [27]	54.21	68.19
GCA+IAL [29]	55.71	69.24
DPCNet [30]	**57.11**	69.24
DSGCN	57.06	**70.57**

Table 3. Comparison with state-of-the-art methods on AFEW.

Methods	Metrics(%)	
	UAR	WAR
C3D	43.75	46.72
I3D-RGB	41.86	45.41
R(2+1)D	42.89	46.19
3DR18	42.14	45.67
R18+LSTM	43.96	48.82
Former-DFER [24]	47.42	50.92
EST [28]	49.57	54.26
STT [26]	49.11	54.23
NR-DFERNet [27]	48.37	53.54
DPCNet [30]	47.86	51.67
DSGCN	**60.46**	**65.49**

categories, regardless of the instances of each category) and weighted average recall (WAR, i.e. accuracy) as the metrics.

4.3 Comparison with State-of-the-Arts

In this section, we compare our best results with current state-of-the-art methods on the FERV39k, DFEW and AFEW benchmarks.

As shown in the Table 1, we compare our method with other state-of-the-art methods on the FERV39k dataset, including C3D, P3D, R2Plus1, 3DR18, R18+LSTM, VGG13+LSTM, Two C3D [6], Two R18+LSTM [6], Two VGG13+LSTM [6], Former-DFER [24], STT [26], NR-DFERNet [27]. DSGCN improvements of 9.88% and 11.77% in UAR and WAR than the previous state-of-the-art method STT. Moreover, we show the performances on each expression in the Table. As can be seen, our method achieve the best results on most of the expressions, only slightly lower than STT on Disgust with a gap of 1.28%. At the same time, in Table 1, we can see that most of the methods perform poorly in "disgust" and "fear", which we believe is caused by insufficient training data in the original datasets.

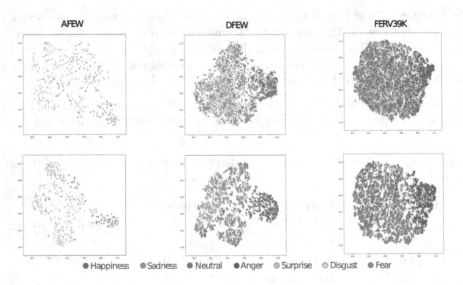

Fig. 2. Illustration of feature distribution learned by the Former-DFER [24] (top) and DSGCN (bottom) on three datasets.

For the DFEW data set, the experiment compared 3DR18, R18+LSTM, R18+GRU, EC-STFL [7], Former-DFER [24], EST [28], STT [26], NR-DFERNet [27], GCA+IAL [29], and DPCNet [30] under 5-fold cross-validation. As shown in Table 2, DSGCN outperforms the comparison methods on the WAR metric, and is very close to the current state-of-the-art method DPCNet on the UAR metric.

Specifically, we have a 1.33% improvement on WAR and only a 0.05% reduction on UAR compared to DPCNet. It should be noticed that DFEW also has a imbalanced data distribution. The proportions of "disgust" and "fear" sequences are 1.22% and 8.14%, which is the reason why our method achieve a relatively low performance in UAR.

For the AFEW dataset, all methods are first pre-trained on DFEW (fd1) and then fine-tuned on AFEW with the same settings. Our method compares C3D, I3D-RGB, R(2+1)D, 3DR18, R18+LSTM, Former-DFER [24], EST [28], STT [26], NR-DFERNet [27], DPCNet [30]. The comparative performance shown in Table 3 shows that DSGCN achieves the best results on both UAR and WAR. In particular, our method shows an improvement of 10.89% and 11.23% in UAR and WAR than the previous state-of-the-art method EST.

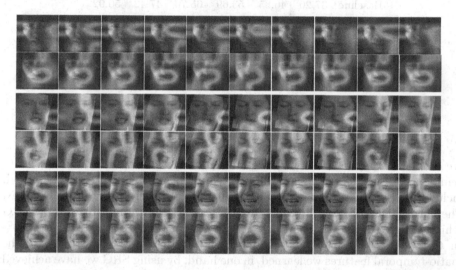

Fig. 3. Visualization of the learned feature maps. There are three sequences are presented, which including the facial expression of Neutral, Anger and Sadness, respectively. For each sequence, the images in the first row are heat-maps generated by the Former-DFER, and the images in the second row are heat-maps generated by DSGCN.

4.4   Visualization Results

We utilize t-SNE [31] to analyze the feature distribution learn by the Former-DFER and DSGCN on three datasets. As shown in Fig. 2, it is obvious that the feature distribution of each category learned by our method is tighter, and the boundaries between different categories are more obvious. This shows that our method can better discriminate different facial expressions in feature level. Furthermore, we conduct experiments to visualize the learned facial feature maps, as shown in Fig. 3, we used neutral, angry, and sad three types of expressions to compare with Fomer-DFER. For the first neutral expression sequence, although

there is no significant expression behavior, our method still pays more attention to facial regions. In the second angry expression sequence and the third
sad expression sequence, our method pays more attention to the facial regions
such as mouth and eyes that contain more emotional information. In the second
sequence where the subject has large head pose changes, our method always
locks on the subject's face region compared to the comparison method.

Table 4. Ablation study to evaluate the effectiveness of different modules in our proposed method.

Methods	FERV39K(%)		DFEW(%)		AFEW(%)	
	UAR	WAR	UAR	WAR	UAR	WAR
Baseline	37.20	46.85	53.69	65.70	47.42	50.92
SRG	44.64	57.47	55.29	67.56	58.95	64.15
DRG	44.35	57.74	56.43	68.38	59.90	64.69
DSGCN	**47.64**	**59.88**	**57.06**	**70.57**	**60.46**	**65.49**

4.5   Ablation Study

We conduct extensive ablation studies on the three video-based FER datasets to
demonstrate the effectiveness of different components of our proposed method.
Including the individual part of SRG and DRG, as well as the final DSGCN.
The Former-DFER is employed as the baseline. As shown in Table 4, our STRG
achieves the WAR and UAR of 40.43%/54.81%, 55.56%/67.04%, 57.67%/62.80%
on three datasets, which outperforms some existing methods because of the
spatio-temporal features we learned. in one hand, by using SRG we have achieved
obvious improvements in WAR and UAR compared to the baseline. This proves
that SRG can effectively enhance facial expressions by learning the similarity of
expressions at different moments in the same video, and provide more robust features for subsequent extraction of temporal information. In other hand, through
the propagation and enhancement of spatio-temporal features, DRG outperforms the baseline to varying degrees on the three datasets. The most significant
improvements are in AFEW, where CRG exceeds baseline by 12.48% and 13.77%
on WAR and UAR. This proves that DRG can indeed capture the correlation
between different sample expressions to strengthen the current expression. We
notice solely using DRG performs slightly better than using SRG, the reason
is dynamic features from other video sequences can better improve the robustness of node features than in the same video sequence. Finally, in the complete
method DSGCN, all indicators in the three datasets reach the highest, the results
prove our method can indeed learn both the spatial and temporal relationship
of the input video facial expressions.

5 Conclusion

This paper proposes a novel dynamic-static graph convolutional network (DSGCN) for dynamic facial expression recognition in-the-wild scenarios. Specifically, the proposed DSGCN mainly consists the static-relational graph (SRG) and the dynamic-relational graph (DRG), it can capture multi-level relationships among the input video sequences, including spatial relationship, temporal relationship, context relationship and sample relationship. Extensive experiments with previous methods show that the proposed DSGCN achieves state-of-the-art results on three popular dynamic FER benchmarks. The abundant ablation studies have validated the effectiveness of each part in DSGCN. Moreover, the visualization results of facial features demonstrate that DSGCN can pay more attention to the salient facial regions. The visualization results of the feature distribution show that the method can better discriminate the learned face features. Furthermore, comparisons with previous methods show that DSGCN achieves state-of-the-art results on three popular dynamic FER benchmarks.

In future work, based on our DSGCN framework, we will further expand it to Micro-Expression Recognition, Pose Prediction, Person Recognition and other fields. Additionally, we plan to integrate our DSGCN framework with self-supervised learning, encouraging the model to learn potential internal relationships in a large amount of unlabeled data, thereby alleviating the impact of imbalances in facial data.

Acknowledgements. This work was supported in part by Zhejiang Provincial Natural Science Foundation of China (No.LDT23F0202, No. LDT23F02021F02, No. LQ22F020013) and the National Natural Science Foundation of China (No. 62036009, No. 62106226).

References

1. Plass-Oude Bos, D., Poel, M., Nijholt, A.: A study in user-centered design and evaluation of mental tasks for BCI. In: Lee, K.-T., Tsai, W.-H., Liao, H.-Y.M., Chen, T., Hsieh, J.-W., Tseng, C.-C. (eds.) MMM 2011. LNCS, vol. 6524, pp. 122–134. Springer, Heidelberg (2011). https://doi.org/10.1007/978-3-642-17829-0_12
2. Lucey, P., Cohn, J.F., Kanade, T., Saragih, J., Ambadar, Z., Matthews, I.: The extended cohn-kanade dataset (ck+): A complete dataset for action unit and emotion-specified expression. In: 2010 IEEE Computer Society Conference on Computer Vision and Pattern Recognition-workshops. pp. 94–101. IEEE (2010)
3. Zhao, G., Huang, X., Taini, M., Li, S.Z., Pietikälnen, M.: Facial expression recognition from near-infrared videos. Image Vis. Comput. **29**(9), 607–619 (2011)
4. Pantic, M., Valstar, M., Rademaker, R., Maat, L.: Web-based database for facial expression analysis. In: 2005 IEEE International Conference on Multimedia and Expo, P. 5. IEEE (2005)
5. Dhall, A., Goecke, R., Lucey, S., Gedeon, T., et al.: Collecting large, richly annotated facial-expression databases from movies. IEEE Multimedia **19**(3), 34 (2012)
6. Wang, Y.: Ferv39k: a large-scale multi-scene dataset for facial expression recognition in videos. In: Proceedings of the IEEE/CVF Conference on Computer Vision and Pattern Recognition, pp. 20922–20931 (2022)

7. Jiang, X.: Dfew: a large-scale database for recognizing dynamic facial expressions in the wild. In: Proceedings of the 28th ACM International Conference on Multimedia, pp. 2881–2889 (2020)
8. Wang, L., He, X., Du, R., Jia, W., Wu, Q., Yeh, W.: Facial expression recognition on hexagonal structure using LBP-based histogram variances. In: Lee, K.-T., Tsai, W.-H., Liao, H.-Y.M., Chen, T., Hsieh, J.-W., Tseng, C.-C. (eds.) MMM 2011. LNCS, vol. 6524, pp. 35–45. Springer, Heidelberg (2011). https://doi.org/10.1007/978-3-642-17829-0_4
9. Huang, X., He, Q., Hong, X., Zhao, G., Pietikainen, M.: Improved spatiotemporal local monogenic binary pattern for emotion recognition in the wild. In: Proceedings of the 16th International Conference on Multimodal Interaction, pp. 514–520 (2014)
10. Chen, J., Chen, Z., Chi, Z., Fu, H.: Emotion recognition in the wild with feature fusion and multiple kernel learning. In: Proceedings of the 16th International Conference on Multimodal Interaction, pp. 508–513 (2014)
11. Liu, M., Shan, S., Wang, R., Chen, X.: Learning expressionlets on spatio-temporal manifold for dynamic facial expression recognition. In: Proceedings of the IEEE Conference on Computer Vision and Pattern Recognition, pp. 1749–1756 (2014)
12. Liu, M., Wang, R., Li, S., Shan, S., Huang, Z., Chen, X.: Combining multiple kernel methods on riemannian manifold for emotion recognition in the wild. In: Proceedings of the 16th International Conference on Multimodal Interaction, pp. 494–501 (2014)
13. Aminbeidokhti, M., Pedersoli, M., Cardinal, P., Granger, E.: Emotion recognition with spatial attention and temporal softmax pooling. In: Karray, F., Campilho, A., Yu, A. (eds.) ICIAR 2019. LNCS, vol. 11662, pp. 323–331. Springer, Cham (2019). https://doi.org/10.1007/978-3-030-27202-9_29
14. Fan, Y., Lu, X., Li, D., Liu, Y.: Video-based emotion recognition using cnn-rnn and c3d hybrid networks,. In: Proceedings of the 18th ACM International Conference on Multimodal Interaction, pp. 445–450 (2016)
15. Zhang, K., Zhang, Z., Li, Z., Qiao, Y.: Joint face detection and alignment using multitask cascaded convolutional networks. IEEE Signal Process. Lett. **23**(10), 1499–1503 (2016)
16. Amos, B., Ludwiczuk, B., Satyanarayanan, M., et al.: Openface: a general-purpose face recognition library with mobile applications. CMU School Comput. Sci. **6**(2), 20 (2016)
17. Fasel, B.: Robust face analysis using convolutional neural networks. In: 2002 International Conference on Pattern Recognition, vol. 2, pp. 40–43. IEEE (2002)
18. Kahou, S.E., et al.: Combining modality specific deep neural networks for emotion recognition in video. In: Proceedings of the 15th ACM on International Conference on Multimodal Interaction, pp. 543–550 (2013)
19. Liu, M., Li, S., Shan, S., Chen, X.: Au-inspired deep networks for facial expression feature learning. Neurocomputing **159**, 126–136 (2015)
20. Baddar, W.J., Ro, Y.M.: Mode variational lstm robust to unseen modes of variation: Application to facial expression recognition. In: Proceedings of the AAAI Conference on Artificial Intelligence, vol. 33(01), pp. 3215–3223 (2019)
21. Lee, J., Kim, S., Kim, S., Park, J., Sohn, K.: Context-aware emotion recognition networks. In: Proceedings of the IEEE/CVF International Conference on Computer Vision, pp. 10143–10152 (2019)
22. Liu, D., Zhang, H., Zhou, P.: Video-based facial expression recognition using graph convolutional networks. In: 2020 25th International Conference on Pattern Recognition (ICPR), pp. 607–614. IEEE (2021)

23. Lee, J., Kim, S., Kim, S., Sohn, K.: Multi-modal recurrent attention networks for facial expression recognition. IEEE Trans. Image Process. **29**, 6977–6991 (2020)
24. Zhao, Z., Liu, Q.: Former-dfer: dynamic facial expression recognition transformer,. In: Proceedings of the 29th ACM International Conference on Multimedia, pp. 1553–1561 (2021)
25. Arnab, A., Sun, C., Schmid, C.: Unified graph structured models for video understanding. In: Proceedings of the IEEE/CVF International Conference on Computer Vision, pp. 8117–8126 (2021)
26. Ma, F., Sun, B., Li, S.: Spatio-temporal transformer for dynamic facial expression recognition in the wild, arXiv preprint arXiv:2205.04749 (2022)
27. Li, H., Sui, M., Zhu, Z., et al.: Nr-dfernet: noise-robust network for dynamic facial expression recognition, arXiv preprint arXiv:2206.04975 (2022)
28. Liu, Y., Wang, W., Feng, C., Zhang, H., Chen, Z., Zhan, Y.: Expression snippet transformer for robust video-based facial expression recognition. Pattern Recogn. **138**, 109368 (2023)
29. Li, H., Niu, H., Zhu, Z., Zhao, F.: Intensity-aware loss for dynamic facial expression recognition in the wild, arXiv preprint arXiv:2208.10335 (2022)
30. Wang, Y., et al.: Dpcnet: dual path multi-excitation collaborative network for facial expression representation learning in videos. In: Proceedings of the 30th ACM International Conference on Multimedia, pp. 101–110 (2022)
31. Van der Maaten, L., Hinton, G.: Visualizing data using t-sne. J. Mach. Learn. Res. **9**(11) (2008)

Hierarchical Supervised Contrastive Learning for Multimodal Sentiment Analysis

Kezhou Chen, Shuo Wang[✉], and Yanbin Hao

MoE Key Laboratory of Brain-inspired Intelligent Perception and Cognition,
University of Science and Technology of China, Hefei, China
chenkezhou@mail.ustc.edu.cn, shuowang.edu@gmail.com

Abstract. Multimodal sentiment analysis (MSA) is dedicated to deciphering human emotions in videos. It is a challenging task due to the semantic disparities among various modalities (e.g., linguistic, visual, and acoustic) present in video content. To bridge these gaps, we leverage contrastive learning and introduce a novel hierarchical multimodal approach termed Hierarchical Supervised Contrastive Learning (HSCL). Initially, we utilize an unimodal fusion combined with a supervised contrastive learning strategy to distill pertinent content from each modality. Subsequently, we combine the bimodal data in pairs and further align them through supervised contrastive learning. This paired data aids in understanding the intricate nuances of multidimensional human emotions. In our method, supervised contrastive learning is tailored to accentuate the significance of label information, facilitating the extraction of sentiment cues from diverse sources. Experimental studies on two benchmark datasets demonstrate the effectiveness of our method. Specifically, compared to the baseline on the CMU-MOSI benchmark, our method achieves 2.59% accuracy improvement for emotion recognition. Codes are released at https://github.com/Turdidae810/HSCL.

Keywords: Multimodal Sentiment Analysis · Hierarchical Alignment · Supervised Contrastive Learning

1 Introduction

Multimodal Sentiment Analysis (MSA) is a complex task that involves interpreting emotions from various modalities such as language, vision, and acoustics [7,27]. In this task, language data is transcribed from spoken words in videos, visual data captures facial expressions, and acoustic data reflects tonal variations. Unlike unimodal sentiment analysis [31], MSA not only examines individual data modalities but also explores the interrelations among them. It provides richer and more accurate emotion recognition since it combines multiple sensory signals. Meanwhile, it is a challenging task since the semantic gaps between these different sensory signals weaken emotional perception [8]. Therefore, how to connect the contents of multimodality is a core problem of this task.

S. Rudinac et al. (Eds.): MMM 2024, LNCS 14555, pp. 56–69, 2024.
https://doi.org/10.1007/978-3-031-53308-2_5

Early work directly stitches and learns the data of the aforementioned modalities [18]. Come to single-modal sentiment analysis, the combination of multiple modalities often complements each other and improves the accuracy of emotion prediction [15]. However, the stitch operation ignores the correlation between different modalities and achieves limited performance improvement. Thus, to improve the perception between different modalities, there are two common solutions used in this task: attention-based methods and alignment-based methods, where the former methods [23] design different attention mechanisms that introduce a high-cost calculation to capture the relations between different modalities' data, and then analyze the sentiment. However, these attention-based methods require a targeted design of the module during the calculation procedure [6]. For alignment-based methods [8,22], they design a cross-modal alignment strategy before feature fusion to learn latent-modality representations. However, they only focused on unimodal representation [8] or bimodal fusion results [22] during the alignment process, which still does not fully address the problem of modality heterogeneity. Meanwhile, the previous methods all neglected the role of supervised signals and underutilized label information during the alignment process, potentially leading to the loss of emotional information in this process.

Inspired by supervised contrastive learning (SCL) [12] that improves both the accuracy and robustness of the matching process with minimal complexity, we introduce SCL into the MSA task and propose a Hierarchical Supervised Contrastive Learning (HSCL) method to align the content from different modalities both unimodal representation and bimodal fusion features. Meanwhile, we use the label to constrain the aligned representations for retaining rich emotional semantics. In our method, we regard language data as the basic knowledge of the whole data. Compared to the other two modalities (video and audio), the text information is recorded more accurately and can be easily analyzed during the calculation. Then we introduce our HSCL training strategy. Specifically, we first employ three different modules to represent the different modalities in features. During the feature extraction, we connect the association between audio-language and video-language by a supervised contrastive learning strategy, where contrastive learning is used to help the feature extraction stage focus on the content between different modalities and provide more precise representations. After the feature extraction, we first fuse these paired data (audio-language and video-language) by designing a cross-modal attention structure. Then, we use the supervised contrastive learning strategy to constrain the attention calculation and filter the sentiment information from different paired data. Meanwhile, we also leverage traditional learning strategies on the fused data to learn the distribution of sentiment from these multimodal data. In our method, we align the low-level feature representations to the high-level fusion features in multiple dimensions and extract task-related information from representations. The contributions of our method can be summarized as follows:

- We introduce supervised contrastive learning and propose a hierarchical training strategy, where the emotion is captured from both the low-level and high-level feature representations.

- We design self- and cross-attention modules to fuse the representations from different modalities data, which provides more effective emotional content.
- The results demonstrate that our method achieves state-of-the-art performance on two publicly available datasets for multimodal sentiment analysis.

2 Related Work

In this section, we first briefly introduce common solutions for sentiment analysis tasks. Then we list the applications of corresponding contrastive learning strategies. Finally, we enumerate the differences between our methods and those of related methods.

2.1 Multimodal Sentiment Analysis

Sentiment analysis was originally developed in natural language processing to discern whether a review was positive or negative, with early efforts focusing solely on language [17]. However, genuine human emotional expression often encompasses multiple modalities. As a result, there has been a growing interest among researchers in multimodal sentiment analysis, which integrates textual, acoustic, and visual cues [18]. Zadeh et al. [27] were the first to employ deep learning techniques to predict sentiment by integrating data from these three modalities. They utilized LSTM [10] to learn intra-modality dynamics and employed a three-fold Cartesian product to fuse the features [27]. Subsequently, the researchers started exploring the effects of different feature-fusion methods. Tsai et al. were the first to introduce cross-modal attention into multimodal sentiment analysis [23]. Rahman et al. proposed MAG-BERT, which incorporated multimodal information into pre-trained language models for fine-tuning [20]. However, these methods mainly focused on fusion techniques and overlooked the influence of multimodal representations on fusion. Tsai et al. were the first to investigate the impact of multimodal representations on model performance. They decomposed the representation into multimodal discriminative and modality-specific generative factors [24]. Hazarika et al. further learned effective modality representations by projecting them into two subspaces to enhance the fusion process [8]. Sun et al. used Deep Canonical Correlation Analysis (DCCA) to learn relationships between different modalities [22]. Lin et al. introduced contrastive learning into multimodal sentiment analysis and captured complex relationships of different modalities by unimodal and multimodal graphs [16].

2.2 Contrastive Learning

Our research has ties to the domain of contrastive learning. Initially introduced as an unsupervised technique for representation learning [3], the essence of contrastive learning lies in gauging distances between samples in feature space. The goal is to minimize the distance between the anchor and positive samples while maximizing the distance between the anchor and negative samples. Owing to its robust capability for representation learning, contrastive methods have found applications in various tasks, including image recognition [3] and sentence embeddings [5]. Subsequent research revealed that the integration of label signals into contrastive learning can enhance representation quality. For instance, Khosla et al. introduced a supervised contrastive loss tailored for encoder training in classification tasks [12]. They expanded the definition of positive pairs to include not only augmented views of the same image but also samples from the same class. Zha et al. tackled the challenge of annotation continuity in regression tasks, leading to the development of supervised contrastive regression [30]. The realm of multimodal learning has also witnessed the adoption of contrastive learning. Radford et al. leveraged inforNCE to derive a visual model guided by language supervision [19]. Similarly, Li et al. harnessed contrastive learning to synchronize image and text representations prior to their fusion [14].

Based on the analysis of the related work, recent methods often treated all modalities equally and ignored the dominant role of linguistic data in multimodal sentiment analysis. In our method, we put textual information in the center during the fusion process. Furthermore, existing methods often only consider relationships between unimodal representations or between bimodal fusion results, without simultaneous consideration of both. In contrast, HSCL employs a hierarchical contrastive learning strategy to address the negative impact of modal heterogeneity. Last but not least, existing methods often underutilize MSA labels when enhancing representations, potentially leading to information loss. To this end, HSCL incorporates supervised contrastive regression in the cross-modal contrastive learning process.

3 Approach

In this section, we dive into the details of our Hierarchical Supervised Contrastive Learning (HSCL) method. Initially, we revisit the preliminaries of multimodal sensitization analysis (MSA) tasks and provide an overview of our framework. Finally, we illustrate the training and inference procedures of our method.

3.1 Preliminaries

The task of multimodal sentiment analysis concentrates on utilizing video clips from textual (t), visual (v), and acoustic (a) modalities as model inputs. Each individual modality is represented as $\boldsymbol{X}_m \in \mathbb{R}^{T_m \times d_m}, (m \in \{t, v, a\})$, where T_m and d_m are used to represent sequence length and feature dimension of modality m, respectively. The goal of our method is to thoroughly examine and merge sentiment-related information from these unaligned multimodal sequences. In this procedure, we aspire to generate a text-driven multimodal representation that can accurately predict the final sentiment analysis results.

An overview of our Hierarchical Supervised Contrastive Learning (HSCL) is shown in Fig. 1. Given three different types of data, we employ distinct feature extractors to convert these data into representative features. Subsequently,

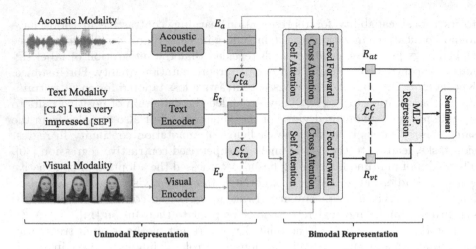

Fig. 1. An overview of our Hierarchical Supervised Contrastive Learning, where $\mathcal{L}_{ta}^C$, $\mathcal{L}_{tv}^C$, and $\mathcal{L}_f^C$ are three supervised contrastive learning modules.

supervised contrast regression [30] is applied to align these unimodal representations. Meanwhile, we design two cross-modal attention structures to fuse the different representations and use high-level supervised contrastive learning to gauge inter-modality relationships. Finally, this process yields task-aligned multimodal representations that can effectively predict the sentiment in a given video.

3.2 Unimodal Representation

Firstly, we encode the low-level multimodal sequence $\mathbf{X}_m$ into multimodal representations $\mathbf{E}_m$. Specifically, we utilize 12-layers pre-trained BERT [11] to extract textual features. For audio and video sequences, we use COVAREP [4] and Facet [2] for the extraction of low-level features and employ two randomly initialized Transformer encoders [25] to capture their temporal relationship:

$$\mathbf{E}_t = \text{BERT}(\mathbf{X}_t; \theta_t^{\text{BERT}}),$$
$$\mathbf{E}_m = \text{Transformer}(\mathbf{X}_m; \theta_m^{\text{Transformer}}), m \in \{a, v\}, \tag{1}$$

where θ_t^{BERT} and $\theta_m^{\text{Transformer}}$ are the parameters of feature extractors, and $\mathbf{E}_m \in \mathbb{R}^{(T_m+1) \times d_m}$. After the feature extraction phase, we select the "[cls]" token at the beginning of each sequence, denoted as R_m, as the global representation of each modality. We then utilized contrastive learning to align the different modalities. To project each unimodal representation into a consistent dimension, we passed it through a fully connected layer:

$$\hat{R}_m = \text{FC}_m(R_m; \theta_m^{\text{FC}}), m \in \{a, v, m\}, \tag{2}$$

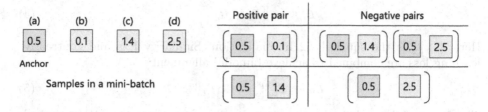

Fig. 2. An example of supervised contrastive regression. In this example, sample (a) is the anchor, and the numbers in the square are the sentiment value of this sample.

where θ_m^{FC} is the parameter of FC_m.

Inspired by the work of [12,30], we incorporate labels during the calculation. Given that sentiment strength is a continuum, we apply supervised contrastive regression as presented in [30]. For a clearer understanding of the process, Fig. 2 provides an illustrative example of how supervised contrastive regression delineates positive and negative samples with a fixed anchor. In this example, we choose a sample with a sentiment strength of "0.5" as an anchor. If a sample labeled "0.1" is deemed positive, then samples (c) and (d) are considered negative due to their greater distance from "0.5" compared to "0.1". Similarly, when sample (c) is selected as the positive sample, we apply the same methodology to identify the corresponding negative sample, which, in this instance, is "2.5". Through this approach, supervised contrastive regression aids the model in defining an embedding space where representation similarities hinge on their annotated values. In practical applications, supervised contrastive regression operates between two modalities, and any given sample in a batch can sequentially serve as the anchor. Thus, based on the analysis of supervised contrastive regression, the operation of contrastive learning between the unimodal representation $\hat{R}_t$ and $\hat{R}_a$ can be calculated as:

$$\mathcal{L}_{ta}^{SupCR} = -\frac{1}{2N}\sum_{i=1}^{2N}\frac{1}{2N-1}\sum_{j=1,j\neq i}^{2N}\frac{\exp\left(\hat{R}_{m,i}\cdot\hat{R}_{m,j}/\tau\right)}{\sum_{k\in\mathbb{A}(k)}\exp\left(\hat{R}_{m,i}\cdot\hat{R}_{m,k}/\tau\right)}, \quad (3)$$

where m represents text modality and audio modality, and N is the batch size. i, j, and k are the indices representing the anchor, the positive sample, and the negative sample, respectively. $\mathbb{A}(k)$ is the set of negative sample indices when i and j are chosen, and $\mathbb{A}(k) = \{k \neq i, |y_i - y_k| \geq |y_i - y_j|; k \in \{1, 2, \ldots, 2N\}\}$. It denotes that only samples that have a greater distance to the anchor than the positive sample can be negative samples. y is the label value here. And τ is the temperature coefficient. In this operation, $\mathcal{L}^{SupCR}$ can learn a multimodal feature space in which the distances between embeddings are organized according to their sentiment intensity, which is beneficial for the remaining calculations.

Due to the heterogeneity of the data from different modalities, $\mathcal{L}^{SupCR}$ may not work. Therefore, a common subspace for different modalities is necessary. We achieve this through the similarity loss function:

$$\mathcal{L}_{ta}^{sim} = \left\| \hat{R}_t - \hat{R}_a \right\|_2^2. \tag{4}$$

Here, $\| \cdot \|_2^2$ is the squared $L2$ normalization. Similarity loss and contrastive learning loss are combined to achieve bimodal alignment:

$$\mathcal{L}_{ta} = \mathcal{L}_{ta}^{SupCR} + \alpha \mathcal{L}_{ta}^{sim}. \tag{5}$$

The forms of $\mathcal{L}_{tv}$ and $\mathcal{L}_{ta}$ is symmetrical. α is a hyper-parameter that controls the impact of $\mathcal{L}^{sim}$.

3.3   Bimodal Representation

Our bimodal fusion process consists of three sublayers: Self-Attention, Cross-Attention, and Feed-Forward. In line with [25], each sublayer incorporates a residual connection as described in [9] and undergoes layer normalization as per [1]. We denoted the procedure of residual connection and normalization as Add&Norm in our method. Then, we describe the bimodal fusion process in detail. First of all, we pass the encoded text through a Self-Attention sublayer:

$$\begin{aligned} \mathbf{H}_t &= \text{Add} \& \text{Norm} \left(\text{Self-Attention}(\mathbf{E}_t) \right) \\ &= \text{Add} \& \text{Norm} \left(\text{softmax} \left(\frac{\mathbf{E}_t \mathbf{W}_Q \mathbf{W}_K^\top \mathbf{E}_t^\top}{\sqrt{d_t}} \right) \mathbf{E}_t \mathbf{W}_V \right) \end{aligned} \tag{6}$$

Here, $\mathbf{W}_Q$, $\mathbf{W}_K$, and $\mathbf{W}_V \in \mathbb{R}^{d_t \times d_t}$ are parameters to project the matrices into local spaces.

Second, we employ Cross-Attention to merge text with other modalities, a technique central to this approach. Previous studies have shown that cross-modal attention effectively enables latent adaptations across different modalities and captures interactions among multimodal sequences over variable time steps [23]. Thus, we harness cross-modal attention to fuse text-audio and text-video sequences. Crucially, text serves as the query in this configuration, positioning it as the target modality, while video and audio information is infused into the to denote either audio or vision, both of which are treated symmetrically during the fusion phase.

$$\begin{aligned} \mathbf{F}_{mt} &= \text{Add} \& \text{Norm} \left(\text{Cross-Attention}_{m \rightarrow t}(\mathbf{H}_t, \mathbf{E}_m) \right), \\ &= \text{Add} \& \text{Norm} \left(\text{softmax} \left(\frac{\mathbf{H}_t \mathbf{W}_{Q_t} \mathbf{W}_{K_m}^\top \mathbf{E}_m^\top}{\sqrt{d_t}} \right) \mathbf{E}_m \mathbf{W}_{V_m} \right), \end{aligned} \tag{7}$$

where $m \in \{a, v\}$. $\mathbf{W}_{Q_t} \in \mathbb{R}^{d_t \times d_t}$, $\mathbf{W}_{K_m} \in \mathbb{R}^{d_m \times d_t}$ and $\mathbf{W}_{V_m} \in \mathbb{R}^{d_m \times d_t}$. The last part of the bimodal fusion is the position-wise feed-forward sublayer, which consists of two linear transformations with a ReLU activation in between:

$$\mathbf{Y}_{mt} = \text{Add} \& \text{Norm} \left(\text{ReLU}(\mathbf{F}_{mt} \mathbf{W}_1 + b_1) \mathbf{W}_2 + b_2 \right), \tag{8}$$

where $\mathbf{W}_1$, $\mathbf{W}_2$ and b_1, b_2 are parameters of Feed-Forward sublayer. After acquiring fusion sequence $\mathbf{Y}_{mt} \in \mathbb{R}^{(T_t+1) \times d_t}$, similar to unimodal representation, we consider the "[cls]" token of them as bimodal representations, denoted as R_{mt}, $m \in \{a, v\}$, and then pass them into a fully-connected layer:

$$\hat{R}_{mt} = \text{FC}(R_{mt}; \theta_{mt}^{FC}), \tag{9}$$

Finally, we apply the supervised contrastive regression in this calculation:

$$\mathcal{L}_{fusion} = -\frac{1}{2N} \sum_{i=1}^{2N} \frac{1}{2N-1} \sum_{j=1, j \neq i}^{2N} \frac{\exp\left(\hat{R}_{mt,i} \cdot \hat{R}_{mt,j}/\tau\right)}{\sum_{k \in \mathbb{A}(k)} \exp\left(\hat{R}_{mt,i} \cdot \hat{R}_{mt,k}/\tau\right)}, \tag{10}$$

where the definition of N, $\mathbb{A}(k)$, and τ is the same as Eq. 3.

3.4 Sentiment Prediction

To predict the sentiment strength, we feed the multimodal representation into a multi-layer perception, consisting of two linear transforms and a tanh in between:

$$\hat{y} = \text{MLP}\left(\frac{1}{2}(R_{at} \oplus R_{vt}); \theta^{MLP}\right), \tag{11}$$

where $\oplus$ is element-wise addition. Then, we calculate the mean absolute error (MAE) between the prediction $\hat{y}$ and the label y:

$$\mathcal{L}_{task} = \frac{1}{N} \sum_{i=1}^{N} |y_i - \hat{y}_i|, \tag{12}$$

N is the batch size here, i represents the index of a sample within the batch. The overall loss function during training is:

$$\mathcal{L} = \mathcal{L}_{task} + \omega_1(\mathcal{L}_{ta} + \mathcal{L}_{tv}) + \omega_2 \mathcal{L}_{fusion}, \tag{13}$$

where ω_1 and ω_2 are hyper-parameters.

4 Experiments

In this section, we conduct experiments to evaluate the performance of our hierarchical supervised contrastive learning (HSCL). We first introduce the experimental settings. And then compare other state-of-the-art methods with ours. Finally, we analyze the ablation studies. Our experiments are intended to address the following research questions (RQ):

RQ1: How does HSCL perform compared to the SOTA MSA methods?
RQ2: What are the effects of different modalities on MSA?
RQ3: What are the effects of hierarchical contrastive learning?
RQ4: How does the MSA label information influence the alignment process?

4.1 Experimental Settings

Datasets. We conducted experiments on two well-known public datasets: CMU-MOSI [28] and CMU-MOSEI [29]. The CMU-MOSI dataset is renowned and comprises 93 videos that span a wide array of topics, featuring 89 distinct YouTube speakers. These videos are further divided into 2,199 short clips, with each being annotated by five evaluators. The annotations in CMU-MOSI are real numbers within the range of $[-3, +3]$; the sign indicates polarity, while the absolute value represents the strength of the sentiment. On the other hand, CMU-MOSEI is an extensive version of CMU-MOSI. It boasts a more substantial dataset with 23,453 annotated clips that encompass over 1,000 speakers and 250 topics. The labeling convention for CMU-MOSEI mirrors that of CMU-MOSI.

Evaluation Metrics. For evaluation, we employed the same metrics used in prior studies [7]. These include mean absolute error (MAE), Pearson correlation coefficient (Corr), seven-class accuracy (ACC-7), and binary classification accuracy (ACC-2) accompanied by the respective F1 score. With the exception of MAE, higher metric values signify superior model performance.

Implementation Details. For a fair comparison, all our experiments utilize the same unaligned raw data as in MMIM [7]. For the MOSI dataset, the learning rate is set at $5e^{-5}$ for BERT fine-tuning and $1e^{-3}$ for the primary model. The parameters α, ω_1, and ω_2 are set to 0.1, 0.01, and 0.1, respectively. For the MOSEI dataset, we establish a learning rate of $1e^{-5}$ for BERT fine-tuning and $1e^{-4}$ for the primary model, with α, ω_1, and ω_2 configured to 0.1, 0.01, and 0.01. We maintain a consistent batch size of 128 across all experiments. The number of transformer encoder layers designated for unimodal feature extraction is fixed at three, and the temperature coefficient, represented as τ, is set at 0.5. Our model utilizes the Adam optimizer [13] and is trained on a single RTX 3090 GPU.

4.2 Comparison with Other Methods (RQ1)

Baselines. To test the performance of our proposed model, we compare it with a variety of previous models in the MSA task. The compared models contain: **TFN** [27]: The tensor fusion network is a method that models uni-, bi-, and tri-modal dynamics using a three-fold Cartesian product. **MFM** [24]: Multimodal Factorization Model learns multimodal discriminative factors and modality-specific generative factors to acquire meaningful multimodal representation and infer missing modalities. **MulT** [23]: Multimodal Transformer is a transformer-based model requiring no alignment assumption by learning a latent crossmodal adaptation through pairwise cross-modal attention mechanism. **MAG-Bert** [20]: MAG-Bert enables the pretrained Bert to accept multimodal nonverbal data during fine-tuning by the Multimodal Adaptation Gate, which can generate a shift to lexical embedding. **MISA** [8]: Modality-invariant and -Specific Representations project unimodal representations into the modality-invariant subspace and

Table 1. Experimental results on CMU-MOSI and CMU-MOSEI

Methods	CMU-MOSI				CMU-MOSEI			
	MAE (↓)	Corr (↑)	ACC-2 (↑)	F1 (↑)	MAE (↓)	Corr (↑)	ACC-2 (↑)	F1 (↑)
TFN	0.901	0.698	80.8	80.7	0.593	0.677	82.5	82.1
MFM	0.877	0.706	81.7	81.6	0.568	0.703	84.4	84.3
MulT	0.861	0.711	84.10	83.90	0.580	0.713	82.5	82.3
MAG-Bert	0.712	0.796	86.10	86.00	–	–	–	–
MISA	0.804	0.764	82.10	82.03	0.568	0.717	84.23	83.97
Self-MM	0.713	0.798	85.98	85.95	0.530	0.765	85.17	85.30
MMIM	0.700	0.800	86.06	85.98	0.526	0.772	85.97	85.94
CubeMLP	0.770	0.767	85.6	85.5	0.529	0.760	85.1	84.5
HGraph-CL	0.717	0.799	86.2	86.2	0.527	0.769	85.9	85.8
DMD	–	–	86.0	86.0	–	–	**86.6**	**86.6**
HSCL	**0.691**	**0.805**	**86.43**	**86.31**	**0.525**	**0.775**	85.86	85.85

Table 2. Role of modalities on CMU-MOSI and CMU-MOSEI

Models	CMU-MOSI				CMU-MOSEI			
	MAE (↓)	Corr (↑)	ACC-2 (↑)	F1 (↑)	MAE (↓)	Corr (↑)	ACC-2 (↑)	F1 (↑)
HSCL	**0.691**	**0.805**	**86.43**	**86.31**	**0.525**	**0.775**	**85.86**	**85.85**
-w/o Audio	0.715	0.797	83.99	84.01	0.532	0.772	85.86	85.83
-w/o Video	0.724	0.786	83.84	83.86	0.536	0.762	83.27	83.46
-w/o Text	1.408	0.004	52.59	51.73	0.816	0.223	63.15	60.82

the modality-specific subspace to reduce the modality gap and capture the characteristic features meanwhile. **Self-MM** [26]: Self-Supervised Multi-task Multimodal sentiment analysis network joint learns one multimodal task and three unimodal subtasks. The labels used in unimodal tasks are auto-generated by the Unimodal Label Generation Module in a self-supervised way. **MMIM** [7]: MultiModal InfoMax maximizes the mutual information between unimodal features and the fusion result to control the information flow from input to fusion representation. **Cube-MLP** [21]: CubeMLP is an MLP-based model mixing multimodal features across three axes in order to reduce computational costs while maintaining high performance. **HGraph-CL** [16]: HGraph-CL is a hierarchical graph contrastive learning framework that builds unimodal and multimodal graphs with a graph contrastive learning strategy to explore sentiment relations. **DMD** [15]: Decoupled Multimodal Distilling decouples representation into modality-irrelevant/-exclusive spaces and uses graph distillation unit to make knowledge transfer between different modalities.

Experimental Results. We evaluate the performance of our proposed HSCL model on the MOSI and MOSEI datasets, as shown in Table 1. Our model outperforms previous methods in most metrics, achieving a mean absolute error (MAE) of 0.691 and 0.525 on the MOSI and MOSEI datasets, respectively. These results confirm the effectiveness of our proposed method.

Table 3. Role of Hierarchical Contrastive Learning on CMU-MOSI

$\mathcal{L}_{at}$	$\mathcal{L}_{vt}$	$\mathcal{L}_{fusion}$	MAE($\downarrow$)	Corr($\uparrow$)	ACC-2($\uparrow$)	F1($\uparrow$)
			0.732	0.782	83.84	83.89
✓			0.711	0.800	83.38	83.38
	✓		0.724	0.792	83.99	83.93
		✓	0.708	0.797	83.38	83.40
✓	✓		0.699	0.803	85.21	85.04
✓		✓	0.703	0.802	84.30	84.28
	✓	✓	0.703	0.791	84.76	84.70
✓	✓	✓	**0.691**	**0.805**	**86.43**	**86.31**

4.3   Ablation Study

To further show the effectiveness of our framework, we conduct a variety of ablation studies on MOSI and MOSEI. We examine the role of HSCL components, including different modalities, contrastive learning at the unimodal representation level, and contrastive learning at the bimodal fusion level.

Effects of Different Modalities (RQ2). Firstly, we remove one modality during the training process. Experiment results are shown in Table 2. We find that removing audio or video leads to performance degradation, which means that non-verbal behaviors are helpful to sentiment analysis. For example, compared to ACC-2 of the experiments with all modalities (86.43%), the experiments without audio and without video are 83.99% and 83.44%, respectively. Moreover, removing the text modality resulted in a large drop in performance that was not observed for other modalities. This shows that in multimodal sentiment analysis, the text modality dominates, which is also in line with our expectations. We can explain this from two aspects. On the one hand, compared to other modalities, language is naturally more capable of expressing emotion-related information. On the other hand, the ability of pre-trained models such as BERT to extract features is more powerful than that of randomly initialized Transformers.

Effects of Hierarchical Contrastive Learning (RQ3). Next, we explore the impact of different levels of contrastive learning loss functions. We mainly observe the effects of audio-text alignment loss, video-text alignment loss, and contrastive loss between fusion results of two modalities, denoted as $\mathcal{L}_{at}$, $\mathcal{L}_{vt}$, and $\mathcal{L}_{fusion}$. Based on the experimental results in Table 3, we can see that removing any of the three losses would lead to performance degradation, and the more losses removed, the greater the performance decline. The removal of all auxiliary loss functions results in an increase of MAE from 0.691 to 0.732, and ACC-2 dropping from 86.43% to 83.84%. This confirms the effectiveness of the hierarchical contrastive learning method proposed in our work.

Table 4. Influence of label Information during Alignment

Method	MAE ($\downarrow$)	Corr ($\uparrow$)	ACC-2 ($\uparrow$)	F1 ($\uparrow$)
$\text{HSCL}_1(\mathcal{L}^{unSup})$	0.715	0.796	84.76	84.76
$\text{HSCL}_2(\mathcal{L}^{SupCon})$	0.714	0.801	84.60	84.63
HSCL	**0.691**	**0.805**	**86.43**	**86.31**

Influence of Label Information During Alignment (RQ4). In the process of cross-modal contrastive learning, we utilize the sentiment intensity annotated for each sample to select positive and negative samples, rather than simply considering different modalities corresponding to the same video as positive samples and others as negative samples. To investigate the impact of label information during the alignment process, we conducted a comparative experiment using different loss functions. Firstly, following [3], we adopted an unsupervised approach for positive and negative sample selection, denoted as $\mathcal{L}^{unSup}$. Secondly, we follow [12], selecting positive and negative samples only based on the sentiment polarity, denoted as $\mathcal{L}^{SupCon}$. The experimental results are shown in Table 4. We observed that removing label information or solely considering sentiment polarity during the contrastive learning process leads to a performance decline. We believe this is because the unsupervised alignment process lacks constraints on the representations, which may result in the loss of emotional semantics.

5 Conclusion

In this paper, we present a supervised contrastive learning approach tailored for multimodal sentiment analysis. We employ a hierarchical contrastive learning structure to effectively align and fuse various modalities. Additionally, through supervised contrastive regression, we aim to define a subspace where embeddings align according to label proximity. This strategy enriches representation at multiple tiers. Empirical results from both the CMU-MOSEI and CMU-MOSI datasets validate the efficacy of our model, and the ablation studies further emphasize the significance of its distinct components.

In our contrastive learning, we primarily focus on the modality-invariant aspect, potentially sidelining unique modality-specific information in the alignment process. Additionally, with the recent surge in performance of large language models (LLMs) in various natural language processing tasks, exploring their application in MSA becomes imperative. In the future, we aim to amplify the modality-specific representation and synergize MSA with LLM.

Acknowledgements. This work was supported by the National Natural Science Foundation of China (Grants No. 62202439).

References

1. Ba, J.L., Kiros, J.R., Hinton, G.E.: Layer normalization. arXiv preprint arXiv:1607.06450 (2016)
2. Baltrušaitis, T., Robinson, P., Morency, L.P.: Openface: an open source facial behavior analysis toolkit. In: 2016 IEEE Winter Conference on Applications of Computer Vision (WACV), pp. 1–10. IEEE (2016)
3. Chen, T., Kornblith, S., Norouzi, M., Hinton, G.: A simple framework for contrastive learning of visual representations. In: International Conference on Machine Learning, pp. 1597–1607. PMLR (2020)
4. Degottex, G., Kane, J., Drugman, T., Raitio, T., Scherer, S.: Covarep-a collaborative voice analysis repository for speech technologies. In: 2014 IEEE International Conference on Acoustics, Speech and Signal Processing (ICASSP), pp. 960–964. IEEE (2014)
5. Gao, T., Yao, X., Chen, D.: SimCSE: simple contrastive learning of sentence embeddings. In: Proceedings of the 2021 Conference on Empirical Methods in Natural Language Processing, pp. 6894–6910 (2021)
6. Han, W., Chen, H., Gelbukh, A., Zadeh, A., Morency, L.P., Poria, S.: Bi-bimodal modality fusion for correlation-controlled multimodal sentiment analysis. In: Proceedings of the 2021 International Conference on Multimodal Interaction, pp. 6–15 (2021)
7. Han, W., Chen, H., Poria, S.: Improving multimodal fusion with hierarchical mutual information maximization for multimodal sentiment analysis. In: Proceedings of the 2021 Conference on Empirical Methods in Natural Language Processing, pp. 9180–9192 (2021)
8. Hazarika, D., Zimmermann, R., Poria, S.: Misa: modality-invariant and-specific representations for multimodal sentiment analysis. In: Proceedings of the 28th ACM International Conference on Multimedia, pp. 1122–1131 (2020)
9. He, K., Zhang, X., Ren, S., Sun, J.: Deep residual learning for image recognition. In: Proceedings of the IEEE Conference on Computer Vision and Pattern Recognition, pp. 770–778 (2016)
10. Hochreiter, S., Schmidhuber, J.: Long short-term memory. Neural Comput. **9**(8), 1735–1780 (1997)
11. Kenton, J.D.M.W.C., Toutanova, L.K.: BERT: pre-training of deep bidirectional transformers for language understanding. In: Proceedings of NAACL-HLT, vol. 1, p. 2 (2019)
12. Khosla, P., et al.: Supervised contrastive learning. In: Advances in Neural Information Processing Systems, vol. 33, pp. 18661–18673 (2020)
13. Kingma, D.P., Ba, J.: Adam: a method for stochastic optimization. arXiv preprint arXiv:1412.6980 (2014)
14. Li, J., Selvaraju, R., Gotmare, A., Joty, S., Xiong, C., Hoi, S.C.H.: Align before fuse: vision and language representation learning with momentum distillation. In: Advances in Neural Information Processing Systems, vol. 34, pp. 9694–9705 (2021)
15. Li, Y., Wang, Y., Cui, Z.: Decoupled multimodal distilling for emotion recognition. In: Proceedings of the IEEE/CVF Conference on Computer Vision and Pattern Recognition, pp. 6631–6640 (2023)
16. Lin, Z., et al.: Modeling intra-and inter-modal relations: hierarchical graph contrastive learning for multimodal sentiment analysis. In: Proceedings of the 29th International Conference on Computational Linguistics, pp. 7124–7135 (2022)

17. Pang, B., Lee, L., Vaithyanathan, S.: Thumbs up? Sentiment classification using machine learning techniques. In: Proceedings of the 2002 Conference on Empirical Methods in Natural Language Processing (EMNLP 2002), pp. 79–86 (2002)
18. Pérez-Rosas, V., Mihalcea, R., Morency, L.P.: Utterance-level multimodal sentiment analysis. In: Proceedings of the 51st Annual Meeting of the Association for Computational Linguistics (Volume 1: Long Papers), pp. 973–982 (2013)
19. Radford, A., et al.: Learning transferable visual models from natural language supervision. In: International Conference on Machine Learning, pp. 8748–8763. PMLR (2021)
20. Rahman, W., et al.: Integrating multimodal information in large pretrained transformers. In: Proceedings of the Conference. Association for Computational Linguistics. Meeting, vol. 2020, p. 2359. NIH Public Access (2020)
21. Sun, H., Wang, H., Liu, J., Chen, Y.W., Lin, L.: CubeMLP: an MLP-based model for multimodal sentiment analysis and depression estimation. In: Proceedings of the 30th ACM International Conference on Multimedia, pp. 3722–3729 (2022)
22. Sun, Z., Sarma, P., Sethares, W., Liang, Y.: Learning relationships between text, audio, and video via deep canonical correlation for multimodal language analysis. In: Proceedings of the AAAI Conference on Artificial Intelligence, vol. 34, pp. 8992–8999 (2020)
23. Tsai, Y.H.H., Bai, S., Liang, P.P., Kolter, J.Z., Morency, L.P., Salakhutdinov, R.: Multimodal transformer for unaligned multimodal language sequences. In: Proceedings of the 57th Annual Meeting of the Association for Computational Linguistics, pp. 6558–6569 (2019)
24. Tsai, Y.H.H., Liang, P.P., Zadeh, A., Morency, L.P., Salakhutdinov, R.: Learning factorized multimodal representations. In: International Conference on Learning Representations (2018)
25. Vaswani, A., et al.: Attention is all you need. In: Advances in Neural Information Processing Systems, vol. 30 (2017)
26. Yu, W., Xu, H., Yuan, Z., Wu, J.: Learning modality-specific representations with self-supervised multi-task learning for multimodal sentiment analysis. In: Proceedings of the AAAI Conference on Artificial Intelligence, vol. 35, pp. 10790–10797 (2021)
27. Zadeh, A., Chen, M., Poria, S., Cambria, E., Morency, L.P.: Tensor fusion network for multimodal sentiment analysis. In: Proceedings of the 2017 Conference on Empirical Methods in Natural Language Processing, pp. 1103–1114 (2017)
28. Zadeh, A., Zellers, R., Pincus, E., Morency, L.P.: Multimodal sentiment intensity analysis in videos: facial gestures and verbal messages. IEEE Intell. Syst. **31**(6), 82–88 (2016)
29. Zadeh, A.B., Liang, P.P., Poria, S., Cambria, E., Morency, L.P.: Multimodal language analysis in the wild: CMU-MOSEI dataset and interpretable dynamic fusion graph. In: Proceedings of the 56th Annual Meeting of the Association for Computational Linguistics (Volume 1: Long Papers), pp. 2236–2246 (2018)
30. Zha, K., Cao, P., Yang, Y., Katabi, D.: Supervised contrastive regression. arXiv preprint arXiv:2210.01189 (2022)
31. Zhang, L., Wang, S., Liu, B.: Deep learning for sentiment analysis: a survey. Wiley Interdiscip. Rev. Data Min. Knowl. Discov. **8**(4), e1253 (2018)

Semantic Importance-Based Deep Image Compression Using a Generative Approach

Xi Gu[1], Yuanyuan Xu[1(✉)], and Kun Zhu[2]

[1] Hohai University, Nanjing, China
yuanyuan_xu@hhu.edu.cn
[2] Nanjing University of Aeronautics and Astronautics, Nanjing, China

Abstract. Semantic image compression can greatly reduce the amount of transmitted data by representing and reconstructing images using semantic information. Considering the fact that objects in an image are not equally important at the semantic level, we propose a semantic importance-based deep image compression scheme, where a generative approach is used to produce a visually pleasing image from segmentation information. A base-layer image can be reconstructed using a conditional generative adversarial network (GAN) considering the importance of objects. To ensure that objects with the same semantic importance have similar perceptual fidelity, a generative compensation module has been designed, considering the varying generative capability of GAN. The base-layer image can be further refined using residuals, prioritizing regions with high semantic importance. Experimental results show that the reconstructed images of the proposed scheme are more visually pleasing compared with relevant schemes, and objects with a high semantic importance achieve both good pixel and semantic-perceptual fidelity.

Keywords: Image Compression · Semantic image coding · Scalable coding

1 Introduction

Image compression is an essential and fundamental topic for efficient image storage and transmission across modern networks. Traditional image codecs, such as WebP [8], JPEG [23], JPEG2000 [22] and BPG [4] use a hybrid image coding framework, where basic coding blocks of an image are transformed, quantized and entropy coded. Block-based coding introduces blocking effects, especially at a low bitrate. Complicate dependency among coding modules makes the codec difficult to optimize as a whole. Due to the strong representation capabilities of deep learning, learned image compression [3] based on deep neural networks, or deep image compression, has received widespread attention in recent years. The learned image compression framework can be optimized end-to-end, showing competitive performance compared with traditional codecs.

To further reduce the amount of transmitted data, image semantic coding [11,12] can be performed. As traditional and learned image codecs, even the lightfiled image codec [14], are typically optimized for pixel fidelity, such as MSE (Mean Square Error) and PSNR (Peak Signal to Noise Ratio), semantic image coding aims at maintaining semantic-perceptual fidelity [11]. According to [21], perceptual fidelity metrics refer to objective metrics of human viewing experience, while semantic fidelity is the semantic difference between original and reconstructed image such as the difference in object detection accuracy. A good choice to optimize towards semantic-perceptual fidelity is to use generative adversarial network (GAN) [7], since GAN can capture both global semantic information and local texture and is powerful to produce a visually pleasing image even with semantic labels only. In recent years, GAN has been used to improve image coding efficiency [1, 2, 11, 12, 17].

Regarding pixel fidelity, the work in [2] has proposed a generative image compression framework based on deep semantic segmentation (DSSLIC), where a down-sampled version of the image and a semantic segmentation map need to be transmitted and used to guide the image generation. Residual can be sent to refine the generated image as well. Similar as DSSLIC, the work in [15] utilizes semantic segmentation map to compress image and the redundancy in the spatial dimension of feature map is addressed by using octave convolution. To avoid using extra bits for segmentation map, the work [10] trained a segmentation network from the up-sampled version of the down-sampled image to obtain an accurate segmentation map at both encoder and decoder.

Targeting for perceptual quality, an image compression system based on GANs has been designed in [1] for extreme low bitrates, where quantized extracted features are fed to GANs. In [17], a neural compression scheme with a GAN has been proposed, which can be optimized to yield reconstructions with high perceptual fidelity that are visually close to the input. The above works optimize towards fidelity for the whole image. However, objects in the same image are different in terms of semantic importance. For example, when viewing portrait photos, humans are more important than buildings semantically. In [1], a selective generative compression scheme has been proposed as well only using GAN to generate unimportant regions, while user-defined regions were preserved with fine details. This scheme can only support two levels of semantic importance. The semantic image coding scheme proposed by Huang et al. [11] quantized features of different semantic concepts adaptively, where bit allocation among semantic concepts was determined using a reinforcement learning algorithm. GAN was used at the decoder to reconstruct images from quantized features. The scheme was optimized in terms of semantic-perceptual loss without considering pixel fidelity.

Pixel fidelity and perceptual fidelity are both important, especially for regions with high semantic importance. Optimizing for both pixel and perceptual loss, Huang et al. [12] proposed a deep image semantic coding scheme which uses both quantized extracted features and the segmentation map for image generation, and residuals were transmitted to restore fine details. However, varying importance of different regions are not considered. In this work, we propose a generative image compression that can support any levels of semantic importance, and both pixel-level and semantic-perceptual distortion are evaluated in the training of the proposed work. In the proposed approach, objects with higher semantic importance are evaluated weighting more on pixel-level accuracy, while the unimportance regions are mainly concerned with perceptual accuracy. The proposed scheme is a scalable coding scheme as the work [24], where the base layer consists of segmentation map and quantized features, and the enhancement layer comprises coded residuals. Like multiple description coding [13,16,26] that combats transmission losses via source coding, a scalable image coding scheme can provide a certain degree of robustness tolerating the loss of enhancement layer.

The main contributions of this work are summarized as follows:

- We propose a generative deep image compression framework considering varying semantic importance of objects. Guided by an importance map of an image of objects, a base-layer image can be reconstructed using a conditional GAN using the extracted feature and segmentation map. Prioritized residual coding is then performed that can be used to refine important regions of the base-layer image.
- Considering the varying generative capability of GAN for objects with different characteristics, a generative compensation module has been designed to ensure that objects with the same semantic importance have similar perceptual fidelity.
- The proposed framework has been optimized and evaluated using both the weighted pixel-level distortion and adversarial loss concerning perceptual loss. As a result, objects with a high semantic importance can achieve both good pixel and semantic-perceptual fidelity, and regions with lower importance are visually pleasing.

2   Semantic Importance-Based Deep Image Compression

Illustration of the proposed framework is shown in Fig. 1. Encoder E extracts the latent representation w from the input image x. w is scaled according to the importance map m and quantized by a quantizer Q. The generator G uses the scaled and quantized latent code $\hat{w}$, and the segmentation map s from the segmentation network F, to produce a base-layer reconstructed image, $\tilde{x}$. In the importance map generation module, a pre-defined semantic importance map, m_s, and a generative compensation map, m_c, are fused to form a final map, m. m_c is generated by the generative-compensation network, A, using s and x as the input. In the residual coding module, a residual processor H conducts prioritized processing for residual, $r = x - \tilde{x}$, according to m_s. The processed residual r' is then compressed and transmitted.

As shown in Fig. 1(a), the image is layered coded. The base layer consists of lossless compressed $\hat{w}$ and s, while the enhancement layer includes lossy compressed r'. In Fig. 1(b), the generator G can use decoded $\hat{w}$ and s to generate an image $\hat{x}$. With an available enhancement layer, the decoded residual $\hat{r}$ is added to $\hat{x}$ to obtain an enhanced image $x' = \tilde{r} + \tilde{x}$.

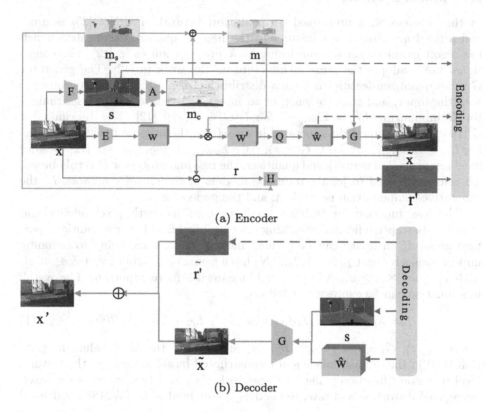

(a) Encoder

(b) Decoder

Fig. 1. The proposed framework. (a) The encoder consists of segmentation network F, encoder network E, generative-compensation network A, quantizer Q, generator G, and residual processor H. (b) The decoder generates a base-layer image using the segmentation map and quantized latent code, while an enhanced image can be obtained combining the residual.

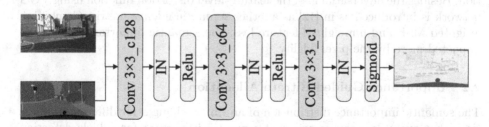

Fig. 2. The network architecture of generative-compensation module.

2.1 Model Learning

In this framework, a pretrained segmentation network, PSPNet [25] is used, and a fixed quantizer maps features to the nearest quantization centers which has a soft quantization version during back propagation as in [12]. The conditional GAN adopts the same architecture as the work in [12]. The generator G takes quantized features $\hat{w}$ form a distribution of $p_{\hat{W}}$ and segmentation map s as the input, and tries to generate an image that can cheat the discriminator D minimizing $L_G = E_{\hat{w} \sim p_{\hat{W}}}[-log(D(G(\hat{w}, s), s))]$ [12]. The discriminator G tries to differentiate the generated images from the original images, minimizing $L_D = E_{\hat{w} \sim p_{\hat{W}}}[-log(1 - D(G(\hat{w}, s), s))] + E_{x \sim p_{X|s}}[-log(D(x, s))]$ [12]. Besides fixed segmentation network and quantizer, the residual processor H is rule-based. Therefore, we need to jointly train the models of the encoder network E, the generative-compensation network A, and the generator G.

The loss function for training needs to consider both pixel fidelity and semantic-perceptual fidelity, weighting more on pixel fidelity for semantic important areas. If an image can be divided into N regions according to semantic importance, a region C_k ($k = 1, 2, ..., N$) has a semantic weight of w_k ($w_k \in [0, 1]$) with $w_1 \geq w_2 \geq ... \geq w_k$. A larger weight means it is more important. The overall loss function can be expressed as follows:

$$L = \lambda_1 WMSE + \lambda_1 d(G(\hat{w}, s), x) + L_G + L_D + \lambda_2 R(\hat{w}), \qquad (1)$$

where $x(i, j)$, $G(\hat{w}, s)(i, j)$, $d()$, $R()$, λ_1 and λ_2 are the pixel values at position (i, j) in the original image and reconstructed base-layer image, the feature-level distortion function using VGG network, rate function, weights for pixel-perceptual distortion and rate, respectively. Weighted MSE (WMSE) is defined as

$$WMSE = \sum_{k=1}^{N} w_k \cdot \frac{1}{|C_k|} \sum_{p(i,j) \in C_k} [x(i, j) - G(\hat{w}, s)(i, j)]^2, \qquad (2)$$

where $p(i, j) \in C_k$ means that the pixel at position (i, j) is in region C_k, and $|C_k|$ are the number of pixles in C_k.

Since the segmentation network is pre-trained, and the optimization is conducted for the base-layer, the rates of s and r' are not included in the loss function. Besides the adversarial loss, the feature-level distortion function using VGG network is introduced as in [12] as another semantic fidelity metric. Combining weighted MSE and unweighted feature loss, the semantic important regions are impacted more by the pixel fidelity.

2.2 Importance Guided Bitrate Allocation

The semantic importance distribution of an image changes for different applications. For example, vehicles are most semantic important for vehicle detection but not for pedestrian detection. Therefore, we assume a predefined semantic importance map m_s is available. Guided by m_s, a bitrate allocation scheme is then designed.

Generative Compensation. Originally, we only use m_s to guide bitrate allocation. However, we observe that objects with the same semantic weight in the generative image do not have the same perceptual quality, due to varying generative capability of GAN. For example, sky and road in an image can have the same semantic weight, but the perceptual quality of sky with simple texture is better than that of a road. Therefore, a generative-compensation module has been added to compensate the generative capability.

The network architecture of the generative-compensation module is shown in Fig. 2. This generative-compensation module, A, is a network with three convolutional layers generating a pixel-level importance map m_c, whose values are within $[0, 1]$. The input of the model is the original image and segmentation map. Including semantic labels can make m_c reflect the semantic and edge information of the image better.

Fused Importance Map. Two maps, m_s and m_c can be fused into one map m to guide the bitrate allocation of different regions, with

$$m = \lambda * m_s + (1 - \lambda) * m_c, \tag{3}$$

where λ is a weight factor. The value range for m is within $[0, 1]$ as well.

For latent code w with C feature maps with a dimension of $H * W$, m is rescaled to $H*W$. Each element in a feature map is multiplied by a corresponding important weight in m. The multiplication is performed at the element level for each channel. By scaling the latent code with a weight between $[0, 1]$ followed by quantization with fixed quantization centers, the coefficients with a small weight are represented with a lower quantization precision and thus a lower bitrate. In this way, bitrate is allocated according to importance.

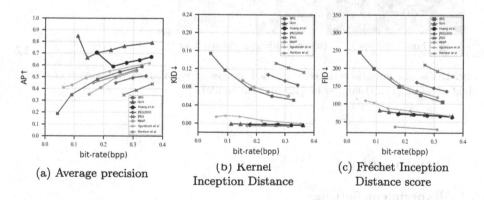

(a) Average precision

(b) Kernel Inception Distance

(c) Fréchet Inception Distance score

Fig. 3. Comparsion of JPEG, JPEG2000, WebP, BPG, Huang *et al.* [12], Agustsson *et al.* [1], and our model in terms of objective metrics. The arrow (↓) on the y-axis indicates that a lower value is better, and (↑) indicates that a higher value is better.

2.3 Prioritized Residual Coding

Residual coding is conducted considering prioritizes as well. A residual processor skips part of residuals according to m_s yielding a processed residual r'. For example, with a low bit rate, we only keep the residual of most important area. If the ω is the semantic weight threshold for residual skipping, r' can be expressed as

$$r'(i,j) = \begin{cases} r(i,j), & if \ x(i,j) \in C_k \ AND \ w_k \geq \omega \\ 0, & otherwise \end{cases} \tag{4}$$

where $r(i,j)$ and $r'(i,j)$ are the values of residual and processed residual at position (i,j), respectively. Different ω is used for varying targeting bitrates. The residual image r' is encoded by the lossy BPG encoder with different quality factors for varying bitrate.

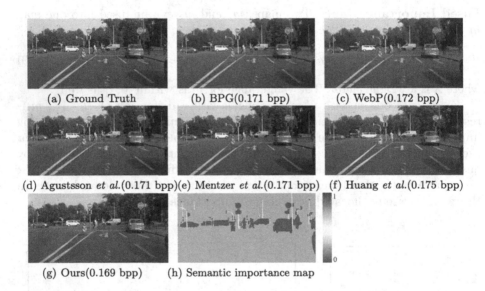

(a) Ground Truth   (b) BPG(0.171 bpp)   (c) WebP(0.172 bpp)

(d) Agustsson *et al.*(0.171 bpp)(e) Mentzer *et al.*(0.171 bpp) (f) Huang *et al.*(0.175 bpp)

(g) Ours(0.169 bpp)   (h) Semantic importance map

Fig. 4. Visual comparison of reconstructed images of WebP, BPG,Agustsson *et al.* [1], Mentzer *et al.* [17], Huang *et al.* [12] and our model.

3 Experiment

3.1 Experiment Settings

The proposed model is trained with the Cityscapes dataset [6] with images down-sampled to $256 * 512$. The training set consists of 2975 images, and our performance is evaluated using the test set. The latent representation w is represented using 128 channels. The semantic importance maps are generated according to

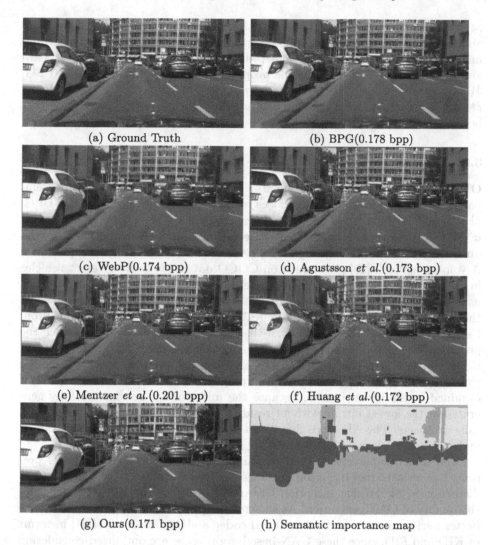

(a) Ground Truth

(b) BPG(0.178 bpp)

(c) WebP(0.174 bpp)

(d) Agustsson *et al.*(0.173 bpp)

(e) Mentzer *et al.*(0.201 bpp)

(f) Huang *et al.*(0.172 bpp)

(g) Ours(0.171 bpp)

(h) Semantic importance map

Fig. 5. Example of using reconstructed images of WebP, BPG,Agustsson *et al.* [1], Mentzer *et al.* [17], Huang *et al.* [12] and our model for vehicle detection.

annotations. Three levels of semantic importance are considered, whose values are set to 1, 0.67, and 0.33, respectively. The first level includes vehicles, objects, and humans, while the second one contains constructions and flats. The remaining belong to the third level. λ_1 and λ_2 in the loss function of (1) are set to 10 and 1, respectively. λ for the importance map in (3) is set to 0.6. The model is trained with an initial learning rate of $3*10^{-4}$ which gradually reduces to 0 after 50 epochs.

The segmentation map and latent code are lossless compressed using vector graph and arithmetic coding consuming 0.01bpp and 0.1bpp, respectively. ω in

(4) is set to 0.9, 0.6, 0.3 for bitrates that are below 0.2bpp, [0.2, 0.4]bpp, and above 0.4bpp, respectively. Residual is compressed by the lossy BPG encoder under quality factors of {40, 35} and {40, 35, 30} for cases with a bitrate below and above 0.2bpp, respectively. The proposed scheme is compared with JPEG, JPEG2000, WebP, and BPG, Agustsson et al. [1], Mentzer et al. [17] and Huang et al.'s generative semantic image compression scheme in [12] which considers both pixel and perceptual fidelity.

3.2   Results

Objective Evaluation. For semantic fidelity, the proposed scheme is compared with the latest traditional codecs, which are JPEG2000, WebP and BPG, JPEG, and learning based codecs including Agustsson et al. [1] and Mentzer et al. [17] for performance on the object detection task. The simulation was conducted on an NVIDIA GeForce RTX 3080, based on the PyTorch [19] platform. A yolo-v3 network [20] pre-trained on COCO dataset is used to detect vehicles in reconstruction images of different methods. Average Precision (AP) [18] is used to measure the performance of vehicle detection at different bitrates, where the IOU threshold is set to 0.5. Figure 3 show performance comparison. From Fig. 3, the proposed scheme achieves the highest AP at all bit-rates, since the proposed scheme is designed to support image compression with different semantic importance. Vehicles in the images are assigned with the highest semantic weight, resulting in both good pixel and perceptual fidelities. The highest AP is obtained at the lowest bitrate, because the unimportant regions are fully generated at the lowest bitrate without residual coding which further differentiates salient objects. The AP drops due to the blurry effect of BPG residual coding, and increases with more bitrate used for residual coding.

The semantic-perceptual quality comparison for the whole images is presented as well. Kernel Inception Distance (KID) [5] evaluates the feature space distortion, while Fréchet Inception Distance score (FID) [9] is used to measure the quality of GAN generation. As shown in Fig. 3, the generative schemes have better performance than the traditional codec and Agustsson et al. [1] in terms of KID and FID, since these GAN-based approaches are optimized considering semantic-perceptual quality. The performance of the proposed and the scheme in [12] are almost the same. Mentzer et al. [17] performs better on FID than our method, since the proposed method focuses on maintaining fidelity for important regions only which does not show competitive generating capability for the whole image.

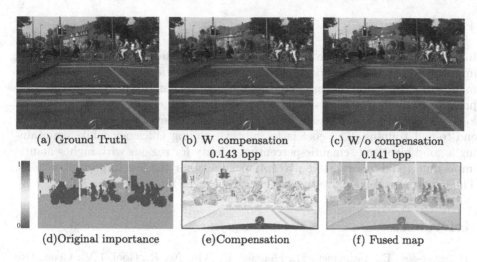

(a) Ground Truth	(b) W compensation 0.143 bpp	(c) W/o compensation 0.141 bpp
(d)Original importance	(e)Compensation	(f) Fused map

Fig. 6. Visual comparison with/without the generative compensation module.

Subjective Evaluation. Except the AP, the objective evaluation above do not differentiate regions with varying importance, and thus subjective evaluation is conducted as well. For perceptual quality, Fig. 4 shows the visual comparison. It can be observed that, even with a lower bitrate, the reconstructed images of the generative approaches are more visual pleasing than those of traditional codecs, Agustsson*et al.* [1] and Mentzer *et al.* [17]. Compared with Huang *et al.*'s method, the regions with high semantic importance in the proposed scheme exhibit a lower level of degradation from the original content, such as the stop sign. In our scheme, semantically unimportant area, such as trees in the background, deviates more from the original content, but is still visually pleasing. For semantic quality, an example for vehicle detection is present in Fig. 5. As shown in the semantic importance map in Fig. 5(h), regions of vehicles have high semantic importance in the proposed scheme. These regions are coded with more bitrate resulting in both good pixel and semantic-percentual fidelity. Compared with other schemes, more correct vehicles are detected using the reconstructed image of the proposed scheme with a lower bitrate.

Effectiveness of Generative Compensation Module. Reconstructed images of the proposed scheme with and without the generative compensation module are shown in Fig. 6. In the original importance map in Fig. 6(d), sky and road have the same semantic weight. However, generative capability of GAN for these objects are different, where sky is more visually pleasing than the road without compensation. With compensation module, the regions with more texture details are compensated as shown in Fig. 6(e), achieving similar perceptual quality for regions with the same semantic importance.

4   Conclusion

In this paper, a generative image compression framework based on semantic importance has been proposed. A predefined semantic importance map and a map from the generation compensation module are combined to guide the compression of the latent representation extracted by the encoder. Both adversarial loss, perceptual loss and semantically weighted pixel-level loss are considered for end-to-end training. The goal is to improve coding efficiency while maintaining a good pixel and semantic-percentual fidelity for regions with high semantic importance, and a reasonable perceptual fidelity for the less important regions. The experimental results verify the effectiveness of the proposed scheme.

References

1. Agustsson, E., Tschannen, M., Mentzer, F., Timofte, R., Gool, L.V.: Generative adversarial networks for extreme learned image compression, pp. 221–231 (2019)
2. Akbari, M., Liang, J., Han, J.: DSSLIC: deep semantic segmentation-based layered image compression. In: IEEE International Conference on Acoustics, Speech and Signal Processing (ICASSP), pp. 2042–2046 (2019)
3. Balle, J., Laparra, V., Simoncelli, E.P.: End-to-end optimization of nonlinear transform codes for perceptual quality. In: Picture Coding Symposium (PCS), pp. 1–5. IEEE, Nuremberg, Germany (2016). https://doi.org/10.1109/PCS.2016.7906310
4. Bellard., F.: BPG Image format
5. Binkowski, M., Sutherland, D.J., Arbel, M., Gretton, A.: Demystifying MMD GANs. ArXiv:1801.01401 (2018)
6. Cordts, M., Omran, M., Ramos, S., Rehfeld, T., Schiele, B.: The cityscapes dataset for semantic urban scene understanding. IEEE (2016)
7. Goodfellow, I., et al.: Generative adversarial nets. In: Neural Information Processing Systems (2014)
8. Google: WebP Image format (2010). https://developers.google.com/speed/webp/
9. Heusel, M., Ramsauer, H., Unterthiner, T., Nessler, B., Hochreiter, S.: GANs trained by a two time-scale update rule converge to a local nash equilibrium. In: Advances in Neural Information Processing Systems 30 (2017)
10. Hoang, T.M., Zhou, J., Fan, Y.: Image compression with encoder-decoder matched semantic segmentation. In: IEEE/CVF Conference on Computer Vision and Pattern Recognition Workshops (CVPRW), pp. 160–161 (2020)
11. Huang, D., Gao, F., Tao, X., Du, Q., Lu, J.: Towards semantic communications: deep learning-based image semantic coding. IEEE J. Selected Areas Commun. **41**(1), 55–71 (2022)
12. Huang, D., Tao, X., Gao, F., Lu, J.: Deep learning-based image semantic coding for semantic communications. In: IEEE Global Communications Conference (GLOBECOM), pp. 1–6 (2021)
13. Liu, M., Zhu, C., Wu, X.: Index assignment design for three-description lattice vector quantization. In: 2006 IEEE International Symposium on Circuits and Systems (ISCAS), pp. 4-pp. IEEE (2006)
14. Liu, Y.Y., Zhu, C., Mao, M.: Light field image compression based on quality aware pseudo-temporal sequence. Electron. Lett. **54**(8), 500–501 (2018)

15. Liu, Z., Meng, L., Tan, Y., Zhang, J., Zhang, H.: Image compression based on octave convolution and semantic segmentation. Knowl.-Based Syst. **228**, 107254 (2021)
16. Meng, L., Li, H., Zhang, J., Tan, Y., Ren, Y., Zhang, H.: Convolutional auto-encoder based multiple description coding network. KSII Trans. Internet and Inform. Syst. (TIIS) **14**(4), 1689–1703 (2020)
17. Mentzer, F., Toderici, G.D., Tschannen, M., Agustsson, E.: High-fidelity generative image compression. Adv. Neural. Inf. Process. Syst. **33**, 11913–11924 (2020)
18. Padilla, R., Netto, S.L., Silva, E.: A survey on performance metrics for object-detection algorithms. In: International Conference on Systems, Signals and Image Processing (IWSSIP) (2020)
19. Paszke, A., et al.: Automatic differentiation in pytorch (2017)
20. Redmon, J., Farhadi, A.: Yolov3: An incremental improvement. arXiv e-prints (2018)
21. Shi, J., Chen, Z.: Reinforced bit allocation under task-driven semantic distortion metrics. In: IEEE International Symposium on Circuits And Systems (ISCAS), pp. 1–5 (2020)
22. Skodras, A., Christopoulos, C., Ebrahimi, T.: The JPEG 2000 still image compression standard. IEEE Signal Process. Mag. **18**(5), 36–58 (2001)
23. Wallace, Gregory, K.: The JPEG still picture compression standard. Communications ACM **34**(4), 30–44 (1991)
24. Zhang, D., et al.: Exploring resolution fields for scalable image compression with uncertainty guidance. IEEE Trans. Circ. Syst. Video Technolpp. (2023). https://doi.org/10.1109/TCSVT.2023.3307438
25. Zhao, H., Shi, J., Qi, X., Wang, X., Jia, J.: Pyramid scene parsing network. In: Proceedings of the IEEE Conference on Computer Vision and Pattern Recognition, pp. 2881–2890 (2017)
26. Zhao, L., Bai, H., Wang, A., Zhao, Y.: Multiple description convolutional neural networks for image compression. IEEE Trans. Circuits Syst. Video Technol. **29**(8), 2494–2508 (2018)

Drive-CLIP: Cross-Modal Contrastive Safety-Critical Driving Scenario Representation Learning and Zero-Shot Driving Risk Analysis

Wenbin Gan[✉], Minh-Son Dao, and Koji Zettsu

Big Data Integration Research Center, National Institute of Information and Communications Technology (NICT), Tokyo, Japan
{wenbingan,dao,zettsu}@nict.go.jp

Abstract. Driving risk analysis, especially in safety-critical driving scenarios (SCDSs), plays a paramount role in providing subsequent driving assistance to mitigate potential traffic hazards. Previous studies predominantly relied on unimodal event data for risk analysis, overlooking the valuable source of information embedded in text narratives. This limitation hindered the full utilization of available data for representing SCDSs effectively. Capitalizing on the success of large language models in natural language processing, text-based models offer new potential for enhancing the performance of this task. In this paper, we introduce a novel framework named Drive-CLIP, designed for cross-modal contrastive SCDS representation learning, incorporating both text narratives and event data as two modalities. Through cross-modal analysis, Drive-CLIP distills SCDS event embeddings from natural language supervision, enabling text-guided zero-shot driving risk analysis on event data. Experiments conducted on a naturalistic driving dataset demonstrate that Drive-CLIP surpasses the performance of current best-performing methods, underscoring its effectiveness and superiority. Furthermore, we highlight that cross-modal analysis yields advantages over using a single data modality, and the cross-modal contrastive SCDS representation learning remains beneficial even in scenarios with limited data.

Keywords: Driving Risk Analysis · Safety-Critical Driving Scenario · Naturalistic Driving Study · Intelligent Vehicles · Driving Assistance · Cross-modal Contrastive Learning · CLIP

1 Introduction

Mitigating traffic hazards stands as a paramount public safety priority on a global scale. It is well acknowledged that a significant proportion of traffic accidents can be alleviated (prevented) by executing suitable driving maneuvers within these SCDSs [2,9]. For this proactive crash mitigation purpose, some advanced driving assistant systems (ADASs) that offer driving assistance have been proposed within this collaborative driver-vehicle co-driving context [2,10].

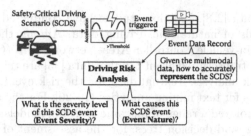

Fig. 1. The research questions solved in this paper.

ADASs assist drivers by reducing their control demands, monitoring the vehicle dynamics, proactively forecasting the driving risk, issuing alerts in the SCDS, offering necessary support and coaching, and have shown potential for promoting safe and comfortable driving [9,17]. As the initial and critical step for offering driving assistance in ADASs, driving risk prediction and analysis detects (evaluates) the driving risks and provides sufficient insights for follow-up action-taking methods in SCDS to play their roles. Hence, it is essential to precisely establish the model that links numerous driving factors in SCDSs to driving risks.

As shown in Fig. 1, the ADASs in vehicles will be generally activated by some triggering events (e.g., the longitudinal/lateral acceleration/deceleration is greater than some thresholds). These events are usually regarded as being critical to the driving safety, and are also called SCDS events. Given the data record of these SCDS events, analyzing the driving risks by **evaluating their severity levels (event severity)** and **analyzing their causes (event nature)** is the research question that we are concerned about in this paper. This holds significance in practical applications: for one thing, predicting and identifying driving risks that could result in accidents during driving provides the essential insights to deliver immediate on-board driving assistance to drivers to offer necessary support, with the ultimate goal of diminishing the chances of accidents. For another, the risk event classification is important for the management and retrieval of individual driving record data, such as the driving risk-based self-coaching [10,30], fleet management [21] and "pay-how-you-drive" car insurance paradigm [3].

To achieve this objective, great effort has been dedicated to improving the efficacy of driving risk analysis. This has involved expanding the consideration of various driving-related factors to represent the SCDS and refining the methods used for analysis [6,33,34]. The studies in [2,9,18] have yielded valuable insights into the variables that impact the probability of a driving risk. Recent advancements in intelligent vehicle technology and transportation have introduced rich sources of data related to drivers, vehicles, and contextual information. These developments have made it both technologically viable and economically practical to construct a comprehensive understanding of natural driving events on roadways [33], particularly in SCDSs. These naturalistic driving studies enable a more precise observation and measurement of SCDS events by generating rich and multi-modal data [2,8]. Existing studies have conducted the risk analysis using different kinds of naturalistic driving data. Among them, the recorded video data [4], the in-vehicle Controller Area Network-BUS (CAN-BUS) sensor data [6,34], and the transcripted event tabular data [33] are the most widely used driving data for risk analysis. Textual data for the risk event narratives is also used in a few studies for text mining analytics [19,20]. For the analysis methods, researchers have employed a range of machine learning models, including logistic regression, boosting, and decision trees, for the assessment of driving risks and shown them to be effective for risk prediction [29,33]. Some work explored the modern deep learning models for the feature learning from data [6,29,34].

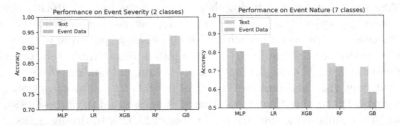

Fig. 2. Comparisons of performance on textual narratives and event data for event severity and event nature classification. Text-based models generally outperform the event-data-based models but text data is rarely used for driving risk analysis, which motivates this paper to use the cross-modal text and event data for driving risk study.

Despite significant advancements in current research, challenges persist with these methods. **Firstly**, most of these studies only use *uni-modal* data to analyze the risk, and the problem of making use of *cross-modal data between multiple different modalities* to achieve a more precise assessment of driving risks remains under-explored. **Secondly**, text data is used in some descriptive statistical analysis and thematic analysis for driving studies, but is rarely used in driving risk assessment. Actually, with the success of large language models (LLMs) in natural language processing (NLP) tasks, text-based models provide the new potential for improving performance in this task. As shown in Fig. 2, the text-based models built on a LLM (see Sect. 2.3) generally outperform the event-data-based models in both event severity and event nature prediction tasks, which motivates this paper to combine the text and event data for cross-modal analysis. **Thirdly**, most existing studies predict the risk levels without spotting the causes of these risks (event nature). From the perspective of intelligent driving assistance, only after the risk causes are detected can follow-up actions be taken to avoid the potential risks. **Fourthly**, the representation learning for the SCDS in existing studies generally subjectively selects the feature set or just concatenates all the raw features together, making it difficult to fully harness the wealth of information embedded in the available data [11,34]. Deep learning models (e.g., Convolutional Neural Network) show effectiveness in feature extraction [29,34], but in contrast to normal driving, SCDS events, such as crashes and near-crashes

(CNC), are quite rare and occur infrequently [29], i.e., the data collection and labeling are not esay and the data amount is usually not big enough. Hence, how to effectively leverage the deep learning models on a small dataset for the SCDS representation learning should be further investigated.

To deal with these issues, this paper proposes a novel framework termed **Drive-CLIP** for cross-modal contrastive SCDS representation learning and zero-shot driving risk analysis. Specifically, to learn more distinguishable representations for the SCDSs, we propose a cross-modal contrastive learning method in an unsupervised fashion by crossing the text and event data. Inspired by the famous CLIP (Contrastive Language-Image Pre-Training) model that jointly learns text and image embeddings [26], we learn the SCDS representation embeddings from two perspectives by jointing text and event data. By leveraging the state-of-the-art LLMs in NLP for text embedding learning, the Drive-CLIP freezes the text encoder and distills the event embeddings using a new event encoder (i.e., **learning event embedding from natural language supervision**), which maps the event embeddings into the same hidden space with the text embeddings. The intuition here is that natural language can serve as a flexible prediction space, facilitating the generalization and transfer of knowledge [16,35]. We want to utilize the power of LLMs to improve the performance of representation learning on small event data (as in real settings, textual narratives are difficult to obtain in real-time while the accessible data is only event data). After obtaining these embeddings, we conduct the driving risk analysis by classifying the SCDS events based on both the event severity and the event nature, not only predicting the risk levels but also spotting the causes of such risks. For the above-mentioned finding in Fig. 2, we train the classifiers for event severity and event nature tasks based on the text embeddings, and freeze these trained classifiers for classifying the event embeddings in a zero-shot manner.

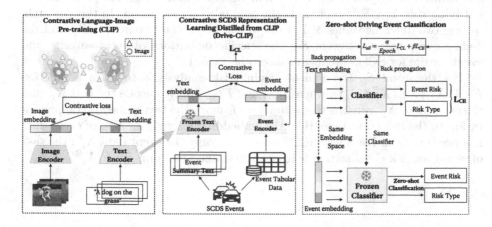

Fig. 3. The Drive-CLIP framework.

The main contributions are summarized as follows:

- To the best of our knowledge, Drive-CLIP is the first method to use cross-modal contrastive learning for the task of driving risk analysis, which learns the SCDS event representations from natural language supervision to achieve good performance.
- We conduct the text-guided zero-shot driving risk analysis for the classification tasks of event severity and event nature based on the event data.
- We evaluate the performance of the Drive-CLIP framework through extensive experiments conducted on a naturalistic driving dataset and demonstrate its effectiveness. Especially, the cross-modal analysis on text and event data outperforms existing models on event data single-modality, and the cross-modal contrastive SCDS representation learning remains beneficial for the baseline models even in scenarios with limited data.

2   Methodology

As illustrated in Fig. 3, our proposed framework mainly consists of two steps: the cross-modal contrastive SCDS representation learning by distilling from CLIP and the zero-shot driving event classification. Hereafter, we first give the task definition and revisit CLIP. Then, we introduce each parts in order.

2.1   Task Definition

Given an SCDS event $\mathcal{E}_i = (X_i, Y_i)$, where $X_i = \{X_i^T, X_i^E\}$ contains the pair of text narrative and event data modalities. Text data $X_i^T = \langle X_{i1}^T, ..., X_{iN}^T \rangle$ with N words is the qualitative description of event $\mathcal{E}_i$ while $X_i^E = \langle X_{i1}^E, ..., X_{iM}^E \rangle$ is the quantitative description with M different variables that impact the probability of a driving risk. $Y_i \in \mathbb{R}^{N_d}$ represents the gold label for the event. In this research, we use two different values of Y_i to represent the 2-class event severity and the 7-class event nature of event $\mathcal{E}_i$. The specific definitions of event severity and event nature for SCDS events are shown in Table 1.

For the event $\mathcal{E}_i$, the goal of driving risk analysis is to learn a function $F : T(X_i^T) \times E(X_i^E) \to Y_i$, where $T(*)$ and $E(*)$ jointly encode the text X_i^T and the event data X_i^E constractively, function F maps the text embedding representation to the risk event type. Here the embedding learning $E(X_i^E)$ of event data is a pre-text task. For a new SCDS event with only the event data X_i^E, the task of zero-shot risk classification is to use the trained function $F : E(X_i^E) \to Y_i$.

Table 1. Definition of event severity and event nature for SCDS events.

Event Category	Risk Event Type	Description
Event Severity (2 classes)	0: Crash	Any collision between the subject vehicle (SV) and another item
	1: Near-Crash	A conflict scenario that necessitates a quick evasive maneuver to prevent a crash
Event Nature (7 classes)	0: Conflict with a lead vehicle	Interaction with a vehicle in front of the SV
	1: Conflict with vehicle in adjacent lane	Interaction with a vehicle traveling in the same direction in the next lane or merging into the same lane
	2: Single vehicle conflict	Any non-motor vehicle conflict occurring on or off the roadway
	3: Conflict with a following vehicle	Interaction with a vehicle behind the SV
	4: Conflict with vehicle moving/turning across/into another vehicle path	Interaction involving a vehicle crossing or turning its lane that makes their driving paths intersecting with each other
	5: Conflict with obstacle/object in roadway	Interaction with any type of animate or inanimate obstacle or object in the roadway
	6: Conflict with oncoming traffic	Interaction with a vehicle traveling toward the SV

2.2 Revisiting CLIP

CLIP [26], learning directly from raw text about images, has recently gained unprecedented attention due to its remarkable ability for contrastive text and image embedding and its exceptional transferability to downstream tasks [23]. As shown in the left part of Fig. 3, CLIP model includes a text encoder and an image encoder for extracting features from two modalities. The interaction layer in the top uses a contrastive learning strategy to map the learned image and text embeddings into the same embedding space, while enabling the embeddings of matched image-text pairs close to each other and the non-matched ones apart. In the implementation, CLIP uses the prompt engineering to transform the image caption of specific category "{label}" to the text description "a photo of a {label}".

2.3 Drive-CLIP

(1) Contrastive SCDS Representation Learning Distilled from CLIP. SCDS Representation Learning is the initial and critical step for assessing the driving risk level and spotting the risk causes. Inspired by the CLIP [26,35], we use a text encoder and an event encoder to extract features from different modalities of the SCDS events, as shown in the middle part of Fig. 3. As CLIP is trained on pairs of text and images, we cannot directly use it for our setting of text and event data pairs, but we can still leverage the pre-trained text encoder on our text narratives. Hence the focus of Drive-CLIP is to build the event encoder by distilling knowledge from the text encoder, i.e., train the event encoder from natural language supervision. Text narratives and event tabular data describe the same SCDS events from two perspectives, and these pairs of data are used for the cross-modal contrastive learning to obtain more distinguishable representations for the SCDSs. Hereafter, we introduce each part of them.

Text Encoder: Text encoder learns the embeddings from text data and transforms text inputs into embedding vectors. In the CLIP, the text encoder is a transformer (or its variants) [31] with 63M-parameter 12 layer 512-wide model with 8 attention heads (for the CLIP-ViT-B/32 model), making it powerful for the text embedding. However, as the text encoder mainly takes the short image caption as input, the max text sequence length is capped at 76. This makes it not applicable for our specific cases where text lengths vary between 10 and 140 words (see Fig. 5). To cope with this, we adopt a sentence BERT [27] (more specifically, the "all-MiniLM-L6-v2" model) to acquire textual features at the sentence level as $\mathcal{V}_i^T \in \mathbb{R}^{d_T}$. It is pre-trained on more than 1 billion sentence pairs and is designed for sentence embedding, with the max sequence length of 256 and the embedding dimension of 384. In the preprocessing phase, we carefully remove all stop words and line breaks from the text to enhance the feature extraction process.

Event Encoder: Event encoder acquires the event data X_E and represents it as the event embedding $\mathcal{V}_i^E \in \mathbb{R}^{d_E}$, where $d_E = d_T$ in the same embedding space. Here we adopt a multi-layer perceptron (MLP) as the event encoder. In the training phase, we freeze the text encoder and use it as natural language supervision to distill the event encoder.

Cross-Modal Contrastive Learning: Existing work reported the issue of representation degeneration [12] particularly in scenarios involving small sample sizes. To solve this, we conduct cross-modal contrastive learning to learn more distinguishable representations across the two considered modalities. Specifically, utilizing the natural language supervision of LLMs, we want to improve the performance of representation learning on small event data. Following CLIP [26], we use the InfoNCEloss [25] as the loss function to conduct the interaction between text and event representations. The loss for the event encoder is defined as:

$$\mathcal{L}_E = -\frac{1}{N} \sum_i^N log \frac{exp(sim(\mathcal{Z}_i^E, \mathcal{Z}_i^T)/\tau)}{\sum_j^N exp(sim(\mathcal{Z}_i^E, \mathcal{Z}_j^T)/\tau)} \tag{1}$$

where $\mathcal{Z}_i^T$ and $\mathcal{Z}_i^E$ are the normalized forms of text embedding $\mathcal{V}_i^T$ and event embedding $\mathcal{V}_i^E$, N is the total number of event and text pairs, the *sim* function is calculated by dot product. Following the original setting in CLIP, the learnable temperature variable τ is initialized to 0.07 to scale the logits.

Asymmetrical loss is defined from the part of text $\mathcal{L}_T$, thus the overall loss function for contrastive learning $\mathcal{L}_{CL}$ is the average of $\mathcal{L}_E$ and $\mathcal{L}_T$, as $\mathcal{L}_{CL} = \frac{1}{2}(\mathcal{L}_T + \mathcal{L}_E)$. The objective of $\mathcal{L}_{CL}$ is to making the event and text embeddings of the N real pairs close to each other in the hidden space while simultaneously the embeddings for the $N^2 - N$ incorrect pairings apart. Note that we did not utilize the annotated labels for the SCDS events during the pre-training phase.

Instead, we solely acquired event data embeddings contrastively through natural language supervision.

(2) Zero-Shot Driving Event Classification. Given the obtained text and event embeddings, we predict the driving event categories. Here we separately train two classifiers using the text embeddings for classification of the event severity and the event nature, respectively. In the test phase, as shown in Fig. 4, we freeze the trained classifier and conduct zero-shot classification of driving events based on the event embeddings. This is consistent with the real settings, as textual narratives are labor-intensive and usually difficult to obtain in real-time while the accessible data is only event data. Another reason for using the text-guided classification is that benefited from the LLM for the text embedding [16], the obtained text embeddings synthesized classifier shows better performance than training directly on the event embeddings, as stated in the experimental results.

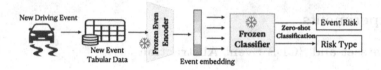

Fig. 4. Zero-shot driving event classification for a new SCDS event.

For the Drive-CLIP, we use a MLP as the classifier for the final zero-shot driving event classification. The classifier is optimized using the cross-entropy loss; specifically, we minimized the following objective function between the true label $t_{i,j}^T$ and the predicted performance $s_{i,j}^T$ at each interaction:

$$\mathcal{L}_{CE} = -\frac{1}{N} \sum_i^N \sum_j^C t_{i,j}^T log\,(f(s_{i,j}^T)), \qquad (2)$$

$$f(s_{i,j}^T) = \frac{e^{s_{i,j}^T}}{\sum_k^C e^{s_{i,k}^T}} \qquad (3)$$

where $C = 2$ for the event severity classification and $C = 7$ for the event nature classification task.

2.4   Objective Function

The final objective function for our Drive-CLIP is formulated as:

$$\mathcal{L}_{all} = \frac{\alpha}{Epoch}\mathcal{L}_{CL} + \beta\mathcal{L}_{CE} \qquad (4)$$

Here the contrastive loss is progressively reduced by dividing it by the ordinal epoch number, starting from 1. Our preference is for the network to prioritize enhancing its feature embedding capabilities in the early stages, while allowing the final driving event classification task to take a more prominent role in later stages as the feature embedding branch stabilizes. The hyper-parameters α and β are to balance these two parts.

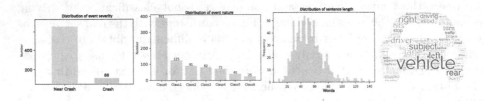

Fig. 5. The distributions of statistical information for our dataset.

3   Experiments

3.1   Dataset and SCDS Event Abstraction

Dataset: Real-world driving data for naturalistic driving study (NDS) from Virginia Tech were used in this study. The dataset known as "100-Car NDS" [8] stands as the first instrumented vehicle study that was carried out in the Northern Virginia/Washington, D.C. area, spanning a duration of two years. This naturalistic driving data was collected unobtrusively from vehicles equipped with an array of sensors and five channels of video cameras. Remarkably, this dataset encompasses around 2,000,000 vehicle miles, accounting for almost 43,000 h of recorded data. It features 102 primary drivers, with data collection spanning a duration of 12 to 13 months for each vehicle involved in the study.

SCDS Event Abstraction: From the data, an event database is created with two types of SCDS events identified: crashes and near-crashes, which contain the potential driving risks. These SCDS events are triggered by some dependent variables, e.g., the lateral acceleration $\geq 0.7\,g$, the longitudinal acceleration/deceleration $\geq 0.6\,g$, and some events are activated by drivers by pressing the event buttons (see Table 2.3 in [8]). After filtering the events based on the event category defined in Table 1, 68 crashes and 758 near-crashes are used to test our framework. The distributions of the statistical information based on the event severity and event nature are shown in Fig. 5.

The collected kinematic and video records allow viewing of all of the pre-event and during-event parameters, based on which a event summary (event tabular data) is created for each event including the event variables, contributing factors, driving environmental factors, road infrastructure, subject driver state variables, etc. [8]. A total of 57 variables in the original dataset are used in this study to comprehensively describe a driving event. Each variable contains a series of possible values, following [9], a data-transcription protocol is used to transform the variables for model input.

Moreover, each event is manually labeled with a text narrative, for example, *"Vehicle in left lane ahead of subject vehicle signals and changes lane in front of subject vehicle. Subject take a brake to avoid hitting lead vehicle in the rear"* for a near-crash event caused by conflicting with vehicle in adjacent lane. As shown in Fig. 5, the length of text narrative ranges from 10 to 140, and the text analysis of word cloud is shown in the rightmost of Fig. 5. Therefore, every event within the dataset consists of a pair of text narrative and event data, representing two distinct modalities. This type of text-event dataset is highly suitable for evaluating the effectiveness of our Drive-CLIP framework.

3.2   Experimental Setting

We implemented Drive-CLIP with the open resource framework PyTorch Lightning. For the text encoder, we used the sentence BERT [27] (the "all-MiniLM-L6-v2" model) as the textual features extractor. For the event encoder and event classifiers, we adopted different architectures of MLPs with dense, dropout and ReLU layers. We adopt the Adam optimizer in our experiments, the model hyper-parameters, i.e., the dropouts, learning rates, batch sizes, and epochs, as well as the different layers of event encoder and classifiers, are obtained using grid search. For the parameters for balancing the two loss parts in Eq. 4, we set α to 5 and β to 1 to progressively prioritize different parts.

For the event data with numerical and categorical types, we use one-hot encoding to represent the categorical variables and min-max normalization to process the numerical ones. The dataset was split into train and test sets with a 8:2 ratio. To balance the data, we applied the synthetic minority oversampling technique (SMOTE) [5] to augment the train data to make it balance between each categories. After obtaining the best set of hyper-parameters for each tasks, we re-trained the models using all the text data, and conducted zero-shot classification on the event data.

We select the best-performing and the most representative baselines that have been shown to be stable and effective for the same task of driving risk classification in previous researches [1,13,28]: the MLP [7,15,28], logistic regression (LR) [14,32], XGBoost [29], Random Forests (RF) [33], Gradient Boosting (GB) [36]. Here some time-series prediction models [6,22] are not compared as such models highlighted the need for larger training sets [13] and are not suitable for the data types used in this paper.

We conduct experiments to answer the following questions:

– How does the Drive-CLIP model perform for the two tasks of zero-shot predicting the event severity and event nature?
– Leveraging the pre-trained LLM for natural language supervision, whether the contrastive SCDS representation learning from event data improves the performance of existing models?

– Whether the cross-modal analysis on text and event data improves the performance on event data single-modality?

3.3 Performance Comparisons on Event Severity Classification

With identical data partitioning, we compare the performance of binary event severity classification between our Drive-CLIP model and several state-of-the-art methods to show the effectiveness of our model, as shown in Table 2. We use different settings for the experiments, the Drive-CLIP can take the original text (OT) as input, while other models cannot directly process the text, hence the original event (OE) data, text embedding (TE) and event embedding (EE) are used instead. These diverse training and testing strategies allow us to evaluate our model from multiple perspectives[1].

Table 2. Performance comparisons of binary classification on event severity. OT: original text data, OE: original event data, EE: event embedding, TE: text embedding. The underlines mean top values and best performances are shaded.

Method	Train	Test	Transfer	ACC	AUC	F1
Drive-CLIP (SCDS-CL + Zero-shot)	OT	OE	✓	0.946	0.904	0.706
Drive-CLIP (SCDS-CL)	OE	OE		0.918	0.915	0.65
MLP	OE	OE		0.828	0.701	0.45
MLP + SCDS-CL	EE	EE		0.886	0.735	0.463
MLP + SCDS-CL + Zero-shot	TE	EE	✓	0.896	0.836	0.539
LR	OE	OE		0.822	0.665	0.435
LR + SCDS-CL	EE	EE		0.887	0.743	0.567
LR + SCDS-CL + Zero-shot	TE	EE	✓	0.912	0.788	0.608
XGBoost	OE	OE		0.831	0.616	0.453
XGBoost + SCDS-CL	EE	EE		0.885	0.701	0.548
XGBoost + SCDS-CL + Zero-shot	TE	EE	✓	0.902	0.813	0.66
RF	OE	OE		0.848	0.571	0.45
RF + SCDS-CL	EE	EE		0.899	0.653	0.538
RF + SCDS-CL + Zero-shot	TE	EE	✓	0.874	0.71	0.55
GB	OE	OE		0.824	0.619	0.459
GB + SCDS-CL	EE	EE		0.898	0.62	0.522
GB + SCDS-CL + Zero-shot	TE	EE	✓	0.814	0.674	0.555

From the results, we have made the following observations: (1) the Drive-CLIP model achieves the best performance in ACC (accuracy), AUC (Area Under Curve) and F1. Using contrastive learning for the SCDS representation (SCDS-CL) and zero-shot classification with OT for training and OE for testing, it gets the best ACC of 0.946 and best F1 of 0.706, while the Drive-CLIP model with only SCDS-CL on the OE data for both training and testing has a slight higher AUC. They both outperform the other baselines for binary event severity classification, among which the LR and XGBoost are high ranked than others.

---

[1] For the Drive-CLIP, we use the setting OT→OE and OE→OE; for other baselines, OE→OE, EE→EE and TE→EE are considered. Additional settings by combining them are not feasible for this task, as in real settings, the accessible data is only event data.

(2) the baselines are generally improved by being equipped with the SCDS-CL. Using the EE from the SCDS-CL, all the models shows better performance than directly on the OE data. This makes sense because the OE data is quite rough and in high dimension, some of the considered variables have low correlations with the driving risk [9], while the event encoder extracts much higher-level features from OE, making it better represent the SCDS, as acknowledged in computer vision field. At the same time, this also validates the effectiveness of event embedding using contrastive learning from natural language supervision. (3) the text-guided event classification does perform better than using the event data single-modality. The zero-shot classification by training on the TE and testing on the EE generally have better performance than training and test both on the EE data. All the baselines with zero-shot show such improvements while the Drive-CLIP obtains slightly competitive results for these two cases. This is consistent with our observation in Fig. 2, as TE-based classifications, benefiting from the excellent power of LLM, generally have better performance than the OE data based classifications. Actually the event embedding is distilled from the supervision of text embedding, an intrinsic reason why the text-guided event classification works better.

3.4   Performance Comparisons on Event Nature Classification

Table 3 shows the performance comparisons of multi-class classification on event nature. We use the similar training and testing strategies with the event severity models. As shown in Table 3, the Drive-CLIP models with/without the zero-shot averagely achieve the best F1 performance of 0.901 and 0.886 on the whole classes. The good performance is also visually evident in Fig. 6, where the micro-average AUC score for all classes reaches 0.94. The Drive-CLIP model accurately predicts the majority of SCDS events, as indicated by the diagonal of the confusion matrix. The complete Drive-CLIP model gets the best F1 scores for class

Table 3. Performance comparisons of multi-class classification on event nature. The values in the seven-class columns are F1 values for each class, while the last column is the micro-average F1 values (equals to the micro ACC) on the whole classes.

Method	Train	Test	Transfer	Class0	Class1	Class2	Class3	Class4	Class5	Class6	micro-ave. F1
Drive-CLIP (SCDS-CL + Zero-shot)	OT	OE	✓	0.94	0.87	0.92	0.9	0.85	0.84	0.68	0.901
Drive-CLIP (SCDS-CL)	OE	OE		0.94	0.87	0.89	0.896	0.846	0.81	0.672	0.886
MLP	OE	OE		0.89	0.78	0.76	0.86	0.67	0.5	0.5	0.807
MLP + SCDS-CL	EE	EE		0.92	0.81	0.76	0.87	0.64	0.79	0.48	0.833
MLP + SCDS-CL + Zero-shot	TE	EE	✓	0.91	0.84	0.86	0.893	0.65	0.67	0.8	0.855
LR	OE	OE		0.9	0.78	0.77	0.76	0.8	0.79	0	0.826
LR + SCDS-CL	EE	EE		0.9	0.82	0.85	0.8	0.68	0.77	0.57	0.837
LR + SCDS-CL + Zero-shot	TE	EE	✓	0.93	0.88	0.81	0.895	0.74	0.8	0.798	0.88
XGBoost	OE	OE		0.93	0.84	0.53	0.59	0.61	0.8	0.51	0.812
XGBoost + SCDS-CL	EE	EE		0.91	0.82	0.75	0.79	0.7	0.71	0.49	0.873
XGBoost + SCDS-CL + Zero-shot	TE	EE	✓	0.92	0.85	0.9	0.83	0.84	0.836	0.676	0.881
RF	OE	OE		0.8	0.76	0.51	0.71	0.65	0.57	0.5	0.724
RF + SCDS-CL	EE	EE		0.86	0.73	0.81	0.35	0.7	0.4	0	0.759
RF + SCDS-CL + Zero-shot	TE	EE	✓	0.89	0.81	0.77	0.71	0.72	0.67	0.5	0.813
GB	OE	OE		0.74	0.62	0.4	0.48	0.5	0.53	0.24	0.586
GB + SCDS-CL	EE	EE		0.88	0.78	0.75	0.58	0.76	0.57	0.29	0.795
GB + SCDS-CL + Zero-shot	TE	EE	✓	0.91	0.79	0.72	0.8	0.74	0.77	0.5	0.831

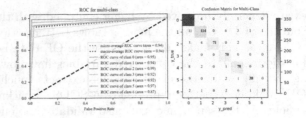

Fig. 6. ROC and confusion matrix for multi-class classification.

0,2,3,4 and 5, while the MLP with SCDS-CL and zero-shot achieves best in class 6, and LR with SCDS-CL and zero-shot does best in class 1. Similar with the observations from Table 2, we can also find that baselines show improvements in varying degrees from the benefits of the SCDS-CL for the event embedding, compared with using the OE data. This can be observed from the micro-average F1 scores. But for some specific classes, the F1 scores fluctuate, e.g., the MLP for class 4 changes from 0.67 to 0.64 while the class 5 from 0.5 to 0.79. Hence the benefits of SCDS-CL vary for each class. Furthermore, zero-shot classification guided by text embedding generally demonstrates superior performance for all models compared to relying solely on single-modality event data.

3.5   Visualization Analysis

In Fig. 7, we show the learning curves of loss and accuracy of the Drive-CLIP model for the event nature classification. The training and testing curves for both cases go steady quickly after about 10 epochs, a good property of our model. Moreover, to illustrate how our cross-modal contrastive learning contributes to the SCDS representation, we employ t-SNE [24] to visualize the embedding space based on modality type and classification task, as depicted in Fig. 8. The text embeddings benefiting from LLM shows good results in both two-class event severity and seven-class event nature cases, as event embeddings belonging to different classes are relatively dispersed. Moreover, in contrast to the original event representations that mixed together, the event embeddings exhibit a lower presence of overlapping confusion zones for both tasks, underscoring the effectiveness of the cross-modal contrastive learning for SCDS event representation. The visualization effectively illustrates why our model improves the performance of existing models, highlighting its superior capabilities for SCDS representation learning using natural language supervision and cross-modalality analysis.

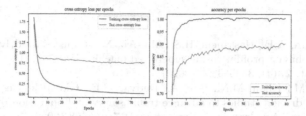

Fig. 7. The learning curves for loss and accuracy per epoch of the Drive-CLIP model.

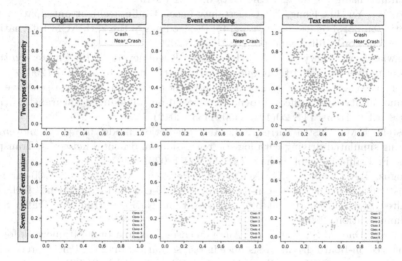

Fig. 8. The visualization of latent embedding space. From left to right: OE, EE and TE, different colors represent different classes. Note that different random states in t-SNE were utilized to generate the upper and lower figures for two tasks.

4 Conclusion

This paper introduced a novel framework called Drive-CLIP for cross-modal contrastive representation learning of Safety-Critical Driving Scenarios (SCDS) and zero-shot analysis of driving risk. In contrast to existing methods that relied on unimodal data for driving risk analysis, we incorporated cross-modal analysis involving textual narratives and event data to tackle two tasks: event severity and event nature classification. By employing cross-modal contrastive learning, we distilled SCDS event embeddings from natural language supervision. Subsequently, we conducted text-guided zero-shot driving risk analysis on event data using these embeddings. We validated the effectiveness of our Drive-CLIP model on a real-world naturalistic driving dataset, and the results demonstrated its superiority over current best-performing baseline models for both tasks. Notably, the text-guided zero-shot event classification outperformed using event data as a single modality. Furthermore, the cross-modal contrastive SCDS representation proved beneficial even in scenarios with limited data.

References

1. Abdelrahman, A.E., Hassanein, H.S., Abu-Ali, N.: Robust data-driven framework for driver behavior profiling using supervised machine learning. IEEE Trans. Intell. Transp. Syst. **23**(4), 3336–3350 (2020)
2. Ahmed, M.M., Khan, M.N., Das, A., Dadvar, S.E.: Global lessons learned from naturalistic driving studies to advance traffic safety and operation research: a systematic review. Accident Anal. Prevention **167**, 106568 (2022)
3. Arumugam, S., Bhargavi, R.: A survey on driving behavior analysis in usage based insurance using big data. J. Big Data **6**, 1–21 (2019)
4. Bao, W., Yu, Q., Kong, Y.: Uncertainty-based traffic accident anticipation with spatio-temporal relational learning. In: Proceedings of the 28th ACM International Conference on Multimedia, pp. 2682–2690 (2020)
5. Chawla, N.V., Bowyer, K.W., Hall, L.O., Kegelmeyer, W.P.: Smote: synthetic minority over-sampling technique. J. Artif. Intell. Res. **16**, 321–357 (2002)
6. Chen, J., Wu, Z., Zhang, J.: Driving safety risk prediction using cost-sensitive with nonnegativity-constrained autoencoders based on imbalanced naturalistic driving data. IEEE Trans. ITS **20**(12), 4450–4465 (2019)
7. Costela, F.M., Castro-Torres, J.J.: Risk prediction model using eye movements during simulated driving with logistic regressions and neural networks. Transport. Res. F: Traffic Psychol. Behav. **74**, 511–521 (2020)
8. Dingus, T.A., Klauer, S.G., Neale, V.L., Petersen, A., et al.: The 100-car naturalistic driving study, phase ii-results of the 100-car field experiment. Tech. rep., United States. Department of Transportation. National Highway Traffic Safety Administration (2006)
9. Gan, W., Dao, M.S., Zettsu, K.: An open case-based reasoning framework for personalized on-board driving assistance in risk scenarios. In: 2022 IEEE International Conference on Big Data (Big Data), pp. 1822–1829. IEEE (2022)
10. Gan, W., Dao, M.S., Zettsu, K.: Procedural driving skill coaching from more skilled drivers to safer drivers: A survey. In: Proceedings of the 4th ACM Workshop on Intelligent Cross-Data Analysis and Retrieval, pp. 10–18 (2023)
11. Gan, W., Dao, M.S., Zettsu, K., Sun, Y.: Iot-based multimodal analysis for smart education: Current status, challenges and opportunities. In: Proceedings of the 3rd ACM ICDAR, pp. 32–40 (2022)
12. Gao, J., He, D., Tan, X., Qin, T., Wang, L., Liu, T.Y.: Representation degeneration problem in training natural language generation models. arXiv preprint arXiv:1907.12009 (2019)
13. Gatteschi, V., Cannavò, A., Lamberti, F., Morra, L., Montuschi, P.: Comparing algorithms for aggressive driving event detection based on vehicle motion data. IEEE Trans. Veh. Technol. **71**(1), 53–68 (2021)
14. Guo, M., et al.: A study of freeway crash risk prediction and interpretation based on risky driving behavior and traffic flow data. Accident Anal. Prevent. **160**, 106328 (2021)
15. Halim, Z., Sulaiman, M., Waqas, M., Aydın, D.: Deep neural network-based identification of driving risk utilizing driver dependent vehicle driving features: A scheme for critical infrastructure protection. J. Ambient. Intell. Humaniz. Comput. **14**(9), 11747–11765 (2023)
16. Jeong, Y., Park, S., Moon, S., Kim, J.: Zero-shot visual commonsense immorality prediction. arXiv preprint arXiv:2211.05521 (2022)

17. Khan, M.Q., Lee, S.: A comprehensive survey of driving monitoring and assistance systems. Sensors **19**(11), 2574 (2019)
18. Kong, X., Das, S., Zhang, Y., Wu, L., Wallis, J.: In-depth understanding of near-crash events through pattern recognition. Transp. Res. Rec. **2676**(12), 775–785 (2022)
19. Kwayu, K.M., Kwigizile, V., Lee, K., Oh, J.S.: Discovering latent themes in traffic fatal crash narratives using text mining analytics and network topology. Accident Anal. Prevent. **150**, 105899 (2021)
20. Lee, S., Arvin, R., Khattak, A.J.: Advancing investigation of automated vehicle crashes using text analytics of crash narratives and Bayesian analysis. Accident Anal. Prevent. **181**, 106932 (2023)
21. Levi-Bliech, M., Kurtser, P., Pliskin, N., Fink, L.: Mobile apps and employee behavior: an empirical investigation of the implementation of a fleet-management app. Int. J. Inf. Manage. **49**, 355–365 (2019)
22. Li, P., Abdel-Aty, M., Yuan, J.: Real-time crash risk prediction on arterials based on LSTM-CNN. Accident Anal. Prevent. **135**, 105371 (2020)
23. Li, Y., et al.: Supervision exists everywhere: a data efficient contrastive language-image pre-training paradigm. arXiv preprint arXiv:2110.05208 (2021)
24. Van der Maaten, L., Hinton, G.: Visualizing data using T-SNE. J. Mach. Learn. Res. **9**(11) (2008)
25. Oord, A.v.d., Li, Y., Vinyals, O.: Representation learning with contrastive predictive coding. arXiv preprint arXiv:1807.03748 (2018)
26. Radford, A., et al.: Learning transferable visual models from natural language supervision. In: ICML, pp. 8748–8763. PMLR (2021)
27. Reimers, N., Gurevych, I.: Sentence-BERT: Sentence embeddings using siamese bert-networks. In: EMNLP. Association for Computational Linguistics (11 2019)
28. Shangguan, Q., Fu, T., Wang, J., Luo, T., et al.: An integrated methodology for real-time driving risk status prediction using naturalistic driving data. Accident Anal. Prevent. **156**, 106122 (2021)
29. Shi, L., Qian, C., Guo, F.: Real-time driving risk assessment using deep learning with xgboost. Accident Anal. Prevent. **178**, 106836 (2022)
30. Takeda, K., Miyajima, C., et. al.: Self-coaching system based on recorded driving data: Learning from one's experiences. IEEE Trans. Intell. Transp. Syst. **13**(4), 1821–1831 (2012)
31. Vaswani, A., et al.: Attention is all you need. In: Advances in Neural Information Processing Systems 30 (2017)
32. Wang, J., Huang, H., Li, Y., Zhou, H., Liu, J., Xu, Q.: Driving risk assessment based on naturalistic driving study and driver attitude questionnaire analysis. Accident Anal. Prevent. **145**, 105680 (2020)
33. Wang, J., Zheng, Y., Li, X., Yu, C., Kodaka, K., Li, K.: Driving risk assessment using near-crash database through data mining of tree-based model. Accident Anal. Prevent. **84**, 54–64 (2015)
34. Wang, Y., Xu, W., Zhang, W., Zhao, J.L.: Safedrive: a new model for driving risk analysis based on crash avoidance. IEEE Trans. ITS **23**(3), 2116-2129 (2020)
35. Wu, H.H., Seetharaman, P., Kumar, K., Bello, J.P.: Wav2clip: learning robust audio representations from clip. In: ICASSP, pp. 4563–4567. IEEE (2022)
36. Zheng, Z., Lu, P., Lantz, B.: Commercial truck crash injury severity analysis using gradient boosting data mining model. J. Safety Res. **65**, 115–124 (2018)

MRHF: Multi-stage Retrieval and Hierarchical Fusion for Textbook Question Answering

Peide Zhu[1]([envelope]), Zhen Wang[2], Manabu Okumura[2], and Jie Yang[1]

[1] Delft University of Technology, Delft, Netherlands
{p.zhu-1,j.yang-3}@tudelft.nl
[2] Tokyo Institute of Technology, Tokyo, Japan
wzh@lr.pi.titech.ac.jp, oku@pi.titech.ac.jp

Abstract. Textbook question answering is challenging as it aims to automatically answer various questions on textbook lessons with long text and complex diagrams, requiring reasoning across modalities. In this work, we propose **MRHF**, a novel framework that incorporates dense passage re-ranking and the mixture-of-experts architecture for TQA. MRHF proposes a novel query augmentation method for diagram questions and then adopts multi-stage dense passage re-ranking with large pretrained retrievers for retrieving paragraph-level contexts. Then it employs a unified question solver to process different types of text questions. Considering the rich blobs and relation knowledge contained in diagrams, we propose to perform multimodal feature fusion over the retrieved context and the heterogeneous diagram features. Furthermore, we introduce the mixture-of-experts architecture to solve the diagram questions to learn from both the rich text context and the complex diagrams and mitigate the possible negative effects between features of the two modalities. We test the framework on the CK12-TQA benchmark dataset, and the results show that MRHF outperforms the state-of-the-art results in all types of questions. The ablation and case study also demonstrates the effectiveness of each component of the framework.

Keywords: Textbook Question Answering · Information Retrieval · Mixture-of-Experts

1 Introduction

The Textbook Question Answering (TQA) task [13] aims at automatically answering questions designed for multimodal textbook lesson materials. Unlike the text-based machine reading comprehension and visual question answering (VQA) tasks, where the context is text or image only, TQA aims to answer multiple types of multimodal scientific questions with scientific knowledge contained in both the text context and scientific diagrams. The requirement to answer multiple types of questions by understanding both the long context and complex diagrams makes TQA a challenging task.

P. Zhu, Z. Wang—Equal Contribution.

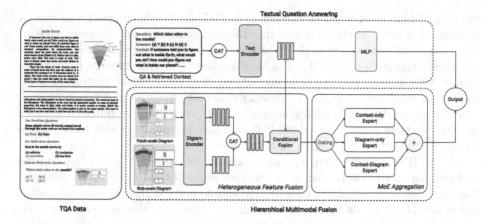

Fig. 1. The pipeline of our proposed MRHF. If a question does not contain a diagram, the upper textual question answering module will be activated to generate the answer directly. In contrast, if the question is a diagram question, the lower hierarchical multimodal fusion module will be activated.

The TQA task has attracted a lot of research efforts [3, 4, 14, 16, 17, 19, 20, 29]. Despite previous progress, TQA remains a challenge. First, previous research retrieves several sentences from the whole corpus using the questions as queries. However, this approach overlooks two critical aspects. First, it fails to account for the fact that many questions are often related to sentences within the same paragraph, where a single sentence may not provide sufficient information for deducing the correct answer. Second, the method is inadequate for retrieving the correct context for diagram questions that tend to be vague and frequently depend on information within the diagrams, e.g., *which of the following labels is correct?*. Secondly, some diagram questions can be answered with the text context or diagram knowledge only, and features of the other modality may be negative for prediction [29]. However, previous research either ignores it or uses manually defined hyper-parameters as a solution, which are not adjustable and learnable for different instances.

In this work, we systematically address these challenges and propose **MRHF**, a novel framework for TQA with multi-stage context retrieval and hierarchical multimodal fusion (Fig. 1). To address the noisy context selection problem, we first apply the pretrained neural passage retriever for paragraph-level multi-stage context retrieval in the TQA task and show its effectiveness. We augment diagram questions with keywords extracted from texts associated with the question diagram and its related teaching diagrams. With the augmented queries, the retrieved paragraphs are more related to the question. Moreover, since the diagram questions can be related to certain specific regions of interest (RoI) or related to knowledge represented by the whole diagram, we propose a heterogeneous feature fusion (HFF) module to learn from different forms of diagram representation, including patch-level features extracted by visual-transformers

(ViT) and blob-level features extracted by YOLO. Furthermore, given that a question may relate exclusively to either the text context or the diagram, to suppress the negative effects from different modalities [29], we introduce the mixture-of-experts aggregation (MoEA). The MoEA consists of three experts, namely the context-only expert, the diagram-only expert, and the context-diagram expert; each is an MLP neural network following different encoding and fusion results, together with a trainable gating network which learns to give different weights to each expert to compose final prediction. Thus, the model is able to rely more on specific experts according to its features.

With MRHF, we perform extensive experiments on the CK12-TQA dataset and compare its performance with previous state-of-the-art (SOTA) methods. The experiment result shows that MRHF significantly surpasses previous methods in the TQA task on both text and diagram questions. We then conduct ablation studies and demonstrate the effectiveness of different components in MRHF. Our contributions in this paper can be summarized as follows: 1) we propose a multi-stage context retrieval method integrated with query augmentation and dense re-ranking, making the context we retrieved more relevant to questions; 2) we propose a hierarchical fusion method that includes the heterogeneous multimodal feature fusion and MoE, surpassing previous methods' performance on the diagram question; 3) detailed experiments and ablation studies prove the efficiency of different components in our method.

2   Related Work

As a complex multimodal QA task, TQA has attracted considerable research interest, particularly following the introduction of the (CK12-TQA) dataset [13]. Most efforts in TQA research can be categorized into three groups: context retrieval, diagram understanding, and question reasoning. Context retrieval is applied for gathering context knowledge related to the question. Most works extract sentences with lexical retrieval methods like TF-IDF [17], Elastic-Search [3], and Solr [4]. Some works like IGMN [16] propose to build essay-level contradiction entity-relationship graphs for reasoning in the long context. In addition to text lexical-based retrieval, ISAAQ [3], MoCA [29] further use different independent semantic-based methods for context retrieval. However, these retrieval methods suffer from noisy results because the sentence-level extraction cannot maintain enough information to answer the questions, and they lack the information from the diagram for retrieval. In recent years, dense passage retrieval methods based on pretrained models [11] and sentence transformers [22] have achieved great progress and shown competent performance in zero-shot retrieval scenarios. Our approach is the first to leverage the capabilities of dense retrievers for TQA.

Diagrams contain complex objects, text, and relations. A lot of attention has been put into effectively leveraging both text and diagram features for answering diagram questions. Some works explore the fine-grained relations among diagram components multimodal graphs for diagram QA, e.g., [13] translates the parsed diagram graphs to factual sentences, [20] builds graphs for reasoning. Some other research leverages attention among the context text and the diagrams for diagram QA, e.g., [3] pretrain the diagram QA model on VQA datasets and leverage bottom-up and top-down attention for multimodal fusion, and [29] proposes to use patch-level diagram features generated by large pretrained visual transformers [1]. However, these methods cannot effectively leverage different-scale knowledge in the diagram, and there are possible negative effects between different modalities. Therefore, we introduce the Mixture-of-Experts (MoE) architecture to the TQA task. MoE has been proposed over two decades ago [7], designed to allow different sub-networks (experts) of a model to specialize for different samples with a learnable gating function. Different types of MoE have been proposed and applied to a range of tasks, including NLP and visual applications [2,24]. This is the first research that adopts MoE for the TQA task.

3 Method

3.1 The TQA Problem

Given a textbook QA dataset that consists of paragraphs $\mathcal{P} = \{P_1, P_2, \ldots, P_N\}$, a list of instructional diagrams $\mathcal{D} = \{d_1, d_2, \ldots, d_M\}$ and a list of questions $\mathcal{Q} = \{Q_1, Q_2, \ldots, Q_K\}$, where d_i denotes the i-th diagram and Q_i is the i-th question. A text question contains one question sentence q_i and its answer options A_i, where $A_i = \{a_{i,j}\}_{j=1}^{O}$ is the list of options. If Q_i is a multiple choice question, A_i is a list of O options, whereas it contains True or False if Q_i is a T/F question. If Q_i is a diagram question, then we represent it as $Q_i = \{q_i, A_i, \delta_i\}$ where δ_i is its corresponding question diagram. Then the answer inference of Q_i using a QA model with trainable parameters θ can be formulated as follows:

$$\hat{a}_i = arg \max_{a_{i,j} \in A_i} \Pr(a_{i,j}|q_i, C_i, [d_k, \delta_i]; \theta) \tag{1}$$

where $C_i \subset \mathcal{P}$ is the retrieved text context from the text contents and $d_k \in \mathcal{D}$ is the retrieved instructional diagram if Q_i is diagram question.

3.2 Multi-stage Context Retrieval

Although TQA lessons are extremely long (over 75% of them have at least 50 sentences), most (about 80%) questions require only several sentences from the same paragraph, and only some questions require information spread across the entire lesson. Instead of retrieving sentences like in previous work, we perform paragraph-level retrieval. We split paragraphs longer than a certain number (128, since most paragraphs are shorter than 128 words) of words into separate shorter paragraphs. In this way, the proposed method can gather cross-paragraph context and keep syntactic and semantic properties in each paragraph. Diagram questions pose distinct challenges for context retrieval that are neglected by previous research. Therefore we develop an extra query augmentation method for

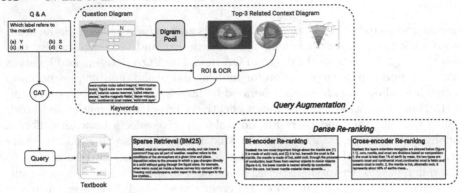

Fig. 2. Our multi-stage context retrieval pipeline includes query augmentation, sparse retrieval, and dense re-ranking.

diagram questions. We would introduce context retrieval pipelines for both diagram and text questions.

➤ **Query Augmentation for DQ.** The context knowledge of diagram questions includes both text and diagrams. As shown in Fig. 2, some diagram questions do not contain enough information to retrieve the text context that can answer it. The texts in the question diagram are replaced with option letters, which makes it difficult to leverage diagram annotations as extra vocabulary for retrieval. Therefore, instead of directly performing context retrieval with questions, we propose performing query augment first by using question diagrams as a bridge and retrieving related teaching diagrams as well as texts associated with them, and then performing context extraction using the same pipeline as text questions with the augmented query.

In detail, we first parse the diagrams that contain complex components like images, arrows, and text that convey critical information for understanding. To extract these components, we fine-tune YOLO(V5) [9] on AI2d [12] to recognize the positions and types of all these components. The AI2D dataset contains diagrams in the same style and annotations for diagrams, including each component's type and position information. Then we use OCR to recognize the text in each text block. Besides text contained in diagrams, we also extract extra text associated with them, including diagram captions, textbook sentences that reference them, and detailed introductions for teaching diagrams.

We then perform instructional diagram retrieval to find related diagrams in textbooks. As components representing the same concepts often have different colors in various diagrams, we first convert the diagram to grayscale to mitigate distractions introduced by colors. We then encode each diagram with the visual transformers model (ViT) and use the embedding of [CLS] token as its representation and rank the instructional diagrams according to the cosine similarity to the question diagram.

$$sim(d_j, \delta_i) = \cos(\text{ViT}_{[\text{CLS}]}(d_j), \text{ViT}_{[\text{CLS}]}(\delta_i)) \tag{2}$$

We choose top-3 instructional diagrams and combine all texts associated with them as associated text. Then we augment r_i with extracted keywords from the associated text. Then we apply the same context knowledge retrieval pipeline introduced in Sect. 3.2 with the augmented query.

➤ **Sparse Retrieval.** With the original QA pairs for text questions or augmented queries for diagram QA, we construct queries (r_i) for context retrieval by combining the questions and all answer options (except the *true, false, none,* and *all* options) with white spaces [W]:

$$r_i = q_i \, [\text{W}] \, a_{i,1} \, [\text{W}] \, a_{i,2} \, [\text{W}] \ldots a_{i,|A_i|}$$

and then perform multi-stage retrieval to obtain the QA context, as shown in Fig. 2. We apply **sparse retrieval** with BM25 [25], a well-adopted space vector-based probabilistic text retrieval method. In this step, we choose top-K_1 paragraphs (C_i^{BM25}) that have the most lexical similarity with the queries.

➤ **Dense Re-ranking.** Different from previous research that uses sparse models such as TF-IDF only, we use two-stage **dense re-ranking** to refine the context based on both lexical and semantic similarity to the question-answer pairs, as large pretrained language model-based dense passage retrieval has demonstrated substantial improvements in retrieval performance. In our dense re-ranker, we first employ a standard pretrained neural IR architecture [11] for a semantic **bi-encoder re-ranking** (BI). It uses the pretrained transformer encoder E_C which encodes the context paragraphs and E_Q on the queries into separate m-dimensional real-valued vectors and retrieves top$-K_2$ paragraphs C_i^{BI} in terms of cosine similarity:

$$sim(P_{i,j}, r_i) = \cos(E_C(P_{i,j}), E_Q(r_i)). \tag{3}$$

where $P_{i,j} \in C_i^{BM25}$. Then, we further leverage **cross-encoder re-ranking** (CE), which is used for matching text pairs by concatenating the query and target paragraph together, treating it as a sequence classification task, and performing full self-attention over the entire sequence. As reported in some research [23], the cross-encoder could have better performance than the bi-encoders with the sacrifice of efficiency. Therefore, we re-rank the semantic retrieval results C_i^{BI} with a pretrained cross-encoder.

$$sim(P_{i,j}, r_i) = \text{MLP}(E([\text{CLS}] P_{i,j} [\text{SEP}] r_i [\text{SEP}])) \tag{4}$$

where $\text{MLP}(h_i) = W_2(W_1 h_i + b_1) + b_2$ is a multilayer perceptron network that takes the encoder E's output h_i as the input for calculating the final matching score, and W_1, W_2, b_1, b_2 are trainable parameters. We choose the top-3 paragraphs as the context passage.

3.3 Textual Question Answering

The process of choosing the correct answer to Text-MC questions is similar to answering the T/F questions, which can be interpreted as verifying whether the context can support the claim of the question and the option pair. Inspired by [8], we transfer multiple-choice questions as a single-choice decision problem and treat T/F questions and Text-MC questions as a sequence bi-classification problem.

$$I_{i,j} = \texttt{[CLS]}\, C_i\, \texttt{[SEP]}\, \hat{a}_{i,j}\, \texttt{[SEP]}$$
$$f_{i,j}^t = \mathrm{MLP}(\mathbf{E}_{\texttt{[CLS]}}(I_{i,j})) \tag{5}$$

where $f_t = \Pr(\hat{a}_{i,j}|q_i, C_i)$ represents the predicted probability of the correctness of the j-th answer option $\hat{a}_{i,j}$ in A_i, and we use softmax to normalize the MLP's output of True/False probability to $[0-1]$. For Text-T/F questions and the option like *none* or *none of the above*, $\hat{a}_{i,j}$ is an empty string. For answer options such as *all* or *all of the above*, we concatenate all other options as the option text. To train the model, we label the sequence with the correct answer option as True, and others as False. We label all answer options as True when the correct answer is *all*. We use Cross-Entropy loss to train the model. For prediction, we chose the option of highest probability on True as the correct answer. To the questions with option *none*, we predict the correct answer is *none* when 1) the *none* option has the highest probability, or 2) all other options' probabilities are below 0.5. For the questions with option *all*, we predict the correct answer is *all* when 1) the *all* option has the highest probability, or 2) the probabilities of all other options are greater than 0.5. Similar to previous TQA researches [3,4,29], we perform pretraining with some extra datasets such as RACE [15] and SQuAD [21].

3.4   Hierarchical Multimodal Fusion

To answer the diagram questions, the solver should be able to reason over both the text (context, questions, and answer options) and the diagrams (question diagram and instructional diagrams). Many (40%) diagram questions require complex diagram parsing [3] and are relevant to certain regions of interest (RoI) or the relation and knowledge represented by the whole diagram. As pointed out in previous work, text contexts of a considerable proportion of diagram questions are rich enough to answer them, and features of the other modality may have negative effects. Therefore, we first propose a heterogeneous feature fusion module to learn from different contextualized diagram representations. Then we adopt the mixture-of-experts architecture to learn from different modality and their interaction.

➤ **Heterogeneous Feature Fusion (HFF).** To better leverage the features from the diagram, here we propose the HFF module. As mentioned above, we first parse the question diagram δ_i and get a list of blobs B_i. Then, we create the patch-level features V_i^p, and the blob-level features V_i^b of B_i using the ViT encoder, where $V_i^b = [\mathrm{ViT}_{\texttt{[CLS]}}(B_i^j)]_0^{|B_i|}$. We concatenate them to generate a heterogeneous representation V_i^d of the diagram. For the j-th option $a_{i,j}$ of the question, we create the text features $V_{a_{i,j}}^t$ for the QA pair, and the text features $V_{c_{i,j}}^t$ for QA pair and the context with the trained text model. Instead of using all the features for further processing, here we use gated attention [30] to find the most important features. The dual fusion of the text feature and the heterogeneous diagram features is calculated as follows:

$$U_{i,j} = V_{c_{i,j}}^t W_u V_i^d$$
$$S_{i,j} = \text{softmax}(U_{i,j}/\sqrt{d_{V_i^d}})$$
$$Z_{i,j} = W_s[V_{c_{i,j}}^t : S_{i,j}^T V_i^d]$$
$$z_{i,j} = \text{MaxPooling}(Z_{i,j}) \tag{6}$$
$$v_i = \text{MaxPooling}(V_{c_{i,j}}^t)$$
$$g_{i,j} = \text{sigmoid}(W_g z_{i,j})$$
$$h_{i,j} = g_{i,j} \cdot \tanh(z_{i,j}) + (1 - g_{i,j}) \cdot v_i$$

where W_u, W_s, W_g are trainable weights.

We then obtain the full-text ([question & answer options & context]) guided representation $h_{i,j}^c$ by using the equation. Similarly, we calculate the qa-text ([question & answer options]) guided representation $h_{i,j}^a$. To calculate the diagram guided representation, we substitute V_i^d and $V_{c_{i,j}}^t$ in the equation obtain $j_{i,j}^d$, and then obtain $j_{i,j}^a$ in the similar way.

➤ **Mixture-of-Experts Aggregation (MoEA).** Since diagram questions can be answered by using only context, or only diagram, or must leverage both diagram and context, we design different experts to handle different situations and use the mixture-of-experts to aggregate the results of those experts.

We first combine $h_{i,j}^c$ and $j_{i,j}^d$ to form $u_{i,j}^c$ representing the fusion of diagram and text with context, and also combine $h_{i,j}^a$ and $j_{i,j}^a$ to form $u_{i,j}^a$ representing the fusion of diagram and text without context, to input different experts in the next step. Based on the text features and the dual fusion representations, we design three question solvers, namely the text question solver $f_{i,j}^t$ which is MLP for $V_{c_{i,j}}^t$, the diagram-only solver $f_{i,j}^a$ which is MLP for $u_{i,j}^a$, and the context-diagram solver $f_{i,j}^c$ which is MLP for $u_{i,j}^c$. We utilize the MoEA with a learnable gating function G to automatically learn to put different weights on these different solvers (experts). We adopt the simple yet widely used gating function [10] to calculate the weights by multiplying a trainable matrix W_γ with the input and then normalize the weights by softmax. We concatenate $u_{i,j}^a$ and $u_{i,j}^c$ as input to the gating function.

$$\mu_{i,j} = G(u_{i,j}^a, u_{i,j}^c)$$
$$G = softmax(W_\gamma[u_{i,j}^a : u_{i,j}^c]) \tag{7}$$
$$f_{i,j}^{MoE} = \mu_{i,j} \cdot [f_{i,j}^t, f_{i,j}^a, f_{i,j}^c]$$

where the output of the gating function μ is a 3-dimension vector $[\mu_0, \mu_1, \mu_2]$ which represents weights for the two experts. The weighted sum of the outputs of the three experts ($f_{i,j}^{MoE}$) is the final prediction for the j−th answer option of diagram multiple-choice question i.

4   Experiment

4.1   Experimental Settings

➤ **Datasets.** We conduct the model evaluation on the CK12-TQA dataset, which contains textbook lessons, different types of questions, and rich textbook diagrams and has become the benchmark dataset for TQA research. Data samples in CK12-TQA can be categorized by the type of questions into three groups: true/false (T/F), text multiple choice (T-MC), and diagram multiple choice (D-MC) questions. The details of the datasets we use are shown in Table 1.

Table 1. Statistics of TQA dataset.

Dataset	Train	Dev	Test	Total	Options
CK12-TQA	15,154	5,309	5,797	26.260	–
–T/F	3,490	998	912	5,400	2
–T-MC	5,163	1,530	1,600	8,293	4–7
–D-MC	6,501	2,781	3,285	12,567	4

➤ **Implementation Details.** To create the dataset for pretraining, we perform named entity recognition and POS tagging using SpaCy [6]. For text context retrieval, we use the pretrained bi-encoders (MPNet [27]) and cross-encoders (MiniLM [28]) provided by Sentence-Transformers library [22]. We extract keywords from texts associated with retrieved instructional diagrams with RAKE [26] and use the top-5 extracted keywords for query augmentation. We use RoBERTa-large [18] for the sequence classification model. We train YOLO [9] on the AI2D dataset for 50 epochs for diagram parsing. For diagram retrieval and encoding, we use the visual transformers pretrained via masked autoencoders [5]. We finetune the model on the CK12-TQA dataset for 10 epochs on one NVIDIA A40 GPU with an initial learning rate at $1e^{-6}$, and the batch size is 4.

4.2   Experiment Results

We evaluate the proposed framework's performance in terms of its accuracy ($\frac{\#correct}{\#questions}$) on T/F, text, and diagram MC questions in both validation and test splits. We compare its performance with the previous state-of-the-art (SOTA) models, including single model approaches **IGMN** [16], **XTQA** [19] and **MHTQA** [4], as well as the ensemble approaches: **ISAAQ** [3] and **MoCA** [29]. The main results are shown in Table 2.

Table 2. Experimental results on the CK12-TQA validation and test splits in terms of accuracy. ⋆ means we choose the best single model from the ensemble solutions for comparison. We train these models with the context extracted using methods introduced in this paper.

Model	Val Set					Test Set				
	T/F	T-MC	T-All	D-MC	All	T/F	T-MC	T-All	D-MC	All
Random	50.86	23.66	34.40	25.83	29.91	50.37	22.93	32.89	24.80	28.31
Single Model										
IGMN	57.41	40.00	46.88	36.35	41.36	–	–	–	–	–
XTQA	58.24	30.33	41.32	32.05	36.46	56.22	33.40	41.67	33.34	36.95
ISAAQ-IR⋆	78.26	67.52	71.76	53.83	62.37	77.74	68.94	72.13	50.50	59.87
MHTQA	82.87	69.22	74.61	54.87	64.27	–	–	–	–	–
MoCA-IR⋆	–	73.33	–	54.15	–	–	–	–	52.12	–
MRHF	**87.48**	**76.80**	**81.01**	**56.27**	**67.90**	**86.51**	**79.19**	**81.85**	**53.97**	**66.05**
w/o CE	84.37	75.36	78.92	55.09	66.29	82.90	77.44	79.42	52.15	63.97
w/o BI&CE	85.47	74.84	79.04	50.38	63.88	82.24	78.25	79.70	51.08	63.48
Ensemble Model										
ISAAQ	81.36	71.11	75.16	55.12	64.66	78.83	72.06	74.52	51.81	61.65
MoCA	81.56	76.14	78.28	56.49	66.87	81.36	76.31	78.14	53.33	64.08
MRHF	87.88	78.95	82.48	56.67	68.80	86.62	80.00	82.40	54.55	66.62

We first compare our method with previous methods in a single-model setting. After further fine-tuning on the CK12-TQA dataset, **MRHF** achieves 81.01% in overall accuracy on the text questions of the validation set, and 81.85% in the test set, which outperforms the previous best single-model MoCA by a margin of about 6.4% and 9.7% on validation and test sets separately in all text questions according to available data. MRHF also outperforms the SOTA single method on diagram multi-choice questions by a margin of 2.12% and 1.85% on the validation set and test set, respectively. Since previous SOTA methods ISAAQ and MoCA are both ensemble models which ensemble multiple models trained on different retrieved results, we also compare MRHF's performance in the ensemble-model setting. It can be found that the accuracy of ensemble MRHF is further improved and achieves new SOTA performance. Moreover, even the single model MRHF exceeds the ensemble MoCA model. These results demonstrate the effectiveness of our proposed MRHF framework.

5   Ablation Studies

5.1   Quantitative Analysis

➢ **Query Augmentation.** We first investigate the impact of query augmentation. Results are shown in Table 3. For the setting without text context, we replace the $I_{i,j}$ with answer option text. As the result shows, the performance on the validation set deteriorates to 46.46% with about 9.8% decline and 3.04% decline on the test set, which first demonstrates the importance of context features. Then we remove the query augmentation, the performance of MRHF declines 1.6% and 1.46% on the validation and test set, respectively, which can demonstrate the effectiveness of query augmentation in context retrieval in Sect. 3.2.

Table 3. Ablations study on the impact of context text, query augment (AGM), and mixture-of-experts on the performance of answering diagram MC questions in CK12-TQA validation and test splits.

Split	MRHF	w/o Text	w/o AGM	w/o HFF	w/o MoEA
Val	**56.27**	46.46	54.67	55.34	54.95
Test	**53.97**	49.99	51.57	51.18	51.96

➤ **Dense Re-ranking.** We then finetune MRHF on CK12-TQA data with context extracted using different retrieval methods, and report MRHF-BM25, MRHF-BI, and MRHF-CE results in Table 2. First, we observe that with BM25 only, it outperforms previous SOTA models that utilize similar lexical retrieval methods, revealing the effectiveness of paragraph-level retrieval. Second, applying Bi-Encoder and Cross-Encoder Re-ranking can further improve the performance, demonstrating that applying the neural ranking models in a zero-shot setting on the TQA task is effective. Third, we observe that the model achieves over 53% accuracy on diagram questions with a text context only, showing the importance of text context in answering diagram questions.

➤ **Hierarchical Multimodal Fusion.** We examine the HFF and MoEA employed in our hierarchical multimodal fusion module, as demonstrated in Table 3. When we remove the HFF, the overall performance drops to 55.34% and 51.18% on the validation and test sets, which demonstrates the usefulness of HFF. We then eliminate the MoEA, and the performance decreases by 1.32% on the validation set and 2.01% on the test set, indicating the effectiveness of MoEA.

5.2   Case Studies

Fig. 3 illustrates the impacts of multi-stage context retrieval with query augmentation for diagram questions. We show the question diagrams with the bounding box of all blobs detected by the diagram parsing step. The example in the first row shows that without AGM, the retrieved context is about *food* topic, which totally drifts off the actual topic on *rain*. By contrast, with AGM, the multi-stage context retrieval method can retrieve the exact context paragraph. Examples in other rows show that AGM can still help improve the retrieval results even for diagram questions where there is rich text information. The last column reports

Question	Answer Options	Question Diagram	Keywords	w/o-AGM	AGM	Option Probs
Which label refers to rains?	(a) V (b) T (c) U (d) E	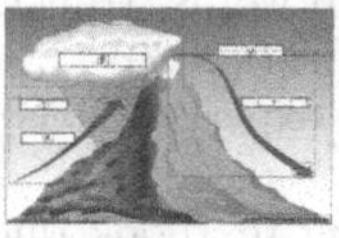	'wind moves particles suspension', '21 wind transports particles', 'oxidiz- ing agent', '23 sand dunes form', 'wind deposits sand', 'desert sand dunes', 'sand dunes line', 'sand dunes', 'ways depending', 'dust storm'	look at the nutrition facts label in figure 17.7. instructions at the right of the label tell you what to look for. at the top of the label, look for the serving size. the serving size tells you how much of the food you should eat to get the nutrients listed on the label. for this food, 1 cup is a serving. the calories in one serving are listed next. in this food, there are 250 calories per serving. next on the nutrition facts label, look for the percent daily values (% dv) of several nutrients.	the diagram is a representation of how a rain shadow is formed. a rain shadow is a dry region of land on the side of a mountain range that is protected from the prevailing winds. prevailing winds are the winds that occur most of the time in a particular location on the earth. the protected side of a mountain range is also called the lee side or the down-wind side. prevailing winds carry air toward the mountain range. as the air rises up over a mountain range, the air cools, water vapor condenses, and clouds form.	Ans, Tex, Dia, T&D (a), 0.087, 0.145, 0.102, 0.265 (b), 0.294, 0.325, 0.334, 0.172 (c), 0.122, 0.177, 0.148, 0.276 (d), 0.501, 0.354, 0.416, 0.287 W, 0.036, 0.048, 0.917
From the diagram, what happens after the presence of moist prevailing winds?	(a) rain shadow desert (b) air cools and condenses (c) dry air is warmed (d) zone of precipitation'	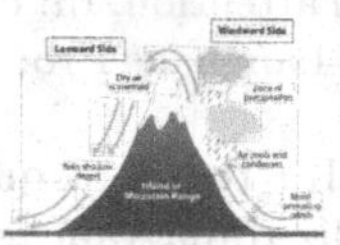	'wind moves particles suspension', '21 wind transports particles', 'physical models liquid', '23 sand dunes form', 'wind deposits sand', 'desert sand dunes', 'sand dunes line', 'sand dunes', 'physical model', 'ways depending'	this diagram shows the effect of rains on hills. the moist air from the green side of the hills rising up in the air and condenses as water vapor. this is called precipitation. the other side of the hill from the rain shadow region. rain shadow region is a region having little rainfall because it is sheltered from prevailing rain-bearing winds by a range of hills. the dry air descends from this region.	this diagram shows the effect of rains on hills. the moist air from the green side of the hills rising up in the air and condenses as water vapor. this is called precipitation. the other side of the hill from the rain shadow region. rain shadow region is a region having little rainfall because it is sheltered from prevailing rain-bearing winds by a range of hills. the dry air descends from this region.	Ans, Tex, Dia, T&D (a), 0.010, 0.247, 0.195, 0.017 (b), 0.681, 0.294, 0.423, 0.471 (c), 0.012, 0.218, 0.188, 0.020 (d), 0.297, 0.241, 0.195, 0.493 W, 0.094, 0.005, 0.902
Identify the part of the plant which functions as nourishment for the embryo, dispersal to a new location, and dormancy during unfavorable conditions.	(a) Seed (b) Fruit (c) Flower (d) Seedling	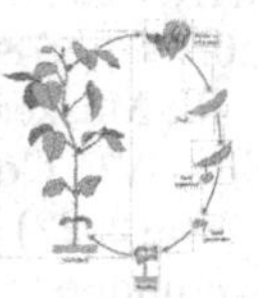	'tremendous feather displays', 'unfavorable mutations selected', 'sand dunes form', 'mutation creates variation', 'namib desert beetle', 'mutations reproduction', 'favorable mutations', 'sand dune', '11 genetic mutation', 'biological diversity'	the diagram shows the internal structure of a corn kernel or corn seed. the seed has three main parts seed coat, endosperm and the embryo. the seed coat is the outermost part of the seed and it protects both the endosperm and the embryo. the endosperm is below the seed coat. it stores food in the seed. the embryo is what gives rise to a new plant. early growth and development of a plant embryo inside a seed is called germination. the embryo consists of the embryonic leaves, cotyledon and the primary root.	..............a seed is a reproductive structure that contains an embryo and a food supply, called endosperm. both the embryo and endosperm are enclosed within a tough outer coating, called a hull (or shell). you can see these parts of a seed in figure 10.14. an embryo is a zygote that has already started to develop and grow. early growth and development of a plant embryo inside a seed is called germination. the seed protects and nourishes the embryo and gives it a huge head start in the "race" of life. both a parent plant and its offspring are better off if they dont grow too closely together..............	Ans, Tex, Dia, T&D (a), 0.390, 0.251, 0.380, 0.201 (b), 0.290, 0.255, 0.218, 0.370 (c), 0.114, 0.256, 0.210, 0.163 (d), 0.206, 0.238, 0.192, 0.266 W, 0.071, 0.004, 0.924

Fig. 3. Case study of AGM and MoEA from test set. Under "Option Probs", "Ans" means answer candidates, and red is the correct answer, "Tex" means context-only expert, "Dia" means diagram-only expert, "T&D" means context-diagram expert, "W" means the weight of different experts.

the prediction by each expert as well as the learned weights for them. The results first suggest that the context-diagram expert has the largest weight. Second, through the mixture, although the single context-diagram expert makes a wrong prediction, the overall results adjusted by text and diagram-only experts successfully choose the correct answer.

6 Conclusion

In this paper, we propose a concise framework MRHF to address the challenges in the textbook question answering task, especially for diagram-related QA pairs. Experiment on CK12-TQA shows that our proposed framework can effectively solve the TQA problem, outperforming previous SOTA results on all types of questions. Even single-model MRHF can achieve considerable performance compared to previous ensemble models, which can significantly simplify the training, maintenance, and deployment of the TQA systems. Ablation studies further demonstrate the effectiveness of its components, including multi-stage context retrieval with query augmentation for diagram questions, multimodal conditional fusion, and the mixture-of-experts architecture. In the future, we will further study and unravel the challenges in multimodal question answering.

References

1. Dosovitskiy, A., et al.: An image is worth 16×16 words: transformers for image recognition at scale. arXiv preprint arXiv:2010.11929 (2020)
2. Fedus, W., Dean, J., Zoph, B.: A review of sparse expert models in deep learning. arXiv preprint arXiv:2209.01667 (2022)
3. Gómez-Pérez, J.M., Ortega, R.: ISAAQ-mastering textbook questions with pre-trained transformers and bottom-up and top-down attention. In: Proceedings of the 2020 Conference on Empirical Methods in Natural Language Processing (EMNLP), pp. 5469–5479 (2020)
4. He, J., Fu, X., Long, Z., Wang, S., Liang, C., Lin, H.: Textbook question answering with multi-type question learning and contextualized diagram representation. In: Farkaš, I., Masulli, P., Otte, S., Wermter, S. (eds.) ICANN 2021. LNCS, vol. 12894, pp. 86–98. Springer, Cham (2021). https://doi.org/10.1007/978-3-030-86380-7_8
5. He, K., Chen, X., Xie, S., Li, Y., Dollár, P., Girshick, R.: Masked autoencoders are scalable vision learners. In: Proceedings of the IEEE/CVF Conference on Computer Vision and Pattern Recognition, pp. 16000–16009 (2022)
6. Honnibal, M., Montani, I.: spaCy 2: natural language understanding with Bloom embeddings, convolutional neural networks and incremental parsing (2017). to appear
7. Jacobs, R.A., Jordan, M.I., Nowlan, S.J., Hinton, G.E.: Adaptive mixtures of local experts. Neural Comput. $3(1)$, 79–87 (1991)
8. Jiang, Y., et al.: Improving machine reading comprehension with single-choice decision and transfer learning. arXiv preprint arXiv:2011.03292 (2020)
9. Jocher, G., et al.: ultralytics/yolov5: v4.0 - nn.SiLU() activations, Weights & Biases logging, PyTorch Hub integration. Zenodo, January 2021. https://doi.org/10.5281/zenodo.4418161
10. Jordan, M.I., Jacobs, R.A.: Hierarchical mixtures of experts and the EM algorithm. Neural Comput. $6(2)$, 181–214 (1994)
11. Karpukhin, V., et al.: Dense passage retrieval for open-domain question answering. arXiv preprint arXiv:2004.04906 (2020)
12. Kembhavi, A., Salvato, M., Kolve, E., Seo, M., Hajishirzi, H., Farhadi, A.: A diagram is worth a dozen images. In: Leibe, B., Matas, J., Sebe, N., Welling, M. (eds.) ECCV 2016. LNCS, vol. 9908, pp. 235–251. Springer, Cham (2016). https://doi.org/10.1007/978-3-319-46493-0_15
13. Kembhavi, A., Seo, M., Schwenk, D., Choi, J., Farhadi, A., Hajishirzi, H.: Are you smarter than a sixth grader? Textbook question answering for multimodal machine comprehension. In: Proceedings of the IEEE Conference on Computer Vision and Pattern Recognition, pp. 4999–5007 (2017)
14. Kim, D., Kim, S., Kwak, N.: Textbook question answering with multi-modal context graph understanding and self-supervised open-set comprehension. In: Proceedings of the 57th Annual Meeting of the Association for Computational Linguistics, pp. 3568–3584 (2019)
15. Lai, G., Xie, Q., Liu, H., Yang, Y., Hovy, E.: RACE: large-scale reading comprehension dataset from examinations. In: Proceedings of the 2017 Conference on Empirical Methods in Natural Language Processing, pp. 785–794 (2017)
16. Li, J., Su, H., Zhu, J., Wang, S., Zhang, B.: Textbook question answering under instructor guidance with memory networks. In: Proceedings of the IEEE Conference on Computer Vision and Pattern Recognition, pp. 3655–3663 (2018)

17. Li, J., Su, H., Zhu, J., Zhang, B.: Essay-anchor attentive multi-modal bilinear pooling for textbook question answering. In: 2018 IEEE International Conference on Multimedia and Expo (ICME), pp. 1–6. IEEE (2018)
18. Liu, Y., et al.: Roberta: a robustly optimized Bert pretraining approach. arXiv preprint arXiv:1907.11692 (2019)
19. Ma, J., Chai, Q., Liu, J., Yin, Q., Wang, P., Zheng, Q.: XTQA: span-level explanations for textbook question answering (2023)
20. Ma, J., Liu, J., Wang, Y., Li, J., Liu, T.: Relation-aware fine-grained reasoning network for textbook question answering. IEEE Transactions on Neural Networks and Learning Systems (2021)
21. Rajpurkar, P., Zhang, J., Lopyrev, K., Liang, P.: Squad: 100,000+ questions for machine comprehension of text. arXiv preprint arXiv:1606.05250 (2016)
22. Reimers, N., Gurevych, I.: Sentence-BERT: sentence embeddings using Siamese BERT-Networks. In: Proceedings of the 2019 Conference on Empirical Methods in Natural Language Processing. Association for Computational Linguistics, November 2019
23. Reimers, N., Gurevych, I.: Sentence-BERT: sentence embeddings using Siamese BERT-Networks. arXiv preprint arXiv:1908.10084 (2019)
24. Riquelme, C., et al.: Scaling vision with sparse mixture of experts. Adv. Neural Inf. Process. Syst. **34**, 8583–8595 (2021)
25. Robertson, S.E., Walker, S.: Some simple effective approximations to the 2-Poisson model for probabilistic weighted retrieval. In: Croft, B.W., van Rijsbergen, C.J. (eds.) SIGIR '94, pp. 232–241. Springer, London (1994). https://doi.org/10.1007/978-1-4471-2099-5_24
26. Rose, S., Engel, D., Cramer, N., Cowley, W.: Automatic keyword extraction from individual documents. Text Min. Appl. Theory **1**(1–20), 10–1002 (2010)
27. Song, K., Tan, X., Qin, T., Lu, J., Liu, T.Y.: MPNet: masked and permuted pretraining for language understanding. Adv. Neural Inf. Process. Syst. **33**, 16857–16867 (2020)
28. Wang, W., Wei, F., Dong, L., Bao, H., Yang, N., Zhou, M.: MINILM: deep self-attention distillation for task-agnostic compression of pre-trained transformers. Adv. Neural Inf. Process. Syst. **33**, 5776–5788 (2020)
29. Xu, F., et al.: MoCA: incorporating domain pretraining and cross attention for textbook question answering. Pattern Recognit. **140**, 109588 (2023)
30. Zhao, Y., Ni, X., Ding, Y., Ke, Q.: Paragraph-level neural question generation with maxout pointer and gated self-attention networks. In: Proceedings of the 2018 Conference on Empirical Methods in Natural Language Processing, pp. 3901–3910 (2018)

Multi-scale Decomposition Dehazing with Polarimetric Vision

Tongwei Ma[1,2,3] [iD], Lilian Zhang[2(✉)] [iD], Bo Sun[3(✉)] [iD], and Chen Fan[2] [iD]

[1] Xinjiang University, Ürümqi 830047, China
tongwei@stu.xju.edu.cn
[2] National University of Defense Technology, Changsha 410073, China
lilianzhang@nudt.edu.cn
[3] Chinese Academy of Sciences, Quanzhou 362000, China
sunbo@fjirsm.ac.cn

Abstract. In this paper, the problem of simultaneous image dehazing of near and far scenes in hazy weather is addressed. We propose the multi-scale decomposition dehazing with polarimetric vision (Pol-MSD) algorithm to solve this problem. First, the contrast limited adaptive histogram equalization method is presented to obtain the contrast enhanced polarization image as the channel of near scene. Meanwhile, based on stokes theory, the incomplete normalized degree of polarization image is obtained as the channel of far scene. Then, the dual channels of near and far scene are decomposed into the base layers and the detail layers. To fusion the base layers, we use sobel gradient map (SGM) as the fusion weight of the base layers, and we proposed a fusion rule for the base layers by taking account of the SGM. To preserve the texture details of near and far scenes in detail layer fusion, we design a novel weighted least squares optimization scheme to fuse the detail layers. To evaluate the proposed Pol-MSD, we create a polarized hazy image datasets. Experiment results demonstrate that the proposed algorithm has significant advantages in qualitative and quantitative evaluations compared with other state-of-the-art dehazing methods.

Keywords: Image dehazing · Polarimetric dehazing · image restoration

1 Introduction

Outdoor hazy images taken in hazy weather usually lose contrast [39] and fidelity [31,33]. It affects many advanced computer vision tasks, such as visual navigation [32], scene understanding [5,25], and autonomous driving [10]. Also, the problem that hazy weather reduces the quality of images and videos taken outdoors has brought great security risks to intelligent transportation. Additionally, the outdoor monitoring and remote sensing system will also be paralyzed due to the degradation of hazy images. Therefore, it is urgent to improve the quality of hazy images. Scholars have developed various methods for dehazing, which

S. Rudinac et al. (Eds.): MMM 2024, LNCS 14555, pp. 112–126, 2024.
https://doi.org/10.1007/978-3-031-53308-2_9

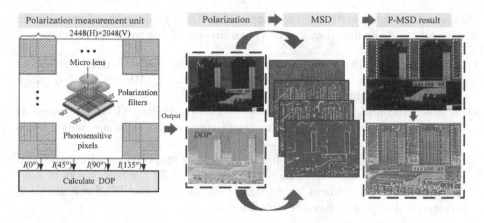

Fig. 1. The overall scheme of Pol-MSD. It first to obtain the contrast enhanced polarization CLAHE (PCLAHE) image as the channel of near scene by using the CLAHE method. Meanwhile, the incomplete normalized degree of polarization image (INDoP) is utilized as the second input to restore the visibility of the far scene. Then, the RGF framework is introduced to perform the MSD of the two inputs. In the fusion of base layers, the sobel gradient map (SGM) is introduced as the fusion weight map of base layers, and a weight measurement method is constructed based on SGM to fuse the base layers. In the fusion of detail layers, a novel WLS optimization method is designed to fuse the detail layers. Finally, the visibility of the far scene can be significantly improved by performing inverse MSD on the fused base layers and detail layers.

are mainly divided into prior-based, fusion-based, and learning-based methods. 1) Prior-based dehazing: These methods are based on the atmospheric scattering model (ASM) [20, 21]. To solve the ill-posed problems in the model, these methods exploit the prior to provide constraints for ill-posedness, thus achieving significant dehazing performance [3, 7]. However, the prior conditions are hypothetical, and these methods cannot realize dehaze effectively outside the prior conditions. 2) Fusion-based dehazing: These methods do not rely on the ASM to dehaze, and the performance of these methods [16, 38] are similar to the prior-based methods. These methods use the Laplacian pyramid or Gaussian pyramid method to decompose hazy images into multi-scales. Then, dehazing and noise reduction are conducted for images of different scales, and the corresponding fusion rules are designed for the base layers and detail layers respectively. Finally, the generated pyramid is folded to restore the haze-free image. Additionally, the multiscale fusion method retains a small amount of object texture details [15, 18]. 3) Learning-based dehazing: Recently, learning-based methods have attracted wide attention. These methods directly restore images using cGAN [23] or encoderdecoder-based networks [4, 9] without using an ASM. However, the training data of these methods are almost synthesized images, so the recovered image usually contains distortion when the scene changes, especially for realistic images.

This paper proposes the multi-scale decomposition dehazing with polarimetric vision (Pol-MSD) algorithm. An overview of Pol-MSD is shown in Fig. 1.

Different from most methods, the proposed method could realize near and far scenes dehazing simultaneously. The contributions of our work are as follows:

1) The contrast limited adaptive histogram equalization (CLAHE) method is used to obtain the contrast enhanced polarization CLAHE (PCLAHE) image as the channel of near-scene.
2) A novel incomplete normalized degree of polarization (INDoP) image is obtained based on stokes theory as the channel of the far scene, which helps to improve the visibility of the hazy image.
3) A novel multi-scale detail layer fusion method is proposed based on weighted least squares (WLS), and a fusion rule is proposed for the base layers by considering the sobel gradient map (SGM). Further, a novel WLS optimization scheme is designed to fuse the detail layers.

2   Related Work

Our work mainly contains two topics: polarimetric dehazing and MSD. In this section, we will briefly review the important theories of these two topics.

2.1   Polarimetric Dehazing

Image dehazing is a challenging task in the field of computer vision. Schechner et al. [21,30] proposed the polarimetric difference dehazing technology firstly. The algorithm assumes that the light reflected by the object does not have remarkable polarization, it collects two polarimetric images with the highest and lowest image contrast. Image dehazing can be realized by differential imaging of two polarimetric images. Polarimetric difference dehazing technology that depends on the selection of the sky region [28]. Researchers discuss methods for selecting sky regions in [8,19]. For some scenes, it is difficult to obtain two orthogonal polarimetric images because the change of the scene may be faster than the rotation of the polarization filter. Liang et al. [12–14] propose the polarimetric dehazing technology of the stokes vector, which uses the angle of polarization to estimate the degree of polarization. Researchers propose a large number of polarimetric dehazing methods based on deep learning to eliminate the impact of haze [27,40]. Some scholars propose a dehazing method for polarization information fusion, which can improve the quality of dehazed images and enrich image details [26,34,41]. The polarization-based dehazinga method has more robust and stable characteristics in scene changes.

2.2   Multi-scale Decomposition (MSD)

Multi-scale decomposition is another important dehazing method. Codruta et al. [2] proved firstly the effectiveness of fusion methods in image dehazing by calculating three weight maps to obtain important image features and merging the input laplacian and weighted gaussian images in a multi-scale fusion. Xue et al. [35] designed a multi-scale feature extraction module that can detect rain bands of different lengths for clear imaging in the rain or hazy background. Liu et al. [17] proposed GridDehazeNet, and the backbone module implements attention-based multi-scale estimation on the grid network to improve the feature extraction performance of traditional multi-scale decomposition. Li et al. [11] used Laplacian pyramid and Gaussian pyramid to decompose the hazy image into different levels, and they exploited different dehazing and noise reduction methods to restore scene brightness at different levels to restore haze-free images. DehazeNet [4] realizes image dehazing by learning the mapping relationship between hazy and transmission images. The algorithm is composed of the coarse-scale net and fine-scale net, where the former predicts the overall transmission based on the image, and the latter refines the dehazing results locally.

3   Proposed Method

When applying the fusion algorithm, the key to achieving the final good visibility result is the carefully customized input. The idea of image fusion is that the processed result combines multiple input images by retaining only the important features of the near and far scenes in the input image. The two inputs of the fusion process are extracted from the original polarization degraded image, where the first input is the PCLAHE image, and second input is the INDoP image.

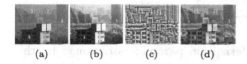

| (a) | (b) | (c) | (d) | (a) | (b) | (c) | (d) |

Fig. 2. (a) Input. (b), (c), (d) and (e) are the enhancing results by histogram equalization (HE), adaptive histogram equalization (AHE) and CLAHE method.

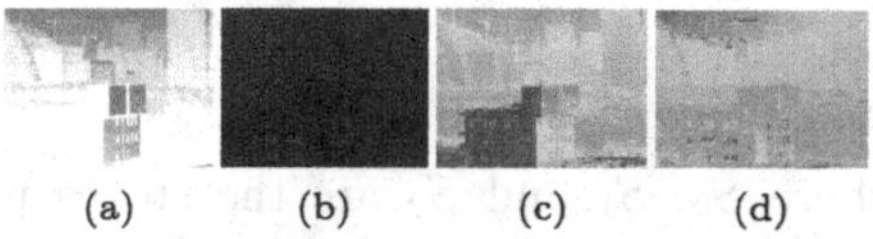

Fig. 3. (a) Incomplete normalized polarimetric image (b) polarimetric image. (a) and (b) are displayed in the HSV space, as shown in (c) and (d).

3.1   Histogram Equalization of the Inputs

The derived input is intended to reduce the degradation due to volume scattering. To achieve the optimal contrast level of the image, it is found that CLAHE [42] is an appropriate strategy in the experiment. The CLAHE produces satisfactory results, and the overall contrast of the image is improved. As shown by marks "1", "2", "3", and "4" in Fig. 2(d), the contrast of the near scene of the image is further improved compared with the corresponding positions in Fig. 2(b). Meanwhile, compared with Fig. 2(c), the problem of distortion due to over-enhancement of the image is avoided. Compared with the input image in Fig. 2(a), there is no "white spot" in the PCLAHE image obtained, and a more pleasing image of the near scene is produced. However, only the PCLAHE image cannot solve the problem of visibility of the far scene, so this paper derives an additional input (described in the next subsection) to enhance the visual visibility of the far scene in the hazy image.

(a)   (b)   (c)   (d)

Fig. 4. (a) is the base layer B_1 of PCLAHE image, (b) is the SGM1 of B_1, (c) is base layer B_2 of INDoP image, and (d) is the SGM2 of B_2.

3.2 The Incomplete Normalized Degree of Polarization Image

In our fusion framework, the second input is an INDoP image derived from the original hazy image. This input is intended to dehaze the far scene, thereby enhancing the texture information of the far scene. This paper starts with the ASM that can accurately describe the degradation process of the image in a hazy environment. Polarization dehazing and most image restoration techniques utilize this model for dehazing processing, and good effect can be achieved. However, there is a great deviation when judging the "brightest" and "darkest" images manually. Therefore, some scholars have proposed polarization-based dehazing method [8,13]. DoP information can be obtained from stokes vector, and the haze can be removed by polarization information:

$$DoP = \frac{\sqrt{S_1^2 + S_2^2}}{S_0} \tag{1}$$

where S_0, S_1 and S_2 are the stokes polarization parameters obtained by the polarimetric images.

In the experiment, the numerator of the DoP in Eq. 1 is normalized, and the denominator is not normalized. The obtained polarimetric images are shown in Fig. 3(a). The polarimetric images obtained by simultaneously normalizing the numerator and denominator of the DoP is shown in Fig. 3(b). Since the polarimetric images obtained by the two methods are not uniform visually, the two types of polarimetric images are displayed in the HSV space. The polarimetric images of the HSV space corresponding to Fig. 3(a) and Fig. 3(b) are shown in Fig. 3(c) and Fig. 3(d), respectively. By comparing the two results obtained by normalization and incomplete normalization, it was found that the INDoP image contains more scene information. Particularly, Fig. 3(c) shows that compared with Fig. 3(d), the former highlights the profile of the far scene behind the two buildings, and the visibility of the far scene and the sky area of the latter is insufficient. So the INDoP image is taken as the second input.

3.3 MSD Based on RGF and Bilateral Filter

The design of the fusion layers needs to consider the desired appearance of the restored output. However, PCLAHE and INDoP images have quite different characteristics: PCLAHE image often contain fine-scale visual detail information in the near scene, while INDoP image is dehazing for the far scene. It can be seen from Fig. 3(a) that the image contains abundant scene information in the far scene, the coarse-scale structure profile of the near scene, and many incompatible "camera quantum noise" [14]. We need to retain the features of near and far scenes represented by PCLAHE and INDoP images to the greatest extent. Recently, Zhang et al. [37] proposed a state-of-the-art filtering frame of rolling guidance filter (RGF), which has both scale-aware and edge-preserving properties. RGF includes two main steps: small structure removal and edge recovery. In Pol-MSD, the coarsest base layer is processed by a bilateral filter to obtain the base layers needed in the fusion process after the RGF of the two derived images.

$$B^j = Bilateral\left(B^{j-1}, \sigma_s^{j-1}, \sigma_r\right) \tag{2}$$

where j is equal to the maximum number of decomposition layers.

Then, the calculation of the detail layer is defined as:

$$D^j = B^{j-1} - B^j \tag{3}$$

where B^j denotes the jth-level filtering image, D^j is the jth-level detail layer.

3.4   Polarization Dehazing Based on WLS Optimization

This paper proposes a novel WLS optimization method in the experiment to fuse the detail layers. The purpose of this optimization is to retain the features of near and far scenes represented by PCLAHE and INDoP images to the greatest extent. Finally, the corresponding inverse MSD is performed to produce the fused image.

1) Fusion of base layers: The traditional average fusion rule of the base layer will lose the coarse information of the near and far scenes, which usually results in the loss of contrast in the fused image. To solve this problem, this paper uses the weighted average technology based on SGM to fuse the base layers. For image fusion, since SGM can reflect salient image features, it is usually used as the activity level measurement to fuse the base layers. Let M_1 and M_2 represent the SGM of the input PCLAHE and INDoP images in Fig. 4, respectively. The fused base layer B_F can be obtained by the following weighted average:

$$B_F = W_b B_1 + (1 - W_b)B_2 \tag{4}$$

where the weight W_b is defined as

$$W_b = 0.5\left(\frac{M_1}{max\left(M_1, M_2\right)}\right) \tag{5}$$

2) Fusion of detail layers: The detail layers of the PCLAHE and INDoP images in Fig. 4 indicate that the difference between the two is not obvious, and the texture information in it needs to be merged into the fused image. However, the traditional maximum absolute rule cannot retain the texture details

of the near and far scenes at the same time. Additionally, the low-pass filtering algorithm for detail layers fusion may lead to the loss of smoothing of important fine-scale structures in the detail layers. Therefore, this paper obtains the jth level fusion detail layer d^j by minimizing the WLS cost function:

$$\sum_p \left(\left(d_p^j - \left(D_1^j \right)_p \right)^2 + \lambda a_p^j \left(d_p^j - \left(D_2^j \right)_p \right)^2 \right) \tag{6}$$

where a_p^j defines the coefficient with spatial variation weight. When selecting the window size of the coefficient, a larger window will blur the fused image and increase the calculation cost, and a small window cannot eliminate the influence of noise. The first term $\left(d_p^j - \left(D_1^j \right)_p \right)^2$ is to minimize the Euclidean distance between the fused detail layer d_p^j and the base layer $\left(D_1^j \right)_p$. The second term $\left(d_p^j - \left(D_2^j \right)_p \right)^2$ intends to make the fused detail layer d_p^j close to the INDoP detail layers $\left(D_2^j \right)_p$. λ is a parameter that controls the trade-off between these two items globally. Eq. 6 is rewritten in a matrix form as follows:

$$d_p^j = \left(I + \lambda A^j \right)^{-1} \left(\left(D_1^j \right)_p + \lambda A^j \left(D_2^j \right)_p \right) \tag{7}$$

Finally, by combining the fused base layer B_F and the fused detail layers d^1, $d^2, \cdots, d^N$, the fused image F can be reconstructed as follows:

$$F = B_F + d^1 + d^2 + \cdots + d^N \tag{8}$$

4   Experiments

In this selection, a serious of experiments are carried out to comprehensively evaluate the performance of the proposed dehazing method.

4.1   Model and Performance

To investigate the best trade-off between performance and parameter size, the parameters of Pol-MSD are progressively modified. We take the rate of new visible edges e, the gradients of visible edges r, the saturated σ [6] as the objective function to adaptively select the best value.

The test values of σ_s are set to 2, 4, and 6, respectively. Meanwhile, the sensitivity of the algorithm to different sizes of σ_r is observed under each σ_s. The value of σ_r is set to [0.01, 0.09] and [0.1, 0.9]. Then, three types of blind evaluation indexes e, r, and σ are used to conduct the experimental verification. The experimental results are shown in Fig. 5, where the break-point is set between 0.09 and 0.1. The convergence of the algorithm for σ_s is analyzed. The index e

Fig. 5. The sensitivity of Pol-MSD to the spatial weights σ_s and range weights σ_r in different sizes. We utilize three kinds of blind evaluation indexes e, r and σ to conduct the experimental verification. (a), (b) and (c) are the variation curves of three evaluation indexes e, r and σ with the increase of space weights σ_s and σ_r, respectively. (d)-(i) are result using different size of σ_s and σ_r.

in Fig. 5(a) and the index σ in Fig. 5(c) are optimal when $\sigma_s = 2$. Based on the above analysis, we further demonstrate the experimental results in Figs. 5(d)-(i). From the three groups of comparative experiments, when σ_s is fixed and σ_r is smaller, the texture details of the near scene in the dehazing image will contain artifacts and distortions. When the value of σ_r is determined, and the smaller the value of σ_s, the clearer the structure of the haze-free image. Therefore, in our article, the default value $\sigma_s = 2$ and $\sigma_r = 0.9$.

4.2   Ablation Study

To better demonstrate the effectiveness of the architecture of our method, we conduct an ablation study by considering the combination of three factors: 1) **P**: contrast enhanced polarization CLAHE (PCLAHE) image, 2) **I**: incomplete normalized degree of polarization (INDoP) image, 3) **W**: weighted least squares (WLS) optimization. The evaluation is performed on the polarimetric hazy images shown in Fig. 6. The quantitative results on dehazed images for the various configurations are tabulated in Table 1. From Fig. 6, the visibility of the restored image is seriously insufficient in the absence of contrast stretching to the near scene. Compared with Fig. 6(c), after adding the texture information of the far scene, the dehazing image obtained by the Pol-MSD method is visually pleasant. After using the max-absolute fusion rule to fuse the detail layer in Fig. 6(d), the image contains the coarse-scale structure profile of the near scene and many incompatible "camera quantum noise". The detail information of the image is seriously lost. The quantitative performance evaluated on polarimetric images also demonstrate the effectiveness of each module in Table 1. The dehazing image by Pol-MSD is improved in terms of visual visibility and saturation.

4.3   Evaluation of Pol-MSD

To demonstrate the capability of the proposed Pol-MSD, Fig. 7 presents the comparison with other polarization-based dehazing methods, including P-MSVD [41], PSDNet [27] and a recent method LTD [40]. As shown in Fig. 7(a), three hazy examples with rich information in structures are used for the comparison. The dehazing results are given in Figs. 7(c)-(f). It is observed in Fig. 7(d)

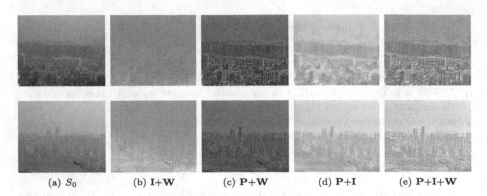

(a) S_0 (b) I+W   (c) P+W   (d) P+I   (e) P+I+W

Fig. 6. The dehazing results using different combination of three factors: **P, I, P**.

Table 1. Ablation study by considering the combination of three factors: **P, I, P**, the best and second-best scores are highlighted in red and blue.

Method	S_0-1			S_0-2		
	r	VCM	En	r	VCM	En
I+W	3.16	37.7	6.10	4.29	21.8	6.06
P+W	5.65	99.7	7.21	7.12	54.1	6.58
P+I	2.73	33.3	6.37	2.82	50.2	6.22
P+I+W(Pol-MSD)	5.84	100	7.26	7.15	54.5	6.77

and Fig. 7(f) that both PSDNet and P-MSVD fail to unveil the details in long distances, and the overall contrast of the image is insufficient. The far scene information is not recovered in Fig. 7(e), In contrast, the proposed Pol-MSD method can restore the texture information of near scene and far scene, demonstrating its remarkable improvement in dehazing.

4.4   Quantitative Evaluation

To demonstrate the capability of the proposed Pol-MSD, Fig. 8(a) shows the polarimetric hazy images we collected. Bedsides, the Blackfly polarization camera equipped with a Sony IMX250 chip [1] is adopted. The camera has several polarization measurement units, each of which consists of 4 pixels. The corresponding pixels have polarization sensitive directions of $0°$, $45°$, $90°$, and $135°$, and polarimetric images in different polarization states (I_0, I_{45}, I_{90}, and I_{135}) are obtained through one exposure.

The dehazing results obtained by using various methods are shown in Figs. 8(c)–(i). Seven typical or recent dehazing methods are adopted to facilitate the comparison, including polarization-based dehazing methods P-MSVD [41], PSDNet [27], Stokes [14] and a recent method LTD [40]. We use three polarization channels (I_0, I_{45}, and I_{90}) as inputs to our method with P-MSVD, Stokes

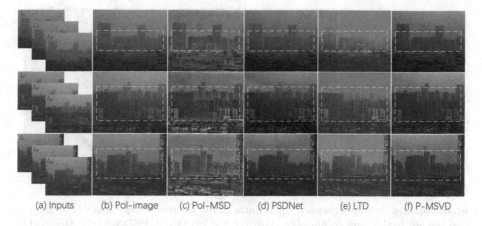

(a) Inputs   (b) Pol-image   (c) Pol-MSD   (d) PSDNet   (e) LTD   (f) P-MSVD

Fig. 7. The comparison with other polarization-based dehazing methods. (a) shows the three polarization channels (I_0, I_{45}, and I_{90}) as inputs, (b) is polarized hazy image(Pol-image). The proposed Pol-BCT method can improve the visibility of near and far scenes simultaneously (see the red and yellow rectangles). Please amplify figures to view the detail differences in bounded regions.

Table 2. Quantitative comparisons in Fig. 8, the best and second-best scores are highlighted in red and blue.

Method	IM-1			IM-2			IM-3			IM-4		
	r	VCM	En	r	VCM	En	r	VCM	En	r	VCM	En
P-MSVD [41]	1.63	54.2	5.98	1.65	58.1	6.22	1.53	41.5	6.62	1.63	51.3	6.26
PSDNet [27]	2.05	52.2	6.05	2.09	32.1	6.58	2.24	46.2	7.21	1.74	48.1	5.97
Stokes [14]	0.84	26.0	4.52	1.23	0.63	3.84	1.43	25.2	3.85	1.86	38.3	6.25
FFA-Net [22]	1.51	48.3	5.8	1.45	12.3	6.43	1.77	24.8	6.42	1.77	8.33	5.96
DehazeFormer [29]	2.92	39.8	6.15	2.92	50.0	6.86	2.49	52.5	6.23	2.83	39.8	6.15
DehazeNet [4]	1.75	33.8	6.33	1.42	38.5	6.63	1.44	33.3	6.88	1.47	36.1	6.61
LBT [24]	2.25	48.3	6.29	2.03	48.1	6.66	1.97	48.3	6.45	2.07	48.5	6.62
Pol-MSD (Ours)	5.31	50.6	6.45	11.4	89.7	7.33	8.72	71.8	7.19	6.37	52.9	7.56

Table 3. Quantitative comparisons in Fig. 9, the best and second-best scores are highlighted in red and blue.

Method	IM-5			IM-6			IM-7		
	$r \uparrow$	VCM $\uparrow$	En $\uparrow$	$r \uparrow$	VCM $\uparrow$	En $\uparrow$	$r \uparrow$	VCM $\uparrow$	En $\uparrow$
LTD [40]	3.81	47.1	7.04	4.21	36.1	7.60	1.43	45.3	7.62
P-MSVD [41]	1.25	12.3	5.83	1.25	4.57	6.46	0.57	12.8	6.38
PSDNet [27]	1.91	9.83	5.34	2.66	11.4	6.22	0.64	12.5	6.17
Stokes [14]	1.54	12.5	6.39	2.47	3.89	7.25	1.22	18.3	7.33
FFA-Net [22]	3.78	7.66	6.18	4.05	29.1	6.41	1.91	23.2	6.66
DehazeFormer [29]	2.51	27.3	5.47	2.74	26.7	6.01	1.88	28.1	7.21
LBT [24]	4.47	24.3	3.92	6.02	25.4	5.75	2.17	32.0	6.07
Pol-MSD (Ours)	5.52	38.2	7.11	5.84	38.2	7.62	2.71	34.7	7.69

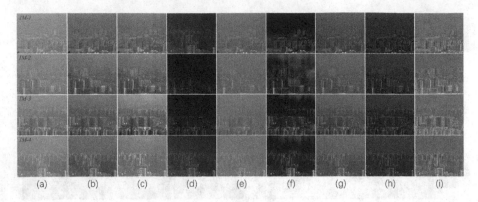

Fig. 8. Qualitative comparisons on real data. (a) is polarized hazy image(Pol-image). From the left to the right column (i.e., (b)-(i)), P-MSVD [41], PSDNet [27], Stokes [14], FFA-Net [22], DehazeFormer [29], DehazeNet [4], LBT [24] and Pol-MSD (ours).

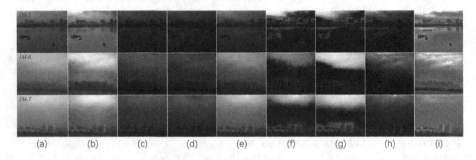

Fig. 9. Dehazing results of different methods in dataset [40]. (a) is polarized hazy image(Pol-image). From the left to the right column (i.e., (b)-(i)), LTD [40], P-MSVD [41], PSDNet [27], Stokes [14], FFA-Net [22], DehazeFormer [29], LBT [24] and Pol-MSD (ours). Pol-MSD generates more realistic images than the other methods in far scene. Please amplify figures to view the details.

and LTD. The PSDNet uses two polarization channels (I_0 and I_{90}) as inputs. Learning-based dehazing methods FFA-Net [22], DehazeNet [4], and Dehaze-Former [29], Prior-based dehazing methods LBT [24]. The three polarized channels (I_0, I_{45}, and I_{90}) are combined as RGB image to satisfy the input requirement of the compared methods: FFA-Net, DehazeNet, DehazeFormer, and LBT. These algorithms with publicly available codes are selected. The performance of Pol-MSD is quantitatively evaluated by using three quantitative indexes, including the blind assessment indicator: the gradients of visible edges r [6], image visibility measurement (IVM) [36], and Entropy (En).

For the hazy scenes in Fig. 8(a), P-MSVD, PSDNet, and DehazeNet is able to remove most haze, but the visibility of far scenes is insufficient. Large dim areas appeared in the dehazing scene by Stokes, DehazeFormer, and LBT. FFA-Net performs haze removal at the cost of introducing pseudo-color. In comparison, the proposed Pol-MSD effectiveness highlights the main information of the scene, especially the overall contrast has been improved. The scores for the dehazing results of various methods shown in Fig. 8 are presented in Table 2. It is observed that the proposed Pol-MSD achieves the best r scores in almost all examples and gains the best or second-best VCM and En scores in most examples. Pol-MSD is mainly reflected in that it can improve the texture features and saturation information of the scenes.

Considering the generalization performance of the learning-based dehazing methods, we further use three images shared by LTD [40] in Fig. 9. For P-MSVD, PSDNet, Stokes, FFA-Net, DehazeFormer, and LBT seem to be dim and fail to reveal the details. For LTD, it successfully remove haze with satisfactory details preservation, however, its dehazing performances drops in the scenes at long distance, as shown in Fig. 9(c). In comparison, the proposed Pol-MSD is able to restore clearer texture details and performs better saturation preservation, as shown in Fig. 9(i). The scores for the dehazing results of various methods shown in Figs. 9 are presented in Table 3. It is observed that the proposed Pol-MSD achieves the best En scores in almost all examples and gains the best or second-best VCM and r scores in most examples, which further indicates its boosted dehazing performance.

5   Conclusion

In this paper, a novel polarization dual channel multi-scale decomposition image dehazing algorithm is proposed to solve the problem of recovering near and far scenes information simultaneously in hazy weather. Firstly, the polarization CLAHE (PCLAHE) image is obtained from the original polarization image as the first derivative image. Meanwhile, according to stokes theory, the INDoP image is obtained from the original polarization image as the second derivative image. To combine the two derived images in different feature sub-bands, RGF and bilateral filter are proposed to decompose the input image into base layers and detail layers. Besides, the SGM is used as the fusion weight of the base layer, and a base layer fusion method based on the SGM is constructed. In detail layer fusion, different characteristics of the PCLAHE and INDoP images are considered, and a novel multi-scale detail layer fusion method based on least squares is proposed. Through this optimization, both the texture details of near and far scenes can be transferred to the fused image. The experimental results indicate that the Pol-MSD can improve the visibility of near and far scenes simultaneously, and the dehazing effect is significant.

Acknowledgements. This work is supported in part by National Natural Science Foundation of China under Grant (62103429), Major Project of Natural Science Foundation of Hunan Province (2021JC0004), Outstanding doctoral research innovation project of Xinjiang University (XJU2022BS082), and Robot and intelligent equipment technology innovation team (2022D14002).

References

1. Imx250 cmos sensor. https://www.sony-semicon.co.jp/e/products/IS/polariza tion/
2. Ancuti, C.O., Ancuti, C.: Single image dehazing by multi-scale fusion. IEEE Trans. Image Process. **22**(8), 3271–3282 (2013)
3. Berman, D., Treibitz, T., Avidan, S.: Non-local image dehazing. In: Conference on Computer Vision and Pattern Recognition (CVPR), pp. 1674–1682 (2016)
4. Cai, B., Xu, X., Jia, K., Qing, C., Tao, D.: Dehazenet: an end-to-end system for single image haze removal. IEEE Trans. Image Process. **25**(11), 5187–5198 (2016)
5. Hahner, M., Dai, D., Sakaridis, C., Zaech, J.N., Gool, L.V.: Semantic understanding of foggy scenes with purely synthetic data. In: 2019 IEEE Intelligent Transportation Systems Conference (ITSC), pp. 3675–3681 (2019)
6. Hautiere, N., Tarel, J.P., Aubert, D., Dumont, E.: Blind contrast enhancement assessment by gradient ratioing at visible edges. Image Analy. Stereology **27**(2), 87–95 (2008)
7. He, K., Sun, J., Tang, X.: Single image haze removal using dark channel prior. IEEE Trans. Pattern Anal. Mach. Intell. **33**(12), 2341–2353 (2010)
8. Kaftory, R., Schechner, Y.Y., Zeevi, Y.Y.: Variational distance-dependent image restoration. In: Proceedings of Conference on Computer Vision and Pattern Recognition (CVPR), pp. 1–8 (2007)
9. Li, B., Peng, X., Wang, Z., Xu, J., Feng, D.: Aod-net: All-in-one dehazing network. In: Proceedings of IEEE Conference on Computer Vision (ICCV), pp. 4770–4778 (2017)
10. Li, N., Zhao, Y., Pan, Q., Kong, S.G., Chan, J.C.-W.: Full-time monocular road detection using zero-distribution prior of angle of polarization. In: Vedaldi, A., Bischof, H., Brox, T., Frahm, J.-M. (eds.) ECCV 2020. LNCS, vol. 12370, pp. 457–473. Springer, Cham (2020). https://doi.org/10.1007/978-3-030-58595-2_28
11. Li, Z., Shu, H., Zheng, C.: Multi-scale single image dehazing using laplacian and gaussian pyramids. IEEE Trans. Image Process. **30**, 9270–9279 (2021)
12. Liang, J., Ren, L., Ju, H., Qu, E., Wang, Y.: Visibility enhancement of hazy images based on a universal polarimetric imaging method. J. Appl. Phys. **116**(17) (2014)
13. Liang, J., Ren, L., Ju, H., Zhang, W., Qu, E.: Polarimetric dehazing method for dense haze removal based on distribution analysis of angle of polarization. Opt. Express **23**(20), 26146–26157 (2015)
14. Liang, J., Zhang, W., Ren, L., Ju, H., E., Q.: Polarimetric dehazing method for visibility improvement based on visible and infrared image fusion. Appl. Opt. **55**(29), 8221–8226 (2016)
15. Liu, W., Zhou, F., Duan, J., Qiu, G.: Image defogging quality assessment: real-world database and method. IEEE Trans. Image Process. **30**, 176–190 (2020)
16. Liu, X., Li, H., Zhu, C.: Joint contrast enhancement and exposure fusion for real-world image dehazing. IEEE Trans. Multimedia **24**, 3934–3946 (2021)
17. Liu, X., Ma, Y., Shi, Z., Chen, J.: Griddehazenet: attention-based multi-scale network for image dehazing. In: Proc. IEEE Conference on Computer Vision (ICCV), pp. 7314–7323 (2019)
18. Ma, J., Zhou, Z., Wang, B., Zong, H.: Infrared and visible image fusion based on visual saliency map and weighted least square optimization. Infrared Phys. Techn. **82**, 8–17 (2017)
19. Namer, E., Schechner, Y.Y.: Advanced visibility improvement based on polarization filtered images. In: Polarization Science and Remote Sensing II, vol. 5888, pp. 36–45 (2005)

20. Narasimhan, S.G., Nayar, S.K.: Contrast restoration of weather degraded images. IEEE Trans. Pattern Anal. Mach. Intell. **25**(6), 713–724 (2003)
21. Nayar, S.K., Narasimhan, S.G.: Vision in bad weather. In: Proc. IEEE Conference on Computer Vision (ICCV), vol. 2, pp. 820–827 (1999)
22. Qin, X., Wang, Z., Bai, Y., Xie, X., Jia, H.: Ffa-net: feature fusion attention network for single image dehazing. In: AAAI, vol. 34, pp. 11908–11915 (2020)
23. Qu, Y., Chen, Y., J., H., Xie, Y.: Enhanced pix2pix dehazing network. In: Proceedings of IEEE Conference on Computer Vision Pattern Recognition (CVPR). pp. 8160–8168 (2019)
24. Raikwar, S.C., Tapaswi, S.: Lower bound on transmission using non-linear bounding function in single image dehazing. IEEE Trans. Image Process. **29**, 4832–4847 (2020)
25. Sakaridis, C., Dai, D., Hecker, S., Van Gool, L.: Model adaptation with synthetic and real data for semantic dense foggy scene understanding. In: Ferrari, V., Hebert, M., Sminchisescu, C., Weiss, Y. (eds.) ECCV 2018. LNCS, vol. 11217, pp. 707–724. Springer, Cham (2018). https://doi.org/10.1007/978-3-030-01261-8_42
26. Shen, L., Zhao, Y., Peng, Q., Chan, J., Kong, S.G.: An iterative image dehazing method with polarization. IEEE Trans. Multimedia **21**(5), 1093–1107 (2018)
27. Shi, Y., Guo, E., Bai, L., Han, J.: Polarization-based haze removal using self-supervised network. Front. Phys. **9**, 789232 (2022)
28. Shwartz, S., Namer, E., Schechner, Y.Y.: Blind haze separation. In: Proceedings of IEEE Conference on Computer Vision and Pattern Recognition (CVPR), vol. 2, pp. 1984–1991 (2006)
29. Song, Y., He, Z., Qian, H., Du, X.: Vision transformers for single image dehazing. IEEE Trans. Image Process. **32**, 1927–1941 (2023)
30. Treibitz, T., Schechner, Y.Y.: Polarization: beneficial for visibility enhancement? In: Proceedings of IEEE Conference on Computer Vision and Pattern Recognition (CVPR), pp. 525–532 (2009)
31. Ullah, H., et al.: Light-dehazenet: a novel lightweight cnn architecture for single image dehazing. IEEE Trans. Image Process. **30**, 8968–8982 (2021)
32. Wan, Z., Zhao, K., Chu, J.: Robust azimuth measurement method based on polarimetric imaging for bionic polarization navigation. IEEE Trans. Instrum. Meas. **69**(8), 5684–5692 (2019)
33. Wang, H., Wang, Y., Cao, Y., Zha, Z.: Fusion-based low-light image enhancement. In: International Conference on Multimedia Modeling, pp. 121–133 (2023)
34. Wang, Y., et al.: Polarimetric dehazing based on fusing intensity and degree of polarization. Optics Laser Technol. **156**, 108584 (2022)
35. Xue, X., Hao, Z., Ding, Y., Jia, Q., Liu, R.: Multi-scale features joint rain removal for single image. In: Proceedings of IEEE Conference on Image Processing (ICIP), pp. 933–937 (2020)
36. Yu, X., Xiao, C., Deng, M., Peng, L.: A classification algorithm to distinguish image as haze or non-haze. In: Proceedings of IEEE International Conference on Image Graphics, pp. 286–289 (2011)
37. Zhang, Q., Shen, X., Xu, L., Jia, J.: Rolling guidance filter. In: Fleet, D., Pajdla, T., Schiele, B., Tuytelaars, T. (eds.) ECCV 2014. LNCS, vol. 8691, pp. 815–830. Springer, Cham (2014). https://doi.org/10.1007/978-3-319-10578-9_53
38. Zheng, M., Qi, G., Zhu, Z., Li, Y., Wei, H., Liu, Y.: Image dehazing by an artificial image fusion method based on adaptive structure decomposition. IEEE Sensors J. **20**(14), 8062–8072 (2020)

39. Zheng, Y., Su, J., Zhang, S., Tao, M., Wang, L.: Dehaze-aggan: unpaired remote sensing image dehazing using enhanced attention-guide generative adversarial networks. IEEE Trans. Geosci. Remote Sens. **60**, 1–13 (2022)
40. Zhou, C., Teng, M., Han, Y., Xu, C., Shi, B.: Learning to dehaze with polarization. Adv. Neural. Inf. Process. Syst. **34**, 11487–11500 (2021)
41. Zhou, W., Fan, C., Hu, X., Zhang, L.: Multi-scale singular value decomposition polarization image fusion defogging algorithm and experiment. Chin. Optics **14**(2), 298–306 (2021)
42. Zuiderveld, K.: Contrast limited adaptive histogram equalization. Graphics Gems, pp. 474–485 (1994)

CLF-Net: A Few-Shot Cross-Language Font Generation Method

Qianqian Jin, Fazhi He$^{(\boxtimes)}$ (ID), and Wei Tang

School of Computer Science, Wuhan University, Wuhan, Hubei, China
fzhe@whu.edu.cn

Abstract. Designing a font library takes a lot of time and effort. Few-shot font generation aims to generate a new font library by referring to only a few character samples. Accordingly, it significantly reduces labor costs and has attracted many researchers' interest in recent years. Existing works mostly focus on font generation in the same language and lack the capability to support cross-language font generation due to the abstraction of style and language differences. However, in the context of internationalization, the cross-language font generation task is necessary. Therefore, this paper presents a novel few-shot cross-language font generation network called CLF-Net. We specifically design a Multi-scale External Attention Module (MEAM) to address the issue that previous works simply consider the intra-image connections within a single reference image, and ignore the potential inter-image correlations between all reference images, thus failing to fully exploit the style information in another language. The MEAM models the inter-image relationships and enables the model to learn essential style features at different scales of characters from another language. Furthermore, to solve the problem that previous approaches usually generate characters with missing or duplicated strokes and blurry stroke edges, we define an Edge Loss to constrain the model to focus more on the edges of characters and make the outlines of generated results clearer. Experimental results show that our CLF-Net is outstanding for cross-language font generation and generates better images than the state-of-the-art methods.

Keywords: Cross-language Font generation · External attention · Few-shot learning

1 Introduction

Font style is the art of visual representation of text and plays a crucial role in conveying information. It can even deliver deeper meaning, such as whether the current content is delightful or horrible. Designing a font is very time-consuming and requires the highly professional ability of the designer. The designer has to make proper artistic effects for strokes so that the font not only conveys the artistic style but also guarantees the original content of the character. In addition, when designing a large font library of multiple languages, the designer needs to

© The Author(s), under exclusive license to Springer Nature Switzerland AG 2024
S. Rudinac et al. (Eds.): MMM 2024, LNCS 14555, pp. 127–140, 2024.
https://doi.org/10.1007/978-3-031-53308-2_10

spend a lot of time and effort to keep the characters of different languages in the same style, which not only demands professional knowledge and skills but also requires the designer to be proficient in different languages.

Therefore, automatic font generation via neural networks has attracted the attention of researchers, and many GAN [5]-based models for automatic font generation have been proposed. Early models [13, 14, 24, 25] need to be pre-trained on large datasets and then fine-tuned for specific tasks, which requires many computational resources and much effort to collect training samples. Recently, many few-shot learning methods [3, 9, 11, 19, 20, 23, 31, 32] have been proposed specifically for the font generation task, and these models can generate complete font libraries of the same language based on a small number of samples.

Nevertheless, in many scenarios, such as designing novel covers in different translations, movie promotional posters for different countries, and user interfaces for international users, it is necessary to keep characters of different languages having the same font style. At the same time, characters of different languages vary greatly in their glyph structure, e.g., the strokes and structures of English letters are very different from those of Chinese characters. Specifically, many components of Chinese characters have no counterparts in English letters, which leads to the fact that learning the style of characters from another language is difficult and requires the model to learn high-level style characteristics. Thus, some efforts [15] attempt to use self-attention mechanism to capture style patterns in another language. However, they ignore the potential inter-image correlations between different reference images and thus fail to learn sufficiently essential features in another language. Therefore, we propose to learn better style representation in another language by analyzing the inter-image relationships between all reference images rather than simply considering the intra-image connections.

In this paper, we propose a novel model named CLF-Net. Its core idea is to learn essential style features in another language by modeling the inter-image relationships between all reference images. Specifically, we design a Multi-scale External Attention Module (MEAM) to capture style features at different scales. The MEAM not only considers the intra-image connections between different regions of a single reference image but also implicitly explores the potential inter-image correlations between the overall style images, which makes it possible to extract the geometric and structural patterns that are consistently present in the style images and thus learn the unified essential style information at different scales in another language. In addition, considering that boundary pixels play a key role in determining the overall style of Chinese characters, we define an Edge Loss to compel the model to preserve more edge information and ensure the generated characters have sharper edges with less blur. Combining these components, we have achieved high-quality cross-language font generation.

Our contributions can be summarized as follows:

1) We first implicitly consider the inter-image associations and propose a novel few-shot cross-language font generation network called CLF-Net instead of simply considering the intra-image connections.

2) We design a Multi-scale External Attention Module (MEAM) to learn the unified essential style information at different scales of characters from another language, which solves the problem that the existing font generation models can not fully exploit style information in another language.
3) We introduce an Edge Loss function to make the model generate characters with sharper edges.
4) By modeling the inter-image relationships, our approach achieves significantly better results than state-of-the-art methods.

2   Related Works

2.1   Image-to-Image Translation

Image-to-image (I2I) translation aims to learn a mapping function from the target domain to the source domain. Pix2pix [12] uses a conditional GAN-based network that requires a large amount of paired data for training. To alleviate the problem of obtaining paired data, the CycleGAN [33] introduces cycle consistency constraints, which allow I2I methods to train cross-domain translations without paired data. FUNIT [16] proposes a few-shot unsupervised image generation method to accomplish the I2I translation task by encoding content images and style images separately and combining them with Adaptive Instance Normalization (AdaIN) [10]. Intuitively, font generation is a typical I2I translation task that maps a source font to a target font while preserving the original character structure. Therefore, many font generation methods are based on I2I translation methods.

2.2   Automatic Font Generation

We categorize automatic font generation methods into two classes: many-shot and few-shot font generation methods. Many-shot font generation methods [13, 14, 24, 25] aim to learn the mapping function between source fonts and target fonts. Although these methods are effective, they are not practical because these methods often first train a translation model and fine-tune the translation model with many reference glyphs, e.g., 775 for [13, 14].

Based on different kinds of feature representation, few-shot font generation methods can be divided into two main categories: global feature representation [1, 4, 27, 31] and component-based feature representation [3, 9, 11, 19, 20, 28, 32]. The global feature representation methods, such as EMD [31] and AGIS-Net [4], synthesize a new glyph by combining a style vector and a content vector together, but they show worse synthesizing quality for unseen style fonts. Since the style of glyphs is highly complex and fine-grained, it is very difficult to generate the font utilizing global feature statistics. Instead, works related to component-based feature representation focus on designing a feature representation that is associated with glyphs' components or localized features. LF-Font [19] designs a component-based style encoder that extracts component-wise features from reference images. MX-Font [20] designs multiple localized encoders and utilizes component labels as weak supervision to guide each encoder to obtain different local style patterns. DFS [32] proposes the Deep Feature Similarity architecture to calculate the feature similarity between the input content images and style images to generate the target images.

In addition, some efforts [9,11,15,21,23] attempt to use the attention mechanism [26] for the font generation task. RD-GAN [11] utilizes the attention mechanism to extract rough radicals from content images. FTransGAN [15] captures the local and global style features based on self-attention mechanism [29]. Our Multi-scale External Attention Module (MEAM), motivated by external attention mechanism [7], extracts essential style features at different scales for cross-language font generation.

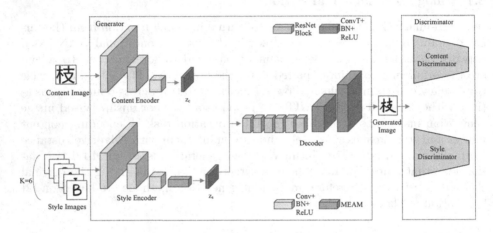

Fig. 1. Architecture overview of the CLF-Net. z_c/z_s denotes the content/style latent feature. Conv denotes a convolutional layer. BN denotes BatchNorm. MEAM denotes the Multi-scale External Attention Module. ConvT denotes a transposed convolutional layer.

3   Method Description

This section describes our method for few-shot cross-language font generation, named CLF-Net. Given a content image and several stylized images, our model aims to generate the character of the content image with the font of the style images. The general structure of CLF-Net is shown in Fig. 1. Like other few-shot font generation methods, CLF-Net adopts the framework of GAN, including a Generator G and two discriminators: content discriminator D_c and style discriminator D_s. Moreover, to make the model show enough generalization ability to learn both local and global essential style features in another language, we propose a Multi-scale External Attention Module (MEAM). More details are given in Sect. 3.2.

3.1 Network Overview

We regard the few-shot font generation task as solving the conditional probability $p_{gt}(x|I_c, I_s)$, where I_c is a content image in the standard style (e.g., Microsoft YaHei), I_s is a few style images having the same style but different contents, and x denotes the target image with the same character as I_c and with the similar style as I_s. Considering that our task is cross-language font generation, I_c and I_s should be from different languages. Therefore, we choose a Chinese character as the content image and a few English letters as the style images to train our CLF-Net. The generator G consists of two encoders and a decoder. The content encoder e_c is used to capture the structural features of the character content. The style encoder e_s is used to learn the style features of the given stylized font. Two encoders extract the style latent feature and content latent feature, respectively. Then the decoder d will take the extracted information and generate the target image $\hat{x}$. The generation process can be formulated as:

$$z_c = e_c(I_c), z_s = e_s(I_s), \tag{1}$$

$$\hat{x} = G(I_c, I_s) = d(z_c, z_s), \tag{2}$$

where z_c and z_s represent the content latent feature and style latent feature.

The content encoder consists of three convolutional blocks, each of which includes a convolutional layer followed by BatchNorm and ReLU. The kernel sizes of the convolutional layers are 7, 3, and 3, respectively.

The style encoder has the same structure as the content encoder, including three convolutional blocks. Moreover, inspired by FTransGAN [15] and external attention [7], we design a Multi-scale External Attention Module (MEAM) after the above layers to capture essential style features at different scales. More details are given in Sect. 3.2.

The decoder takes the content feature z_c and style feature z_s as input and outputs the generated image $\hat{x}$. The decoder consists of six ResNet blocks [8] and two transposed convolutional layers that upsample the spatial dimensions of the feature maps. Each transposed convolutional layer is followed by BatchNorm and ReLU.

The discriminators include a content discriminator and a style discriminator, which are used to check the matching degree from the style and content perspective separately. Following the design of PatchGAN [12], two patch discriminators utilize image patches to check the features of the real images and the fake images both locally and globally.

3.2 Multi-scale External Attention Module

Since self-attention mechanism [29] is applicable to the GAN [5] framework, both generators and discriminators are able to model relationships between spatial regions that are widely separated. However, self-attention only considers the relationships between elements within a data sample and ignores the potential relationships between elements in different references, which may limit the ability and flexibility of self-attention. It is not difficult to see that incorporating correlations between different style reference images belonging to the same font helps to contribute to a better feature representation for cross-language font generation.

External attention [7] has linear complexity and implicitly considers the correlations between all references. As shown in Fig. 2a, external attention calculates an attention map between the input pixels and an external memory unit $M \in \mathbb{R}^{S \times d}$ by:

$$A = (\alpha)_{i,j} = \text{Norm}(FM^T), \tag{3}$$

$$F_{out} = AM, \tag{4}$$

and $\alpha_{i,j}$ in Eq. (3) is the similarity between the i-th pixel and the j-th row of M, where M is an input-independent learnable parameter that is a memory of the whole training dataset. A is the attention map inferred from the learned dataset-level prior knowledge.

External attention separately normalizes columns and rows using the double-normalization method proposed in [6]. The formula for this double-normalization is:

$$(\tilde{\alpha})_{i,j} = FM_k^T, \tag{5}$$

$$\hat{\alpha}_{i,j} = \exp(\tilde{\alpha}_{i,j}) / \sum_k \exp(\tilde{\alpha}_{k,j}), \tag{6}$$

$$\alpha_{i,j} = \hat{\alpha}_{i,j} / \sum_k \hat{\alpha}_{i,k}. \tag{7}$$

Finally, it updates the input features of M according to the similarities in A. In practice, it uses two different memory units, M_k and M_v, as the key and value to improve the capability of the network. This slightly alters the computation of external attention to

$$A = \text{Norm}(FM_k^T), \tag{8}$$

$$F_{out} = AM_v. \tag{9}$$

As mentioned above, the style of glyphs is complex and delicate. When designing the fonts, experts need to consider multiple levels of styles, such as component-level, radical-level, stroke-level, and even edge-level. Therefore, to improve the attention modules in FTransGAN [15], we design a Multi-scale External Attention Module (MEAM) to capture style features at different scales.

In particular, our method can model relationships between all style reference images from another language with the presence of the MEAM. With the MEAM, we can obtain high-quality essential style features at different scales. Specifically, when the style reference images go into the style encoder, whose architecture is shown in Fig. 2b, they will first go through three convolution blocks. Afterward, we feed the feature map outputted by the last convolutional block in the above layers into the MEAM. The MEAM first further extracts two

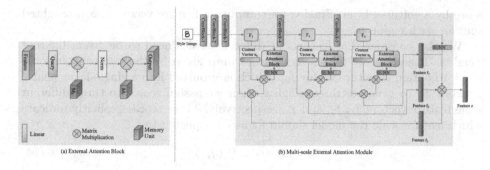

(a) External Attention Block

(b) Multi-scale External Attention Module

Fig. 2. The architecture of External Attention Block and the style encoder with Multi-scale External Attention Module. NN denotes a single-layer neural network. ConvBlock-1, ConvBlock-2, ConvBlock-3, ConvBlock-4, and ConvBlock-5 denote convolutional blocks, each of which includes a convolutional layer followed by BatchNorm and ReLU. F_3, F_4, and F_5 denote the feature maps with receptive fields of 13×13, 21×21, and 37×37, respectively.

feature maps separately through two consecutive convolutional blocks, each of which has a convolutional layer with kernel sizes of 3, and each convolutional layer is followed by BatchNorm and ReLU. Then the MEAM uses three juxtaposed External Attention Blocks to process the above three feature maps with receptive fields of 13×13, 21×21, and 37×37, respectively. Thus, the feature maps with different receptive fields contain the multi-scale features. The context information is obtained and incorporated into the feature map through an External Attention Block, which is computed as:

$$h_r = EA(v_r), \tag{10}$$

where EA denotes the External Attention Block, $\{v_r\}_{r=1}^{H \times W}$ denotes each region of the feature map and the new feature vector h_r contains not only the information limited to their receptive field but also the context information from other regions of other reference images.

Then, considering that not all regions contribute equally, we assign scores to each region. Specifically,

$$u_r = S_1(h_r), \tag{11}$$

$$a_r = \text{softmax}(u_r^T u_c), \tag{12}$$

$$f = \sum_{r=1}^{H \times W} a_r v_r. \tag{13}$$

That is, we input the feature vector h_r into a single-layer neural network S_1 and get u_r as the latent representation of h_r. Next, the importance of the current region is measured using the context vector u_c, which is randomly initialized and co-trained with the whole model. After that, we can obtain the normalized score by a softmax layer. Finally, we compute a feature vector f as a weighted sum for each region v_r.

We also consider that features at different scales need to be given different weights. Therefore, we flatten the feature map given by the last convolutional block to obtain a feature vector f_m, which is inputted into a single-layer neural network S_2 to generate three weights, then we assign scores to three different scale feature vectors f_1, f_2, and f_3, respectively. These scores explicitly indicate which feature scale the model should focus on. Specifically,

$$w_1, w_2, w_3 = S_2(f_m), \tag{14}$$

$$z = \sum_{i=1}^{3} w_i f_i, \tag{15}$$

where w_1, w_2, and w_3 are the three normalized scores given by the neural network and z is the weighted sum of three feature vectors. Note that each time the style encoder will accept K images. Thus, the final latent feature z_s is the average of all vectors:

$$z_s = \frac{1}{K} \sum_{K} z^k. \tag{16}$$

Besides, we copy the style latent feature z_s seven times to match the size of the content latent feature z_c.

3.3   Loss Function

To achieve few-shot cross-language font generation, our CLF-Net employs three kinds of losses: 1) Pixel-level loss to measure the pixel-wise mismatch between generated images and the ground-truth images. 2) Edge Loss to make the model pay more attention to the edge pixels of characters and make the edges of generated images sharper. 3) Adversarial loss to solve the minimax game in the GAN framework.

Pixel-Level Loss: To learn pixel-level consistency, we use L1 loss between generated images and the ground truth images:

$$\mathcal{L}_{L1} = \mathbb{E}_{x, \hat{x} \in P_{(x, \hat{x})}}[\| x - \hat{x} \|_1]. \tag{17}$$

Edge Loss: Pixel-level loss is widely used in existing font generation models. They all estimate the consistency of the distribution of the two domains based on the per-pixel difference between the generated and real characters. However, in the font generation task, the weights of pixels in the images of Chinese characters are different. Different from pixels used as background or fill, boundary pixels play a key role in the overall style of Chinese characters. Therefore, our model needs to pay more attention to the edges of each Chinese character. To preserve more edge information of Chinese characters, we define an Edge Loss to limit our model to generate results with sharper edges inspired by [21]. We utilize Canny algorithm [2] to extract the edges of generated images and the target images and utilize L1 loss function to measure the pixel distance between the two edges:

$$\mathcal{L}_{edge} = \mathbb{E}_{x,\hat{x} \in P_{(x,\hat{x})}}[\|\ Canny(x) - Canny(\hat{x})\ \|_1].$$ (18)

Adversarial Loss: Our proposed method uses a framework based on GAN. The optimization of GAN is essentially a game problem, and its goal is to allow generator G to generate examples that are indistinguishable from the real data to deceive the discriminator D. In CLF-Net, the generator G has to extract the information from the style images I_s and the content image I_c, and generate an image with the same content as I_c and the similar style as I_s, and then the discriminators D_c and D_s are used to determine whether the generated image has no difference with the reference images in terms of content and style. We use hinge loss [18] function to compute the adversarial loss as:

$$\mathcal{L}_{adv} = \mathcal{L}_{advc} + \mathcal{L}_{advs},$$ (19)

$$\mathcal{L}_{advc} = \max_{D_c} \min_{G} \mathbb{E}_{I_c \in P_c, I_s \in P_s}[\log D_c(I_c) + \log(1 - D_c(\hat{x}))],$$ (20)

$$\mathcal{L}_{advs} = \max_{D_s} \min_{G} \mathbb{E}_{I_c \in P_c, I_s \in P_s}[\log D_s(I_s) + \log(1 - D_s(\hat{x}))],$$ (21)

where $D_c(\cdot)$ and $D_s(\cdot)$ represent the output from the content discriminator and style discriminator respectively.

Combining all losses mentioned above, we train the whole model by the following objective:

$$\mathcal{L} = \lambda_{L1}\mathcal{L}_{L1} + \lambda_{edge}\mathcal{L}_{edge} + \lambda_{adv}\mathcal{L}_{adv},$$ (22)

where λ_{L1}, λ_{edge}, and λ_{adv} are the weights for controlling these terms.

4 Experiments

4.1 Datasets

For a fair comparison, our experiments use the public dataset of FTransGAN [15], which contains 847 grayscale fonts (stylized inputs), each font with about 1000 commonly used Chinese characters and 52 English letters of the same style. The test set consists of two parts: images with known contents but unknown styles and images with known styles but unknown contents. They randomly select 29 characters and fonts as unknown contents and styles and leave the rest as training data.

4.2 Training Details

We trained CLF-Net on Nvidia RTX 3090 with the following parameters on the above dataset. For experiments, we use Chinese characters as the content input and English letters as the style input. We set $\lambda_{L1} = 100$, $\lambda_{edge} = 10$, $\lambda_{adv} = 1$, and $K = 6$.

4.3 Competitors

To comprehensively evaluate the model, we chose the following three models, EMD [31], DFS [32], and FTransGAN [15] as our competitors. As mentioned above, previous works usually focus on font generation for a specific language, and there are few works on cross-language font generation. Therefore, in addition to FTransGAN being specifically designed for the cross-language font generation task, EMD and DFS are both designed for monolingual font generation, and we make them suitable for the cross-language task according to the modifications made by the authors of FTransGAN.

Table 1. Quantitative evaluation on the test set. The bold numbers indicate the best, and the underlined numbers represent the second best.

	Content-aware		Style-aware		Pixel-level	
	Accuracy↑	mFID↓	Accuracy↑	mFID↑	MAE↓	SSIM↑
Evaluation on the unseen character images						
EMD	81.2	116.9	24.4	597.1	**0.117**	0.497
DFS	89.2	150.0	2.7	820.6	0.185	0.303
FTransGAN	97.0	<u>49.8</u>	<u>58.1</u>	<u>308.9</u>	<u>0.121</u>	<u>0.501</u>
Ours	**97.5**	**45.8**	**61.6**	**294.1**	<u>0.121</u>	**0.503**
Evaluation on the unseen style images						
EMD	85.5	184.4	4.4	623.2	**0.166**	**0.384**
DFS	91.7	230.7	0.7	662.4	0.214	0.231
FTransGAN	<u>99.8</u>	<u>97.8</u>	<u>11.7</u>	**418.8**	<u>0.179</u>	0.368
Ours	**99.9**	**96.9**	**11.9**	<u>427.5</u>	0.180	<u>0.369</u>

4.4 Quantitative Evaluation

Quantitative evaluation of generative models is inherently difficult because there are no generalized rules for comparing ground truths and generated images. Recently, several evaluation metrics [30] based on different assumptions have been proposed to measure the performance of generative models, but they remain controversial. In this paper, we evaluate the models using various similarity metrics from pixel-level to perceptual-level. As shown in Table 1, our model outperforms existing methods in most metrics.

Pixel-Level Evaluation. A simple way to quantitatively evaluate the model is to calculate the distance between generated images and the ground truths. The pixel-wise assessment is to compare the pixels that are at the same position in the ground truths and generated images. Here, we use the following two metrics: mean absolute error (MAE) and structural similarity (SSIM).

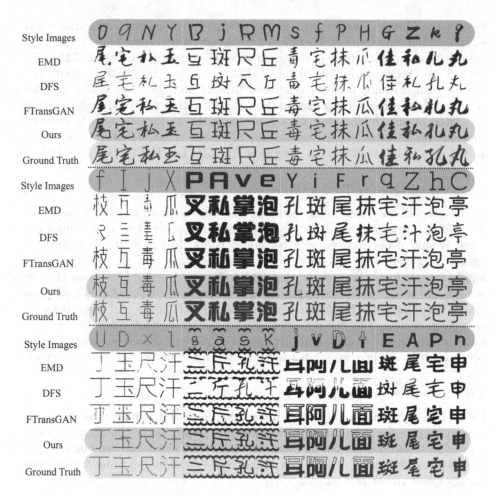

Fig. 3. Visual comparison of our proposed model and its competitor.

Table 2. Effect of different components in our method. The bold numbers indicate the best, and the underlined numbers represent the second best.

	Content-aware		Style-aware		Pixel-level	
	Accuracy ↑	mFID ↓	Accuracy ↑	mFID ↑	MAE ↓	SSIM ↑
Evaluation on the unseen character images						
FM-$\mathcal{L}_{edge}$	_97.4_	_48.4_	_58.3_	_311.7_	**0.121**	_0.502_
FM-$\mathcal{L}_{edge}$-MEAM	97.1	50.7	46.5	361.3	_0.127_	0.482
FM	**97.5**	**45.8**	**61.6**	**294.1**	**0.121**	**0.503**
Evaluation on the unseen style images						
FM-$\mathcal{L}_{edge}$	**99.9**	_98.9_	_10.9_	428.5	**0.180**	_0.367_
FM-$\mathcal{L}_{edge}$-MEAM	_99.7_	106.9	_10.9_	**417.2**	_0.181_	0.360
FM	**99.9**	96.9	11.9	_427.5_	**0.180**	**0.369**

Perceptual-Level Evaluation. However, pixel-level evaluation metrics often go against human intuition. Therefore, we also adopt perceptual-level evaluation metrics to comprehensively evaluate all models. Drawing on FTransGAN [15], we use the Fréchet Inception Distance (FID) proposed in [22] to compute the feature map distance between generated images and the ground truths. This metric evaluates the performance of the network rather than simply comparing generated results. In this way, we can evaluate the performance of the content encoder and the style encoder separately. The score is calculated from the top-1 accuracy and the mean Fréchet Inception Distance (mFID) proposed by [17].

4.5   Visual Quality Evaluation

In this section, we qualitatively compare our method with the above methods. The results are shown in Fig. 3. We have randomly selected some outputs from three groups of our model and other competitors. In Fig. 3, the first group is handwriting fonts, the second group is printing fonts, and the third group is highly artistic fonts. We can see that EMD [31] erases some fonts with thinner strokes and works worse on highly artistic fonts. DFS [32] performs poorly on most fonts. FTransGAN [15] ignores fine-grained local styles and is not detailed enough in dealing with the style patterns of the stroke ends on highly artistic fonts, which causes artifacts and black spots in generated images. Our approach generates high-quality images of various fonts and achieves satisfactory results.

4.6   Ablation Study

Edge Loss. As shown in Table 2, after stripping out the Edge Loss from full model(FM), we find that Edge Loss significantly improves the classification accuracy of style and content labels. From the mFID [17] scores, we can observe that the feature distribution of the images generated by the model trained with Edge Loss is closer to the real images.

Multi-scale External Attention Module. Continue taking out the Multi-scale External Attention Module (MEAM), according to Table 2, both pixel-level and perceptual-level metrics drop rapidly.

5   Conclusion

In this paper, we propose an effective few-shot cross-language font generation method called CLF-Net by learning the inter-image relationships between all style reference images. In CLF-Net, we design a Multi-scale External Attention Module for extracting essential style features at different scales in another language and introduce an Edge Loss function that produces results with less blur and sharper edges. Experimental results show that our proposed CLF-Net is highly capable of cross-language font generation and achieves superior performance compared to state-of-the-art methods. In the future, we plan to extend the model to the task of font generation across multiple languages.

Acknowledgement. This work was supported by China Yunnan province major science and technology special plan project 202202AF080004. The numerical calculations have been done on the Supercomputing Center of Wuhan University.

References

1. Azadi, S., Fisher, M., Kim, V.G., Wang, Z., Shechtman, E., Darrell, T.: Multi-content GAN for few-shot font style transfer. In: Proceedings of the IEEE Conference on Computer Vision and Pattern Recognition, pp. 7564–7573 (2018)
2. Canny, J.: A computational approach to edge detection. IEEE Trans. Pattern Anal. Mach. Intell. **6**, 679–698 (1986)
3. Cha, J., Chun, S., Lee, G., Lee, B., Kim, S., Lee, H.: Few-shot compositional font generation with dual memory. In: Vedaldi, A., Bischof, H., Brox, T., Frahm, J.-M. (eds.) ECCV 2020. LNCS, vol. 12364, pp. 735–751. Springer, Cham (2020). https://doi.org/10.1007/978-3-030-58529-7_43
4. Gao, Y., Guo, Y., Lian, Z., Tang, Y., Xiao, J.: Artistic glyph image synthesis via one-stage few-shot learning. ACM Trans. Graph. (TOG) **38**(6), 1–12 (2019)
5. Goodfellow, I., et al.: Generative adversarial nets. In: Advances in Neural Information Processing Systems, vol. 27 (2014)
6. Guo, M.H., Cai, J.X., Liu, Z.N., Mu, T.J., Martin, R.R., Hu, S.M.: PCT: point cloud transformer. Comput. Vis. Media **7**, 187–199 (2021)
7. Guo, M.H., Liu, Z.N., Mu, T.J., Hu, S.M.: Beyond self-attention: external attention using two linear layers for visual tasks. IEEE Trans. Pattern Anal. Mach. Intell. **45**(5), 5436–5447 (2022)
8. He, K., Zhang, X., Ren, S., Sun, J.: Deep residual learning for image recognition. In: Proceedings of the IEEE Conference on Computer Vision and Pattern Recognition, pp. 770–778 (2016)
9. He, X., Zhu, M., Wang, N., Gao, X., Yang, H.: Few-shot font generation by learning style difference and similarity (2023). https://doi.org/10.48550/arXiv.2301.10008
10. Huang, X., Belongie, S.: Arbitrary style transfer in real-time with adaptive instance normalization. In: Proceedings of the IEEE International Conference on Computer Vision, pp. 1501–1510 (2017)
11. Huang, Y., He, M., Jin, L., Wang, Y.: RD-GAN: few/zero-shot Chinese character style transfer via radical decomposition and rendering. In: Vedaldi, A., Bischof, H., Brox, T., Frahm, J.-M. (eds.) ECCV 2020. LNCS, vol. 12351, pp. 156–172. Springer, Cham (2020). https://doi.org/10.1007/978-3-030-58539-6_10
12. Isola, P., Zhu, J.Y., Zhou, T., Efros, A.A.: Image-to-image translation with conditional adversarial networks. In: Proceedings of the IEEE Conference on Computer Vision and Pattern Recognition, pp. 1125–1134 (2017)
13. Jiang, Y., Lian, Z., Tang, Y., Xiao, J.: DCfont: an end-to-end deep Chinese font generation system. In: SIGGRAPH Asia 2017 Technical Briefs, pp. 1–4 (2017)
14. Jiang, Y., Lian, Z., Tang, Y., Xiao, J.: SCfont: structure-guided Chinese font generation via deep stacked networks. In: Proceedings of the AAAI Conference on Artificial Intelligence, vol. 33, pp. 4015–4022 (2019)
15. Li, C., Taniguchi, Y., Lu, M., Konomi, S., Nagahara, H.: Cross-language font style transfer. Appl. Intell. 1–15 (2023)
16. Liu, M.Y., et al.: Few-shot unsupervised image-to-image translation. In: Proceedings of the IEEE/CVF International Conference on Computer Vision, pp. 10551–10560 (2019)

17. Liu, M.Y., Tuzel, O.: Coupled generative adversarial networks. In: Advances in Neural Information Processing Systems, vol. 29 (2016)
18. Miyato, T., Kataoka, T., Koyama, M., Yoshida, Y.: Spectral normalization for generative adversarial networks. arXiv preprint arXiv:1802.05957 (2018)
19. Park, S., Chun, S., Cha, J., Lee, B., Shim, H.: Few-shot font generation with localized style representations and factorization. In: Proceedings of the AAAI Conference on Artificial Intelligence, vol. 35, pp. 2393–2402 (2021)
20. Park, S., Chun, S., Cha, J., Lee, B., Shim, H.: Multiple heads are better than one: few-shot font generation with multiple localized experts. In: Proceedings of the IEEE/CVF International Conference on Computer Vision, pp. 13900–13909 (2021)
21. Ren, C., Lyu, S., Zhan, H., Lu, Y.: SAfont: automatic font synthesis using self-attention mechanisms. Aust. J. Intell. Inf. Process. Syst. **16**(2), 19–25 (2019)
22. Salimans, T., Goodfellow, I., Zaremba, W., Cheung, V., Radford, A., Chen, X.: Improved techniques for training GANs. In: Advances in Neural Information Processing Systems, vol. 29 (2016)
23. Tang, L., et al.: Few-shot font generation by learning fine-grained local styles. In: Proceedings of the IEEE/CVF Conference on Computer Vision and Pattern Recognition, pp. 7895–7904 (2022)
24. Tian, Y.: Rewrite: neural style transfer for Chinese fonts (2016). https://github.com/kaonashi-tyc/Rewrite
25. Tian, Y.: zi2zi: Master Chinese calligraphy with conditional adversarial networks (2017). https://github.com/kaonashi-tyc/zi2zi
26. Vaswani, A., et al.: Attention is all you need. In: Advances in Neural Information Processing Systems, vol. 30 (2017)
27. Wen, Q., Li, S., Han, B., Yuan, Y.: ZiGAN: fine-grained Chinese calligraphy font generation via a few-shot style transfer approach. In: Proceedings of the 29th ACM International Conference on Multimedia, pp. 621–629 (2021)
28. Xie, Y., Chen, X., Sun, L., Lu, Y.: DG-font: deformable generative networks for unsupervised font generation. In: Proceedings of the IEEE/CVF Conference on Computer Vision and Pattern Recognition, pp. 5130–5140 (2021)
29. Zhang, H., Goodfellow, I., Metaxas, D., Odena, A.: Self-attention generative adversarial networks. In: International Conference on Machine Learning, pp. 7354–7363. PMLR (2019)
30. Zhang, R., Isola, P., Efros, A.A., Shechtman, E., Wang, O.: The unreasonable effectiveness of deep features as a perceptual metric. In: Proceedings of the IEEE Conference on Computer Vision and Pattern Recognition, pp. 586–595 (2018)
31. Zhang, Y., Zhang, Y., Cai, W.: Separating style and content for generalized style transfer. In: Proceedings of the IEEE Conference on Computer Vision and Pattern Recognition, pp. 8447–8455 (2018)
32. Zhu, A., Lu, X., Bai, X., Uchida, S., Iwana, B.K., Xiong, S.: Few-shot text style transfer via deep feature similarity. IEEE Trans. Image Process. **29**, 6932–6946 (2020)
33. Zhu, J.Y., Park, T., Isola, P., Efros, A.A.: Unpaired image-to-image translation using cycle-consistent adversarial networks. In: Proceedings of the IEEE International Conference on Computer Vision, pp. 2223–2232 (2017)

Multi-dimensional Fusion and Consistency for Semi-supervised Medical Image Segmentation

Yixing Lu[1], Zhaoxin Fan[2], and Min Xu[2(✉)]

[1] University of Liverpool, Liverpool, UK
[2] Mohamed bin Zayed University of Artificial Intelligence,
Abu Dhabi, United Arab Emirates
xumin100@gmail.com

Abstract. In this paper, we introduce a novel semi-supervised learning framework tailored for medical image segmentation. Central to our approach is the innovative Multi-scale Text-aware ViT-CNN Fusion scheme. This scheme adeptly combines the strengths of both ViTs and CNNs, capitalizing on the unique advantages of both architectures as well as the complementary information in vision-language modalities. Further enriching our framework, we propose the Multi-Axis Consistency framework for generating robust pseudo labels, thereby enhancing the semi-supervised learning process. Our extensive experiments on several widely-used datasets unequivocally demonstrate the efficacy of our approach.

Keywords: Medical image segmentation · Semi-supervise learning · ViT-CNN fusion · Multi-axis consistency

1 Introduction

Medical image segmentation is a pivotal and intricate process within the realm of intelligent diagnosis, entailing the extraction of regions of interest within medical imagery. This task is of paramount importance for enabling precise diagnosis and tailored treatment. Over recent years, Convolutional Neural Networks (CNNs) [4, 12,15] and Vision Transformers (ViTs) [5,11,32], both endowed with a U-shaped architecture, have witnessed significant advances in the domain of medical image segmentation.

Medical image segmentation literature mainly employs pretrained Convolutional Neural Networks (CNNs) [15] or Transformers [5]. The benefits of using both CNNs and Vision Transformers (ViTs) haven't been thoroughly explored. Interestingly, CNNs and ViTs seem to complement each other for medical image understanding. CNNs excel in local feature recognition [3], while ViTs are superior in comprehending long-range dependencies [7,14]. For medical image segmentation, combining these strengths is crucial to understanding the organ

---

Z. Fan—Equal Contribution.
We thank Bowen Wei for helpful discussions on this work.

S. Rudinac et al. (Eds.): MMM 2024, LNCS 14555, pp. 141–155, 2024.
https://doi.org/10.1007/978-3-031-53308-2_11

and its interrelations with others. This raises the question: *Can we fuse the strengths of CNNs and ViTs into a single framework for medical image segmentation?* An additional noteworthy observation is that both Convolutional Neural Networks (CNNs) and Vision Transformers (ViTs) tend to necessitate extensive quantities of annotated data for effective training. While this may be manageable in the realm of natural image segmentation, it poses a formidable hurdle in the context of medical image segmentation, where the process of annotation is both laborious and costly. As a result, a second question emerges: *Is it feasible to reduce the model's dependency on annotated medical image-mask pairs without undermining the performance of the ViT/CNN?*

In response to the aforementioned questions, we introduce a novel semi-supervised learning framework for medical image segmentation in this study. We first propose a simple yet efficacious Multi-scale ViT-CNN Fusion scheme. The underlying principle of this approach is that the integration of ViT and CNN features can equip the model with the ability to capture both intricate local details and extensive global long-range dependency information. Additionally, given that both the CNN and ViT are pretrained on large-scale networks, they can retain both abstract natural features and domain-specific medical features during fusion, thereby further enhancing the segmentation task. Moreover, inspired by vision-language models [24,30,36,39], and considering the relative ease of obtaining text descriptions for medical images, we introduce a text-aware language enhancement mode to further enrich the learned features. This effectively addresses the first question. Subsequently, we incorporate a Multi-Axis Consistency framework in our study to extend our approach to scenarios where annotated labels are limited. Within this framework, we unify and formulate multiple consistency regularizations under a single framework, dubbed Multi-AXIs COnsistency (MaxiCo). This framework combines intra-model, inter-model, and temporal consistency regularizations to generate robust probability-aware pseudo-labels, thereby enabling the use of a large corpus of unlabeled data for semi-supervised network training. Furthermore, we design a voting mechanism within this module, whereby each intermediate output can contribute to the final pseudo-label. This mechanism further enhances the trustworthiness of pseudo-labels and bolsters the final model's performance, therefore providing a satisfactory answer to the second question.

To deliver a comprehensive demonstration of the efficacy of our proposed method, we have undertaken extensive experimentation using the MoNuSeg [18] dataset and QaTa-COV19 [9] dataset. The empirical results obtained from these experiments substantiate that our method establishes a new benchmark in fully-supervised settings, outperforming existing state-of-the-art methodologies. Moreover, within semi-supervised scenarios, our strategy shows remarkable superiority over other leading-edge techniques.

Our contribution can be summarized as: 1) We pioneer a semi-supervised framework that harnesses the power of textual information to support fused ViT-CNN networks for medical image segmentation, representing a unique approach to this problem. 2) We propose a novel Multi-scale Text-aware ViT-CNN Fusion methodology that adroitly amalgamates CNNs and ViTs to boost segmentation accuracy. 3) We introduce a novel Multi-Axis Consistency Learning module that capitalizes on consistency regularizations to generate reliable pseudo-labels for semi-supervised learning, effectively addressing the issue of data scarcity.

2   Related Work

Transformers in Medical Image Segmentation. The success of Vision Transformer (ViT) [10] in various computer vision tasks has led to its integration into medical image segmentation [11,22,38,41,42]. Some studies use transformers for image representation [14], while others propose hybrid encoders combining transformers and convolutional neural networks (CNNs) [7]. Cao et al. [5] proposed a pure transformer network replacing convolutional layers with Swin Transformer blocks [23] and the traditional Unet skip-connection with a transformer-based channel-attention module [34]. However, these methods are typically trained in a fully-supervised manner, which may be impractical due to the scarcity of annotated medical data. To address this, we introduce a semi-supervised approach that fuses ViT-CNN networks for medical image segmentation, aiming to overcome the challenge of limited annotated data.

Semi-supervised Medical Image Segmentation. In light of the challenge posed by the dearth of annotated data in medical image segmentation, semi-supervised learning has come to the fore as a promising solution [33,35,37,40]. Predominant strategies for semi-supervised learning encompass pseudo labeling [8], deep co-training [43], and entropy minimization [13]. In our work, we adopt a consistency learning framework to generate pseudo labels for unmarked images. There are several versions of consistency regularization methods, including temporal consistency [19], model-level consistency [25], and pyramid consistency [26], in existing literature. However, most of these methods only depend on a single type of regularization, which will limit the power of the model. In contrast, our approach amalgamates multiple consistency regularizations and employs a voting mechanism to produce more robust pseudo labels, which demonstrates better performance.

Vision-Language Fusion for Dense Predictions. In recent years, the fusion of vision and language in large-scale pretraining has garnered significant attention. The CLIP model [29] showcases this, demonstrating impressive transfer learning across multiple datasets. Building on this, researchers have explored fine-tuning the CLIP model for dense prediction tasks [20,30,36], framing it as a pixel-text matching problem. Vision-language models also enable zero-shot inference, bridging the gap between seen and unseen categories [39]. Some research has further explored the potential of visual prompts and their interpolation with text prompts during training [24]. In this paper, we use textual information to enhance ViT-CNN training for medical image segmentation, showcasing a novel application of vision-language fine-tuning.

3 Methodology

3.1 Overview

Given an image $I \in X$, where X represents the space of all possible medical images, and a text input $T \in T$, where T is the space of all possible text inputs (e.g., medical notes or labels), the task of medical image segmentation in our study is to learn a mapping function $F_\theta : X \times T \to Y$:

$$F_\theta : X \times T \to Y \tag{1}$$

where Y represents the segmentation masks corresponding to the input medical image, and θ denotes the parameters of our model. The goal is to train the model such that the mapping function F_θ can accurately predict the segmentation mask $y \in Y$ for any given input image and text (I, T).

To address the first challenge, our Multi-scale Text-aware ViT-CNN Fusion scheme integrates a pretrained ViT and CNN, incorporating text features for increased prediction accuracy. We perform vision-language pretraining to obtain vision and text features, aligning them to formulate ViT features. These are then fused with CNN features at various resolutions, enabling the efficient use of local and global features.

To facilitate semi-supervised training, we introduce a Multi-Axis Consistency framework to generate pseudo labels, leveraging inter-model, multi-scale intra-model, and temporal consistency. Our network makes multiple predictions in a single pass, generating probabilistic pseudo labels via a voting mechanism, supporting semi-supervised training.

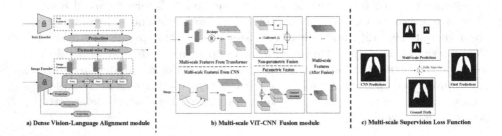

Fig. 1. The illustration of Multi-scale Text-aware ViT-CNN Fusion.

3.2 Multi-scale Text-Aware ViT-CNN Fusion

In this section, we present a novel architectural design named Multi-scale Text-aware ViT-CNN Fusion, as depicted in Fig. 1. This scheme is primarily composed of three major components: *Dense Vision-Language Alignment module*, *Multi-scale ViT-CNN Fusion module*, and a *Supervised Loss function* for joint training. The Dense Vision-Language Alignment model is responsible for aligning the vision and text features into a common embedding space. By performing this alignment, we can effectively exploit the complementary information from both modalities to enhance the feature representations. The second component, Multi-scale ViT-CNN Fusion module, facilitates the fusion of the features extracted by the ViT and the CNN. This fusion is carried out at multiple scales, allowing the model to capture abstract features, domain-specific features, local details, and global long-range dependencies at different resolutions. Finally, the Multi-scale Supervision Loss Function. By optimizing this loss, the network learns to predict segmentation masks in a multi-scale and progressive manner. Next, we introduce them in detail.

Dense Vision-Language Alignment Module. In our work, we incorporate both image and text information as inputs for segmentation. The inclusion of text allows us to capture the strengths of transformers more effectively and bolsters the fusion of Vision Transformers (ViTs) and Convolutional Neural Networks (CNNs). This approach leverages contextual cues from text to enhance segmentation precision. To align the image and text features, we adopt a progressive approach. Visual features are extracted from a sequence of layers $(3, 6, 9, 12)$ of a pretrained visual encoder, forming a set $L = x_1, x_2, ..., x_{\ell-1}, x_\ell$ where x_1 and x_ℓ represent the features from the shallowest and deepest layers, respectively. We obtain text embeddings, denoted by y, from a pretrained clinical text encoder.

We use transformer layers with skip connections to compactly represent visual features, as shown in the equation below:

$$X_i = \begin{cases} \mathsf{TransLayer}_i(W_i^x x_i), & i = \ell \\ \mathsf{TransLayer}_i(W_i^x x_i + X_{i+1}), & i \in L - \{\ell\} \end{cases} \quad (2)$$

Here, W_i^x is a linear layer for dimension reduction, and $\mathsf{TransLayer}_i$ denotes a Transformer layer. We reduce the dimension of the text embeddings and transfer them to different layers using simple MLP blocks:

$$Y_i = \begin{cases} W_i^y y, & i = \ell \\ W_i^y Y_{i+1}, & i \in L - \{\ell\} \end{cases} \quad (3)$$

Here, W_i^y signifies a linear layer for dimension reduction, while W_i^y denotes an MLP block for transferring text embeddings. We perform Vision-Text alignment using element-wise multiplication so that the alignment is "dense":

$$Z_i = W_i^X X_i \odot W_i^Y Y_i, \quad i \in L \quad (4)$$

In the equation above, $\odot$ represents element-wise multiplication, and W_i^X and W_i^Y are tensor reshape operations for X_i and Y_i, respectively.

Multi-scale ViT-CNN Fusion Module. ViTs and CNNs each have their unique strengths in image analysis tasks. ViTs excel in capturing global dependencies, while CNNs are particularly adept at extracting local features. However, when dealing with complex tasks such as medical image segmentation, a combination of these two can be beneficial, leveraging global contextual understanding and local feature extraction. Addressing this, we propose a dense fusion of ViT and CNN features at different resolutions. This approach is designed to enhance local interactions and preserve global knowledge. Our fusion method follows two guiding principles: 1) it should improve model performance, and 2) it should maintain the robustness of each individual feature to avoid over-dependence on either.

We begin with a non-parametric fusion method, where the fusion parameter β is uniformly sampled from $[0, 1]$. A Unet CNN processes the input $I \in \mathbb{R}^{H \times W \times 3}$, projecting it initially to C_1 and then applying $(N-1)$ down/up-sampling operations to yield multi-scale features $F_j^{CNN} \in \mathbb{R}^{H_j \times W_j \times C_j}$ at N different resolutions ($N = 4$ in our case).

ViT features Z_i are projected to match the size of the corresponding CNN features F_j^{CNN}, resulting in the ViT features F_j^{ViT}. These are then fused with the CNN features as follows:

$$F_j = \beta F_j^{CNN} + (1 - \beta) F_j^{ViT}, \quad j = 1, 2, ..., N \tag{5}$$

β is sampled from $[r_1, r_2](0 \leq r_1 < r_2 \leq 1)$, and F_j is the fused feature map at level j.

Beyond non-parametric fusion, we explore parametric fusion, employing a channel attention mechanism [1] at each scale. This mechanism is defined as:

$$\hat{X} = W \cdot \mathsf{Attention}(\hat{Q}, \hat{K}, \hat{V})$$

$$\mathsf{Attention}(\hat{Q}, \hat{K}, \hat{V}) = \hat{V} \cdot \mathsf{Softmax}(\frac{\hat{K} \cdot \hat{Q}}{\alpha}) \tag{6}$$

Here, $\hat{X} \in \mathbb{R}^{H \times W \times C}$ denotes the output feature map; $\hat{Q} \in \mathbb{R}^{HW \times C}, \hat{K} \in \mathbb{R}^{C \times HW}, \hat{V} \in \mathbb{R}^{HW \times C}$ are tensors reshaped from Q, K, V, respectively; W is a 1×1 convolution for output projection; α is a learnable parameter to control the magnitude of $\hat{K} \cdot \hat{Q}$. The definition of Q, K, V is $Q, K, V = W^Q X, W^K X, W^V X$ and $X = \mathsf{LayerNorm}([F^{CNN}, F^{VIT}])$, where $[\cdot, \cdot]$ denotes feature concatenation and $W^{(\cdot)}$ denotes point-wise 1×1 convolutions.

Multi-scale Supervision Loss Function. In our proposed network architecture that combines ViT and CNN in a multi-scale manner, multiple predictions are generated in a single forward pass. However, relying exclusively on the final output for training can lead to convergence issues. To address this, we propose an end-to-end optimization of multiple predictions, which we term the Multi-scale Supervision Loss function.

We denote the network's final prediction as P and its multi-scale predictions as $S = Q_1, Q_2, ..., Q_{s-1}, Q_s$, where Q_s represents the prediction at the s-th scale. For simplicity, we exclude the final prediction P from the set S. The CNN branch prediction is represented by R. Then, we utilize multi-scale predictions $S + P$ and the CNN output R to compute the loss. The Multi-scale Supervision Loss function is formulated as follow:

$$\mathcal{L}_{ms} = \alpha_1 \mathcal{L}(P, T) + \alpha_2 \mathcal{L}(R, T) + \alpha_3 \frac{1}{|S|} \sum_{s=1}^{|S|} \mathcal{L}(Q_s, T) \tag{7}$$

Here, $\mathcal{L}$ refers to the average of Dice loss and Cross Entropy loss. T denotes the ground truth label, and $|S|$ is the cardinality of set S. The weights α_1, α_2, and α_3 are used to balance each term in the loss function, and are set to $\alpha_1 = \alpha_2 = 1$ and $\alpha_3 = 0.6$ for all our experiments.

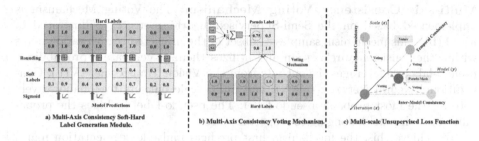

a) Multi-Axis Consistency Soft-Hard   b) Multi-Axis Consistency Voting Mechanism.   c) Multi-scale Unsupervised Loss Function
Label Generation Module.

Fig. 2. The illustration of Multi-Axis Consistency Framework.

3.3   Multi-axis Consistency Framework

In our pursuit to accomplish semi-supervised learning, we present a novel Multi-Axis Consistency framework as illustrated in Fig. 2. This all-encompassing framework is made up of three main components: the *Multi-Axis Consistency Soft-Hard Label Generation Module, the Multi-Axis Consistency Voting Mechanism, and the Multi-scale Unsupervised Loss Function.* The Soft-Hard Label Generation Module generates robust labels, taking into consideration intra-model and inter-model consistency, as well as temporal consistency. The Consistency Voting Mechanism selects the most probable predictions across different models and scales, thereby enhancing the robustness and accuracy of the learning process. The Multi-scale Unsupervised Loss Function provides a metric for model optimization in scenarios where ground truth labels are absent, promoting the extraction of valuable features from unlabeled data. Next, we introduce them in detail.

Multi-axis Consistency Soft-Hard Label Generation Module. This innovative module is designed based on the Coordinate Systems we established to model different consistency paradigms, as depicted in Fig. 2 (a). We represent the input as x and the consistency condition as $\theta = [m, s, t]^T$, indicating that the output is generated by model m, scale s, and training iteration t. The module's objective is to minimize the distance between two outputs under consistency regularization from multiple axes. The module achieves this by applying an augmentation σ to the input x and generating a modified input $\hat{x}$ while ensuring a small consistency relaxation ϵ. This process is expressed as follows:

$$\min \| f(x, \theta) - f(\hat{x}, \theta + \epsilon) \|$$
$$s.t. \hat{x} = \sigma(x), \|\epsilon\| \to 0 \tag{8}$$

Then, the module generates robust labels by predicting multiple segmentation maps $P_\theta \in \mathbb{R}^{H \times W \times K}, \theta \in \Theta$, where K denotes the number of segmentation classes. These are the soft labels. The module then converts these soft labels into binary hard labels using a threshold. Next, we introduce the whole process in our, multi-axis consistency voting mechanism.

Multi-axis Consistency Voting Mechanism. The Voting Mechanism is implemented based on the Semi-Supervised Learning strategy, illustrated in Fig. 2 (b). This mechanism samples a subset of the predicted segmentation maps within the consistency relaxation and utilizes them to generate a probabilistic pseudo-label. It leverages the outputs from the Vision Transformer (ViT), Convolutional Neural Network (CNN), and multi-scale outputs to collaboratively vote for the most probable pseudo-label. The pseudo-label includes the probability of each pixel belonging to a specific class.

To achieve this, the mechanism first predicts multiple segmentation maps, which can be considered as "soft labels" indicating the probability of each class. These soft labels are then converted into binary "hard labels" using a threshold of 0.5, as shown in the first part of the equation.

$$M_\theta(h, w, k) = \begin{cases} 1 \text{ for } P_\theta(h, w, k) \geq 0.5 \\ 0 \text{ for } P_\theta(h, w, k) < 0.5 \end{cases} \tag{9}$$

Here, $M_\theta(h, w, k)$ denotes the binary hard label of pixel (h, w) for class k under condition θ and $P_\theta(h, w, k)$ represents the soft label corresponding to the same. The final pseudo-label M_{pseu} is then generated by taking the average of these binary hard labels across all conditions $\theta \in \Theta$, as expressed in the second part of the equation.

$$M_{pseu} = \frac{1}{|\Theta|} \sum_{\theta \in \Theta} M_\theta \tag{10}$$

In this equation, M_{pseu} represents the final probabilistic pseudo-label and $|\Theta|$ denotes the cardinality of set Θ. In this way, the Multi-Axis Consistency Voting Mechanism generates robust probabilistic pseudo-labels that reflect the consensus among different models (ViT, CNN) and multi-scale outputs, embodying the concept of multi-axis consistency.

Multi-scale Unsupervised Loss Function. The unsupervised loss function is incorporated in our semi-supervised training via an unsupervised loss term, as is shown in Fig. 2 (c). For unlabeled images, the function first generates pseudo-labels according to multiple network outputs. Multiple outputs from the current training iteration as well as previous iterations all contribute to the generation of the pseudo-label. The function aims to minimize the distance between contributors from the current training iteration to the pseudo-label:

$$\mathcal{L}_{unsup} = \frac{1}{|\Theta|} \sum_{\theta \in \Theta} \mathcal{L}(P_\theta, M_{pseu}) \tag{11}$$

Table 1. Results under Fully-supervised Learning and Comparison with the state-of-the-arts. "PF" and "NPF" represent Parametric fusion and Non-Parametric Fusion, respectively. Different values of β during inference are also included.

Method	MoNuSeg		QaTa-COV19	
	Dice (%)	mIoU (%)	Dice (%)	mIoU (%)
Unet	76.45	62.86	79.02	69.46
Unet++	77.01	63.04	79.62	70.25
AttUnet	76.67	63.74	79.31	70.04
nnUnet	80.06	66.87	80.42	70.81
MedT	77.46	63.37	77.47	67.51
TransUnet	78.53	65.05	78.63	69.13
GTUnet	79.26	65.94	79.17	69.65
Swin-Unet	77.69	63.77	78.07	68.34
UCTransNet	79.87	66.68	79.15	69.60
Ours+PF	79.91	66.74	**82.29**	**72.87**
Ours+NPF	**80.60**	**67.66**	82.03	72.80

Table 2. Results under Semi-Supervised Setting. "PF" and "NPF" represent Parametric fusion and Non-Parametric Fusion, respectively.

Setting	Labels (%)	MoNuSeg	
		Dice (%)	mIoU (%)
PF	25	78.59	64.99
	50	78.85	65.36
	100	79.91	66.74
NPF	25	78.47	64.88
	50	79.26	65.94
	100	80.16	67.06

Here, $\mathcal{L}$ represents the average of Dice loss and Cross-Entropy loss. The final loss for semi-supervised learning $\mathcal{L}_{final}$ is represented by the weighted sum of L_{sup} and L_{unsup}, as shown below:

$$\mathcal{L}_{final} = \mathcal{L}_{sup} + \lambda \mathcal{L}_{unsup} \qquad (12)$$

Here, λ is a weight factor, defined by a time-dependent Gaussian warming-up function to control the balance between the supervised loss and unsupervised loss.

4 Experiments

4.1 Experiment Setup

We use pretrained vision and language transformers, which remain frozen during training. We use a ViT pretrained on the ROCO dataset [28] via DINO [6] as our vision backbone, and Clinical BERT [2] as our language backbone. We adopt U-Net [31] as the CNN branch. We set the batch size to 4 and the initial learning rate to 10^{-3}, using the Adam optimizer [17] with cosine annealing cyclic schedule. Data augmentation includes random flips and 90° rotations. All experiments are conducted on an A5000 GPU. We use Dice and mIoU metrics as our evaluation metrics. The experiments are conducted on MoNuSeg [18] and QaTa-COV19 [9] datasets. The MoNuSeg dataset includes images of tissue from various patients with tumors and approximately 22,000 nuclear boundary annotations across 30 training images and 14 test images. And the QaTa-COV19 dataset includes 9258 annotated COVID-19 chest radiographs. The text annotations for both datasets are derived from [22].

4.2 Quantitative Results

In this section, we conduct main experiments on MoNuSeg dataset on both fully-supervised and semi-supervised settings. We also include the fully-supervised results on QaTa-COV19 dataset.

Results on Fully-Supervised Setting. In our research, we compare our methodology to state-of-the-art methods in a fully-supervised setting, these methods include Unet [31], Unet++ [44], AttUnet [27], nnUnet [16], MedT [32], transUnet [7], GTUnet [21], Swin-Unet [5], and UCTransNet [34]. The results of our approach and the other state-of-the-art methods in a fully-supervised learning setting are presented in Table 1. Notably, our method demonstrates a significant improvement over existing approaches. With the employment of parametric ViT-CNN fusion, our method achieves results that are not only comparable with TransUnet [7] on the MoNuSeg dataset but also surpasses it under specific conditions. More notably, our approach exhibits superior performance under non-parametric feature fusion, namely, random fusion with a uniformly sampled β during training. In this case, our method sets a new benchmark on the MoNuSeg dataset, outperforming all other state-of-the-art methods. This remarkable performance demonstrates the robustness of the random fusion strategy, where both the ViT and CNN branches can learn strong representations.

Results on Semi-supervised Setting. In Table 2, we present our results under a Semi-Supervised setting. These results are achieved by using 25% and 50% labels for model evaluation, under our proposed Multi-Axis Consistency framework. The performance of our method stands out in several ways: 1) Most notably, our method delivers a comparable result to TransUnet in a fully-supervised setting even with only 25% labels. Moreover, when we possess 50% labels, the result improves significantly. This clearly showcases the potential of our proposed method. It illustrates how our method can effectively reduce the reliance on labeled data by learning from limited data and large-scale unlabeled data, thereby alleviating the cost of labels. 2) Our method with ViT-CNN random fusion and parametric channel attention consistently produces strong results across all semi-supervised settings. While the version with ViT-CNN random fusion outperforms the version with parametric channel attention by a small margin (less than 0.5%) when using 50% and 100% labels, the results are fairly comparable. This highlights the advantages of our multi-scale ViT-CNN fusion, underscoring its ability to capture both global and local interactions and retain pre-trained knowledge. All these findings reinforce the effectiveness and efficiency of our method in Semi-Supervised settings, demonstrating that it is a promising approach for future research and applications.

4.3 Qualitative Comparisons

Figure 3 provides a visual comparison between our method and our baseline method [24]. Both sets of results are obtained under a fully-supervised training setting. To facilitate a clearer distinction between the two, we have highlighted

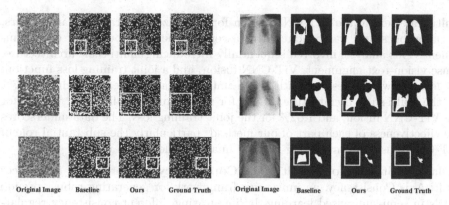

Fig. 3. Qualitative comparison. Left: Visualization results on MoNuSeg dataset. Right: Visualization results on QaTa-COV19 dataset.

specific areas in each sample within a red box. Upon close examination, it's evident that our proposed method offers superior results with respect to the precision of boundary delineation and the accuracy of shape representation. These improvements are most apparent within the highlighted regions, where our method's predictions exhibit finer detail and higher fidelity to the original structures. A key factor contributing to this enhanced performance is the introduction of our multi-scale text-aware ViT-CNN fusion. This innovative approach significantly improves local feature extraction within the target medical domain, allowing for more accurate and detailed segmentation results. This clearly demonstrates the advantage of our method over traditional approaches, and its potential for providing superior outcomes in complex medical image analysis tasks.

4.4 Ablation Study

Table 3. Ablation studies on proposed modules.

Method	ViT-CNN Dense Fusion				MoNuSeg	
	Multi-scale Arch.	Text	ViT-CNN Fusion	Loss	Dice (%)	mIoU (%)
Baseline					68.03	52.05
	✓				76.30	61.92
Ours	✓	✓			78.47	64.67
	✓	✓	✓		79.14	65.78
	✓	✓	✓	✓	80.16	67.06

(a) Ablation study on proposed modules under fully-supervised learning.

Setting	Multi-Axis Consistency			MoNuSeg
	Intra-Model	Inter-Model	Temporal	Dice (%)	mIoU (%)
Sup. Only				77.94	64.12
	✓			78.68	65.11
		✓		78.52	64.92
Ours			✓	78.01	64.15
	✓	✓		78.88	65.37
	✓		✓	78.84	65.31
		✓	✓	78.57	64.97
	✓	✓	✓	79.26	65.94

(b) Ablation study on Multi-Axis Consistency framework under semi-supervised learning with 50% training labels.

Ablation Studies on Mutli-scale Text-aware ViT-CNN Fusion. To evaluate our ViT-CNN Fusion, we employed the state-of-the-art vision-language transformer dense finetuning method as our baseline, which didn't perform well in medical image segmentation due to over-reliance on the pretrained backbone and lack of multi-scale dense features. To counter these issues, we propose a multi-scale text-aware ViT-CNN fusion for optimized pretrained transformers. An ablation study was conducted to ascertain the contribution of each component. This analysis involved sequentially introducing multi-scale architecture, dense vision-text alignment, ViT-CNN fusion, and a joint training loss function. Table 3a shows the results, with significant Dice gains for each module: 8.27% for the multi-scale architecture, 2.17% for the vision-text alignment, 0.67% for the ViT-CNN fusion, and 1.02% for the joint training loss. The data underscores the effectiveness of each part of our method, particularly the substantial role of ViT-CNN Fusion in improving medical image segmentation tasks.

Ablation Studies on Multi-axis Consistency. Our research introduces Multi-Axis Consistency, an innovative framework for generating robust pseudo labels in semi-supervised learning, by integrating different consistency regularization types. Table 3b displays our results: each consistency regularization type improves semi-supervised performance compared to a supervised-only setting, highlighting their importance in semi-supervised learning. Notably, peak performance is achieved when all three types are combined, demonstrating the effectiveness of the Multi-Axis Consistency framework. This comprehensive approach leads to superior performance in semi-supervised learning, marking a significant advancement in generating pseudo labels and improving model performance.

In-depth Discussion on Multi-scale ViT-CNN Fusion. In this section, we address two key questions experimentally: 1Why does ViT-CNN fusion work in semi-supervised settings? Our results (Table 3a and Fig. 4) demonstrate this module's effectiveness in fully-supervised learning. Fig. 5 further shows that ViT-CNN fusion is crucial in semi-supervised settings, with performance increasing as β decreases from 0.8 to 0.2. This suggests that both Transformer and CNN branches can independently perform well in such settings. 2) Why is multi-scale fusion important? We conducted ablation studies on fusion levels using different approaches: Non-Parametric Random Fusion and Parametric Channel attention. Figure 6 shows that increased feature fusion levels improve model performance, underscoring the importance of multi-scale dense features in medical image segmentation and the effectiveness of our proposed multi-scale ViT-CNN fusion method.

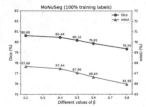 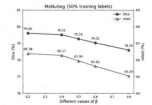 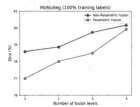

Fig. 4. Impact of different β (full).   **Fig. 5.** Impact of different β (semi).   **Fig. 6.** Impact of fusion level numbers.

5 Conclusion

In this paper, we propose a novel semi-supervised learning framework for medical image segmentation. In our work, a Text-aware ViT-CNN Fusion scheme is proposed to take advantages of both pretrained ViTs and CNNs as well as extracting both abstract features and medical domain specific features. Besides, a novel Multi-Axis Consistency framework is proposed to vote for pseudo label to encourage semi-supervised training. Experiments on several widely used datasets have demonstrated the effectiveness of our method.

References

1. Ali, A., et al.: Xcit: cross-covariance image transformers. In: Advances in NeurIPS (2021)
2. Alsentzer, E., et al.: Publicly available clinical BERT embeddings. arXiv preprint (2019)
3. Baker, N., et al.: Local features and global shape information in object classification by deep convolutional neural networks. Vision. Res. **172**, 46–61 (2020)
4. Cai, S., et al.: Dense-unet: a novel multiphoton in vivo cellular image segmentation model based on a convolutional neural network. Quant. Imaging Med. Surg. **10**(6), 1275 (2020)
5. Cao, H., et al.: Swin-unet: Unet-like pure transformer for medical image segmentation. arXiv:2105.05537 (2021)
6. Caron, M., et al.: Emerging properties in self-supervised vision transformers. In: Proceedings of IEEE/CVF ICCV (2021)
7. Chen, J., et al.: Transunet: transformers for medical image segmentation. arXiv:2102.04306 (2021)
8. Chen, X., et al.: Semi-supervised segmentation with cross pseudo supervision. In: Proceedings of IEEE/CVF CVPR (2021)
9. Degerli, A., et al.: OSEGnet: operational segmentation network for COVID-19 detection using chest x-ray images. In: Proceedings of ICIP, pp. 2306–2310. IEEE (2022)
10. Dosovitskiy, A., et al.: Transformers for image recognition at scale. arXiv:2010.11929 (2020)
11. Gao, Y., Zhou, M., Metaxas, D.N.: UTNet: a hybrid transformer architecture for medical image segmentation. In: de Bruijne, M., et al. (eds.) MICCAI 2021. LNCS, vol. 12903, pp. 61–71. Springer, Cham (2021). https://doi.org/10.1007/978-3-030-87199-4_6
12. Guo, C., et al.: SA-unet: Spatial attention u-net for retinal vessel segmentation. In: Proceedings of ICPR, pp. 1236–1242. IEEE (2021)
13. Hang, W., et al.: Local and global structure-aware entropy regularized mean teacher model for 3D left atrium segmentation. In: Martel, A.L., et al. (eds.) MICCAI 2020. LNCS, vol. 12261, pp. 562–571. Springer, Cham (2020). https://doi.org/10.1007/978-3-030-59710-8_55
14. Hatamizadeh, A., et al.: Unetr: transformers for 3d medical image segmentation. In: Proceedings of IEEE/CVF WACV (2022)
15. Huang, H., et al.: Unet 3+: a full-scale connected unet for medical image segmentation. In: Proceedings of ICASSP, pp. 1055–1059. IEEE (2020)

16. Isensee, F., et al.: nnu-net: a self-configuring method for segmentation. Nat. Methods (2021)
17. Kingma, D., Ba, J.: Adam: a method for stochastic optimization. arXiv preprint (2014)
18. Kumar, N., et al.: A multi-organ nucleus segmentation challenge. IEEE Trans. Med. Imaging **39**(5), 1380–1391 (2020). https://doi.org/10.1109/TMI.2019.2947628
19. Laine, S., Aila, T.: Temporal ensembling for semi-supervised learning. arXiv:1610.02242 (2016)
20. Li, B., et al.: Language-driven semantic segmentation. arXiv preprint arXiv:2201.03546 (2022)
21. Li, Y., et al.: GT u-net: a u-net like group transformer network for tooth root segmentation. In: Proceedings of MLMI (2021)
22. Li, Z., et al.: LVIT: language meets vision transformer in medical image segmentation. IEEE Trans. Med. Imaging (2023)
23. Liu, Z., et al.: Swin transformer: hierarchical vision transformer. In: Proceedings of IEEE/CVF ICCV (2021)
24. Lüddecke, T., et al.: Image segmentation using text and image prompts. In: Proceedings of IEEE/CVF CVPR (2022)
25. Luo, X., et al.: Semi-supervised medical image segmentation via cross teaching. arXiv:2112.04894 (2021)
26. Luo, X., et al.: Semi-supervised medical image segmentation via uncertainty rectified pyramid consistency. Med. Image Anal. (2022)
27. Oktay, O., et al.: Attention u-net: learning where to look for the pancreas. arXiv preprint (2018)
28. Pelka, O., Koitka, S., Rückert, J., Nensa, F., Friedrich, C.M.: Radiology objects in COntext (ROCO): a multimodal image dataset. In: Stoyanov, D., et al. (eds.) LABELS/CVII/STENT -2018. LNCS, vol. 11043, pp. 180–189. Springer, Cham (2018). https://doi.org/10.1007/978-3-030-01364-6_20
29. Radford, A., et al.: Learning transferable visual models from natural language supervision. In: Proceedings of ICML (2021)
30. Rao, Y., et al.: Denseclip: language-guided dense prediction with context-aware prompting. In: Proceedings of IEEE/CVF CVPR (2022)
31. Ronneberger, O., et al.: U-net: convolutional networks for biomedical image segmentation. In: Proceedings of International Conference on Medical image computing and computer-assisted intervention (2015)
32. Valanarasu, J.M.J., Oza, P., Hacihaliloglu, I., Patel, V.M.: Medical transformer: gated axial-attention for medical image segmentation. In: de Bruijne, M., et al. (eds.) MICCAI 2021. LNCS, vol. 12901, pp. 36–46. Springer, Cham (2021). https://doi.org/10.1007/978-3-030-87193-2_4
33. Wang, G., et al.: Semi-supervised segmentation with multi-scale guided dense attention. IEEE Trans. Med. Imaging (2021)
34. Wang, H., et al.: Uctransnet: rethinking the skip connections in u-net with transformer. In: Proceedings of AAAI (2022)
35. Wang, K., et al.: Tripled-uncertainty guided mean teacher model for segmentation. In: Proceedings of International Conference on Medical Image Computing and Computer-Assisted Intervention (2021)
36. Wang, Z., et al.: Cris: clip-driven referring image segmentation. In: Proceedings of IEEE/CVF CVPR (2022)
37. Wu, Y., et al.: Mutual consistency learning for semi-supervised segmentation. Med. Image Anal. (2022)

38. Xie, Y., Zhang, J., Shen, C., Xia, Y.: CoTr: efficiently bridging CNN and transformer for 3D medical image segmentation. In: de Bruijne, M., et al. (eds.) MICCAI 2021. LNCS, vol. 12903, pp. 171–180. Springer, Cham (2021). https://doi.org/10.1007/978-3-030-87199-4_16

39. Xu, M., et al.: A simple baseline for zero-shot semantic segmentation with pretrained vision-language model. arXiv preprint (2021)

40. You, C., et al.: SimCVD: contrastive voxel-wise representation distillation for semi-supervised medical image segmentation. IEEE Trans. Med. Imaging (2022)

41. Zhang, Y., et al.: A multi-branch hybrid transformer network for corneal endothelial cell segmentation. In: de Bruijne, M., et al. (eds.) MICCAI 2021. LNCS, vol. 12901, pp. 99–108. Springer, Cham (2021). https://doi.org/10.1007/978-3-030-87193-2_10

42. Zhang, Y., Liu, H., Hu, Q.: TransFuse: fusing transformers and CNNs for medical image segmentation. In: de Bruijne, M., et al. (eds.) MICCAI 2021. LNCS, vol. 12901, pp. 14–24. Springer, Cham (2021). https://doi.org/10.1007/978-3-030-87193-2_2

43. Zhou, Y., et al.: Semi-supervised multi-organ segmentation via deep multi-planar co-training. arXiv preprint (2018)

44. Zhou, Z., et al.: Unet++: a nested u-net architecture for medical image segmentation. In: Deep Learning in Medical Image Analysis and Multimodal Learning for Clinical Decision Support (2018)

Audio-Visual Segmentation by Leveraging Multi-scaled Features Learning

Sze An Peter Tan⑩, Guangyu Gao(✉)⑩, and Jia Zhao⑩

School of Computer Science and Technology, Beijing Institute of Technology,
Beijing 100081, China
guangyugao@bit.edu.cn

Abstract. Audio-visual segmentation with semantics (AVSS) is an advanced approach that enriches Audio-visual segmentation (AVS) by incorporating object classification, making it a more challenging task to accurately delineate sound-producing objects based on audio-visual cues. To achieve successful audio-visual learning, a model must accurately extract pixel-wise semantic information from images and effectively connect audio-visual data. However, the robustness of such models is often hindered by the limited availability of comprehensive examples in publicly accessible datasets. In this paper, we introduce an audio-visual attentive feature fusion module designed to guide the visual segmentation process by injecting audio semantics. This module is seamlessly integrated into a widely adopted U-Net-like model. Meanwhile, to enhance the model's ability to capture both high and low-level features, we implement double-skip connections. Besides, in the exploration of intra- and inter-frame correspondences, we also propose an ensemble model proficient in learning two distinct tasks: frame-level and video-level segmentation. To address the task's diverse demands, we introduce two model variants, one based on the ResNet architecture and the other based on the Swin Transformer model. Our approach leverages transfer learning and employs data augmentation techniques. Additionally, we introduce a custom regularization function aimed at enhancing the model's robustness against unseen data while simultaneously improving segmentation boundary confidence through self-supervision. Extensive experiments demonstrate the effectiveness of our method as well as the significance of cross-modal perception and dependency modeling for this task.

Keywords: Audio-Visual Segmentation · Double-Skip Connection · Multi-Scale · Self-supervision

1 Introduction

In our surroundings, a myriad of audio and visual stimuli coexist, and humans demonstrate a remarkable proficiency in establishing associations and extracting semantic meaning from these unprocessed inputs. For example, we possess the exceptional ability to swiftly discern the source of an alarm or the call of an animal, accurately pinpointing its location. However, automating this process presents a formidable challenge, highlighting the significance of research in audio-visual perception as an essential and imperative area of investigation. A nascent challenge in this domain is known as audio-visual segmentation (AVS) [37], which entails generating a pixel-level map of objects producing sound corresponding to an image frame. Building upon AVS, audio-visual segmentation with semantics (AVSS) [38] furthers the task by incorporating object classification. Both AVS and AVSS are pivotal in enhancing our comprehension and analysis of audio-visual scenes, with wide-ranging applications in areas such as surveillance, robotics, and human-computer interaction. The core challenge involves the precise identification and classification of sound-producing objects within the intricate and dynamic realm of audio-visual data.

In previous work [37,38], the authors introduced methods incorporating temporal pixel-wise audio-visual interaction (TVAPI) modules to achieve integration between audio and images. These modules established a shortcut connection between the encoder and decoder, facilitating the fusion of audio features with image sequence embeddings. The training approach employed by this method combined a segmentation loss with a custom loss referred to as audio-visual mapping (AVM) loss [37]. Regrettably, this combination failed to impose a robust constraint for learning joint audio-visual representations. We argue that this approach assumes that the audio and visual objects are always related in a one-to-one manner, which does not hold true in real-world scenarios. The unconstrained attention fusion learning of the audio-visual representations is highly likely to overfit to any biases present in the training set. Hence, this results in poor generalization beyond the training set scenes.

On another note, the method proposed in [37,38] approaches audio-visual segmentation as a frame-level segmentation problem. While image segmentation models excel at delineating objects or regions within individual frames, they often struggle to maintain consistency and precision across consecutive frames in a video sequence. The inclusion of video data introduces temporal dynamics encompassing object motion, alterations in appearance, and potential occlusions, all of which can significantly influence the performance of segmentation algorithms. It is currently unclear to what extent video perturbations will pose a robustness challenge. However, there have been reports highlighting performance degradation attributed to video perturbations [4,11].

In this paper, we propose a novel AVS/AVSS method by incorporating double skip connections. Firstly, drawing inspiration from [26], we incorporate double skip connections to amalgamate the capability of capturing multi-scale information from both low-level and high-level features. This multi-scale learning in segmentation is advantageous as it enables models to comprehend images at various levels of detail, ultimately leading to more precise and robust object segmentation across different scenarios [5,17]. Meanwhile, to address video object tracking, we design an independent video-level segmentation model that takes video frames and their corresponding audio as inputs. Leveraging the collective knowledge of frame-level and video-level segmentation models enhances overall predictive performance and robustness, in comparison to relying solely on a single model. It's worth noting that audio and visual objects are not always related in a one-to-one manner. Toward this, we propose a self-supervised loss function called Patch Classification Loss (P-Loss). We use automatically generated pseudo-labels that suggest regions likely to contain an object, encouraging the model to "group" pixels together when there is a high likelihood they belong to a specific object. In summary, our main contributions can be summarized as:

- **Double Skip Connections Framework**: We introduce a framework with double skip connections to effectively harness multi-scale information from both low-level and high-level features.
- **Ensemble Model for Temporal Consistency**: To promote the learning of temporal consistency and mitigate the challenges posed by video perturbations, we design an ensemble model comprising both frame-level and video-level segmentation models.
- **Self-supervised Loss Function (P-Loss)**: We design a self-supervised loss function, referred to as P-Loss, to further improve the model's performance, achieving new state-of-the-art performance over baselines [37,38].

2   Related Work

2.1   Audio-Visual Tasks

In recent years, the field of audiovisual learning has gained significant momentum, aiming to replicate human perceptual abilities. Researchers have proposed various approaches to explore and utilize the relationships between audio and visual signals.

Audio-visual temporal correspondence [15] is a challenging task where a model determines whether an audio sample and a video sequence are synchronized or asynchronous. Unlike audio-visual correspondence (AVC) [1–3], this task focuses on learning time-sensitive features in audio and video streams to recognize synchronization rather than relying solely on semantic matches. Additionally, audio-visual event localization (AVEL) [24,36] aims to pinpoint the timing of audible or visual events using predefined labels. Accurate localization for short- and long-term temporal interactions requires effective multimodal feature alignment, making it a highly demanding task. Similarly, audio-visual video parsing (AVVP) [16,33] involves segmenting videos into temporal event segments and categorizing them as audible events, visual events, or a combination of both. This challenge is complicated by audio from sources outside the visual frame, lacking a visual counterpart but contributing to overall comprehension.

While progress has been made in various audio-visual tasks, they primarily operate at the level of timeframes or frames for comprehension. In contrast, sound source localization (SSL) [10,30] focuses on scene understanding at a patch level. As a result, SSL results are often visualized as heatmaps, generated either by comparing audio features with visual feature maps or by using class activation mapping (CAM) [35]. CAM, however, does not capture the precise shape of sound sources. Audio-visual segmentation (with semantics) (AVS/AVSS) sets a higher expectation, necessitating models to predict whether each pixel corresponds to incoming audio signals. This approach enables finer-grained, pixel-level understanding, taking into account the shape and boundaries of sound objects. AVS/AVSS scenarios present additional challenges due to their dynamic nature, involving varying numbers of sound sources.

2.2   Semantic Segmentation

Semantic segmentation involves partitioning an image into segments and assigning meaningful labels to each of these segments. While traditional image segmentation may rely on color, texture, or other visual properties, semantic segmentation takes it a step further by assigning a meaningful class label to every pixel in an image.

Commonly used architectures for semantic segmentation, include Fully Convolutional Networks (FCN) [18], U-Net [25] and Deeplab [7]. FCNs utilize convolutional layers to generate pixel-wise predictions, making them adaptable for handling images of various sizes. U-Net comprises a contracting path for context capture and an expanding path for precise localization. Architectures with encoder-decoder structures and skip connections, like DeepLab [7], are also widely used. To bolster feature extraction capabilities, these segmentation models often leverage backbones or encoders pretrained on extensive image classification datasets, including ResNet [12], ResNeXt [31], and EfficientNet [28].

It is noteworthy that recent research has shown a growing interest in multimodal semantic segmentation. Notably, SAM [14] has emerged as a recent advancement, enabling networks to comprehend a wide range of cues, from sparse elements like dots, boxes, and text, to dense cues such as masks.

3   Methodology

As an overview of our proposed method, to tackle the AVS/AVSS problem, we build a UNet [25] -like architecture multimodal model with an attentive feature fusion module, as shown in Fig. 1. The model encodes audio and frames into representative features, injects the categorical audio information into the visual features via the fusion module, and then decodes the pixel-wise masks by a decoder. Unlike the single skip connections in UNet, we formulated double skip connections with ASPP and GAB modules. Furthermore, our model leverages deep supervision [39] to generate mask predictions of multiple scales, which are utilized for loss function and serve as one of the inputs to GAB.

To further promote the learning of inter-frame correspondence, long short-term memory modules are inserted in between the base model. This model is a video-level segmentation model, as shown in Fig. 2.

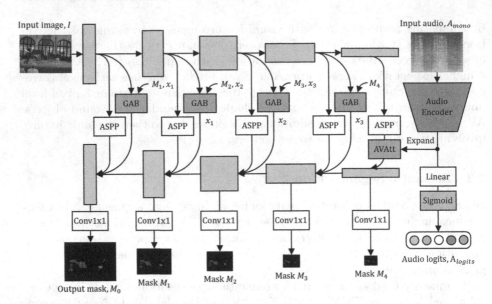

Fig. 1. Overview of our proposed framework.

3.1 Audio Feature Encoder

We deploy a pre-trained VGGish [13] trained on AudioSet [9], which is a variant of the VGG [27] model. The encoder takes log mel spectrogram audio inputs $A_{mono} \in \mathbb{R}^{N \times F \times T}$, and outputs audio features $A_S \in \mathbb{R}^{N \times D}$, where $D = 256$ is the feature dimension. The 1000-wide fully connected layer at the end has been replaced with a 256-wide fully connected layer, serving as a compact embedding layer. In the context of AVSS, for probability prediction, the model employs a linear classifier, followed by a sigmoid activation function applied to A_S:

$$P_S = sigmoid\left(g_1\left(A_S\right)\right) \tag{1}$$

where $g_1 \in \mathbb{R}^{N \times K}$ is a linear classifier, and K is the number of classes.

3.2 Visual Sub-network

The visual sub-network comprises an *image feature encoder* and a *decoder*. During encoding, spatial information decreases while feature information intensifies, enabling the network to comprehend the global image context and generate hierarchical visual feature maps denoted as $F_i \in \mathbb{R}^{N \times C_i \times h_i \times w_i}$, where h_i, w_i. Here, h_i and w_i represent the dimensions of the output features at stage i, determined by the specific backbone model.

The network incorporates skip connections between the encoder and decoder, comprising Atrous Spatial Pyramid Pooling (ASPP) [7] and Group Aggregation Bridge (GAB) [26]. ASPP process the visual features F_i to produce

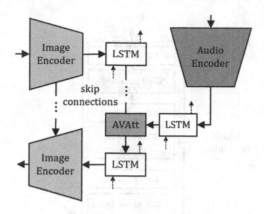

Fig. 2. Video-Level Segmentation Model.

$V_i \in \mathbb{R}^{N \times C \times h_i \times w_i}$. These modules encourage the recognition of visual objects with different receptive fields by employing multiple parallel filters with different rates. GAB output $J_i \in \mathbb{R}^{N \times C \times h_i \times w_i}$. They are crucial for capturing context from low and high-level features, enhancing the model's ability for accurate segment predictions. The GAB module takes in low-level and high-level features, along with a refining mask, enabling it to produce comprehensive predictions.

At the j-th stage, where $j = 1, 2, 3, 4$, the outputs from stage V_{5-j}, J_{5-j} of the encoder, and the last stage Z_{6-j} are utilized in the decoding process.

$$Z_{6-j} = \begin{cases} AVAtt(V_5, A_S), & j = 0 \\ g_2(concat(V_{5-j}, J_{5-j}, Z_{6-j})), & j > 0 \end{cases} \qquad (2)$$

where $g_2 \in \mathbb{R}^{C_1 \times C}$ is a 1×1 channel-wise convolution, $C_1 = 768$ and $C = 256$. The decoded features are then upsampled to the next stage. The output of the decoder is a list of mask predictions of multiple scales, $M_i \in \mathbb{R}^{N \times h_i \times w_i}$. M_i is scaled up bilinearly to match the size of ground truth mask M_{true}.

3.3 Audio-Visual Attentive Feature Fusion (AVAtt)

The visual information in $A_v = V_5$ contains crucial contextual details for image segmentation, while A_S offers categorical audio insights regarding detected objects based on input audio. Our aim is to integrate features from both encoders to utilize audio cues as supplementary guidance for visual features. This approach ensures that the predicted segmentation mask aligns with the audible objects in the frame. To achieve this, we utilize a multi-head attention mechanism. This method combines these features, optimizing computational efficiency and effectiveness by striking a balance between the two.

Specifically, the multi-modal fusion module consists of a stack of $L = 3$ Transformer Decoder layers [29], as depicted in Fig. 3. To align both audio and visual features, A_S is expanded to match the dimensions of A_v: $A_S \in$

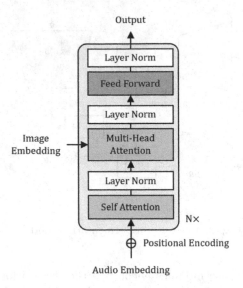

Fig. 3. Audio-Visual Attentive Feature Fusion.

$\mathbb{R}^{N \times D \to N \times 1 \times D \to N \times C_i \times D}$. In i-th layer, AVAtt takes $A_v \in \mathbb{R}^{N \times W \times H}$ and $A_S \in \mathbb{R}^{N \times C_i \times D}$ as input, and outputs new improved visual features $\widehat{A_v}$, by following the procedure as stated below. AVAtt uses Self Attention (SA) and Multi-head Attention (MHA) to find sounding regions in A_v by A_S:

$$\widehat{A_S} = LN \left(SA \left(Q_{AS}, \ K_{AS}, \ V_{AS} \right) + A_S \right) \tag{3}$$

$$\widehat{A_v} = LN \left(MLP \left(LN \left(MHA \left(Q_{\widehat{A_S}}, \ K_{Av}, \ V_{Av} \right) + A_v \right) \right) \right) \tag{4}$$

where LN denotes layer normalization. SA and MHA can be formulated as:

$$SA \left(Q_A, \ K_A, \ V_A \right) = A + V_A^T \times softmax \left(K_A^T \times Q_A \right) \tag{5}$$

while MHA is an extension of this mechanism by using multiple parallel heads to capture different patterns of relationships within the input sequence.

3.4 Objective Function

The primary loss function for the AVS task is a multi-staged binary-cross-entropy-Dice (BCE-Dice) loss, while for the AVSS task we select cross-entropy-Dice (CE-Dice) loss instead, where the function at stage-i is as formulated as:

$$L_i = \left\{ \begin{array}{ll} BCEDice \left(M_{true}, \ M_i \right), & if \ AVS \\ CEDice \left(M_{true}, \ M_i \right), & if \ AVSS \end{array} \right. \tag{6}$$

where $M_i \in \mathbb{R}^{N \times H \times W}$ is the predicted map at stage-i, $M_{true} \in \mathbb{R}^{N \times H \times W}$ is the ground truth, and *BCEDice, CEDice* are simply the summation of two losses.

We employ combined losses to enhance robustness [23,32], measuring pixel-wise differences between predicted segmentation and ground truth labels.

The ground truth M_{true} annotates pixels related to sounding objects, not all visible objects M_{all} in the frame. To encourage clustering of pixels belonging to the same object, we introduce the "Patch Classification Loss" (P-Loss). This loss utilizes the pre-trained Segment Anything Model (SAM) [14] with an Automatic Mask Generator to generate pseudo-labels, indicating potential object regions by assigning unique pixel-wise patch IDs ranging from 0 to m. The introduced P-Loss minimizes the sum of intra-region Kullback-Leibler (KL) divergence loss between pixel-wise probabilities within the same region and their average value. This objective encourages effective grouping of visual representations, strengthening the model's confidence in identifying segmentation boundaries. The mathematical expression for P-Loss is as:

$$P_i = P \times (M_{sam} == i) \tag{7}$$

$$L_p\left(P,\ M_{sam}\right) = \frac{1}{m}\sum_{i=1}^{m} KL\left(\log softmax\left(P_i\right)\middle| softmax\left(\frac{1}{n_i}\sum_{j=1}^{n_i} P_{ij}\right)\right) \tag{8}$$

where P is the output logits, m represents the number of patches in each batch, and n_i represents the number of elements in patches i.

We incorporate an additional regularization of audio-visual mapping (AVM) loss as [37]. This loss serves to enforce a strong correlation between audio and visual signals, ensuring that the masked visual features maintain similar distributions to their corresponding audio features. Specifically, AVM loss is computed using KL divergence between the average pooled visual features and the corresponding audio features, as formula:

$$L_1 = L_{AVM} = sum\left(KL\left(avg\left(M_p \odot Z_p\right),\ A_p\right)\right) \tag{9}$$

where M_p is the ground truth label, Z_p is the visual feature map and A_p is the audio feature map of p-th pixel.

In place of AVM, we introduced a multi-label classification (BCE) loss on the audio logits A_{logits}, as we noticed AVM contributes relatively little contribution in improving evaluation results as for AVSS task. Ground truth class labels are retrieved from the ground truth mask, encouraging the class-label-guided audio encoder to share its learned semantic cues with the visual subnetwork.

$$L_1 = L_{BCE}\left(A_{logits},\ A_{true}\right) \tag{10}$$

Deep supervision [39] is employed to calculate the loss function for different stages, in order to generate more accurate mask information. Hence, the total objective function is computed as the weighted sum of three losses, as follows:

$$L = \sum_{i=0}^{S} \lambda_i L_i + \lambda_1 L_1 + \lambda_2 L_p \tag{11}$$

where λ_i, λ_1, λ_2 are constant weights. In this paper, we set λ_i to 1, 0.5, 0.4, 0.3, ... from $i = 0$ to $i = S$. Considering the expensive computation cost when training with multiple loss functions, as a balance, we set $\S = 4$ and $\S = 2$ in the case of AVS and AVSS tasks respectively.

3.5 Data Augmentation

We apply several transformations to both image and audio files in the original data: (a) Audio: Gaussian noise, gain, gain transition, loudness normalization, pitch shift, resample, and time stretch; (b) Image: Blur, brightness/contrast, and Gaussian noise. We apply 20 random combinations of transformations to each sample. All the transformations are made in a very small amount to prevent the potential removal of key features that render the augmented data useless.

4   Experiments

4.1 AVSBench Dataset

Our method was assessed using AVSBench-v2 [38], an extensive dataset comprising around 12,000 videos, each lasting 5 or 10 s, and encompassing 71 classes of the "background", and diverse sounds, animal noises, vehicle sounds, and musical instruments. This dataset is divided into three subsets. The first subset contains a single sound source in each video, forming the semi-supervised Single Sound Source Segmentation (S4) task. The second subset involves multiple sound sources, resulting in the fully supervised Multiple Sound Source Segmentation (MS3) task. In this subset, each video includes two or more categories from the single-source subset, and all sounding objects are visible within the frames. Ground truths for these two subsets comprise binary masks indicating the pixels emitting the sounds. The third subset, Audio-Visual Semantic Segmentation (AVSS), is a semantic-labels subset introducing semantic labels for the sounding objects, exploring a fully supervised approach. Each video in the dataset is further divided into five equal 1-second clips, and the last frame of each clip is manually pixel-level annotated. Specially, for videos in the training split of single-source, only the first sampled frame is annotated.

4.2 Implementation Details

We utilize three image encoders: ResNet-34, ResNet-50, and SwinV2. Due to computational constraints, we employ ResNet-34 as the image encoder for our video-level segmentation model. Our ensemble model combines outputs from both frame- and video-level models. For initializing our image and audio encoders, we use default pre-trained weights: ImageNet1k for ResNet and SwinV2, and AudioSet [9] for VGGish. Our training strategy involves initial pre-training on all datasets, followed by fine-tuning on individual datasets, following an *abundance* approach. We employ the AdamW optimizer [19] with an initial learning rate and weight decay of $1e - 4$, as well as hyperparameters $\lambda_1 = 0.25$ and $\lambda_2 = 0.1$. The pre-training and fine-tuning phases run for 10 and 20 epochs, respectively. We manage the learning rate using the ReduceLROn-Plateau schedule, decreasing it by a factor of 0.8 with patience of 1 step per epoch. Early stopping is enabled for training termination. During fine-tuning, the initial learning rate is adjusted to 1e-5 and the weight decay to 1e-3. All model training occurs on a single A100 GPU with 40 GB memory, using a batch size of 16. Our method's performance is assessed on the validation set, and we compare it with other approaches on the test set.

Table 1. Comparisons on audio-visual segmentation under the MS3 and S4 settings. Results of mIoU (%) and F-score are reported.

Metric		mIoU		F-score	
Setting		S4	MS3	S4	MS3
SSL	LVS [6]	37.94	29.45	0.510	0.330
	MSSL [22]	44.89	26.13	0.663	0.363
VOS	3DC [20]	57.10	36.92	0.759	0.503
	SST [8]	66.29	42.57	0.801	0.572
SOD	iGAN [21]	61.59	42.89	0.778	0.544
	LGVT [34]	74.94	40.71	0.873	0.593
Baseline [37]	R50	72.79	47.88	0.848	0.578
	PVT-v2	78.74	54.00	0.879	0.645
Ours	R34	76.56	55.90	0.827	0.590
	R50	74.25	59.00	0.815	0.593
	R101	77.15	58.68	0.869	0.644
	SwinV2-b	**80.53**	**66.66**	**0.882**	**0.683**

4.3 Comparison with State-of-the-Arts

In our evaluation, we follow the methodology outlined in [38] and utilize Mean Intersection over Union (mIoU) and F-score as metrics, with a weighting factor $\beta = 0.3$. We present quantitative comparison results for related tasks in audio-visual segmentation under both the S4 and MS3 settings, as summarized in

Table 2. Comparative analysis with approaches on audio-visual segmentation under the AVSS setting. Results of mIoU (%) and F-score are reported.

Metric	VOS		Baseline [38]		Ours			
	3DC	AOT	R50	PVT-v2	R34	R50	R101	SwinV2-b
mIoU	17.27	25.40	20.18	29.77	30.97	50.74	50.30	**54.58**
F-score	0.216	0.310	0.252	0.352		0.634	0.616	**0.657**

Table 1. The methods we compare include those designed for tasks such as sound source localization (SSL), video object segmentation (VOS), and salient object detection (SOD). In the case of the AVSS setting, we conduct a comparison with the baseline method and two additional methods derived from the VOS task, as indicated in Table 2.

4.4   Ablation Study

We performed several ablation studies to evaluate our method's performance. To ensure fairness, all models are trained from scratch in various settings. The results are presented using the AVSS test split, unless otherwise specified.

Impact of Double Skip Connections. We study the effect of difference structures of skip connections on the overall results in Table 3.

Table 3. Impact of different structures of skip connections. The double skip connection structure brings the highest improvement.

	Skip connections		mIoU	F-score
	ASPP	GAB		
SwinV2-b	✗	✗	24.82	0.301
	✓	✗	43.09	0.523
	✗	✓	39.51	0.485
	✓	✓	**54.58**	**0.657**

Impact of Proposed loss Functions. Table 4 demonstrates that the combination of P-Loss and audio multi-class classification loss functions contribute the most to the overall improvement of our model.

Table 4. Impact of different combinations of loss functions. All our proposed loss functions are found to contribute to the final model.

	Loss functions					mIoU	F-score
	CE	Dice	P-Loss	Audio BCE	AVM [37]		
Baseline: PVT-v2 [38]	✓				✓	30.21	0.401
SwinV2-b	✓					28.50	0.386
	✓	✓				31.14	0.399
	✓	✓	✓			43.72	0.525
	✓	✓	✓	✓		**54.58**	**0.657**
	✓	✓	✓		✓	46.49	0.573

Impact of Ensemble Model. In Table 5, we examine the influence of employing multiple learners within our methodology. Notably, we observe performance enhancements for smaller backbone architectures, such as ResNet-34 and

ResNet-50. In contrast, larger models like ResNet-101 and SwinV2 exhibit little improvement. We infer that the integration of multiple learners within ResNet-34 and ResNet-50 ensemble models leads to improved representation learning, as these learners complement each other. Conversely, the larger ResNet-101 and SwinV2 models independently achieve superior representation learning, rendering the incorporation of the video-level segmentation model less contributory, as it cannot provide representations commensurate with the larger models.

Table 5. Impact of ensemble model. The additional video-level segmentation model brings little improvement than none.

	w/ ensemble model				w/o ensemble model			
	R34	R50	R101	SwinV2-b	R34	R50	R101	SwinV2-b
mIoU	30.97	50.74	50.30	54.58	23.51	46.07	50.36	53.90
F-score	0.432	0.634	0.616	0.657	0.314	0.542	0.588	0.654

5 Conclusion

In this paper, our primary focus is on addressing the challenge of pixel-wise audio-visual (semantic) segmentation. To comprehensively tackle this problem, we introduce a model with dual skip connections, emphasizing the exploration of multi-scale features. Our audio encoder is guided by classification tasks, and we propose a fusion mechanism to incorporate categorical audio information into visual features. To enhance temporal consistency, we introduce an ensemble model that combines frame-level and video-level segmentation approaches. Furthermore, we introduce a self-supervised loss function designed to promote the model's ability to identify all objects present in an input image, regardless of whether they are generating sound or not. We evaluate our method on the AVSBench-v2 dataset, and the quantitative results demonstrate that our model outperforms existing benchmarks, establishing new state-of-the-art performance. Particularly noteworthy are the substantial enhancements observed in the AVSS task. These findings underscore the promise of our approach in integrating audio information into visual data and enhancing pixel-wise visual semantics.

Acknowledgment. This work was supported by the National Natural Science Foundation of China under Grant No. 61972036 and the Industry-University-Institute Cooperation Foundation of the Eighth Research Institute of China Aerospace Science and Technology Corporation (No. SAST2022-049).

References

1. Arandjelovic, R., Zisserman, A.: Look, listen and learn. In: Proceedings of the IEEE International Conference on Computer Vision, pp. 609–617 (2017)
2. Arandjelovic, R., Zisserman, A.: Objects that sound. In: Proceedings of the European Conference on Computer Vision (ECCV), pp. 435–451 (2018)
3. Aytar, Y., Vondrick, C., Torralba, A.: Soundnet: learning sound representations from unlabeled video. Adv. Neural Inf. Processing Syst. **29** (2016)
4. Azulay, A., Weiss, Y.: Why do deep convolutional networks generalize so poorly to small image transformations? arXiv preprint arXiv:1805.12177 (2018)
5. Cai, Z., et al.: A unified multi-scale deep convolutional neural network for fast object detection. In: 14th European Conference on Computer Vision, pp. 354–370 (2016)
6. Chen, H., et al.: Localizing visual sounds the hard way. In: Proceedings of the IEEE/CVF conference on computer vision and pattern recognition. pp. 16867–16876 (2021)
7. Chen, L.C., et al.: Deeplab: semantic image segmentation with deep convolutional nets, atrous convolution, and fully connected CRFS. IEEE Trans. Pattern Anal. Mach. Intell. **40**(4), 834–848 (2017)
8. Duke, B., et al.: Sstvos: sparse spatiotemporal transformers for video object segmentation. In: Proceedings of the IEEE/CVF Conference on Computer Vision and Pattern Recognition, pp. 5912–5921 (2021)
9. Gemmeke, J.F., et al.: Audio set: an ontology and human-labeled dataset for audio events. In: 2017 IEEE International Conference on Acoustics, Speech and Signal Processing (ICASSP), pp. 776–780 (2017)
10. Geng, T., et al.: Dense-localizing audio-visual events in untrimmed videos: a large-scale benchmark and baseline. In: Proceedings of the IEEE/CVF Conference on Computer Vision and Pattern Recognition, pp. 22942–22951 (2023)
11. Gu, K., Yang, B., Ngiam, J., Le, Q., Shlens, J.: Using videos to evaluate image model robustness. arXiv preprint arXiv:1904.10076 (2019)
12. He, K., et al.: Deep residual learning for image recognition. In: Proceedings of the IEEE Conference on Computer Vision and Pattern Recognition, pp. 770–778 (2016)
13. Hershey, S., Chaudhuri, S., Ellis, D.P., Gemmeke, J.F., et al.: CNN architectures for large-scale audio classification. In: 2017 IEEE International Conference on Acoustics, Speech and Signal Processing (ICASSP), pp. 131–135 (2017)
14. Kirillov, A., et al.: Segment anything. arXiv preprint arXiv:2304.02643 (2023)
15. Korbar, B., Tran, D., Torresani, L.: Cooperative Learning of Audio and Video Models from Self-supervised Synchronization, vol. 31 (2018)
16. Lin, Y.B., Tseng, H.Y., Lee, H.Y., Lin, Y.Y., Yang, M.H.: Exploring cross-video and cross-modality signals for weakly-supervised audio-visual video parsing. Adv. Neural. Inf. Process. Syst. **34**, 11449–11461 (2021)
17. Liu, J., et al.: Multi-scale triplet CNN for person re-identification. In: Proceedings of the 24th ACM International Conference on Multimedia, pp. 192–196 (2016)
18. Long, J., Shelhamer, E., Darrell, T.: Fully convolutional networks for semantic segmentation. In: Proceedings of the IEEE Conference on Computer Vision and Pattern Recognition, pp. 3431–3440 (2015)
19. Loshchilov, I., Hutter, F.: Decoupled weight decay regularization. arXiv preprint arXiv:1711.05101 (2017)

20. Mahadevan, S., et al.: Making a case for 3d convolutions for object segmentation in videos. arXiv preprint arXiv:2008.11516 (2020)
21. Mao, Y., et al.: Transformer transforms salient object detection and camouflaged object detection. arXiv preprint arXiv:2104.10127 **1**(2), 5 (2021)
22. Qian, R., et al.: Multiple sound sources localization from coarse to fine. In: 16th European Conference on Computer Vision, pp. 292–308 (2020)
23. Rajput, V.: Robustness of different loss functions and their impact on network's learning. Available at SSRN 4065778
24. Ramaswamy, J., Das, S.: See the sound, hear the pixels. In: Proceedings of the IEEE/CVF Winter Conference on Applications of Computer Vision, pp. 2970–2979 (2020)
25. Ronneberger, O., Fischer, P., Brox, T.: U-net: convolutional networks for biomedical image segmentation. In: Navab, N., Hornegger, J., Wells, W., Frangi, A. (eds.) MICCAI 2015. LNCS, vol. 9351, pp. 234–241. Springer, Cham (2015). https://doi.org/10.1007/978-3-319-24574-4_28
26. Ruan, J., Xie, M., Gao, J., Liu, T., Fu, Y.: EGE-UNET: an efficient group enhanced UNET for skin lesion segmentation. arXiv preprint arXiv:2307.08473 (2023)
27. Simonyan, K., Zisserman, A.: Very deep convolutional networks for large-scale image recognition. arXiv preprint arXiv:1409.1556 (2014)
28. Tan, M., Le, Q.: Rethinking model scaling for convolutional neural networks. In: International Conference on Machine Learning, pp. 6105–6114 (2019)
29. Vaswani, A., et al.: Attention is all you need. Adv. Neural Inf. Process. Syst. **30** (2017)
30. Wu, X., et al.: Binaural audio-visual localization. AAAI **35**(4), 2961–2968 (2021)
31. Xie, S., Girshick, R., Dollár, P., Tu, Z., He, K.: Aggregated residual transformations for deep neural networks. In: Proceedings of the IEEE Conference on Computer Vision and Pattern Recognition, pp. 1492–1500 (2017)
32. Yeung, M., Sala, E., Schönlieb, C.B., Rundo, L.: Unified focal loss: generalising dice and cross entropy-based losses to handle class imbalanced medical image segmentation. Comput. Med. Imaging Graph. **95**, 102026 (2022)
33. Yu, J., et al.: Mm-pyramid: multimodal pyramid attentional network for audio-visual event localization and video parsing. In: Proceedings of the 30th ACM International Conference on Multimedia, pp. 6241–6249 (2022)
34. Zhang, J., Xie, J., Barnes, N., Li, P.: Learning generative vision transformer with energy-based latent space for saliency prediction. Adv. Neural. Inf. Process. Syst. **34**, 15448–15463 (2021)
35. Zhou, B., Khosla, A., Lapedriza, A., Oliva, A., Torralba, A.: Learning deep features for discriminative localization. In: Proceedings of the IEEE Conference on Computer Vision and Pattern Recognition, pp. 2921–2929 (2016)
36. Zhou, J., Zheng, L., Zhong, Y., Hao, S., Wang, M.: Positive sample propagation along the audio-visual event line. In: Proceedings of the IEEE/CVF Conference on Computer Vision and Pattern Recognition, pp. 8436–8444 (2021)
37. Zhou, J., et al.: Audio-visual segmentation. In: European Conference on Computer Vision, pp. 386–403 (2022)
38. Zhou, J., et al.: Audio-visual segmentation with semantics. arXiv preprint arXiv:2301.13190 (2023)
39. Zhou, Z., et al.: Unet++: a nested u-net architecture for medical image segmentation. In: Deep Learning in Medical Image Analysis (DLMIA), pp. 3–11 (2018)

Multi-head Hashing with Orthogonal Decomposition for Cross-modal Retrieval

Wei Liu[1], Jun Li[1(✉)], Zhijian Wu[2], Jianhua Xu[1], and Bo Yang[3]

[1] School of Computer and Electronic Information, Nanjing Normal University, Nanjing 210023, China
lijuncst@njnu.edu.cn
[2] School of Data Science and Engineering, East China Normal University, Shanghai 200062, China
[3] School of Artificial Intelligence, Nanjing University of Information Science and Technology, Nanjing 210044, China

Abstract. Recently, cross-modal hashing has become a promising line of research in cross-modal retrieval. It not only takes advantage of complementary multiple heterogeneous data modalities for improved retrieval accuracy, but also enjoys reduced memory footprint and fast query speed due to efficient binary feature embedding. With the boom of deep learning, convolutional neural network (CNN) has become the de facto method for advanced cross-model hashing algorithm. Recent research demonstrates that dominant role of CNN is challenged by increasingly effective Transformer architectures due to their advantages of long-range modeling by relaxing local inductive bias. However, the absence of inductive bias shatters the inherent geometric structure, which inevitably leads to compromised neighborhood correlation. To alleviate this problem, in this paper, we propose a novel cross-modal hashing method termed Multi-head Hashing with Orthogonal Decomposition (MHOD) for cross-modal retrieval. More specifically, with the multi-modal Transformers used as the backbones, MHOD leverages orthogonal decomposition for decoupling local cues and global features, and further captures their intrinsic correlations through our designed multi-head hash layer. In this way, the global and local representations are simultaneously embedded into the resulting binary code, leading to a comprehensive and robust representation. Extensive experiments on popular cross-modal retrieval benchmarking datasets demonstrate the proposed MHOD method achieves advantageous performance against the other state-of-the-art cross-modal hashing approaches.

Keywords: Cross-modal Retrieval · Transformer · Orthogonal Decomposition · Multi-head Hashing · Aggregated Hash Codes

Supported by the National Natural Science Foundation of China under Grant 62173186, 62076134, 62303230 and Jiangsu provincial colleges of Natural Science General Program under Grant 22KJB510004.

S. Rudinac et al. (Eds.): MMM 2024, LNCS 14555, pp. 170–183, 2024.
https://doi.org/10.1007/978-3-031-53308-2_13

1   Introduction

With the rapid growth of multi-modal data including images, texts, videos and audios, cross-modal retrieval [6,9,15,30] aims to perform fast and accurate retrieval among different modalities, e.g., text-to-image or image-to-text retrieval. By fully exploiting the complementarity among multiple data modalities, inter-modality correlation can be uncovered in depth for significantly improved retrieval accuracy. With the dramatic expansion of multi-modal data volume, achieving efficient cross-modal retrieval is becoming increasingly urgent. Emerging as a popular line of research in cross-modal retrieval, cross-modal hashing [2,10,24] aims to project data of different modalities onto a Hamming space, yielding compact hash codes of maintained similarity for binary feature embedding.

As deep learning has prospered in recent years, the most representative convolutional neural network (CNN) has considerably advanced the cross-modal hashing for unprecedented performance improvements [1,2,10]. In particular, with the rise of Transformer architecture and large-scale pre-trained models [14,19,22], there is a major shift from the CNN-based to the Transformer-based methods [7,8,24], since the latter demonstrates superior performance in cross-modal hashing. In particular, the Transformer-based vision-language models including BERT [4,14] and CLIP [19] can well interpret and encode semantic representations of both images and text for accurate cross-modal retrieval. Aiming to capture global dependencies in sequential data, the Transformer model [25] learns global contextual information by employing a self-attention mechanism to assess the relative importance of each position with respect to other positions, making the individual local contents downplayed in the feature embedding. Although massive efforts are devoted to combining global and local information for generating more discriminative hashing codes, most studies [16,18,26] perform feature fusion prior to hashing process without exploring intrinsic correlation between global and local cues in the process of binary embedding. In this sense, the feature fusion is relatively independent of the binary embedding, leading to the hashing codes which lack global-local perception capability.

To address the above-mentioned drawback, in this study, we propose a novel cross-modal hashing method which is multi-head hashing with orthogonal decomposition (MHOD) for cross-modal retrieval. In MHOD, an orthogonal decomposition module is imposed on local tokens and global features derived from multi-modal Transformer backbones, resulting in a set of tokens serving as the local features. Next, both the local and global features are delivered to our designed multi-head hashing layer to generate separate hashing codes. The resulting binary codes are aggregated using a pooling-like operation, enabling the combination of local and global information in a unified binary representation. To summarize, the contributions of our work are threefold as follows:

- We leverage orthogonal decomposition for decoupling the local cues and the global features, such that both local and global information can be fully encoded in our cross-modal hashing framework.

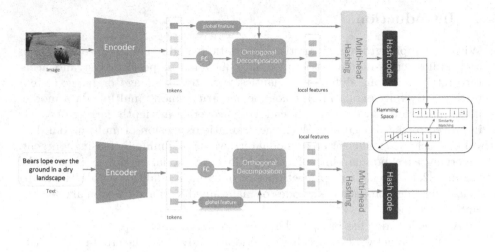

Fig. 1. The network architecture of the proposed MHOD. It consists of three primary blocks including feature encoder used as Transformer backbone, orthogonal decomposition module used for decoupling local cues and global features along with multi-head hashing layer for generating aggregated hash codes. The resulting hash codes can be exploited for accurate and fast cross-modal retrieval in the Hamming space. Different from the existing methods, our method is capable of simultaneously integrating local and global features into binary hashing within Transformer-based cross-modal hashing framework, and considerably benefits mining the intrinsic correlation among different modalities for accurate retrieval.

- To further explore the intrinsic correlation between the local and global features, we simultaneously integrate them into our designed multi-head hashing layer to generate aggregated hash codes with preferable global-local perception capability. This is in contrast to the previous methods in which feature fusion and binary hashing are separately handled.
- Extensive experiments on two public benchmarking cross-modal retrieval datasets demonstrate the superiority of our proposed MHOD against the other state-of-the-art cross-modal hashing models.

The remainder of this paper is organized as follows. We elaborate on our proposed MHOD model in Sect. 2 and carry out extensive experimental evaluations in Sect. 3. The paper is finally concluded in Sect. 4.

2   The Proposed Method

2.1   The Model Framework

While Transformer-based cross-modal hashing methods have achieved considerable success, they either overlook the local clues during the hashing process or perform local-global coupling independent of binary embedding, leading to hash codes with degraded local-global perception. To address the drawback, we propose a cross-modal hashing method termed MHOD for cross-modal retrieval in this study. As shown in Fig. 1, our MHOD first leverages feature encoder modules for generating global features from class tokens across data modalities. Subsequently, local clues are decoupled from the local tokens via orthogonal decomposition and combined with global features in the multi-head hashing layer, producing aggregated hash codes for cross-modal retrieval. Next, We will discuss these key modules and the training loss functions in details.

2.2   Feature Encoder

For notation, $D = \{X_i, Y_i\}_{i=1}^N$ denotes a batch of pairwise data modalities, while N is the number of instances. Since we mainly focus on two different modalities of image and text in cross-modal hashing, X_i and Y_i indicate the original i^{th} image and text instance respectively. In each training batch, $F \in R^{N \times d}$ denotes the feature embedding extracted from the feature encoder where d is the dimension of feature embedding. In addition, F^I and F^T respectively represent the feature of image and text modality.

With the help of the successful pre-trained large models, we adopt the pre-trained CLIP model as our feature encoder for both image and text input. More specifically, image encoder is used as the vision Transformer structure ViT [5], while the text encoder is GPT2 [20] which is a modified architecture developed from Transformer. For brevity, the CLIP encoders are denoted as $CLIP$ which takes the cross-modal image and text as input. Mathematically, the feature encoding process of raw data can be formulated as:

$$F_g, F_t = CLIP(D) \tag{1}$$

where F_g is essentially the object-specific class token obtained by feature encoding from a global perspective, while F_t denotes the remaining local-aware tokens.

2.3   Orthogonal Decomposition

Since local tokens F_t in Eq. (1) also characterize certain global information via self-attention mechanism for exploring local interaction, it is necessary to decouple the local clues and the global information for deriving local features from F_t. Following [27], consequently, we leverage the orthogonal decomposition module for decoupling local cues and the global features. Specifically, F_g and F_t are treated as the input, while F_t^k denotes the kth token in F_t. At first, the tokens pass through consecutive Fully-Connected layers (FCs) such that they have the same dimension as the global features. Afterwards, each token F_t^k is projected onto the global feature F_g, which can be mathematically expressed as follows:

$$F_{t,proj}^k = \frac{F_t^k \cdot F_g}{\|F_g\|_2^2} F_g \tag{2}$$

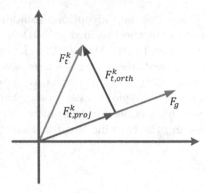

Fig. 2. Illustration of orthogonal decomposition for deriving the local features from local tokens. For notation, F_g represents global feature and F_t^k indicates a local token. In order to decouple F_t^k and F_g, we first project F_t^k onto F_g to obtain $F_{t,proj}^k$ which encodes the global component related to F_g. Thus, the orthogonal component $F_{t,orth}^k$ obtained by subtracting $F_{t,proj}^k$ from F_t^k can be treated as the local feature that is independent of F_g.

where $F_t^k \cdot F_g$ indicates dot product operation and $\|\cdot\|_2$ is ℓ_2 norm. As demonstrated in Fig. 2, the orthogonal component can be calculated as the difference between F_t^k and its projection onto F_g, which is formulated as:

$$F_{t,orth}^k = F_t^k - F_{t,proj}^k \qquad (3)$$

In this way, $F_{t,orth}^k$ is independent of the global feature F_g, and can be treated as the local feature F_l that is separated from the global clues.

2.4   Multi-head Hashing

To leverage both the global and local information effectively, the decoupled global and local features are forwarded to hashing layer for generating and aggregating efficient binary hashing codes. In MHOD, the hashing layer includes multiple heads as shown in Fig. 3. Each head comprises a MLP, a *tanh* activation function, and a *sign* function. The MLP maps high-dimensional features to low-dimensional ones. Using the *tanh* function, the low-dimensional features are rescaled from -1 to 1. Finally, the *sign* function converts these features into discrete binary embeddings for generating the hash code b:

$$b = sign(tanh(MLP(feat))) \qquad (4)$$

Each hashing head within our multiple heads receives different input features $feat$, with the first head taking global features. Each of the remaining local feature groups is averaged, producing averaged local features delivered to individual hashing head. Thus a hashing matrix $H \in R^{L \times M}$ can be derived, where M is the number of hashing heads and L denotes the length of hash code. We generate multiple hash codes from global and local vectors, but only one hash code

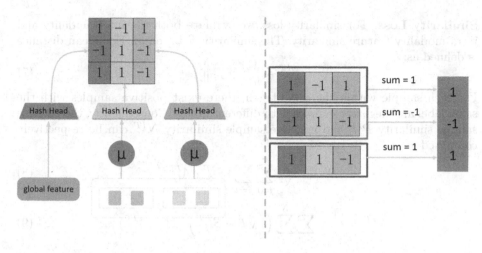

Fig. 3. The structure of our designed multi-head hashing module (left) and accumulation mechanism for aggregating multi-head output into the final hash code (right). Each hashing head contains a MLP, a *tanh* activation function, and a *sign* function. μ represents the mean value operation which is used to average the local features. With the hash codes derived from multiple hashing heads, a simple accumulation strategy is adopted to generate the aggregated hash code.

contributes to the final feature matching. The matching is refined at each bit, handling each bit of the final hash code via a voting mechanism. This ensures that each bit of the hash code is optimized. As illustrated in Fig. 3, we adopt a simple fusion mechanism to aggregate the hashing codes resulting from different heads. Mathematically, it can be formulated as:

$$B_i = \begin{cases} 1 & \sum_{m=1}^{M} O(i,m) > 0 \\ -1 & otherwise \end{cases} \tag{5}$$

Specifically, i^{th} element of the hash code B_i is set as 1 when the sum of ith row in H is greater than 0, and -1 conversely.

2.5 Loss Function

The loss function of our MHOD network for model training includes two critical components, namely similarity loss and hashing loss. Let f_I^i denote the feature for image i and f_T^j for the corresponding text j. With label a_i for each sample, the semantic relation of two different samples can be defined as:

$$A_{ij} = \begin{cases} 1 & a_i \cdot a_j > 0 \\ 0 & a_i \cdot a_j = 0 \end{cases} \tag{6}$$

Similarity Loss. For similarity loss, we evaluate both the inter-modality and intra-modality feature similarity. The similarity S based on Euclidean distance is defined as:

$$S_{ij}^{IT} = \left\| f_I^i - f_T^j \right\|_2 \tag{7}$$

For each sample within the same batch, there exist positive samples with the same label and negative samples with different labels. Consequently, the positive sample similarity P^{IT} and negative sample similarity N^{IT} can be respectively computed as:

$$P^{IT} = \frac{1}{N^2} \sum_{i=1}^{N} \sum_{j=1}^{N} \left(S_{ij}^{IT} \cdot A_{ij}^{IT} \right)^2 \tag{8}$$

$$N^{IT} = \frac{1}{N^2} \cdot \sum_{i=1}^{N} \sum_{j=1}^{N} \left(\left(\sqrt{L} - S_{ij}^{TT} \right) \cdot \left(1 - A_{ij}^{IT} \right) \right)^2 \tag{9}$$

Therefore, cross-modal similarity can be formulated as:

$$L_{sim}^{IT} = P^{IT} + N^{IT} \tag{10}$$

Similarly, the image-related and text-specific intra-modality similarity L_{sim}^{II} and L_{sim}^{TT} can be obtained respectively, and the complete similarity loss L_{sim} is:

$$L_{sim} = L_{sim}^{IT} + L_{sim}^{II} + L_{sim}^{TT} \tag{11}$$

Hashing Loss. In addition to similarity loss, hashing loss L_{hash} aims to calculate the information loss of binary embedding:

$$L_{hash} = \frac{1}{M} \left(\sum_{m=1}^{M} H_I^m + \sum_{m=1}^{M} H_T^m \right) \tag{12}$$

where H represents modality-specific hashing loss. More specifically, image-related hashing loss H_I can be formulated as:

$$H_I = \frac{1}{N} \sum_{n=1}^{N} \sqrt{\sum_{l=1}^{L} \left(f_I^{(n,l)} - h_I^{(n,l)} \right)^2} \tag{13}$$

where h is computed as $sign(f)$. Besides, H_T can be calculated analogously. Notably, the modality-specific hashing loss is computed independently for each hashing head.

Overall Loss. The overall loss function is the weighted sum of the above-mentioned similarity loss and hashing loss:

$$L = L_{sim} + \lambda L_{hash} \tag{14}$$

where λ is a balancing hyper-parameter to compromise between the two terms. It will be discussed in the parameter analysis in the following section of experiments.

Table 1. Comparison of different cross-modal hashing methods in the two public datasets (mAP@all). The best results are highlighted **in bold** and the second best are <u>underlined</u>. The results demonstrate that our method is superior to the other state-of-the-art models in both MS-COCO and NUS-WIDE.

Dataset	Method		I to T			T to I	
		16bit	32bit	64bit	16bit	32bit	64bit
MS-COCO	DCMH [10]	0.5533	0.5540	0.5667	0.5272	0.5467	0.5521
	SCAHN [12]	0.6095	0.6502	0.6435	0.6035	0.6403	0.6435
	MSSPQ [31]	0.5710	0.5862	0.5881	0.5472	0.5630	0.5985
	DADH [1]	0.6388	0.6668	0.6812	0.6027	0.6334	0.6528
	DCHMT [24]	<u>0.6447</u>	<u>0.6757</u>	<u>0.6915</u>	<u>0.6531</u>	<u>0.6832</u>	<u>0.7025</u>
	MHOD (Ours)	**0.6595**	**0.6870**	**0.7056**	**0.6713**	**0.6940**	**0.7130**
NUS-WIDE	DADH [1]	0.6492	0.6662	0.6664	0.6501	0.6679	0.6808
	DMFH [18]	0.6065	0.6212	0.6396	0.6307	0.6468	0.6798
	TEACH [28]	0.6512	0.6643	0.6704	0.6732	0.6871	0.6893
	DCHMT [24]	<u>0.6799</u>	<u>0.6992</u>	<u>0.7038</u>	<u>0.6876</u>	<u>0.7104</u>	<u>0.7253</u>
	SCAHN [12]	0.6155	0.6403	0.6662	0.6446	0.6702	0.6980
	MHOD (Ours)	**0.6992**	**0.7083**	**0.7168**	**0.7041**	**0.7135**	**0.7299**

3   Experiments

3.1   Datasets and Experimental Settings

We have evaluated our approach in two popular benchmarking datasets for cross-modal retrieval, i.e., MS-COCO [13] and NUS-WIDE [3]. On both datasets, we randomly select 10,000 samples for training data, 5,000 samples as queries and the rest as retrieval database. We initialize both the image and text encoders with a pre-trained CLIP(ViT-B/32) model. In our proposed MHOD, Adam optimizer [11] is used for model training. The initial learning rate is set as 0.001 in MS-COCO, 0.0001 in NUS-WIDE, and 1e-7 for the Transformer encoders. The batch size is set to 64 and the number of hashing heads is 3. In terms of evaluation metric, mAP@all and mAP@50 are used for performance measures. All the experiments are conducted on a server with Intel i9-10900K CPU and one NVIDIA RTX3090 GPU using PyTorch framework.

3.2   Results

Comparative Studies. As demonstrated in Table 1, we have compared our MHOD model with recent eleven state-of-the-art cross-modal hashing methods including DADH [1], DMFH [18], TEACH [28], DCHMT [24], SCAHN [12],

Table 2. Comparison of different cross-modal hashing methods in NUS-WIDE dataset (mAP@50). The best results are highlighted **in bold**.

Method	I to T			T to I		
	16bit	32bit	64bit	16bit	32bit	64bit
DJSRH [23]	0.724	0.773	0.798	0.712	0.744	0.771
HNH [29]	0.582	0.789	0.800	0.423	0.747	0.781
DUCH [17]	0.753	0.775	0.814	0.726	0.758	0.781
DAEH [21]	0.766	0.789	0.809	0.718	0.751	0.766
MHOD (Ours)	**0.806**	**0.817**	**0.836**	**0.766**	**0.773**	**0.783**

Table 3. Ablation studies in MS-COCO using 64bit hashing code (mAP@all). MH denotes our designed multi-head hashing component for generating aggregated hashing codes. It should be noted that the first two methods corresponding to the first two rows directly employ a straightforward *sign* function mapping instead of MH for binary embedding.

Global	Local	MH	I to T	T to I
✓			0.6998	0.7010
	✓		0.6912	0.6970
✓		✓	0.7013	0.7076
	✓	✓	0.6934	0.7012
✓	✓	✓	0.7056	0.7130

DCMH [10], MSSPQ [31], DUCH [17], DAEH [21], DJSRH [23] and HNH [29] in the two benchmarking datasets for different cross-modal retrieval tasks. More specifically, for image-to-text retrieval task, our MHOD reports the highest mAP@all scores at 65.95%, 68.70% and 70.56% in MS-COCO, surpassing DCHMT model by 1.5%, 1.1% and 1.4% when the length of the hashing code is 16, 32 and 64, respectively. In NUS-WIDE, analogous advantages against DCHMT can also be observed with respective performance gains of 1.9%, 0.9% and 1.3% for various hashing codes of different lengths. Similar results are also shown for text-to-image retrieval task, suggesting that our method beats the other competing models in both MS-COCO and NUS-WIDE. In terms of the mAP@50 metric, our MHOD also reports the highest accuracies of 80.6%, 81.7% and 83.6% in NUS-WIDE when the length of hash code is 16, 32 and 64 respectively, revealing consistent advantages against the state-of-the-arts as shown in Table 2.

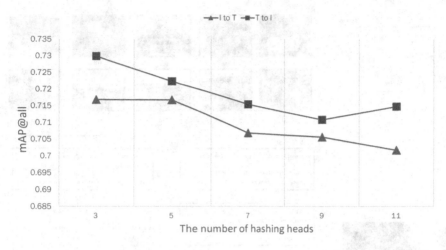

Fig. 4. Performance of the proposed MHOD with different numbers of hashing head in NUS-WIDE using 64bit hashing code.

3.3   Ablation Study

To gain an insight into the effectiveness of each module within our MHOD network, we have carried out comprehensive ablation studies for exploring the effect of individual module on our model. Firstly, we investigate the feature fusion module and compare different strategies of feature embedding. As illustrated in Table 3, combining both local and global features contributes to further performance boost. For text-to-image retrieval task, To be specific, our complete model with local-global fusion provides respective performance gains of 0.5% and 1.2% over the approach with single global feature embedding and single local feature embedding. While the global features plays a dominant role in feature embedding, local contents are supplementary to global feature and conducive to boosting model performance. On the other hand, with our multi-head hashing layer (which is MH for short in Table 3), the retrieval accuracy increases from 70.10% to 70.76% when only global feature embedding is performed. When only considering local cues, similar improvement can also be observed, which substantially suggests the beneficial role of our MH module.

In our MHOD model, the multi-head hashing module consists of multiple hashing heads. In addition to the above ablation studies, we discuss the effect of different head numbers on the model performance. As shown in Fig. 4, the overall declined performance is observed with increasing head number. This can be explained by less tokens assigned to each head when the hashing head increases. Consequently, the amount of information available for each individual hashing head decreases. This reduction in token allocation can adversely affect the retrieval accuracy of the hashing code generated from each head. As a result, the aggregated hashing code combining the outputs of all the heads may still suffer from degraded retrieval accuracy.

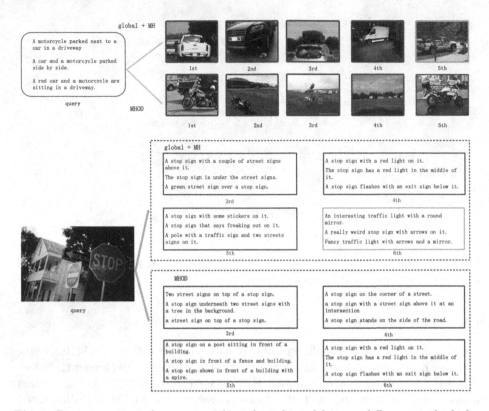

Fig. 5. Demonstration of top returned results achieved by two different methods for different cross-modal retrieval tasks. The query-related ground-truths are highlighted in a green box, whereas the mismatched ones are annotated in a red box. Compared to the method only considering global information, our MHOD model can retrieve more positive images or texts that better match the query by leveraging both local and global contents. (Color figure online)

In addition to the above quantitative results of ablation studies, we also present some qualitative results as shown in Fig. 5. It is shown that compared to only using global feature, our proposed MHOD using both global and local features not only brings a boost in retrieval accuracy, but also helps us find more query-related targets. For instance, for text-to-image retrieval task, the query texts are closely related to two objects, namely car and motorcycle. The model which only focuses on global contents mainly captures the car object while overlooking the other one. In contrast, by taking advantage of both global and local clues, our model can find the images including both car and motorcycle, which implies the beneficial and supplementary role of local information in capturing multiple objects in cross-model retrieval.

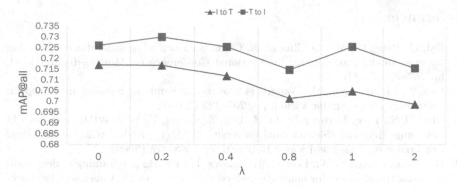

Fig. 6. λ Sensitivity Analysis in NUS-WIDE.

3.4   Parameter Sensitivity Analysis

In the loss function of our MHOD model as formulated in Eq. (14), the hyper-parameter λ needs to be tuned for balancing the similarity loss and hashing loss. Figure 6 demonstrates the performance of our model with varying λ values in NUS-WIDE for different cross-modal retrieval tasks. It can be observed that the best results of 72.99% and 71.68% are reported when λ equals 0.2 for image-to-text retrieval and text-to-image retrieval task. Interestingly, the results even exceed the model performance when λ equals 0 which implies hashing loss is not involved in our model. Profiting from the multi-head hashing module, our model does not severely suffer the information loss resulting from binary embedding of the hashing loss.

4   Conclusions

In this paper, we present a novel cross-modal hashing method termed MHOD for cross-modal retrieval task. More specifically, it leverages the multi-modal Transformer encoders for generating global features and local tokens, which is followed by the orthogonal decomposition module for decoupling the local cues and the global features. Then, the feature fusion is achieved by passing both global and local features through multi-head hashing layer for generating aggregated hash codes. Different from the previous methods in which either local contents are downplayed or the local-global fusion is independent of binary hashing, our method can integrate local-global feature fusion into the hashing process for improved local-global perception capability. Extensive experiments in two public benchmarking datasets show that the proposed MHOD achieves the state-of-the-art performance.

References

1. Bai, C., Zeng, C., Ma, Q., Zhang, J., Chen, S.: Deep adversarial discrete hashing for cross-modal retrieval. In: International Conference on Multimedia Retrieval, pp. 525–531 (2020)
2. Cao, Y., Liu, B., Long, M., Wang, J.: Cross-modal hamming hashing. In: European Conference on Computer Vision, pp. 207–223 (2018)
3. Chua, T.S., Tang, J., Hong, R., Li, H., Luo, Z., Zheng, Y.: NUS-WIDE: a real-world web image database from national university of Singapore. In: ACM International Conference on Image and Video Retrieval, pp. 368–375 (2009)
4. Devlin, J., Chang, M.W., Lee, K., Toutanova, K.: BERT: pre-training of deep bidirectional transformers for language understanding. In: North American Chapter of the Association for Computational Linguistics, pp. 4171–4186 (2019)
5. Dosovitskiy, A., et al.: An image is worth 16x16 words: transformers for image recognition at scale. In: International Conference on Learning Representations, pp. 1–22 (2021)
6. Faghri, F., Fleet, D.J., Kiros, J.R., Fidler, S.: VSE++: improved visual-semantic embeddings with hard negatives. In: British Machine Vision Conference, pp. 1–14 (2018)
7. Hong, J., Liu, H.: Deep cross-modal hashing retrieval based on semantics preserving and vision transformer. In: International Conference on Electronic Information Technology and Computer Engineering, pp. 52–57 (2022)
8. Huo, Y., et al.: Deep semantic-aware proxy hashing for multi-label cross-modal retrieval. IEEE Trans. Circuits Syst. Video Technol. (2023)
9. Jia, C., et al.: Scaling up visual and vision-language representation learning with noisy text supervision. In: International Conference on Machine Learning, pp. 4904–4916 (2021)
10. Jiang, Q.Y., Li, W.J.: Deep cross-modal hashing. IEEE Conference on Computer Vision and Pattern Recognition, pp. 3270–3278 (2016)
11. Kingma, D.P., Ba, J.: Adam: a method for stochastic optimization. In: International Conference for Learning Representations, pp. 1–15 (2015)
12. Liang, M., et al.: Semantic structure enhanced contrastive adversarial hash network for cross-media representation learning. In: ACM International Conference on Multimedia, pp. 277–285 (2022)
13. Lin, T.Y., et al.: Microsoft COCO: common objects in context. In: European Conference on Computer Vision, pp. 740–755 (2014)
14. Lu, J., Batra, D., Parikh, D., Lee, S.: ViLBERT: petraining task-agnostic visiolinguistic representations for vision-and-language tasks. In: Neural Information Processing Systems, pp. 13–23 (2019)
15. Ma, L., Li, H., Meng, F., Wu, Q., Ngi Ngan, K.: Global and local semantics-preserving based deep hashing for cross-modal retrieval. Neurocomputing **312**, 49–62 (2018)
16. Ma, X., Zhang, T., Xu, C.: Multi-level correlation adversarial hashing for cross-modal retrieval. IEEE Trans. Multimedia **22**, 3101–3114 (2020)
17. Mikriukov, G., Ravanbakhsh, M., Demir, B.: Deep unsupervised contrastive hashing for large-scale cross-modal text-image retrieval in remote sensing. arXiv preprint arXiv:2201.08125 (2022)
18. Nie, X., Wang, B., Li, J., Hao, F., Jian, M., Yin, Y.: Deep multiscale fusion hashing for cross-modal retrieval. IEEE Trans. Circuits Syst. Video Technol. **31**, 401–410 (2021)

19. Radford, A., et al.: Learning transferable visual models from natural language supervision. In: International Conference on Machine Learning, pp. 8748–8763 (2021)
20. Radford, A., Wu, J., Child, R., Luan, D., Amodei, D., Sutskever, I., et al.: Language models are unsupervised multitask learners. OpenAI blog **1**(8), 9 (2019)
21. Shi, Y., et al.: Deep adaptively-enhanced hashing with discriminative similarity guidance for unsupervised cross-modal retrieval. IEEE Trans. Circuits Syst. Video Technol. **32**, 7255–7268 (2022)
22. Singh, A., et al.: FLAVA: a foundational language and vision alignment model. In: IEEE Conference on Computer Vision and Pattern Recognition, pp. 15617–15629 (2022)
23. Su, S., Zhong, Z., Zhang, C.: Deep joint-semantics reconstructing hashing for large-scale unsupervised cross-modal retrieval. In: IEEE International Conference on Computer Vision, pp. 3027–3035 (2019)
24. Tu, J., Liu, X., Lin, Z., Hong, R., Wang, M.: Differentiable cross-modal hashing via multimodal transformers. In: ACM International Conference on Multimedia, pp. 453–461 (2022)
25. Vaswani, A., et al.: Attention is all you need. In: Neural Information Processing Systems, pp. 6000–6010 (2017)
26. Wang, H., Zhao, K., Zhao, D.: A triple fusion model for cross-modal deep hashing retrieval. Multimedia Syst. **29**, 347–359 (2022)
27. Yang, M., et al.: DOLG: single-stage image retrieval with deep orthogonal fusion of local and global features. In: IEEE International Conference on Computer Vision, pp. 11752–11761 (2021)
28. Yao, H.L., Zhan, Y.W., Chen, Z.D., Luo, X., Xu, X.S.: TEACH: attention-aware deep cross-modal hashing. In: International Conference on Multimedia Retrieval, pp. 376–384 (2021)
29. Zhang, P., Luo, Y., Huang, Z., Xu, X.S., Song, J.: High-order nonlocal hashing for unsupervised cross-modal retrieval. World Wide Web **24**, 563–583 (2021)
30. Zhen, L., Hu, P., Wang, X., Peng, D.: Deep supervised cross-modal retrieval. In: IEEE Conference on Computer Vision and Pattern Recognition, pp. 10386–10395 (2019)
31. Zhu, L., Cai, L., Song, J., Zhu, X., Zhang, C., Zhang, S.: MSSPQ: multiple semantic structure-preserving quantization for cross-modal retrieval. In: International Conference on Multimedia Retrieval, pp. 631–638 (2022)

Fusion Boundary and Gradient Enhancement Network for Camouflage Object Detection

Guangrui Liu and Wei Wu[✉]

Inner Mongolia University, Hohhot, China
cswuwei@imu.edu.cn

Abstract. The problems of boundary interruption and missing internal texture feature have not been well solved in the current camouflaged object detection model, and the parameters of the model are generally large. To overcome these challenges, we propose a fusion boundary and gradient enhancement networks BGENet, which guides the context features by gradient features and boundary features together. BGENet is divided into three branches, context feature branch, boundary feature branch and gradient feature branch. Furthermore, a parallel context information enhancement module is introduced to enhance the context features. The designed pre-background information interaction module is used to highlight the boundary features of the camouflaged object and guide the context features to compensate for the boundary breaks in the context features, while we use the learned gradient features to guide the context features through the proposed gradient guidance module, and enhances internal information about context features. Experiments on CAMO, COD10K and NC4K three datasets confirm the effectiveness of our BGENet, which uses only 20.81M parameters and achieves superior performance compared with traditional and SOTA methods.

Keywords: Camouflaged object detection · Object gradient · Object boundary

1 Introduction

The main objective of camouflage object detection (COD) [7,8] is to find object with features such as similar color, pattern and texture structure to the background in a particular scene. The realization of camouflage object detection contributes greatly to a number of real-world applications, e.g., polyp segmentation [9], lung infection segmentation [4], and anti-military camouflage [16], etc., which have both scientific and practical value.

Early on, manual features were used to process camouflage object detection (COD), but this approach was only applicable to camouflaged scenes with simple backgrounds. In recent studies [7,14], compelling results have been achieved by using the entire object-level ground-truth mask for supervision, and later on, research has focused on various sophisticated techniques to enhance the COD

© The Author(s), under exclusive license to Springer Nature Switzerland AG 2024
S. Rudinac et al. (Eds.): MMM 2024, LNCS 14555, pp. 184–198, 2024.
https://doi.org/10.1007/978-3-031-53308-2_14

features, e.g., boundary-guided [25,31] and gradient-map-guided [12], and other techniques. However, learning features only from the boundary-guided will lose some features inside the object, especially for some complex scenarios [12]. While learning features only from the gradient-map is able to preserve the internal information of the object and also compensate boundary information, but not all training images have complete gradient maps (see Fig. 1) [12], which makes the method limited. At the same time, we learned from [28] that using multiple forms of feature co-guidance can compensate for the disadvantages of using one form of feature alone, thus improving the performance of the object detection.

Inspired by the above ideas, we propose an effective lightweight network (termed as BGENet) for the COD task, which introduces both boundary and gradient to guide the features of camouflaged object. The underlying assumptions are that the boundary guidance method is used to solve the problem of camouflaged object boundary interruption and that the loss of internal texture features of camouflaged object is compensated by gradient guidance. Our model is divided into three branches, context feature branch, boundary feature branch, and gradient feature branch. In order to fuse the features of different branches, inspired by human observation of camouflaged object, the Pre-Background Information Interaction Module (PBIM) is proposed, which is divided into foreground and background streams, and highlights the boundary features through the interaction of foreground and background streams information, and the Boundary Guidance Module (BG) [31] is introduced in the PBIM, which the rough feature map after the boundary guidance is obtained while enhance boundary learning ability. In addition, a gradient guidance module (GGM) is proposed, which uses a grouping strategy to integrate gradient features and context features. Meanwhile, in order to extract rich context semantic information, a Parallel Context Information Enhancement Module (PCIE) is proposed. Benefiting from these designs, our BGENet is able to better preserve the boundary information and internal information of camouflaged object.

On three challenging COD datasets, our BGENet uses only 20.81M parameters and achieves better performance. Compared with the current cutting-edge COD methods, BGENet is more lightweight and has better performance. Our main contributions are summarized as follows:

1) We propose a fusion boundary and gradient enhancement network BGENet for camouflage object detection tasks. It achieved better results with fewer parameters on three challenging dataset.
2) We design a parallel context information enhancement module. Through four parallel branches, making the features extracted by the backbone network have richer context information.
3) We designed the pre-background information interaction module and the gradient guidance module. The former is used to separate the foreground and background information of the object, and better highlight the boundary information, so as to conduct the boundary guidance of the context features, the latter adopts the grouping strategy and uses the learned gradient features to guide the context features.

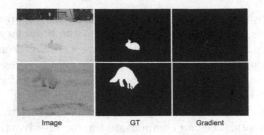

Fig. 1. We observe that not all images have a complete gradient map.

2 Related Work

2.1 Traditional Methods

Traditional techniques for camouflaged object detection involve extracting manual features from the camouflaged area and background. These features include 3D convexity [22], covariance matrix [24], expectation maximization statistics [17], Gaussian mixture models [10], etc. While these methods work well in simple backgrounds, they struggle to perform effectively in complex backgrounds.

2.2 CNN-Based Methods

CNN-based methods are generally categorized into the following strategies: 1) Two-stage strategy: the SINet [8] employs a search-and-identify strategy, which was practiced in the early days of the COD fulfill. SINetV2 [7] further improves performance by introducing the neighbor connection decoder (NCD) and group inversion attention mechanisms in SINet. 2) Joint learning strategies: ANet [14] pioneered joint learning by combining classification and segmentation approaches. JCSOD [15] extends this concept by incorporating salient-to-camouflaged object learning. LSR [18] ranks the camouflage degree of objects, while MFFN [30] leverages multiple observation perspectives to effectively distinguish semantic camouflage features. 3) Guidance-based strategies: The recently proposed BSANet [31] and BGNet [25] use boundary-guided to guide the learning of camouflage features. Gradient maps of camouflaged object were first used to guide the learning of context features by Ji et al. [12] 4) Attention-based strategies: Sun et al. [1] introduced a network with an attention-induced cross-level fusion module for combining multi-scale features, along with a two-branch global context module for extracting diverse context information. Taking cues from predator detection, Mei et al. [21] designed PFNet with a localization and focusing module for recognition. Zhuge et al. [32] proposed a cube-like COD architecture that integrates multi-layer features using attention fusion and X-connections.

2.3   Transformer-Based Methods

Transformer-based techniques are a recent avenue for enhancing detection precision. Mao et al. [19] introduced a Transformer-driven difficulty-aware learning approach for spotting camouflaged and salient objects. UGTR [29] employs probabilistic models within the Transformer framework to grasp the uncertainty surrounding camouflaged objects. Huang et al. [11] presented FSPNet, a novel method that employs local enhancement and hierarchical decoding of Transformer features to progressively detect camouflaged objects.

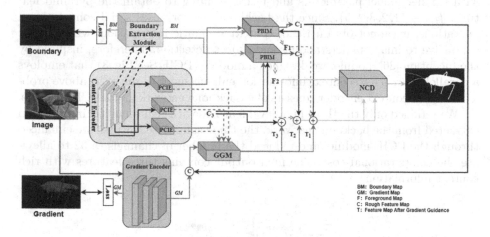

Fig. 2. Overall pipeline of the proposed BGENet. ⊕ represents element-by-element addition, ⓒ represents a concatenation.

3   Proposed BGENet

It is understood from [28] that multiple features that complement each other can improve the accuracy of object detection. Inspired by this, we propose a fusion boundary and gradient enhancement network (BGENet), the overall structure of the network is shown in Fig. 2. For an input $I \in R^{W \times H \times 3}$, where W and H represent the width and height of the image, we use EfficientNetV2 [27] as the backbone to extract multi-level features $f_i, i \in \{1, 2, 3, 4, 5\}$. Firstly, we input f_3, f_4, f_5 into the PCIE module to obtain rich context features, and input f_3, f_4, f_5 into Boundary Extraction Module (BEM) to obtain boundary information [31]. At the same time, the image I is input into the gradient encoder to obtain the required gradient features [12]. Then, through the interaction of the foreground and background information of the context feature through the Pre-Background Information Interaction Module (PBIM) designed by us, the foreground map of the object and the rough feature map after boundary guidance is obtained, the boundary information of the object is more prominent. In addition, we also design the Gradient Guidance Module (GGM), which uses the gradient features to guide the context features through the grouping strategy, so that the context features have rich internal texture features. Finally, the guided features and the foreground map are added fusion, and input into the NCD [7] to obtain the final detection result.

3.1 Context Encoder

For the input camouflaged image $I \in R^{W \times H \times 3}$, we use EfficientNetV2 [27] with smaller model parameters and faster training to obtain the pyramid features $f_i, i \in \{1, 2, 3, 4, 5\}$. Since the backbone network uses serial convolutional operations, it cannot obtain richer context semantic information, which is not conducive to image understanding and object detection. Therefore, inspired by the Inception [26] module, we designed a module (PCIE See Fig. 3) that employs a parallel approach to context information enhancement to solve the above problem. So, our context encoder consists of a backbone network and a PCIE module.

We retained only the three high-level features with rich semantic information extracted from the backbone network, the three candidate features are enhanced through the PCIE module, and reduced the number of channels to 32 to alleviate the computational cost. The final output contains three features with rich context information.

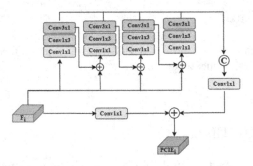

Fig. 3. Structural diagram of the proposed PCIE. Extracting context information through four parallel branches.

3.2 Gradient Encoder

We introduce the gradient map of the camouflag object as supervision for overcoming the problem of missing internal texture features. To reduce the computational burden, we designed a lightweight gradient encoder consisting of only a few residual blocks. Table 1 shows the detailed information of the gradient encoder.

We use the output of the fourth residual block as the supervised computational loss, since the output of the third residual block has a greater resolution and is able to retain more detailed features, the features output from the third residual block are used to fuse with the rough feature map output from the PBIM (Sec. 3.5 Details) to make up for the lack of gradient maps, and the fused features are used to co-guide the context features.

Table 1. Detailed information on gradient encoders. K: kernel size, C: output channels, S: stride and P: padding. CBR represents a residual block consisting of a Convolution layer, a Batch Normalization layer and a ReLU layer.

Layer	Input Size	Output Size	Component	K	C	S	P
1st	$3 \times H \times W$	$64 \times \frac{H}{2} \times \frac{W}{2}$	CBR	7	64	2	3
2nd	$64 \times \frac{H}{2} \times \frac{W}{2}$	$64 \times \frac{H}{4} \times \frac{W}{4}$	CBR	3	64	2	1
3rd	$64 \times \frac{H}{4} \times \frac{W}{4}$	$32 \times \frac{H}{8} \times \frac{W}{8}$	CBR	3	32	2	1
4th	$32 \times \frac{H}{8} \times \frac{W}{8}$	$1 \times \frac{H}{8} \times \frac{W}{8}$	CBR	1	1	1	0

3.3   Gradient Guidance Module

In order to fully integrate the context features with the gradient features, a gradient guidance module (GGM, see Fig. 4) is elaborated. The module uses a grouping strategy [7], where the extracted gradient features G_i and context features X_i are equally divided into 8 groups, it is expressed as follows:

$$G_i \rightarrow G_i^m, m \in \{1, 2, 3, 4, 5, 6, 7, 8\} \tag{1}$$
$$X_i \rightarrow X_i^m, m \in \{1, 2, 3, 4, 5, 6, 7, 8\} \tag{2}$$

then each context feature and gradient feature are alternately connected together. Three parallel branches are used for grouping convolution of the connected features, the number of groups is set to 2, 4, 8, and finally, the features A_j output from the three parallel branches are summed element-by-element with the input context features to obtain the gradient-guided features T_i, it is expressed as follows:

$$T_i = X_i \oplus \sum_{j=1}^{3} A_j \tag{3}$$

where $\oplus$ means the element-wise addition, and $\sum$ denotes a sum of multiple terms, X_i, $i \in \{3, 4, 5\}$ denotes a context feature.

3.4   Boundary Extraction Module

In order to extract the features of the boundary, we designed a simple Boundary Extraction Module (BEM) that can efficiently extract boundary features from

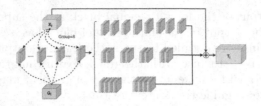

Fig. 4. Structure of the proposed gradient guidance module.

feature $f_i, i \in \{3,4,5\}$ retained by the backbone network. Specifically, boundary feature extraction is performed for $f_i, i \in \{3,4,5\}$ by simple one residual block, and the rough boundary feature of the camouflaged object can be obtained by simply joining the three extracted boundary features (See Fig. 5). Specifically:

$$f_3^B = CBR(f_3) \tag{4}$$

$$f_4^B = CBR(f_4) \tag{5}$$

$$f_5^B = CBR(f_5) \tag{6}$$

$$f_B = Concat(f_3^B, f_4^B, f_5^B) \tag{7}$$

where $f_i^B, i \in \{3,4,5\}$ denotes the ith boundary feature. CBR represents a residual block consisting of a Convolution layer, a Batch Normalization layer and a ReLU layer. f_B represents the final boundary feature.

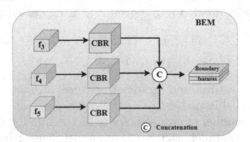

Fig. 5. Structure of the proposed boundary extraction module.

3.5   Pre-background Information Interaction Module

Boundary information at the pre-background junction is crucial for detecting camouflaged objects. Effective fusion of pre-background information enables us to better find the boundary information of an object. Inspired by [2], we introduce the Pre-Background Information Interaction Module (PBIM) with dual streams: one stream emphasizes image foreground information, and the other focuses on image background information. Interaction pre-background information to enhance the learning ability of boundaries.

Specifically, each stream in the PBIM is multiplied by the corresponding attention map, for example, the foreground attention map of layer i is the feature map of the PBIM output of layer $i+1$ calculated by sigmod, denoted by $W_f^i = \sigma(C_{i+1})$, meanwhile, the background attention map of layer i is obtained by subtracting the foreground attention map using 1, written as $W_b^i = 1 - W_f^i$. Before multiplying, the number of channels is expanded to the same number of channels as the PCIE output result, and the calculation formula is as follows:

$$B_i = Conv(PCIE_i \otimes expand(W_b^i)) \tag{8}$$

$$F_i = Out_i = Conv(PCIE_i \otimes expand(W_f^i)) \tag{9}$$

where $PCIE_i$ denotes the ith layer feature map generated by the PCIE module, W_f^i and W_b^i are foreground and background attention maps, and $expand()$ indicates that the number of channels will be expanded, $\otimes$ represents a multiplication operation. Out_i is the feature map of the ith PBIM output, supervised by ground-truth, and F_i and B_i denote the feature maps of the foreground and background outputs.

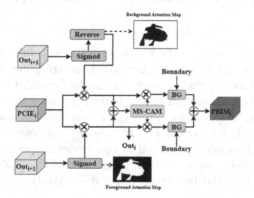

Fig. 6. PBIM structure diagram. The boundary of the camouflaged object is highlighted by coordinating foreground and background information. $\otimes$ represents the element-by-element product, $\oplus$ represents element-by-element addition.

Due to the different contributions of the two streams in PBIM, the Multi-Scale Channel Attention Module (MS-CAM) [3] was introduced in order to calculate the degree of contribution of the two streams. Specifically, we use F_i and B_i as inputs to MS-CAM to compute the weight matrices for the foreground and background flow contributions, denoted as W and $1 - W$, meanwhile, a boundary guidance module [31] is added after each attention module to guide the context features, increase the model's understanding of the boundary, and is the boundary information can be better highlighted. Finally the results of the two streams in PBIM are fused by an addition operation. The formula for PBIM can be written as:

$$PBIMF_i = BG_i(W \otimes F_i, Boundary) \tag{10}$$

$$PBIMB_i = BG_i((1 - W) \otimes B_i, Boundary) \tag{11}$$

$$PBIM_i = PBIMF_i \oplus PBIMB_i, i = 2, 3 \tag{12}$$

where BG_i is the boundary guidance module, $PBIMF_i$ and $PBIMB_i$ are the outputs of the foreground and background streams respectively, $PBIM_i$ is the output rough feature map of the ith PBIM, and $Boundary$ is the boundary map predicted by the BEM. The structure of PBIM is shown in Fig. 6.

3.6 Loss Function

The overall loss function is calculated as follows:

$$L_{\text{Total}} = L_C\left(P^C, Z^C\right) + \sum_{i=1}^{3} L_C\left(C^i, Z^C\right) + L_G\left(P^G, Z^G\right) + L_B\left(P^B, Z^B\right) \tag{13}$$

where L_{Total} stands for total loss, L_C, L_G, and L_B denote the object segmentation loss, object gradient loss, and object boundary loss, respectively. Specifically:

$$L_C = L_{IOU}^W + L_{BCE}^W \tag{14}$$

where L_{IOU}^W and L_{BCE}^W denote the weighted IOU loss and the weighted cross-entropy loss [7,8], respectively, which adaptively assign different weights to each pixel to avoid the model focusing more on the background when there are far fewer foreground pixels than background pixels, resulting in poor model performance. L_G uses a standard mean square error loss function and L_B uses a logits binary cross-entropy loss function. P^C, P^G and P^B represent the segmentation, gradient, and boundary maps predicted by the model, respectively, C^i represents the rough feature maps output by PBIM, and Z^C, Z^G and Z^B denote the real ground-truth, gradient, and boundary maps, respectively.

4   Experiment

4.1   Setting

Dataset. We validate BGENet on challenging COD datasets: CAMO [14], COD10K [7], and NC4K [18]. CAMO has 1000 train samples and 250 test samples. COD10K is the largest dataset, has 6000 training images (3040 positive) and 4000 testing images (2026 positive) across 5 classes and 69 subclasses. NC4K is the largest test dataset, containing 4121 positive samples for testing generalization ability of the model. The same scheme as in [7], we train our model on the hybrid dataset (i.e., COD10K training set + CAMO training set) with 4,040 samples, our method is evaluated on each of the three datasets. Table 2 shows experimental results.

Metrics. We employ standard evaluation metrics for COD tasks: S-measure (S_α) [5], E-measure (E_ϕ) [6], weighted F-measure (F_β^w) [20], and mean absolute error (MAE). S-measure quantifies structural similarity between non-binary predictions and true masks. E-measure takes into account both local similarity and global similarity between two binary images. Weighted F-measure is a weighting of precision and recall on top of the F-measure. Mean absolute error calculates pixel-wise differences between prediction map and ground-truth map.

Implementation Details. We implemented our BGENet using the PyTorch [23] and trained it on a single NVIDIA Tesla P40 GPU. The backbone network was EfficientNetV2-Small [27], pretrained on ImageNet, and we extracted features from three cross-sectional outputs. Our model was trained using the Adam [13] optimizer, employing an SGDR strategy for learning rate adjustment. The initial learning rate was 10^{-4} with 0.1 weight decay, a batch size of 12, and a maximum of 100 epochs. Before training, images were resized to 352×352 and data augmentation included color enhancement, random cropping, flipping, and rotation.

Table 2. Comparison with SOTAs on three challenging COD benchmark datasets. The best results are highlighted in bold and the second best are underlined. ↑ denotes the bigger the better, ↓ denotes the smaller the better.

Model	Param	CAMO-Test				COD10K-Test				NC4K			
		S_α ↑	E_ϕ ↑	F_β^w ↑	MAE ↓	S_α ↑	E_ϕ ↑	F_β^w ↑	MAE ↓	S_α ↑	E_ϕ ↑	F_β^w ↑	MAE ↓
SINet [8]	48.95M	0.751	0.771	0.606	0.100	0.771	0.806	0.551	0.051	0.808	0.871	0.723	0.058
JCSOD [15]	121.63M	0.803	0.853	0.759	0.076	0.809	0.884	0.684	0.035	0.842	0.898	0.771	0.047
PFNet [21]	46.50M	0.782	0.852	0.695	0.085	0.800	0.868	0.660	0.040	0.829	0.887	0.745	0.053
UGTR [29]	48.87M	0.785	0.859	0.686	0.086	0.818	0.850	0.667	0.035	0.839	0.874	0.747	0.052
BSANet [31]	32.58M	0.796	0.851	0.717	0.079	0.818	0.891	0.699	0.034	0.841	0.897	0.771	0.048
BGNet [25]	79.85M	0.812	0.870	0.749	0.073	0.831	<u>0.901</u>	**0.722**	0.033	0.851	0.907	<u>0.788</u>	0.044
SINetV2 [7]	26.98M	0.820	0.882	0.743	0.070	0.815	0.887	0.680	0.037	0.847	0.903	0.770	0.048
DGNet [12]	21.02M	<u>0.839</u>	<u>0.901</u>	<u>0.769</u>	<u>0.057</u>	<u>0.822</u>	0.896	0.693	<u>0.033</u>	<u>0.857</u>	<u>0.911</u>	0.784	<u>0.042</u>
DGNet-S [12]	**8.30M**	0.826	0.893	0.754	0.070	0.810	0.888	0.672	0.036	0.854	0.902	0.764	0.047
Ours	<u>20.81M</u>	**0.854**	**0.907**	**0.792**	**0.054**	**0.853**	**0.909**	<u>0.716</u>	**0.031**	**0.863**	**0.916**	**0.796**	**0.041**

4.2 Results and Analysis

We compared to models that appeared in mainstream computer vision conferences in recent years for comparison, including SINet [8], JCSOD [15], PFNet [21], UGTR [29], BSANet [31], BGNet [25], SINetV2 [7], DGNet [12], DGNet-S [12].

Quantitative Analysis. Table 2 demonstrates the results of our model with the current SOTA model. Our model achieves superior results on all four evaluation metrics. Our model size is the second smallest model so far, comparing to DGNet-S [12], our model parameters are slightly larger than it, but the MAE shrinks by almost 23% on the CAMO [14] dataset, and by about 13% on the COD10K [7] and NC4K [18] datasets. Although the MAE of our method is 0.002 higher and S_α is 0.006 lower than BGNet [25] on the COD10K [7] dataset, it is 59.04M lower on the parameters.

Visual Analysis. A visual comparison of our BGENet with six representative COD baseline is shown in Fig. 7. It is clearly seen that our method is able to better preserve the boundary features and internal gradient features of the camouflaged object, such as the second and sixth rows. And our method also has an improvement for the boundary break problem, as shown in the fifth line.

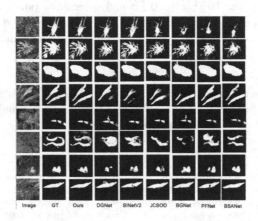

Fig. 7. Visual comparison of our BGENet with other SOTA methods.

Figure 8 shows the detection effect comparison between our method and DGNet [12] when the gradient map is insufficient, it can be seen that our method is still able to accurately detect camouflaged object when the gradient map is severely missing, which can solve the problem that DGNet [12] cannot be accurately detected when the gradient map is severely missing.

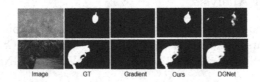

Fig. 8. With fewer gradient maps, our method is visually compared with the detection results of DGNet.

4.3 Ablation Study

To validate the effectiveness of our proposed modules, including PCIE, PBIM and GGM, we perform ablation experiments on two datasets, COD10K-Test [7] and NC4K [18], using different model design schemes. The experimental results are shown in Table 3, where #1 represents a backbone-only network scenario.

Effectiveness of PCIE. Due to limited context in backbone network extracted features, we employ PCIE for enhanced context information. The comparative experimental results in Table 3 highlights PCIE's pivotal role in obtaining context information.

Effectiveness of PBIM. In our approach, PBIM is introduced to realize the separation of foreground and background, to better highlight the boundary features of the camouflaged object through the interaction of foreground and background information, and to realize the boundary features to guide the context features through the boundary guidance module in PBIM. Comparing #2 and #3 in Table 3, we can see that our designed PBIM has some effectiveness.

Table 3. Ablation Study. Validation of each module on the two large test datasets.

No.	PCIE	PBIM	GGM	COD10K-Test				NC4K			
				$S_\alpha \uparrow$	$E_\phi \uparrow$	$F_\beta^w \uparrow$	$MAE \downarrow$	$S_\alpha \uparrow$	$E_\phi \uparrow$	$F_\beta^w \uparrow$	$MAE \downarrow$
#1	-	-	-	0.822	0.897	0.687	0.034	0.856	0.910	0.775	0.044
#2	✓	-	-	0.822	0.898	0.688	0.034	0.855	0.909	0.774	0.045
#3	✓	✓	-	0.825	0.899	0.690	0.034	0.858	0.911	0.778	0.044
#4	✓	-	✓	0.828	0.903	0.704	0.032	0.859	0.914	0.788	0.042
#5	-	✓	-	0.825	0.898	0.689	0.035	0.858	0.912	0.779	0.044
#6	-	-	✓	0.829	0.901	0.706	0.033	0.861	0.915	0.790	0.042
#7	-	✓	✓	0.831	0.904	0.710	0.032	0.862	0.916	0.793	0.041
Ours	✓	✓	✓	**0.853**	**0.909**	**0.716**	**0.031**	**0.863**	**0.916**	**0.796**	**0.041**

Effectiveness of GGM. In order to validate the effectiveness of the proposed GGM, we chose to remove the GGM module for comparison with the final model. Over comparing rows #3 and **Ours** in Table 3, it can be seen that equipped with the GGM module, the MAE decreases by about 8.8% on the COD10K [7] test set and the MAE decreases by about 6.8% on the NC4K [18].

5 Conclusion

In this paper, we propose a fusion boundary and gradient enhancement network (BGENet) for effective camouflage object detection. In our model, the context features are enhanced by the designed PCIE, and the designed PBIM effectively separates the object's pre-background and makes the pre-background information interact effectively to better highlight the boundary information and further guide the context features. In addition, the designed GGM employs a grouping strategy to efficiently utilize the learned gradient features to guide the context features.

We conducted experiments on three challenging COD datasets and our method achieves better performance with fewer parameters compared with some recent methods. However, comparing with SOTA Transformer-based methods [11], our method still needs to be improved. In the future, it is still worth researching how to achieve better detection results with fewer parameters.

Acknowledgements. This work is supported by the Inner Mongolia Science and Technology Project No.2021GG0166.

References

1. Chen, G., Liu, S.J., Sun, Y.J., Ji, G.P., Wu, Y.F., Zhou, T.: Camouflaged object detection via context-aware cross-level fusion. IEEE Trans. Circuits Syst. Video Technol. **32**(10), 6981–6993 (2022)
2. Chen, S., Tan, X., Wang, B., Hu, X.: Reverse attention for salient object detection. In: Ferrari, V., Hebert, M., Sminchisescu, C., Weiss, Y. (eds.) ECCV 2018. LNCS, vol. 11213, pp. 236–252. Springer, Cham (2018). https://doi.org/10.1007/978-3-030-01240-3_15
3. Dai, Y., Gieseke, F., Oehmcke, S., Wu, Y., Barnard, K.: Attentional feature fusion. In: Proceedings of the IEEE/CVF Winter Conference on Applications of Computer Vision, pp. 3560–3569 (2021)
4. Fan, C., Zeng, Z., Xiao, L., Qu, X.: GFNet: automatic segmentation of COVID-19 lung infection regions using CT images based on boundary features. Pattern Recogn. **132**, 108963 (2022)
5. Fan, D.P., Cheng, M.M., Liu, Y., Li, T., Borji, A.: Structure-measure: a new way to evaluate foreground maps. In: Proceedings of the IEEE International Conference on Computer Vision, pp. 4548–4557 (2017)
6. Fan, D.P., Gong, C., Cao, Y., Ren, B., Cheng, M.M., Borji, A.: Enhanced-alignment measure for binary foreground map evaluation. arXiv preprint arXiv:1805.10421 (2018)
7. Fan, D.P., Ji, G.P., Cheng, M.M., Shao, L.: Concealed object detection. IEEE Trans. Pattern Anal. Mach. Intell. **44**(10), 6024–6042 (2021)
8. Fan, D.P., Ji, G.P., Sun, G., Cheng, M.M., Shen, J., Shao, L.: Camouflaged object detection. In: Proceedings of the IEEE/CVF Conference on Computer Vision and Pattern Recognition, pp. 2777–2787 (2020)
9. Fan, D.-P., et al.: PraNet: parallel reverse attention network for polyp segmentation. In: Martel, A.L., et al. (eds.) MICCAI 2020. LNCS, vol. 12266, pp. 263–273. Springer, Cham (2020). https://doi.org/10.1007/978-3-030-59725-2_26
10. Gallego, J., Bertolino, P.: Foreground object segmentation for moving camera sequences based on foreground-background probabilistic models and prior probability maps. In: 2014 IEEE International Conference on Image Processing (ICIP), pp. 3312–3316. IEEE (2014)

11. Huang, Z., et al.: Feature shrinkage pyramid for camouflaged object detection with transformers. In: Proceedings of the IEEE/CVF Conference on Computer Vision and Pattern Recognition, pp. 5557–5566 (2023)
12. Ji, G.P., Fan, D.P., Chou, Y.C., Dai, D., Liniger, A., Van Gool, L.: Deep gradient learning for efficient camouflaged object detection. Mach. Intell. Res. **20**(1), 92–108 (2023)
13. Kingma, D.P., Ba, J.: Adam: a method for stochastic optimization. arXiv preprint arXiv:1412.6980 (2014)
14. Le, T.N., Nguyen, T.V., Nie, Z., Tran, M.T., Sugimoto, A.: Anabranch network for camouflaged object segmentation. Comput. Vis. Image Underst. **184**, 45–56 (2019)
15. Li, A., Zhang, J., Lv, Y., Liu, B., Zhang, T., Dai, Y.: Uncertainty-aware joint salient object and camouflaged object detection. In: Proceedings of the IEEE/CVF Conference on Computer Vision and Pattern Recognition, pp. 10071–10081 (2021)
16. Liu, M., Di, X.: Extraordinary MHNet: military high-level camouflage object detection network and dataset. Neurocomputing. 126466 (2023)
17. Liu, Z., Huang, K., Tan, T.: Foreground object detection using top-down information based on EM framework. IEEE Trans. Image Process. **21**(9), 4204–4217 (2012)
18. Lv, Y., et al.: Simultaneously localize, segment and rank the camouflaged objects. In: Proceedings of the IEEE/CVF Conference on Computer Vision and Pattern Recognition, pp. 11591–11601 (2021)
19. Mao, Y., et al.: Transformer transforms salient object detection and camouflaged object detection. arXiv preprint arXiv:2104.10127 1(2), 5 (2021)
20. Margolin, R., Zelnik-Manor, L., Tal, A.: How to evaluate foreground maps? In: Proceedings of the IEEE Conference on Computer Vision and Pattern Recognition, pp. 248–255 (2014)
21. Mei, H., Ji, G.P., Wei, Z., Yang, X., Wei, X., Fan, D.P.: Camouflaged object segmentation with distraction mining. In: Proceedings of the IEEE/CVF Conference on Computer Vision and Pattern Recognition, pp. 8772–8781 (2021)
22. Pan, Y., Chen, Y., Fu, Q., Zhang, P., Xu, X., et al.: Study on the camouflaged target detection method based on 3d convexity. Mod. Appl. Sci. **5**(4), 152 (2011)
23. Paszke, A., et al.: Pytorch: an imperative style, high-performance deep learning library. In: Advances in Neural Information Processing Systems, vol. 32 (2019)
24. Sengottuvelan, P., Wahi, A., Shanmugam, A.: Performance of decamouflaging through exploratory image analysis. In: 2008 First International Conference on Emerging Trends in Engineering and Technology, pp. 6–10. IEEE (2008)
25. Sun, Y., Wang, S., Chen, C., Xiang, T.Z.: Boundary-guided camouflaged object detection. arXiv preprint arXiv:2207.00794 (2022)
26. Szegedy, C., Vanhoucke, V., Ioffe, S., Shlens, J., Wojna, Z.: Rethinking the inception architecture for computer vision. In: Proceedings of the IEEE Conference on Computer Vision and Pattern Recognition, pp. 2818–2826 (2016)
27. Tan, M., Le, Q.: Efficientnetv2: smaller models and faster training. In: International Conference on Machine Learning, pp. 10096–10106. PMLR (2021)
28. Wang, J., et al.: Multi-feature information complementary detector: a high-precision object detection model for remote sensing images. Remote Sens. **14**(18), 4519 (2022)
29. Yang, F., et al.: Uncertainty-guided transformer reasoning for camouflaged object detection. In: Proceedings of the IEEE/CVF International Conference on Computer Vision, pp. 4146–4155 (2021)

30. Zheng, D., Zheng, X., Yang, L.T., Gao, Y., Zhu, C., Ruan, Y.: MFFN: multi-view feature fusion network for camouflaged object detection. In: Proceedings of the IEEE/CVF Winter Conference on Applications of Computer Vision, pp. 6232–6242 (2023)
31. Zhu, H., et al.: I can find you! boundary-guided separated attention network for camouflaged object detection. In: Proceedings of the AAAI Conference on Artificial Intelligence, vol. 36, pp. 3608–3616 (2022)
32. Zhuge, M., Lu, X., Guo, Y., Cai, Z., Chen, S.: Cubenet: x-shape connection for camouflaged object detection. Pattern Recogn. **127**, 108644 (2022)

Find the Cliffhanger: Multi-modal Trailerness in Soap Operas

Carlo Bretti[1]([envelope]) [iD], Pascal Mettes[1] [iD], Hendrik Vincent Koops[2] [iD],
Daan Odijk[2] [iD], and Nanne van Noord[1] [iD]

[1] Informatics Institute, University of Amsterdam, Amsterdam, Netherlands
{c.bretti,p.s.m.mettes,n.j.e.vannoord}@uva.nl
[2] RTL Nederland, Hilversum, Netherlands
{vincent.koops,daan.odijk}@rtl.nl

Abstract. Creating a trailer requires carefully picking out and piecing together brief enticing moments out of a longer video, making it a challenging and time-consuming task. This requires selecting moments based on both visual and dialogue information. We introduce a multi-modal method for predicting the trailerness to assist editors in selecting trailer-worthy moments from long-form videos. We present results on a newly introduced soap opera dataset, demonstrating that predicting trailerness is a challenging task that benefits from multi-modal information. Code is available at https://github.com/carlobretti/cliffhanger.

Keywords: trailer generation · multi-modal · multi-scale

1 Introduction

Trailers are short previews of long-form videos with the aim of enticing viewers to watch a movie or TV-show [2]. Professional editors create trailers by selecting and editing together moments from long-form video. Determining which moments to select can be a time-consuming task as it involves scanning the entire long-form video and picking moments that match aesthetic or semantic criteria. Those selected moments are considered to have high *trailerness* - they represent the moments that are most suitable to be used in a trailer. By learning to automatically recognize moments that have high trailerness it becomes possible to support editors in creating trailers - as well as boost their creativity by selecting or recommending moments they may not have initially picked.

Trailers are a key component of soap operas, as these follow a regimented format with daily episodes, fixed time slots, and every episode ending with a trailer enticing the viewer to stay tuned. Unlike trailers for movies, which may feature cinematographically spectacular footage, the mainstay for soap opera trailers is the continuation of prominent storylines, i.e., how the events from the current episode unfold in future episodes. As such, the context in which the material was created and the format followed may strongly influence what moments are used in trailers. This context for television programming, and soap operas in

S. Rudinac et al. (Eds.): MMM 2024, LNCS 14555, pp. 199–212, 2024.
https://doi.org/10.1007/978-3-031-53308-2_15

particular, due to their grounding in everyday situations, reflects national and cultural identities [34], therefore requiring learning approaches that take into account both the complexity and the idiosyncrasies of the source material.

Crucially, what determines the trailerness of a moment may thus be conveyed through various modalities, e.g., a shot that is particularly visually attractive or exciting, or a line of dialogue that is particularly funny. While previous works have aimed to extract moments with high trailerness [37], this multi-modal aspect has not received sufficient attention. Next to multi-modality, we find that the time scale of trailer-worthy moments is understudied, with most approaches focusing only on shot-level moments. Yet, trailers are short in duration and typically fast-paced, therefore being able to select shorter-than-shot moments would be highly desirable. To select moments with high trailerness most previous works on trailer generation either rely on dense annotations [28] or external information that is integrated in a hand-crafted manner [25,29,40]. Instead, we propose to leverage existing trailers to learn what moments have high trailerness, based on both visual and subtitle information across multiple time-scales.

Our results show that trailer generation is a highly challenging and subjective task and with our findings, we demonstrate the benefits of a multi-modal approach for trailer generation. In the following, we discuss related work, our proposed multi-scale and multi-modal method, our soap opera trailerness dataset, and present the findings in more detail.

2 Related Work

Movies and TV shows are rich and inherently multimodal sources of information. Within the domain of movie and TV series data, different tasks have been researched [6,9,33,39]. Within this field, we specifically focus on trailer generation, to reduce a long-form video into a short-form video that could be used as a trailer [25,28,37]. A closely related task to trailer generation and a common way to produce shorter videos from longer ones is video summarization [7,41]. In the following, we discuss works from both directions and how they differ.

2.1 Video Summarization

The aim of video summarization is to reduce the length of a long video in a brief and faithful manner. There have been both supervised and unsupervised approaches. Unsupervised methods often focus on generating a summary that is most representative of the original video in terms of its ability to reconstruct the original video [24]. Since our focus is on trailers, and scenes in a trailer are meant to elicit attention rather than provide a comprehensive summary, we focus our discussion of video summarization literature on supervised methods. One relevant line of work in supervised learning consists of modeling videos as sequences of frames [41]. Building on this, another avenue focused on first modeling frames within shots in a video as a sequence and then modeling entire videos as a sequence of shots [42–45]. Various methods have also employed attention to model the relationship between multiple frames [7,17,18,21,23]. Lastly, multimodal summarization has been explored in a few cases. [5] provided a method to identify salient moments in a video using multimodal hand-crafted features. Some have additionally proposed query-dependent video summarization based on a textual query [11,12]. Combining deep features extracted from audio and video at multiple levels using attention has been explored recently [42].

Within video summarization, the goal is to provide a faithful and brief representation of the source material. In contrast, we can characterize trailer generation as a form of biased video summarization aimed at enticing the audience to watch subsequent material. Trailer generation is therefore a more subjective task, an aspect which will be highlighted in the following section.

2.2 Trailer Generation

Trailer generation is not a well-studied topic, with few works focusing on it specifically. Earlier approaches used hand-crafted features, whereas later works focus on learning-based approaches, primarily by incorporating external information.

From the earlier works, the work in [28] is most akin to the learning-based approaches in that a supervised model is trained on hand-crafted visual and audio features to classify shots as trailer worthy. This approach differs from other earlier works which used a more rule-based method, either informed by the typical structure of a Hollywood trailer [36], or by extracting key symbols of a movie (e.g., title logo and theme music) [16].

Among more recent works, there is a wider emphasis on using a learning-based approach, where the majority of approaches rely on external information. For example, in [40] a model of visual attractiveness is learned by leveraging eye-movement data, which is then used to generate a trailer tailored to a piece of music using a graph-based algorithm. Similarly, [25] uses narrative structure and sentiment in screenplays to generate trailers.

In [29] presented an approach that was used to create a trailer for the 2016 sci-fi film *Morgan*. Their approach uses PCA on features extracted from different modalities (e.g. emotions, objects, scenes, and sounds) to select 3 principle components to be used as a scoring mechanism for suggesting shots to editors. Their final trailer is produced by editors who use the system as a selection mechanism.

In contrast to these works, our work is based on learning trailer-worthy sequences through ranking, which is more closely related to recent video summarization approaches. A similar approach to ranking shots rooted in visual learning was proposed in [37]. Here, a model is trained to co-attend pairs of movies and trailers, and obtain correlation scores of shots across end-to-end. This can then be leveraged to learn a ranking model that ranks shots based on whether they are determined to be key moments of a movie. Our proposed method is distinct from previous works in that it consists of a multi-modal learning approach using visual and text inputs at multiple scales. Moreover, we leverage trailers of long videos to directly obtain trailerness annotations for training. In particular, the multi-modal and multi-scale aspects of trailer generation were understudied, whereas we find that these are greatly beneficial for performance.

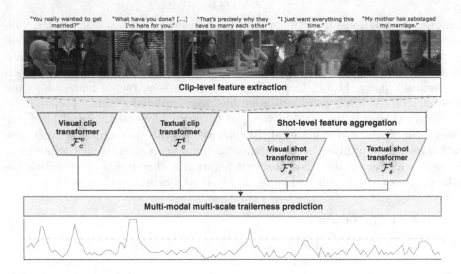

Fig. 1. Estimating trailerness in videos with our Trailerness Transformer. Given a video denoting a movie or tv series episode, we first encode clip-level and shot-level encodings for both the visual and textual video modalities. We then train transformers for each combination of modality and temporal scale, after which we aggregate the trailerness predictions of all transformers.

3   Method

In this section, we will first define a notion of trailerness, a method for obtaining trailer labels from editor selections, and finally a method for predicting trailerness scores as depicted in Fig. 1.

3.1   The Trailerness of Video

We are interested in discovering which parts of a TV episode or movie are most suitable to be used in a trailer. More generally, for all frames x_f in a video m we define *trailerness* as a score that indicates how suitable each frame is to be selected for a trailer. Subsequently, sequences of high trailerness can be used by editors to piece together trailers.

Each video is denoted as $m = \{(v, t)\}$, where $v \in \mathbb{R}^{H \times W \times C \times |x_f|}$ denotes the visual track and $t_i = (f_{\text{start}}, f_{\text{end}}, b)$ denotes the natural language sentence b, and the start and end frame for the ith subtitle, both tracks contain complementary information for determining trailerness.

For learning trailerness, we are given a training collection of videos M and a collection of corresponding trailers G. Each video corresponds to one trailer, hence $|M| = |G|$. Moreover, each trailer $g \in G$ solely relies on content from its corresponding video, *i.e.* $g \subset m$. The trailers are generated by professional editors to obtain high-quality selections. In order to predict which frames of a new video are worthy of being in a trailer, we first need to obtain frame-level annotations for all videos in M by matching them visually to the editor-standard trailers. Then, we approach the problem as a localized scoring problem where we assign each frame $x_{f_j} \in x_f$ a trailerness score and obtain trailerness binary predictions, which can then be evaluated against the editor-standard labels.

3.2 Trailer Labels from Editor Selections

To establish a ground truth for trailerness, we perform visual matching per frame. We obtain annotations for all videos by matching each frame from videos M against the trailers G. Here, we outline a method to obtain frame-level annotations which we later aggregate to obtain clip-level labels for optimization.

First, we utilize a hashing function $\mathcal{F}_{\text{hash}}$ to compute a perceptual hash for each frame in video track v for every video $m \in M$. We then do the same for all trailers $g \in G$. For similarity search, we compute the Hamming distance between the hashes of frames in the trailers and the frames in the long videos. By doing so, we obtain a fast visual matching score between each trailer frame $r \in g$ and long video frames, which we then threshold to obtain editor-standard binary labels. For each individual frame x_{f_j} in each video m, we first determine the following hashing distance:

$$y_{f_j} = [\![\left(\min_{r \in g_i} d_h(\mathcal{F}_{\text{hash}}(v_{f_j}), \mathcal{F}_{\text{hash}}(r)) \right) < \tau]\!], \tag{1}$$

with $d_h(\cdot, \cdot)$ the Hamming distance and τ a hashing distance threshold, stating the minimum required similarity between video and trailer frames. Stacking the labels of all frames in video m results in a label vector $y_f \in \{0, 1\}_{j=1}^{|x_f|}$ indicating which frames were deemed trailer-worthy by a professional editor.

3.3 Multi-scale and Multi-modal Trailerness Transformer

Given dataset M where each video $m = \{(v, t, y_f)\}$ contains video frames, subtitles, and annotations, we seek to obtain an end-to-end model that provides trailerness predictions for all frames in a test video. We hypothesize that such a prediction is best made when considering both visual and textual modalities over both long- and short-term intervals. To that end, we introduce a Trailerness Transformer that performs multi-modal and multi-scale trailerness prediction. Below we outline the three main stages of our approach, namely encoding, multi-modal and -scale transformers, and prediction aggregation.

Visual and Textual Encoding. We operate on four types of sequences, varying in temporal scale and modality. We consider a short clip-level temporal scale of 64 frames (2.56 s at 25 fps) and a longer shot-level temporal scale, with each shot consisting of multiple clips. For the visual modality, we employ a pre-trained video embedding function $\mathcal{F}_V(\cdot)$ to extract features for each 64-frame clip from a [4]. A video v is divided into 64-frame clips and is given as $v_c = \{v_{c_j}\}_{j=1}^{|v_c|}$. Then, for each clip we extract features as $E_{c_j}^v = \mathcal{F}_V(v_{c_j})$, yielding $E_c^v = \{E_{c_j}^v\}_{j=1}^{|v_c|}$ for a long video m. For the textual stream, we concatenate any subtitle with temporal overlap with the clip timeframe for each clip. We then take the corresponding features from a pre-trained language model, $E_c^t = \{\mathcal{F}_T(t_{c_j})\}_{j=1}^{|t_c|}$.

Besides a local clip-level view, we also investigate a longer shot-level scale. We define shots as video sequences for which the boundaries are represented by shot transitions. These shot boundaries are determined by a shot transition detector $\mathcal{F}_{\text{shots}}$ which outputs cut probabilities for each frame [32]. By using shots, we have more coarse sequences that naturally follow from the source material. For visual features at a shot level, we aggregate over the extracted features E_c^v based on the shot boundaries obtained through $\mathcal{F}_{\text{shots}}$ to then obtain shot-level visual features E_s^v. On the other hand, for the textual features, we again first concatenate subtitles as for clips, this time based on the overlap between subtitle boundaries and shot boundaries to obtain t_s. Shot-level textual features can then be computed directly using the pre-trained text embedding function $\mathcal{F}_T$, with $E_s^t = \{\mathcal{F}_T(t_{s_j})\}_{j=1}^{|t_s|}$. Finally, we obtain four streams of features, depending on the combination of scale and modality, $E_c^v, E_c^t, E_s^v, E_s^t$ for a long video m. We will use E_* to refer to multiple scales and $E^\star$ to refer to multiple modalities.

Modality- and Scale-specific Transformers. On the four combinations of modality and temporal scale, we train an individual transformer, where all transformers follow the architecture of Vaswani *et al.* [35]. We conceptualize our problem as a sequence-to-sequence binary classification task, where given a video m as input features $E_*^\star \in \mathbb{R}^{D \times |x_*|}$. D is the number of dimensions in output from video feature extractor $\mathcal{F}_V(\cdot)$ or $\mathcal{F}_T(\cdot)$ for text, while $|x_*|$ the number of sequences for a given video m depending on the scale at play.

We consider a transformer architecture $\mathcal{F}_*^\star(\cdot)$ consisting of a linear layer, positional encodings, two transformer encoder blocks with hidden dimensionality d_k, and finally, an output linear layer followed by a sigmoid to obtain trailerness scores $O_*^\star$. We first map the input features through a linear layer and add the positional encodings [35] as given by

$$PE_{(\text{pos},h)} \begin{cases} \sin\left(\frac{\text{pos}}{10000^{h/d_k}}\right) & \text{if } h \bmod 2 = 0 \\ \cos\left(\frac{\text{pos}}{10000^{(h-1)/d_k}}\right) & \text{otherwise} \end{cases} \tag{2}$$

where pos represents the temporal position of a token (in our case either a clip or a shot) within the full video sequence and h represents the specific dimension. The resulting output is then fed through the transformer encoder blocks. Each transformer encoder block is composed of two sub-layers: a multi-head attention mechanism (MHSA) and a multi-layer perceptron (MLP). A residual connection [10] is added to the output of each sublayer, and the resulting output undergoes layer normalization [1].

The transformer encoder output is then fed through a final linear layer followed by a sigmoid, to obtain trailerness predictions $O_*^\star = \mathcal{F}_*^\star(E_*^\star)$ with $O_*^\star \in \mathbb{R}^{1 \times |x_*|}$. The trailerness labels are provided at frame level, but we investigate clip- and shot-level encodings in our approach. A clip is deemed positive if at least one third of the frames within the clip are positive. Similarly, a shot is deemed positive if at least one third of the clips contained within the shot are positive. For a video m this results in clip-level annotations $y_c \in \{0,1\}_{j=1}^{|x_c|}$ and shot-level annotations $y_s \in \{0,1\}_{j=1}^{|x_s|}$. Focal loss [22] has been found to reliably work well for highly imbalanced data and is therefore used for training:

$$\mathcal{L} = -\alpha(1 - o_\mathrm{p})^\gamma \log(o_\mathrm{p}), \tag{3}$$

with

$$o_\mathrm{p} = \begin{cases} o^\star_{*j} & \text{if } y_{*j} = 1, \\ 1 - o^\star_{*j} & \text{otherwise} \end{cases} \tag{4}$$

where $o^\star_{*j} \in O^\star_*$ is a single prediction for a subsequence for a video m. The parameter α controls the balance between positive and negative examples, emphasizing positive examples in this case. On the other hand, γ determines a modulating factor that allows the loss to distinguish between easy and hard examples.

Fusing Transformer Predictions. We obtain four different trailerness prediction streams for a video, one for each combination of scale and modality. We seek to fuse these predictions to make use of the complementary information across the four transformers. However, the number of sequences $|x_*|$ depends on the size of the sequences, i.e. whether they are clips or shots. At test time, we upsample all predictions from each stream from either clip- or shot-level predictions to frame-level predictions. This allows us to fairly compare outputs from different streams. We can also then perform late fusion of predictions using different possible combinations of the four multi-modal multi-scale streams. This can be done simply by averaging frame-level prediction likelihoods for the considered streams [30].

4   The GTST Dataset

To investigate multi-modal trailerness prediction we introduce the **GTST** dataset. This multi-modal dataset consists of 63 episodes from the long-running Dutch soap opera "Goede Tijden, Slechte Tijden". Each episode is around 20 min in length and for each episode, we have the visual video track and the time-coded subtitles. A typical episode of GTST consists of three blocks: a recap sequence, the body of the episode, and a preview sequence. The recap sequence serves as a way to bring viewers up to speed with the plot. It summarizes the main plot points of relevant earlier episodes, with an emphasis on the preceding episode. On the other hand, the preview sequence serves as a way to persuade viewers to tune in for the next episode. It features clips from the next episode that are meant to catch viewers' attention, i.e. cliffhangers or shocking revelations or events. In the context of soap operas, a preview of an episode represents a very close match to what a trailer represents for a feature film. We will therefore rely on previews to obtain editor-standard trailerness labels for the original long videos, i.e. episodes in the context of a soap opera.

Table 1. Comparing our GTST trailer dataset to popular video summarization datasets and trailer moment detection datasets. * indicates the subset of VISIOC-ITY with only tv shows. Our dataset is unique in its combination of editor-standard labels, multi-modal nature, and availability for open research.

Dataset	#Videos	Avg. length	Goal	Label type	Multimodal	Openly available
SumME [8]	25	2 min	summary	user-generated		✓
TVSum [31]	50	4 min	summary	user-generated		✓
VISIOCITY [20]	67	55 min	summary	user-generated		✓
VISIOCITY* [20]	12	22 min	summary	user-generated		✓
LSMTD [14]	508	N/A	trailers	paired data		
MovieNet [13]	1100	N/A	trailers	paired data	✓	
TMDD [37]	150	N/A	trailers	paired data		
GTST	63	22 min	trailers	editor-standard	✓	✓

Our dataset features long videos from a TV show and features readily available high-quality editor-standard labels compared to user-generated summaries [20,37]. Table 1 compares different datasets for summarization and trailer generation to our dataset. All datasets for summarization are publicly available and they are densely labeled with user-generated annotations, but consist of only video data. The availability of data for trailer generation is much more limited, most datasets consist of commercial movies paired with their trailers. For these datasets no annotations are available, but the trailers could be used for supervision - as we demonstrate with our approach for extracting editor-standard pseudo-labels. Overall, this table highlights the unique trailer-oriented, editor-standard, and multi-modal properties of GTST. The dataset is divided into training/validation/test sets with 60%/20%/20% per set respectively.

5   Experiments

We focus on two types of empirical evaluations: (i) quantitative evaluations of the trailerness transformer compared to editor selections, (ii) qualitative analyses of discovered trailerness.

5.1   Setup

Encoding. For our experiments, we use three pre-trained models to obtain encodings and shot boundaries. Visual features are extracted using I3D [4] as video feature extractor $\mathcal{F}_V$, as provided in [15]. The textual features are computed using a multilingual sentence embedding model $\mathcal{F}_S$ [26,27]. This is a multilingual version of MiniLM [38], a compressed version of a pre-trained transformer. Lastly, TransNetV2 [32] serves as our shot boundary detector $\mathcal{F}_{shots}$. As for the transformer architecture employed, we empirically set the number of attention heads to 4, and we use one transformer encoder block. For the focal loss, we empirically set the parameters $\alpha = 0.95$ and $\gamma = 1$ through validation.

Table 2. Trailerness prediction on individual transformer streams, in percentage %. Overall, clip-level transformers obtain better F1 scores. Across modalities, visual clip-level trailerness boosts precision, while textual shot-level trailerness boosts recall, highlighting their complementary nature.

Clip-level		Shot-level				
Visual	Textual	Visual	Textual	F1	Prec	Rec
✓				6.9 ± 1.0	4.0 ± 0.7	28.0 ± 11.8
	✓			5.5 ± 1.3	3.3 ± 0.9	18.2 ± 2.8
		✓		5.0 ± 2.6	3.0 ± 1.4	26.2 ± 25.0
			✓	5.2 ± 0.5	2.8 ± 0.3	40.9 ± 11.6

Obtaining Trailer Labels. To obtain the per-frame binary trailerness labels for all videos in our dataset, we first split the videos to isolate the body of the episode based on their known structure. We also extract the preview sequence from each episode that corresponds to the next episode. We then remove all the redundant sequences such as opening titles or title cards before and after ad breaks. After having extracted all the frames from the body of an episode and from its corresponding preview we compute perceptual hashes using Image-hash [3]. We perform similarity search using FAISS [19] and match the frames in a preview sequence to the original frames in the episode to obtain editor-standard trailerness labels.

Evaluation. For evaluation, we binarize predictions based on a threshold (0.5) and compute F1, precision, and recall between the frames predicted to be in a trailer and the frames selected by editors. We run the models with 5 different random seeds and report mean and standard deviation estimates for each metric.

5.2 Evaluating Modalities and Temporal Scales

We first evaluate the four individual combinations of modalities and temporal scales for their ability to predict trailerness in unseen test videos. In Table 2, we show the performance of our transformer-based approach across the four combinations. We find that the visual clips stream $\mathcal{F}_c^v$ provides the best performance in terms of F1-score, (6.9%). Moreover, the streams using features extracted at a clip level, *i.e.* $\mathcal{F}_c^v$ (6.9%) and $\mathcal{F}_c^t$ (5.5%), perform better overall than the streams using features at a shot level $\mathcal{F}_s^v$ (5.0%), $\mathcal{F}_s^t$ (5.2%).

Only using the visual stream at a shot level, as is done in the trailer-based approach of Wang et al. [37] leads to lower performance. Within the visual modality, a more localized approach seems to improve performance.

Table 3. Results for late fusion of different trailerness streams, in percentage %. In our results, a triplet, combining modalities at a shot-level with clip-level visual predictions balances precision and recall best, as indicated by the F1 score.

Clip-level		Shot-level				
Visual	Textual	Visual	Textual	F1	Prec	Rec
✓	✓			6.5 ± 2.0	3.9 ± 1.1	20.5 ± 8.1
		✓	✓	7.1 ± 0.6	4.0 ± 0.3	37.5 ± 10.2
✓			✓	7.8 ± 0.4	4.7 ± 0.2	23.4 ± 5.7
	✓		✓	7.2 ± 1.0	4.2 ± 0.5	26.1 ± 7.7
✓			✓	7.9 ± 1.4	4.7 ± 1.0	30.6 ± 12.1
	✓	✓		6.7 ± 2.1	4.5 ± 1.3	17.0 ± 9.6
✓	✓	✓		7.3 ± 1.4	4.6 ± 0.8	18.5 ± 5.7
✓	✓		✓	6.9 ± 2.2	4.2 ± 1.2	22.4 ± 11.0
✓		✓	✓	9.2 ± 0.9	5.6 ± 0.4	30.1 ± 9.7
	✓	✓	✓	8.4 ± 1.1	5.2 ± 0.8	24.6 ± 2.4
✓	✓	✓	✓	8.5 ± 1.4	5.3 ± 0.8	22.5 ± 6.0

5.3   Combining Modalities and Temporal Scales

Through late fusion, we can consider different combinations of the four streams and their effect on predicting trailerness in test videos as we show inTable 3. First, the top two rows in Table 3 showcase that a multi-modal fusion of predictions at a shot level (7.1%) performs better than multimodal fusion at a clip level (6.5%) in terms of F1. The third and fourth rows indicate that fusing clip-level and shot-level predictions for a single modality results in slightly better performance for the visual stream, with 7.8% for visual clips and shots together against 7.2% for predictions based on text fused at a clip and shot level. Overall, multi-modal or multi-scale fusion boosts the performance of the weaker individual streams, with fusing the modalities at a shot level performing the best.

In rows five and six, we consider two pairs with neither matching modality nor matching size. Combining the visual feature at the clip level with text features at the shot level results in the highest F1 score for fusion based on two streams. Fusing clip-level predictions from text with shot-level predictions from visual features also boosts performance compared to the two individual streams as reported in Table 2, albeit not as strongly. Fusing across temporal scales and modalities results generally also results in an increase in performance, indicating that for combinations of two streams this is similarly beneficial.

The most striking result for triplets of streams is that by combining predictions from both modalities at a clip level with predictions from text at a shot-level we achieve our best performance in terms of F1, at 9.2%, above the best performing individual stream (visual clips) at 6.9%. Lastly, we show that a fusion of predictions from all streams does not result in an increase in performance and instead leads to a 0.7% decrease in performance from our best model, from 9.2% to 8.5%.

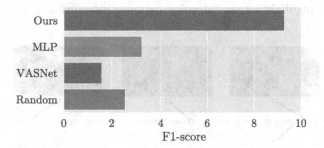

Fig. 2. Baseline comparisons. An MLP-based architecture outperforms the random baseline and a frame-based summarization method (VASNet [7]), and our model outperforms them all by incorporating sequential order and temporal positioning.

5.4 Comparisons to Baselines

In Fig. 2 we compare results from our best-performing model to three baselines: random, a multi-layer perceptron (MLP), and a frame-based video summarization baseline VASNet [7]. We show that our approach performs far better than a random baseline, where clips or shots are randomly assigned a trailerness score. For the second baseline, we train an MLP and separately perform hyperparameter tuning to find the best values for α and γ in the focal loss and we set $\alpha = 0.98$ and $\gamma = 1$. We additionally include early stopping as a regularizing factor. For the MLP, we find that the best results are given by fusing the clip and shot-level visual streams. We show that while an MLP architecture obtains results better than random, our choice of architecture showcases a 2.9x increase in F1, from 3.2% to 9.2%. This is because our transformer-based approach incorporates sequentiality and positioning in its architecture, whereas the MLP baseline treats each subsequence individually, with no knowledge of sequentiality. As denoted by F1, our model provides a better overall performance, making it more suitable for predicting trailerness without overpredicting the positive class. For our third baseline, we separately train VASNet, and tune hyperparameters. For a fair comparison we replace the mean squared error term with the focal loss, and follow the same evaluation procedure as ours. Hyperparameter tuning resulted in setting $\alpha = 0.999$ and $\gamma = 1$. We additionally adapt by splitting up long videos into shorter clips, to avoid complexity issues due to the self-attention operation over too long of a sequence. VASNet results in an F1 of 1.6% below the performance of the other baselines and our model. This highlights the difference between summarisation and trailer generation, despite their similar premise.

5.5 Qualitative Results

To gain more insight into our approach, we show qualitative results for different models in Figs. 3 and 4. In Fig. 3 we relate textual and visual information to trailerness independently, showing how emotionally charged visuals and urgent textual calls to action yield higher trailerness in unimodal settings.

In Fig. 4 we relate trailerness to the results of the best performing multi-stream model highlighting the interplay between modalities.

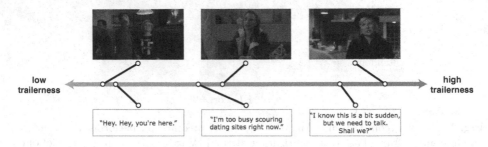

Fig. 3. Qualitative results for visual and text streams at a clip level individually. Emotionally-charged visuals and urgent calls to action in text yield higher trailerness than transitory visuals and playful subtitles.

Fig. 4. Qualitative results for our best-performing model. Scenes with bright visuals and emphatic dialogue yield higher trailerness than scenes with generic visuals and a lack of dialogue.

6   Conclusion

Which moments in a long-form video are suited for a trailer depends on a variety of factors, including the creative style of the editor and narrative aspects (*e.g.* avoiding spoilers). This subjective nature makes selecting moments with high trailerness in a fully automatic manner a challenging task. We presented an approach that leverages existing trailers to generate annotations of trailerness, and use this to train a multi-modal and multi-scale model to predict trailerness underscoring the complexity of trailer generation and how it benefits from contextual information.

Acknowledgements. This research was made possible by the TKI ClickNL grant for the AI4FILM project.

References

1. Ba, J.L., Kiros, J.R., Hinton, G.E.: Layer Normalization (2016)
2. Bordwell, D., Thompson, K., Smith, J.: Film Art: an introduction (2020)
3. Buchner, J.: ImageHash: a Python perceptual image Hashing module. https://github.com/JohannesBuchner/imagehash (2021)
4. Carreira, J., Zisserman, A.: Quo Vadis, Action Recognition? A new model and the kinetics dataset. In: CVPR (2017)
5. Evangelopoulos, G., et al.: Multimodal saliency and fusion for movie summarization based on aural, visual, and textual attention. IEEE Transactions on Multimedia (2013)
6. Everingham, M., Sivic, J., Zisserman, A.: Hello! My name is... Buffy" - Automatic Naming of Characters in TV Video. In: BMVC (2006)
7. Fajtl, J., Sokeh, H.S., Argyriou, V., Monekosso, D., Remagnino, P.: Summarizing Videos with Attention. In: ACCVW (2019)
8. Gygli, M., Grabner, H., Riemenschneider, H., Van Gool, L.: Creating summaries from user videos. In: ECCV (2014)
9. Hanjalic, A., Lagendijk, R.L., Biemond, J.: Automated high-level movie segmentation for advanced video-retrieval systems. IEEE Trans. Circuits Syst. Video Technol. (1999)
10. He, K., Zhang, X., Ren, S., Sun, J.: Deep residual learning for image recognition. In: CVPR (2016)
11. Huang, J.H., Murn, L., Mrak, M., Worring, M.: GPT2MVS: generative pre-trained transformer-2 for multi-modal video summarization. In: ACM ICMR (2021)
12. Huang, J.H., Worring, M.: Query-controllable video summarization. In: ACM ICMR (2020)
13. Huang, Q., Xiong, Y., Rao, A., Wang, J., Lin, D.: MovieNet: a holistic dataset for movie understanding. In: ECCV (2020)
14. Huang, Q., Xiong, Y., Xiong, Y., Zhang, Y., Lin, D.: From trailers to storylines: an efficient way to learn from movies (2018)
15. Iashin, V.: Video Features. https://github.com/v-iashin/video_features (2023)
16. Irie, G., Satou, T., Kojima, A., Yamasaki, T., Aizawa, K.: Automatic trailer generation. In: ACM MM (2010)
17. Ji, Z., Jiao, F., Pang, Y., Shao, L.: Deep attentive and semantic preserving video summarization. Neurocomputing (2020)
18. Ji, Z., Xiong, K., Pang, Y., Li, X.: Video summarization with attention-based encoder–decoder networks. IEEE Trans. Circuits Syst. Video Technol. (2020)
19. Johnson, J., Douze, M., Jégou, H.: Billion-scale similarity search with GPUs. IEEE Trans. Big Data (2021)
20. Kaushal, V., Kothawade, S., Iyer, R., Ramakrishnan, G.: Realistic video summarization through VISIOCITY: a new benchmark and evaluation framework. In: ACM MMW (2020)
21. Li, P., Ye, Q., Zhang, L., Yuan, L., Xu, X., Shao, L.: Exploring global diverse attention via pairwise temporal relation for video summarization. Pattern Recogn. (2021)
22. Lin, T.Y., Goyal, P., Girshick, R., He, K., Dollar, P.: Focal loss for dense object detection. In: ICCV (2017)
23. Liu, Y.T., Li, Y.J., Yang, F.E., Chen, S.F., Wang, Y.C.F.: Learning hierarchical self-attention for video summarization. In: ICIP (2019)

24. Mahasseni, B., Lam, M., Todorovic, S.: Unsupervised video summarization with adversarial LSTM networks. In: CVPR (2017)
25. Papalampidi, P., Keller, F., Lapata, M.: Film trailer generation via task decomposition (2021)
26. Reimers, N., Gurevych, I.: Sentence-BERT: sentence embeddings using Siamese BERT-networks. In: EMNLP (2019)
27. Reimers, N., Gurevych, I.: Making monolingual sentence embeddings multilingual using knowledge distillation. In: EMNLP (2020)
28. Smeaton, A.F., Lehane, B., O'Connor, N.E., Brady, C., Craig, G.: Automatically selecting shots for action movie trailers. In: ACM ICMR (2006)
29. Smith, J.R., Joshi, D., Huet, B., Hsu, W., Cota, J.: Harnessing A.I. for augmenting creativity: application to movie trailer creation. In: ACM MM (2017)
30. Snoek, C.G.M., Worring, M., Smeulders, A.W.M.: Early versus late fusion in semantic video analysis. In: ACM MM (2005)
31. Song, Y., Vallmitjana, J., Stent, A., Jaimes, A.: TVSum: summarizing web videos using titles. In: CVPR (2015)
32. Souček, T., Lokoč, J.: TransNet V2: an effective deep network architecture for fast shot transition detection (2020)
33. Tapaswi, M., Zhu, Y., Stiefelhagen, R., Torralba, A., Urtasun, R., Fidler, S.: MovieQA: understanding stories in movies through question-answering. In: CVPR (2016)
34. Turner, G.: Cultural Identity, Soap Narrative, and Reality TV. Television & New Media (2005)
35. Vaswani, A., et al.: Attention is all you need. In: NeurIPS (2017)
36. von Wenzlawowicz, T., Herzog, O.: Semantic video abstracting: automatic generation of movie trailers based on video patterns. In: SETN (2012)
37. Wang, L., Liu, D., Puri, R., Metaxas, D.N.: Learning trailer moments in full-length movies with co-contrastive attention. In: ECCV (2020)
38. Wang, W., Wei, F., Dong, L., Bao, H., Yang, N., Zhou, M.: MiniLM: Deep self-attention distillation for task-agnostic compression of pre-trained transformers. In: NeurIPS (2020)
39. Wu, C.Y., Krahenbuhl, P.: Towards long-form video understanding. In: CVPR (2021)
40. Xu, H., Zhen, Y., Zha, H.: Trailer generation via a point process-based visual attractiveness model. In: IJCAI (2015)
41. Zhang, K., Chao, W.L., Sha, F., Grauman, K.: Video summarization with long short-term memory. In: ECCV (2016)
42. Zhao, B., Gong, M., Li, X.: Hierarchical multimodal transformer to summarize videos. Neurocomputing (2022)
43. Zhao, B., Li, X., Lu, X.: Hierarchical recurrent neural network for video summarization. In: ACM MM (2017)
44. Zhao, B., Li, X., Lu, X.: HSA-RNN: hierarchical structure-adaptive RNN for video summarization. In: CVPR (2018)
45. Zhao, B., Li, X., Lu, X.: TTH-RNN: tensor-train hierarchical recurrent neural network for video summarization. IEEE Trans. Ind. Electron. (2021)

SM-GAN: Single-Stage and Multi-object Text Guided Image Editing

Ruichen Li[1], Lei Wu[1(✉)], Pei Dong[1], and Minggang He[2]

[1] Shandong University, Jinan, China
202135267@mail.sdu.edu.cn, i_lily@sdu.edu.cn
[2] Shandong Survey and Design Institute of Water Conservancy, Jinan, China

Abstract. In recent years, text-guided scene image manipulation has received extensive attention in the computer vision community. Most of the existing research has focused on manipulating a single object from available conditioning information in each step. Manipulating multiple objects often relies on iterative networks progressively generating the edited image for each instruction. In this paper, we study a setting that allows users to edit multiple objects based on complex text instructions in a single stage and propose the Single-stage and Multi-object editing Generative Adversarial Network (SM-GAN) to tackle problems in this setting, which contains two key components: (i) the Spatial Semantic Enhancement module (SSE) deepens the spatial semantic prediction process to select all correct positions in the image space that need to be modified according to text instructions in a single stage, (ii) the Multi-object Detail Consistency module (MDC) learns semantic attributes adaptive modulation parameter conditioned on text instructions to effectively fuse text features and image features, and can ensure the generated visual attributes are aligned with text instructions. We construct the Multi-CLEVR dataset for CLEVR scene image construction using complex text instructions with single-stage processing. Extensive experiments on the Multi-CLEVR and CoDraw datasets have demonstrated the superior performance of the proposed method.

Keywords: Text-guided image editing · Image generation · Generative adversarial networks · Multi-modal fusion

1 Introduction

With the development of deep learning [8,23], text-guided image editing has become an active research field. It has applications in computer-aided design, intelligent creation and so on.

Most current studies for text-guided image editing focus on using a single text instruction or description to edit a specific object in an image [2,3,13,14,21]. But in practical applications, it is more common that there are multiple interrelated objects in an image, and multiple objects in the scene need to be edited simultaneously. If the user is given a piece of text, the model can extract the key

© The Author(s), under exclusive license to Springer Nature Switzerland AG 2024
S. Rudinac et al. (Eds.): MMM 2024, LNCS 14555, pp. 213–226, 2024.
https://doi.org/10.1007/978-3-031-53308-2_16

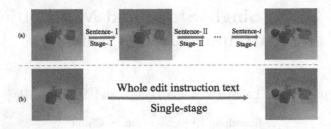

Fig. 1. We aim to address the multi-object text-guided image editing task, as shown in this figure. (a) is the existing method, which uses an iterative approach to editing images, processes only one instruction at a time, and generates an edited image in the middle stage of each iteration. (b) is our method, instead of editing iteratively, we can process multiple objects simultaneously and accurately in a single stage, which can also effectively avoid the artifacts and blurring caused by multiple image editing operations.

information provided by the text and preview the image after editing multiple objects. That will improve the usability and convenience of image editing operations. Some methods attempt to operate on multiple objects and typically rely on an iterative network [5–7,18] that gradually generates an edited image for each instruction. However, the expressive-intensive areas of the image may result in artifacts and blurring due to multiple editing operations. Make certain parts of the resulting image resemble a collage of blurred shapes and details. Furthermore, when these models attempt to process all the objects through a single edit stage, they suffer from missing objects and semantic inaccuracies and confuse multiple features leading to incorrect operations.

In this paper, we investigate a new setting that allows users to edit multiple objects by using complex text instructions through a single stage. We show an example in Fig. 1. Instead of iteratively processing each sentence of the entire text instruction to generate many intermediate images, we want to quickly preview the image by modifying all the parts in a single stage, according to multiple editing instructions. But there are still some challenges. The main challenge in the setting is selecting all the suitable object locations simultaneously in a complex scenario with multiple objects. Furthermore, the different objects have to be accurately edited separately in terms of size or texture according to the semantics of the text instructions.

To solve the above problems, we propose a new model, named SM-GAN, to edit multiple objects in complex scenarios through a single-stage generation module. There are two main modules in our proposed SM-GAN: the Spatial Semantic Enhancement module (SSE) and the Multi-object Detail Consistency module (MDC). Specifically, the SSE module deepens the spatial semantic prediction process to predict a spatial attention mask based on the complex instructions. It can select all areas in the image that need to be edited, effectively solving the problem of missing objects in editing. The MDC module is to align the instruction semantics with various visual attributes that need to be modified for accurate and detailed modification of multiple objects simultaneously. The

MDC module learns semantics attributes and adaptive modulation parameters conditioned on text to fully capture the semantic information related to the edited objects by effectively fusing text and image features. For better training and evaluation of our model, we generated a Multi-CLEVR dataset based on CLEVR [11] dataset. Extensive experiments on Multi-CLEVR and CoDraw [12] datasets show that our method significantly outperforms the state-of-the-art methods, with more accurate editing operations and richer instance-level detail in the generated images. In addition, we conducted a series of analytical experiments to evaluate the importance of each component in our approach and to validate the effectiveness of SM-GAN further.

In summary, our primary contributions are summarized as follows:

- We propose the SM-GAN to deal with a new setting which is editing multiple objects in complex scenarios through single-stage generation based on the complex text instruction.
- We propose the SSE module to select all the correct object locations simultaneously based on the complex text instruction and the MDC module to align the instruction semantics with various visual attributes of these multiple objects.
- To verify the model's validity, we construct the Multi-CLEVR dataset for editing construction of CLEVR scenes with single-stage processing of complex text instructions, and the experimental results on the Multi-CLEVR and CoDraw datasets demonstrate that SM-GAN outperforms the state-of-the-art approaches. The generated images reflect more accurate editing operations and richer instance-level details.

2 Related Works

2.1 Single-Object Text-Guided Image Editing

Text-guided image editing aims to semantically process images based on text instructions or descriptions rather than creating them from scratch. A single object can be edited by a single operation, so there is no multi-stage problem involved. Most of the current research has focused on datasets such as lowers [20] and birds [26] as well as human faces [16], etc., where there is often only one prominent object subject in these images. Similarly to Xu et al., [28], Nam et al. [19] introduce TAGAN, which includes word-level local discriminators. Mani-GAN [17] use two modules to manipulate only corresponding regions that match text descriptions. Some models perform operations based on instructions rather than using descriptive text. [14, 29] can edit the image according to the directive text instructions containing a clear action object and action type. EditGAN [15] embeds an image into the GAN's latent space and performs conditional latent code optimization according to the segmentation edit, effectively modifying the image. Xu et al. propose a novel framework for disentangled text-driven image manipulation, i.e., Predict, Prevent, and Evaluate (PPE) [27]. Some models can edit scenes with multiple subjects, but they deal with only one of the objects at

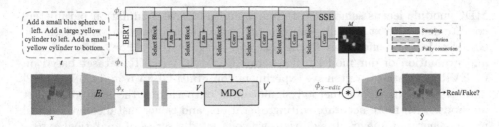

Fig. 2. Overview of the SM-GAN. Given the multimodal input: a source image x and a complex text instruction t, our goal is to synthesize an image $\hat{y}$ according to the text instructions. The green part is our proposed spatial semantic enhancement module(SSE), and the yellow part is the multi-object detail consistency module (MDC). (Color figure online)

a time based on instructions or captions. TIM-GAN [29] divides the operation into where and how phases based on textual instructions, treating text as neural operators to modify image features.

2.2   Multi-object Text-Guided Image Editing

When there are multiple text instructions to edit different objects, most methods rely on an iterative network [5–7,18] that gradually generates an edited image for each instruction, requiring multiple stages to complete the final edit. El-Nouby et al. [6] first introduce the Generative Neural Visual Artist (GeNeVA) task, accompanied by two distinct datasets, CoDraw [12] and iCLEVR [6]. One of their most practical extensions to one-step generation is studying a system for iteratively generating images based on continuous language input or feedback. Similar to the task of GeNeVA, Matsumori et al. [18] extract fine-grained text representations for the generator, and a Text-Conditioned U-Net discriminator architecture, which discriminates both the global and local representations of fake or real images. Jiang et al. [10] propose an interactive editing framework that can edit a certain attribute to different degrees, and the key insight is to model a continual "semantic field" in the GAN latent space.

However, the above methods do not apply to the setting of editing multiple objects in a single stage. When editing multiple objects, using multiple stages to generate intermediate images iteratively brings challenges in maintaining an instruction-independent context and accurately generating high-quality edited images.

3   Method

An overview of the proposed SM-GAN model is shown in Fig. 2. In the entire image editing model, the image to be edited is first encoded using the image encoder E_I, then the edited image embedding is generated by combining the spatial semantic masks and mixing the encoded text information, and finally the decoder G generates the image. Our model mainly consists of a text encoder, image encoder, generator and discriminator, and two processing modules SSE and MDC.

We use the pre-trained Bidirectional Encoder Representations from Transformers (BERT) [4] as the text encoder. In BERT, the complex text instructions t are tokenized and passed through Multi-Head attention. The residual connection and layer normalization are also applied on top of the sentence embeddings. We retain the embedding of the last layer. Ultimately, each token in the instructions is encoded as a 768-dimensional text embedding ϕ_t.

We elaborate on other important parts of our model as follows.

3.1 Spatial Semantic Enhancement Module

Editing multiple objects in a single stage means that the network needs to deal with more complex spatial relationships. In order to predict all locations to be edited in the input image x, we deepen the spatial semantic prediction process. As shown in the Fig. 2, we first input the text features ϕ_t into the SSE module, combined with a noise vector z sampled by Gaussian distribution, and finally generate a spatial attention mask M which serves as a reference to the intended modified position as follows:

$$M = SSE(\phi_t, z) \tag{1}$$

where ϕ_t is reshaped into h, which is a 512-dimensional condition feature, through linear layers.

In the process of predicting spatial attention masks in the SSE module, the spatial select block in the SSE module uses affine transformations to precisely select the desired position matching the given instructions. The ith select block takes two inputs: the hidden features h_i and the regional mask features $v_i \in \mathbb{R}^{c \times h \times w}$. Then, h_i is further processed to produce $W_i(h_i)$ and $b_i(h_i)$ that have the same size as v_i. Finally, we fuse the two modality representations to produce v_{i+1}.

$$v_{i+1} = b_i(h_i) + W_i(h_i) \cdot v_i \tag{2}$$

Between the first few layers of select blocks, we use cross-attention for spatial adaptability selection to enhance expressions of the general location where multiple objects need to be edited. Specifically, we reshape the hidden feature h_i and pass it through a Linear layer $L_q \in \mathbb{R}^{c \times (h \times w)}$ to obtain the query Q. Meanwhile, the mask feature maps $v_i \in \mathbb{R}^{c \times h \times w}$ is reshaped and passed through two convolution layers to obtain key K_i and value V_i.

$$v_{i+1} = L_{proj} \left[softmax \left(\frac{Q \cdot K_i^\top}{\sqrt{c}} \right) V_i^\top \right] \tag{3}$$

where $v_{i+1} \in \mathbb{R}^{c \times h \times w}$ and the $L_{proj} \in \mathbb{R}^{c \times c}$ is a linear layer.

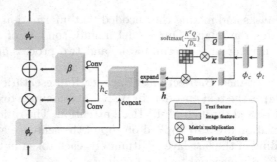

Fig. 3. The MDC module calculates self-attention on the text features ϕ_c, then expands into h and concatenates the input features V in the channel dimension. The model learns to generate modulation parameters that are used to modify the visual feature.

Between the following select blocks, we use a 3×3 convolution filter, aiming to stabilize the positions in the mask found and refine the information of multiple object positions and edges in the mask according to the continuously enhanced spatial semantics.

$$v_{(x,y)} = Conv(v_i) \tag{4}$$

where $v_{(x,y)}$ is the pixel in the currently processed map feature v_i. The SSE module deeply fuses text-image information and progressively optimizes the spatial semantic prediction, eventually can determine the spatial information of multiple objects to be edited in a single stage.

3.2 Multi-object Detail Consistency Module

For multiple objects that need to be edited, the visual attributes that need to be processed are also different. To ensure accurate operation on each object, we propose the MDC module align visual attributes with text instructions that contain multiple object editing information.

As is shown in Fig. 3, our module has two inputs, the image feature ϕ_V processed by convolution and text feature ϕ_t encoded by BERT. First, ϕ_t is reshaped into 512-dimensional features through linear layers. Then, we employ a conditioning augmentation [30] technique that element-wise multiplies the standard deviation of sentence embeddings $\sigma(\phi_t)$ by the noise with a standard Gaussian distribution ϵ_t (e.g., $\epsilon_t \sim \mathcal{N}(0,1)$). Then, we add the mean of sentence embeddings $\mu(\phi_t)$, resulting in the conditioned text features ϕ_c:

$$\phi_c = \mu(\phi_t) + \sigma(\phi_t) \odot \epsilon_t \tag{5}$$

To further understand the semantics of text instructions, we use the scaled dot-product self-attention [25] to summarize the keywords in the instructions. ϕ_c denote the text instructions, and the query, key, and value in the attention are computed by: $Q = \phi_c \cdot W_q$, $K = \phi_c \cdot W_k$, $V = \phi_c \cdot W_v$, where $W_q, W_k, W_v \in R^{d_o \times d}$

are linear weight matrices to learn, and d is the output dimension. We obtain the attended text feature h by:

$$h = Attention(Q, K, V) = softmax(\frac{Q^\top K}{\sqrt{d}})V, \tag{6}$$

in which the softmax function encourages higher attention weights over keywords. Because of the numerous and chaotic operation objects and editing features, it is necessary to further enhance the alignment of text instructions and image editing features to avoid confusing the correspondence between objects and specific instructions. We expand and repeat the instruction feature h in spatial dimension to be h'. h' and image features ϕ_V are concatenated to generate condition feature h_c and serve as the conditional signal for editing. We model the modulation parameters conditioned on h_c through adaptive learning as follows:

$$\beta = W_1 h_c, \quad \gamma = W_2 h_c, \tag{7}$$

where W_1 and W_2 are learnable convolutional filters. The produced modulation parameters γ and β achieve text-image features fusion through scaling and shifting the visual feature map ϕ_V:

$$\phi_{V'} = \gamma \odot \phi_V + \beta, \tag{8}$$

where the $\odot$ is element-wise dot product and $\phi_{V'}$ indicates the edited visual feature map. The edited visual feature map $\phi_{V'}$ combined with the predicted spatial mask M to modify the original image x. Finally, the generator generates the edited image $\hat{y}$ based on the features of the edited image $\phi_{\hat{y}}$.

$$\phi_{\hat{y}} = (1 - M) \odot \phi_x + M \odot \phi_{V'} \tag{9}$$

3.3 Objective Function

For training, we use the standard conditional GAN objective, which consists of an adversarial loss $\mathcal{L}_{GAN}$ and an l_1 reconstruction loss called $\mathcal{L}_{L1}$. Conditional GANs learn a mapping from observed image x to $\hat{y}$, $G : x \rightarrow \hat{y}$, and y represents the ground truth.

$$\mathcal{L}_{cGAN}(G, D) = \mathbb{E}_{x,y}[log D(x, y)]$$
$$+ \mathbb{E}_x[log(1 - D(x, G(x)))], \tag{10}$$

$$\mathcal{L}_{L1}(G) = \mathbb{E}_{x,y}[\|y - G(x)\|_1], \tag{11}$$

where G tries to minimize this objective against an adversarial D that tries to maximize it. After generating the edited image, an adversarial loss $\mathcal{L}_{GAN}$ will be predicted to evaluate the visual realism and semantic consistency of inputs. By distinguishing generated images from real samples, the discriminator promotes the generator to synthesize images with higher quality and text-image semantic consistency.

Our final objective is

$$G^* = arg\underset{G}{min}\underset{D}{max}\mathcal{L}_{cGAN}(G, D) + \lambda\mathcal{L}_{L1}(G). \tag{12}$$

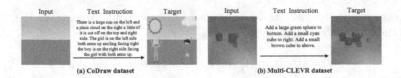

(a) CoDraw dataset   (b) Multi-CLEVR dataset

Fig. 4. (a) and (b) are examples of image instruction pairs in CoDraw and Multi-CLEVR datasets, respectively.

4   Experiments

4.1   Datasets

Thus, we follow previous studies [6, 29] on related text-guided image editing, which used the CLEVR and CoDraw datasets. The CLEVR is a programmatically generated dataset that is popular in the Visual Question Answering (VQA) community. During the construction of the CLEVR dataset, the Blender [31] was used for rendering. Previous work related to ours took advantage of this by improving the rendering process of the CLEVR dataset to generate its own usable dataset. Similarly, we also refer to previous work and construct the Multi-CLEVR dataset.

- **Multi-CLEVR:** An example from the Multi-CLEVR dataset is presented in Fig. 4(a). To generate the text instructions, we follow the previous research [29] and use a simple text template. For example, the instruction will follow the template: *Add a [object shape] [object color] [object] to [relative position].* We ensure that each text-image pair contains text instructions for multiple object editing operations and a variety of visual attributes. The Multi-CLEVR dataset includes 18K training and 4K test examples.
- **CoDraw:** It consists of scenes and sequences of images of children playing in a park. In order to adapt to the task, we intercept some scenes and corresponding text instructions from each sequence. Each sample includes a reference image, a target image, and a text instruction specifying the modification (Fig. 4(b)). We ensure that complex text instructions contain operations on multiple objects and properties to operate on. The dataset includes 8K training and 1.7K test examples.

4.2   Implementation Details

We use 3 down-sampling layers for the image encoder E_I. The encoded image has 256 feature channels, and the attended text embedding dimension is $d = 512$. We construct the generator G using two residual blocks followed by three transposed-convolutional layers. Each layer is equipped with instance normalization and ReLU. The parameters in the image encoder E_I and generator G are initialized by training an image autoencoder for 30 epochs. Then we fix E_I's parameters and optimize the other parts of the network in the end-to-end training for 120 epochs. We set the generator and discriminator learning rates to 0.0002 and exponential rates of $(\beta_1, \beta_2) = (0.5, 0.999)$. We resize the images to 256×256 for all the experiments.

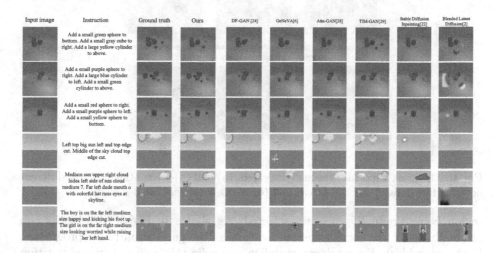

Fig. 5. Example generation results. We show the manipulation results using different approaches on the Multi-CLEVR (top) and CoDraw (bottom) datasets.

4.3   Qualitative Results

To evaluate the visual quality of generated images, we first show some subjective comparisons among the state-of-the-art methods and our proposed SM-GAN in Figure 5. All methods are implemented using their official code or adapted official code. Similar to TIM-GAN, the GeNeVA can make modifications to images but often does not follow the text instructions. When editing, they also suffer from missing objects and confusing features between different target objects. When using text instructions, Attn-GAN and DF-GAN cannot edit the image while maintaining the original image features. They tend to copy the input images and often generate random objects following similar input layouts. This is especially true when DF-GAN handles CoDraw datasets. As can be seen in Fig. 5, although the diffusion model can generate the relevant objects according to the text, it cannot ensure that the objects are at the specific locations described in the text. However, the SM-GAN generates results reflecting richer instance-level details and all the semantic features of the instructions, such as the boy's clothing and the cloud shadows, which are often realized as blurred color patches in the generated results of previous methods.

To verify the validity of spatial selection by attn-blocks and conv-blocks based on spatial semantic information, we output spatial attention masks at different network locations of our proposed SM-GAN, showing the process of the mask

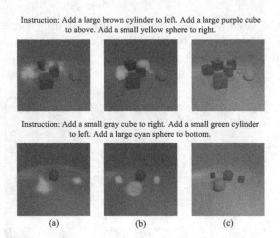

Fig. 6. The mask region are gradually selected. (a) and (b) are spatial masks at different network locations of our SM-GAN, and (c) is the final generated images.

region being gradually correctly selected. As shown in Fig. 6, attn-blocks can gradually select the correct spatial location as the generation phase progresses. In the conv-block, the mask can stabilize and refine the selected spatial position according to the gradually enhanced spatial semantics so that the mask can provide a more accurate and effective editing reference.

Table 1. Quantitative results on Multi-CLEVR and CoDraw with different models. Among them, SD-Inpainting refers to the Stable Diffusion Inpainting model and BL-Diffusion refers to the Blended Latent Diffusion model.

Method	Multi-CLEVR					CoDraw				
	PSNR↑	SSIM↑	FID↓	IS↑	LPIPS↓	PSNR↑	SSIM↑	FID↓	IS↑	LPIPS↓
DF-GAN [24]	26.2625	0.8037	25.82	1.82±0.04	0.1891	16.3929	0.4468	115.36	2.03±0.13	0.3954
GeVeVA [6]	25.7962	0.8114	23.67	2.16±0.03	0.1813	16.0951	0.7068	142.83	1.01±0.00	0.3694
Attn-GAN [28]	25.4123	0.6927	25.65	1.95±0.04	0.1933	16.2525	0.6836	78.23	3.30±0.14	0.4433
TIM-GAN [29]	26.3898	**0.9109**	13.06	2.35±0.02	0.1575	18.1107	0.7435	87.14	3.65±0.18	0.3249
SD-Inpainting [22]	25.1513	0.7027	23.21	2.13±0.04	0.1833	16.0365	0.7031	95.53	1.15±0.12	0.3934
BL-Diffusion [1]	24.4035	0.6517	22.27	2.15±0.02	0.1995	17.8625	0.6878	95.28	2.64±0.03	0.4081
Ours	**27.7071**	0.9041	**10.95**	**3.26±0.06**	**0.1455**	**18.7415**	**0.7738**	**64.37**	**3.67±0.19**	**0.3041**

4.4 Quantitative Results

We use multiple metrics to measure the quality and semantic consistency of the results. All evaluation metrics are computed in the same image size for meaningful and fair comparisons with previous methods. The overall results are summarized in Table 1. Our proposed SM-GAN achieves the best performance compared to previous methods. These results show that SM-GAN can make accurate semantic modifications on the source image based on text instruction to produce a better-quality image. This is because, after the space location selection of the SSE module, the MDC module can ensure that the generated visual attributes are aligned with textual instructions containing multiple object editing information.

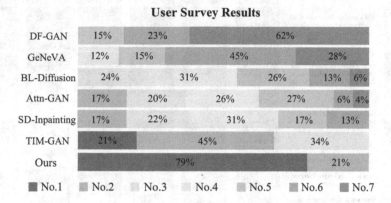

Fig. 7. Human evaluation results. The users are asked to rank these methods by the quality of the generated results. Among them, No. 1 is the best overall effect, and No. 7 results from an unsatisfactory overall effect.

Human Evaluation. We also conduct a human evaluation. There were 20 participants. We experimented on two datasets using seven different methods, each producing 40 pairs of before and after editing images. Each participant was asked to rank the quality of the images generated by the seven methods in terms of visual quality and semantic relevance. Among them, No. 1 is the best overall effect, and No. 7 results from an unsatisfactory overall effect. We count the percentage of each ranking in the results generated by each method in Fig. 7. As shown in the figure, 79% of the answers rated our method as the best overall result, and the remaining 21% of the groups rated our method as the second best overall result.

The above results are consistent with the quantitative results in Table 1. The human evaluation shows that SM-GAN can edit source images more accurately and semantically according to text instructions to produce edited images that align with users' psychological expectations.

Table 2. Ablation study. The performance of the various modules in our model.

Method	Multi-CLEVR					CoDraw				
	PSNR↑	SSIM↑	FID↓	IS↑	LPIPS↓	PSNR↑	SSIM↑	FID↓	IS↑	LPIPS↓
w/o.attn in SSE	25.8174	0.7391	18.30	2.23±0.07	0.2895	15.0329	0.6031	162.28	2.06±0.05	0.3752
w/o.conv in SSE	26.0417	0.8046	21.61	2.71±0.02	0.3052	16.1315	0.6864	102.27	1.76±0.08	0.3551
w/o.MDC	25.4303	0.7579	31.99	2.40±0.05	0.3623	15.6415	0.6235	86.45	2.31±0.12	0.3798
w/o.SSE	**27.7122**	0.8153	**6.91**	2.53±0.02	0.2602	16.8442	0.6834	197.25	1.01±0.00	0.4123
full model	27.7071	**0.9041**	10.95	**3.26±0.06**	**0.1455**	**18.7415**	**0.7738**	64.37	**3.67±0.19**	**0.3041**

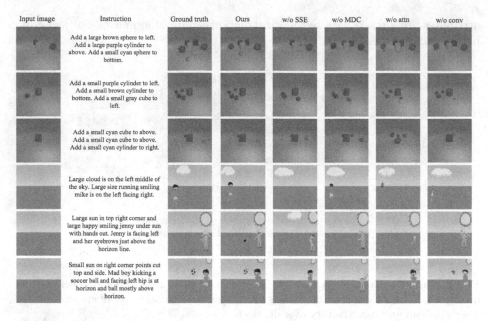

Fig. 8. Ablation study results. The SSE and MDC modules are removed separately, with the other modules retained. The two right columns remove all attn-blocks and conv-blocks from the SSE module, respectively.

4.5   Ablation Studies

In this section, we further analyze the importance of our method's components. The corresponding results are shown in Table 2 and Fig. 8. When the SSE module is removed, all objects to be edited are not correctly selected in space, which is not conducive to accurate modifications later(1st, 4th rows). When the MDC module is removed, even if all the correct edit locations are selected in space, the model cannot correctly distinguish and edit multiple attributes and features, so the multiple objects and edit semantics are inconsistent (3rd row). The MDC module is also important for editing details, such as the little boy in 4th line whose face is blurred and whose limbs are not clearly structured. The figure also shows the instability and ambiguity of the spatial location selection after the removal of the attn-block and conv-block, respectively (2nd, 5th, and 6th rows). This leads to a dramatic deformation and distortion of the final generated edited image, which does not reflect the editing semantics well.

Table 2 shows quantitative comparison results of the ablation study. Our full model achieved better results. An exception to this is the significant drop in FID scores following the removal of the SSE module, which is due to the diversity of the results generated being taken into account in calculating this metric. However, in the image editing task, more accurate spatial position expression and editing operation will reduce diversity.

5   Conclusion

In this paper, we investigate a new setting that allows users to edit multiple objects through a single-stage generation using text instructions. A Single-stage Multi-object Generative Adversarial Network (SM-GAN) is proposed for text-guided image editing. The SSE can effectively utilize spatial semantic information better identify the object locations that need to be modified. To produce correct editing results with multiple editing features, we devise the MDC to assign appropriate manipulation features to each object. Both quantitative and human evaluation studies show that SM-GAN improves the generation compared to the baseline models. We also construct the Multi-CLEVR dataset and hope our work can extend and inspire research on text-guided image editing. Because GAN-based methods usually need to rely on datasets for training, our future work will explore the possibility of using diffusion models for our proposed task.

References

1. Avrahami, O., et al.: Blended latent diffusion. arXiv arXiv:2206.02779 (2022)
2. Avrahami, O., et al.: Blended diffusion for text-driven editing of natural images. In: Proceedings of the IEEE/CVF Conference on Computer Vision and Pattern Recognition, pp. 18208–18218 (2022)
3. Cheng, Y., et al.: Sequential attention GAN for interactive image editing. In: Proceedings of the 28th ACM International Conference on Multimedia, pp. 4383–4391 (2020)
4. Devlin, J., et al.: BERT: pre-training of deep bidirectional transformers for language understanding. arXiv preprint arXiv:1810.04805 (2018)
5. El-Nouby, A., et al.: Keep drawing it: iterative language-based image generation and editing. arXiv arXiv:1811.09845 (2018)
6. El-Nouby, A., et al.: Tell, draw, and repeat: generating and modifying images based on continual linguistic instruction. In: Proceedings of the IEEE/CVF International Conference on Computer Vision, pp. 10304–10312 (2019)
7. Fu, T., et al.: SSCR: iterative language-based image editing via self-supervised counterfactual reasoning. arXiv preprint arXiv:2009.09566 (2020)
8. Goodfellow, I., et al.: Generative adversarial nets. In: Advances in Neural Information Processing Systems, vol. 27 (2014)
9. Heusel M., et al.: GANs trained by a two time-scale update rule converge to a local Nash equilibrium. In: Advances in Neural Information Processing Systems, vol. 30, pp. 6629–6640 (2017)
10. Jiang, Y., et al.: Talk-to-edit: fine-grained facial editing via dialog. In: Proceedings of the IEEE/CVF International Conference on Computer Vision, pp. 13799–13808 (2021)
11. Johnson, J., et al.: CLEVR: a diagnostic dataset for compositional language and elementary visual reasoning. In: Proceedings of the IEEE Conference on Computer Vision and Pattern Recognition, pp. 2901–2910 (2017)
12. Kim, J., et al.: CoDraw: collaborative drawing as a testbed for grounded goal-driven communication. arXiv arXiv:1712.05558 (2017)
13. Kawar, B., et al.: Imagic: text-based real image editing with diffusion models. In: Proceedings of the IEEE/CVF Conference on Computer Vision and Pattern Recognition, pp. 6007–6017 (2023)

14. Liu, Y., et al.: Describe what to change: a text-guided unsupervised image-to-image translation approach. In: 28th ACM International Conference on Multimedia, pp. 1357–1365 (2020)
15. Ling, H., et al.: EditGAN: high-precision semantic image editing. Adv. Neural. Inf. Process. Syst. **34**, 16331–16345 (2021)
16. Liu, Z., Luo, P., et al.: Deep learning face attributes in the wild. In: Proceedings of the IEEE International Conference on Computer Vision, pp. 3730–3738 (2015)
17. Li, B., Qi, X., Lukasiewicz, T., Torr, P.H.: ManiGAN: text-guided image manipulation. In: Proceedings of the IEEE/CVF Conference on Computer Vision and Pattern Recognition, pp. 7880–7889 (2020)
18. Matsumori, S., et al.: LatteGAN: visually guided language attention for multi-turn text-conditioned image manipulation. IEEE Access **9**, 160521–160532 (2021)
19. Nam, S., Kim, Y., Kim, S.J.: Text-adaptive generative adversarial networks: manipulating images with natural language. In: Advances in Neural Information Processing Systems, vol. 31 (2018)
20. Nilsback, M.-E., Zisserman, A.: Automated flower classification over a large number of classes. In: 2008 Sixth Indian Conference on Computer Vision, Graphics & Image Processing, pp. 722–729. IEEE (2008)
21. Patashnik, O., Wu, Z., Shechtman, E., Cohen-Or, D., Lischinski, D.: StyleCLIP: text-driven manipulation of StyleGAN imagery. In: Proceedings of the IEEE/CVF International Conference on Computer Vision, pp. 2085–2094 (2021)
22. Rombach, R., Blattmann, A., Lorenz, D., Esser, P., Ommer, B.: High-resolution image synthesis with latent diffusion models. In: Proceedings of the IEEE/CVF Conference on Computer Vision and Pattern Recognition, pp. 10684–10695 (2022)
23. Shi, J., Xu, N., Xu, Y., Bui, T., Dernoncourt, F., Xu, C.: Learning by planning: language-guided global image editing. In: Proceedings of the IEEE/CVF Conference on Computer Vision and Pattern Recognition, pp. 13590–13599 (2021)
24. Tao, M., Tang, H., Wu, F., Jing, X.-Y., Bao, B.-K., Xu, C.: DF-GAN: a simple and effective baseline for text-to-image synthesis. In: Proceedings of the IEEE/CVF Conference on Computer Vision and Pattern Recognition, pp. 16515–16525 (2022)
25. Vaswani, A., et al.: Attention is all you need. Adv. Neural. Inf. Process. Syst. **30**, 5998–6008 (2017)
26. Wah, C., Branson, S., et al.: The Caltech-UCSD Birds-200-2011 dataset
27. Xu, Z., et al.: Predict, prevent, and evaluate: disentangled text-driven image manipulation empowered by pre-trained vision-language model. In: Proceedings of the IEEE/CVF Conference on Computer Vision and Pattern Recognition, pp. 18229–18238 (2022)
28. Xu, T., Zhang, P., et al.: AttnGAN: fine-grained text to image generation with attentional generative adversarial networks. In: Proceedings of the IEEE Conference on Computer Vision and Pattern Recognition, pp. 1316–1324 (2018)
29. Zhang, T., Tseng, H.-Y., Jiang, L., Yang, W., Lee, H., Essa, I.: Text as neural operator: image manipulation by text instruction. In: Proceedings of the 29th ACM International Conference on Multimedia, pp. 1893–1902 (2021)
30. Zhang, H., et al.: StackGAN: text to photo-realistic image synthesis with stacked generative adversarial networks. In: Proceedings of the IEEE International Conference on Computer Vision, pp. 5907–5915 (2017)
31. Blender Online Community: Blender - a 3D modelling and rendering package (2016). http://www.blender.org/

MAVAR-SE: Multi-scale Audio-Visual Association Representation Network for End-to-End Speaker Extraction

Shilong Yu[1] and Chenhui Yang[2]([⊠])

[1] Department of Computer Science and Technology, Xiamen University, Xiamen 361000, Fujian, China
yushilong@stu.xmu.edu.cn
[2] Intelligent Laboratory, Xiamen University, Xiamen 361000, Fujian, China
chyang@xmu.edu.cn

Abstract. Speaker extraction to separate the target speech from the mixed audio is a problem worth studying in the speech separation field. Since human pronunciation is closely related to lip motions and facial expressions during speaking, this paper focuses on lip motions and their relationship to pronunciation and proposes a multi-scale audio-visual association representation network for end-to-end speaker extraction (MAVAR-SE). Moreover, multi-scale feature extraction and jump connection are used to solve the problem of information loss due to the lack of memory ability of convolution. This method is not limited by the number of speakers in the mixed speech and does not require prior knowledge such as speech features of related speakers, to realize speaker-independent multimodal time domain speaker extraction. Compared with other recent methods on VoxCeleb2 and LRS2 data sets, the proposed method shows better results and robustness.

Keywords: Speaker extraction · Speech separation · Audio-visual · Multi-scale · Attention mechanism

1 Introduction

In the real world, speech communication usually takes place in a complex multi-person environment, and the interference between sound sources leads to the reduction of acceptable information, and people are often only interested in the speech of a specific speaker. In addition, separating mixed speech signals to improve speech articulation and intelligibility is a necessary preprocessing for speech recognition or other speech interactions. Therefore, how to separate speech accurately and extract speech of the target speaker has become a key point in speech processing technology.

Traditional speech separation methods are mainly based on signal processing, independent component analysis (ICA) [1–3], non-negative matrix factorization (NMF) [4–6], and computational auditory scene analysis (CASA) [7–9]. Although the traditional methods have good effects, they are often based on certain assumptions, and the processing effect is not ideal in the real scene. As computer processing speeds increase, deep learning has great advantages in processing large amounts of data [10–12]. In 2012, Wang et al. [13] combined deep neural networks with traditional CASA methods for speech separation, which made a huge breakthrough. Subsequently, deep learning methods became popular in the field of speech separation [14–17].

With the rapid development of short video, audio-visual data can be easily obtained, and there are inextricably linked among speech, mouth shapes, and lip motions. Therefore, multi-modal speech separation methods are becoming more and more popular [18–20]. Hou et al. [21] combined visual and auditory information to propose a multitask-based speech enhancement model, but this approach relies on the speakers and specific models need to be trained for different persons. Then researchers from Google [22] proposed an audio-visual speech separation model which is independent of the speaker and proved that the speech separation effect of multi-mode is better than that of single-mode.

However, most of the methods require short-time Fourier transform (STFT) and its inverse transformation (iSTFT) to complete audio preprocessing and restoration [23–26]. Luo et al. proposed TasNet [27] (called TasNet-LSTM and TasNet-BLSTM in comparative experiments) based on LSTM, which realizes the end-to-end separation of speech in the time domain through encoder and decoder, replacing the conversion of audio in the time-frequency domain.

However, the computational performance of this model in long sequence data is limited by LSTM. Then Luo et al. made improvements on this basis and replaced LSTM with the time-domain convolutional network, and proposed Conv-TasNet [28], which reduced the model volume and computational overhead while modeling long sequences. The above two methods are all audio-only speech separation. Wu et al. [29] proposed the audio-visual speech separation network by adding visual information based on Conv-TasNet (AV-ConvTasnet), and the effect is better than the audio-only method. Pan et al. improved based on AV-ConvTasnet and proposed the speaker extraction model named MuSE [30], which continuously optimized the network during training by repeatedly predicting the audio mask and performing self-feedback. However, such convolution-based methods would suffer from the information loss of long sequences, which is caused by the lack of memory ability of convolution.

Inspired by the work of Pan et al. [30], we use the correlation between audio-visual synchronized information to explore the association representation between lip motions and pronunciation, and propose a multi-scale audio-visual association representation network for end-to-end speaker extraction.

The contributions of this paper are as follows:

1. Three modules are proposed for global feature extraction of lip motions, detail mapping between lip motions and pronunciation, and speaker representation based on facial expressions, mouth shapes, and timbre tone.
2. In this paper, multi-scale feature extraction and multi-scale jump connection are adopted to solve the problem of information loss in the convolution process due to the lack of memory ability of convolution.

3. The network proposed in this paper is to extract the corresponding speech by the visual features of the target speaker when the number of speakers in the mixed audio is unknown, and the end-to-end speaker speech extraction in the time domain is realized irrespective of the speaker.

2 Methods

2.1 Overview

The proposed model consists of encoders, the multi-modal speaker extraction network, and the decoder. As shown in Fig. 1, the model contains two input tensors, which are the mixed audio signal $A_{raw} \in \mathbb{R}^T$ and the video image sequence $V_{raw} \in \mathbb{R}^{H \times W \times F}$ of the target speaker to be extracted, where T represents the number of sampling points of the audio, W and H represent the width and height of the images respectively, and F represents the number of video frames.

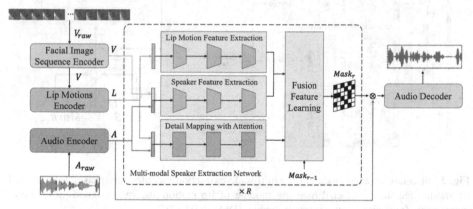

Fig. 1. The architecture of the proposed MAVAR-SE model.

Encoders are used to extract the features of the audio and the video image sequence respectively to generate the audio feature $A \in \mathbb{R}^{N \times M}$ and facial sequence feature $V \in \mathbb{R}^{512 \times F}$, and then calculate the feature of lip motions $L \in \mathbb{R}^{512 \times F}$, according to the facial sequence feature. The multi-modal speaker extraction network consists of four modules, which are multi-scale lip motion feature extraction, detail mapping learning of lip motion and pronunciation, multi-scale speaker feature extraction, as well as fusion feature learning. Finally, the network generates an audio mask $Mask_r$. Using the same strategy as MuSE [30], the speaker extraction network is repeated for R times, and the mask generated by each prediction is used as the input of the next network to form self-feedback to improve the accuracy of network. The initial input of $Mask_{r-1}$ is the original audio feature.

According to the mask, the corresponding part is extracted from the mixed audio, and the extracted audio signal is decoded and restored in the decoder, so as to realize the end-to-end time domain speech extraction.

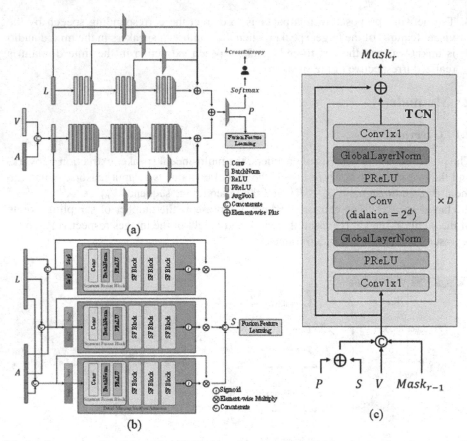

Fig. 2. Structure of each module. (a) Multi-scale extraction modules for speaker features and lip motion features. (b) Detail mapping module of lip motion and speech based on attention mechanism. (c) Fusion feature learning module (D: 8, d: 0 to 7).

2.2 Encoders and the Decoder

The model encodes and decodes audio in the same way as MuSE [30]. A 1D convolution with convolution kernel size K (K is 40 in this paper) and stride size K/2 encodes the mixed audio to generate an audio feature vector $A \in \mathbb{R}^{N \times M}$, where N represents the number of feature channels and M is the time domain feature size after convolution. For feature extraction of the video, a pre-trained speech recognition model [31] is adopted to encode the image sequence to obtain feature $V \in \mathbb{R}^{512 \times F}$. Since the pronunciation of each word in speech is closely related to the motions of the lip, the differences between the frames are important for the study of the correlation between auditory and visual features. Therefore, the image sequence feature V is used to calculate the interframe lip motion feature $L \in \mathbb{R}^{512 \times F}$.

$$L = Cat(V_0, V_i - V_{i-1}), 1 \leq i < F \tag{1}$$

where *Cat* represents the concatenation on the time frame dimension. And the facial sequence feature V and lip motion feature L are extended to the same size as the audio feature A by linear interpolation.

2.3 Multi-modal Speaker Extraction Network

2.3.1 Multi-scale Lip Motion Feature Extraction

Due to the great differences among the features of different lip motions, the convolution and average pooling are adopted to extract and distinguish the global features of different lip motions, while retaining the key information of lip motions and removing redundant information in this module. As shown in Fig. 2(a), multiple serial down-sampling modules are used to realize multi-scale feature extraction. Each down-sampling module consists of convolution, batch normalization, ReLU, and average pooling layer with the window size of 3. However, since convolution pays more attention to the extraction of local features, the dependencies between information are lost in the lip sequence, and the association between the information cannot be established. Therefore, the jump connection is used as the information supplement to solve this problem.

2.3.2 Detail Mapping of Lip Motion and Speech Based on Attention Mechanism

The lip characterization used in the lip motion feature extraction module of Sect. 2.3.1 is the complete feature L of the sequence and extracts the global information of lip motions, while this module is to learn the relationship between lip motion and speech pronunciation in small segments, focusing on local information. Since different motions of the lip correspond to different pronunciation, the study of the mapping relationship between lip motion and speech plays an important role in the separation of speech.

As shown in Fig. 2(b), this module divides the speech feature A and lip motion feature L into three segments of the same length, and after concatenating on the dimension of the feature channel and performing fusion learning, the Sigmoid activation function is used to generate a weight matrix for the fusion feature and multiply with the original one. Thus, weights are assigned to different feature points to form the attention mechanism of local key information, and the output feature S is input into the fusion feature learning module of Sect. 2.3.4.

$$S = Cat\left(\sigma\left(Att\left(Cat\left(A_{seg}^i, L_{seg}^i\right)\right)\right) * Cat\left(A_{seg}^i, L_{seg}^i\right)\right) \tag{2}$$

where i represents the corresponding segment index, σ represents the Sigmoid function, and *Att* represents the feature attention.

2.3.3 Multi-scale Speaker Feature Extraction

Different from the conventional method of separating the speeches according to the number of mixed speakers, this paper realizes the function of extracting the corresponding speech from the mixed audio through the visual information of a single target speaker. Since there are great differences in timbre and intonation between individuals, so this module is proposed based on this. This module realizes speaker extraction through speech features A and facial features V of expressions and mouth shapes. And the multi-scale connection used in the module is also designed to obtain sufficient global context information.

This module uses a similar structure to the module of Sect. 2.3.1. As shown in Fig. 2(a), firstly, the audio feature A and facial sequence feature V are concatenated on the time dimension, and then the high-level abstract information is obtained through three serial multi-scale feature extraction modules. The multi-scale feature extraction module includes convolution, batch normalization, PReLU and average pooling layer with a window size of 3, which pays attention to the differences between different speakers while narrowing the feature map. However, the pooling process will lead to the loss of some details and the convolution will lead to the loss of information dependencies in the sequence. Therefore, the output results are supplemented by the extracted features of different scales in the way of jump connection, so as to retain the integrity of the information. Finally, the extracted speaker features and the lip features of Sect. 2.3.1 are added and pooled to obtain feature P.

2.3.4 Fusion Feature Learning

As shown in Fig. 2(c), After adding the feature P and the mapping feature S, the facial feature V and the previous weight matrix output $Mask_{r-1}$ are fused. A new weight matrix $Mask_r$ is generated by the temporal convolutional net (TCN) of Conv-TasNet [28].

2.4 Two-Stage Training Strategy

We used the same loss functions as Pan et al. [30], part of which is the negative of the scale-invariant signal-to-distortion ratio (SI-SDR) [32], a quantitative measure of time-domain similarity between two audios. And in order to allow the network to better distinguish the differences between different individuals, the output feature P in Fig. 2(a) is used to predict the corresponding speaker in the data set, and the cross-entropy loss function is used to judge the prediction accuracy, thus helping to improve the accuracy of speaker speech extraction.

The loss function during training is defined as:

$$Loss = \alpha L_{SISNR} + \beta L_{CrossEntropy} \tag{3}$$

The proposed model is trained in two stages. In the first stage, $\beta > 0$ in the loss function, and in the second stage, β is adjusted to 0. Compared with the single-stage training, firstly, the two-stage training can solve the problem of overfitting in late iterations. Secondly, it breaks the bottleneck of the speech separation effect of the single-stage training model and improves the accuracy of speech separation. Finally, it makes the model fit the main task better.

3   Experiments

In this paper, training and testing are conducted on the public data set VoxCeleb2 [33], from which 800 speakers are randomly selected as the training set and validation set. Meanwhile, 118 speakers who are unseen and unheard are randomly selected as the test set. The speeches of two different speakers are mixed with a random signal-to-noise ratio of -10 dB to 10 dB to generate 16.67 h of the training set, 4.17 h of the validation set, and 2.5 h of the test set. The audio sampling rate is 16 kHz, and the video frame rate is 25 FPS. Before the experiment, the video frames are input into the pre-trained speech recognition model [31] to generate the facial sequence feature V as the visual information by the network in this paper.

During the experiment, a total of 100 epochs are carried out with an initial learning rate of 0.001. The training is divided into two stages. In the first stage, the parameters α and β in the loss function are 1 and 0.1. When the iteration reaches the 40th epoch, the training enters the second stage and β is adjusted to 0. During training, when the loss function value on the validation set does not decrease for 3 consecutive epochs, the learning rate is reduced to half.

3.1   Comparison with Recent Methods

In this paper, several recent methods are trained and tested on the consistent samples in the VoxCeleb2 data set. For the signal quality assessment of the extracted speech, scale-invariant signal-to-distortion ratio (SI-SDR), signal distortion ratio (SDR), signal interference ratio (SIR), and signal artifact ratio (SAR) [34] are used as evaluation metrics. Perceptual evaluation of speech quality (PESQ) [35] and short-term objective intelligibility (STOI) [36] are used as objective metrics for the auditory perception of extracted speech.

All methods are trained on NVIDIA GeForce RTX 3090 for 100 epochs, about 78 h, and the model with the best effect on the validation set is selected for testing. The test results are shown in Table 1. Our method has significantly improved on most objective metrics, especially the SDR of the extracted speech, which reaches 11.47dB. Compared with Conv-TasNet, an audio-only separation method, our method has an increase of 7.4% in SDR, 7.3% in SI-SDR, 6.6% in PESQ, and STOI is 86.11%, which increases by 1.4%. Compared with MuSE, an audio-visual separation model, our method increases the SDR, SI-SDR, and PESQ by 4.7%, 5.5%, and 1.9% respectively.

3.2   Ablation Experiments

In order to prove the effectiveness of each module, we take MuSE as the baseline to compare the model effects with different modules. As shown in Table 2, M1 represents the multi-scale speaker feature extraction module, M2 represents the multi-scale lip motion feature extraction module, and M3 represents the detail mapping module of lip motion and speech based on the attention mechanism.

Table 1. Comparison of test results on the VoxCeleb2 data set.

Method	SDR	SIR	SAR	SI-SDR	PESQ	STOI (%)
TasNet-LSTM	7.32	14.08	9.05	6.88	1.62	78.53
TasNet-BLSTM	10.02	17.88	11.22	9.60	1.87	83.49
Conv-TasNet	10.68	18.50	11.90	10.14	1.97	84.73
MuSE	10.95	**18.90**	12.43	10.31	2.06	85.21
Ours	**11.47**	18.11	**13.19**	**10.88**	**2.10**	**86.11**

In combination with Fig. 3, SI-SDR on the test set under different epochs indicate that each module is helpful to improve the quality and intelligibility of the extracted speech.

Table 2. Results of the ablation experiments.

Method	M1	M2	M3	SDR	SIR	SAR	SI-SDR	PESQ	STOI (%)
Baseline				10.95	18.90	12.43	10.31	2.056	85.21
Ours	✓			11.06	18.96	12.40	10.46	2.058	85.61
Ours	✓	✓		11.28	**19.28**	12.62	10.70	2.086	85.98
Ours	✓	✓	✓	**11.47**	18.11	**13.19**	**10.88**	**2.101**	**86.11**

3.3 Comparison of Single-Stage and Two-Stage Training

Figure 4 shows the changes of the loss function value on the validation set in single-stage training and two-stage training. Adding the cross-entropy loss of the speaker prediction to the loss function can accelerate the convergence rate of the model. In the middle and later periods of training, due to the limitation of the quantity of speakers in the data set, the model will overfit the speaker, which is manifested as the value of the loss on the validation set increases with the iteration. Overfitting will lead to the reduction of the generalization ability of the model. On the other hand, since the prediction of the speaker is not the main task, in order to make the model perform better in the task of speech extraction, the coefficient of the cross-entropy loss is adjusted to 0 in the second stage.

The differences between different starting epochs of the second stage are also compared in Fig. 4. Starting the second stage of training earlier can reduce the learning rate of the model faster, but the speed of finding the optimal solution is different. We compared the effects of the models entering the second stage from the 30th epoch, 40th epoch, and 50th epoch. The optimal solution can be found the fastest when entering the second stage from the 40th epoch.

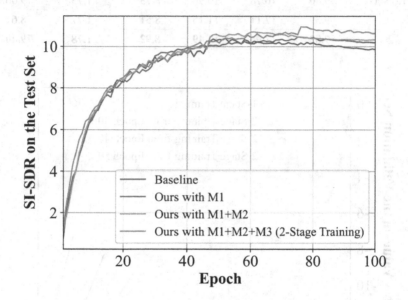

Fig. 3. SI-SDR results on the test set under different epochs.

3.4 Comparison of Effects on Another Data Set

In order to test the generalization ability and robustness of the proposed method, we use the model trained on the VoxCeleb2 data set to test on the LRS2 [31] data set. 6000 clips from different videos are randomly selected from the LRS2 data set for audio mixing with a random signal-to-noise ratio of -10dB to 10dB to generate 3000 test samples. The test results are shown in Table 3. Compared with other time-domain methods, the model proposed in this paper still shows better speech extraction capability even on other unseen data.

Table 3. Comparison of test results on the LRS2 data set.

Method	SDR	SIR	SAR	SI-SDR	PESQ	STOI (%)
TasNet-LSTM	5.83	11.93	8.11	5.29	1.47	73.09
TasNet-BLSTM	8.94	16.20	10.45	8.44	1.69	78.61
Conv-TasNet	9.40	16.78	10.96	8.73	1.73	79.00
MuSE	9.39	17.11	11.15	8.51	1.77	78.67
Ours	**9.82**	**17.96**	**11.49**	**8.92**	**1.78**	**79.40**

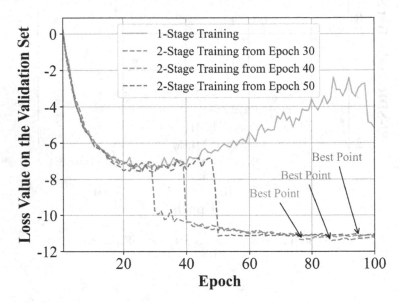

Fig. 4. The change of the loss function value on the validation set with the single-stage and two-stage training.

4 Conclusions

In this paper, we propose a multi-scale audio-visual association representation network for the time domain end-to-end speaker extraction (MAVAR-SE), which can realize the task of extracting the target speaker's speech from mixed audio based on the input visual information of the target speaker. The proposed method pays more attention to the correlation between lip motion and pronunciation and we propose three modules to extract and represent their features and associations. What's more, the proposed method can achieve speaker-independent speaker extraction without knowing the pre-recorded speech of the relevant speaker and the number of speakers in the mixed audio. We have tested on different data sets, and the results show that our method is superior to other time-domain speech separation methods in terms of audio separation effect and robustness. In the future, we will explore the reduction of computing costs.

References

1. Comon, P.: Independent component analysis, a new concept? Signal Process. **36**(3), 287–314 (1994)
2. Davies, M.E., James, C.J.: Source separation using single channel ICA. Signal Process. **87**(8), 1819–1832 (2007)
3. Zhang, K., Wei, Y., Wu, D.: Adaptive speech separation based on beamforming and frequency domain-independent component analysis. Appl. Sci. **10**, 7 (2020)
4. Lee, D.D., Seung, H.S.: Learning the parts of objects by non-negative matrix factorization. Nature **401**(6755), 788–791 (1999)
5. Virtanen, T.: Monaural sound source separation by nonnegative matrix factorization with temporal continuity and sparseness criteria. IEEE Trans. Audio Speech Lang. Process. **15**(3), 1066–1074 (2007)
6. Hsu, C.C., Chi, T.S., Chien, J.T.: Discriminative layered nonnegative matrix factorization for speech separation. In: Proceedings of the 17th Annual Conference of the International Speech Communication Association (INTERSPEECH 2016), pp. 560–564 (2016)
7. Parsons, T.W.: Separation of speech from interfering speech by means of harmonic selection. J. Acoust. Soc. Am. **60**(4), 911–918 (1976)
8. Li, H., Wang, Y., Zhao, R., Zhang, X.: An unsupervised two-talker speech separation system based on CASA. Int. J. Pattern Recognit. Artif. Intell. **32**, 7 (2018)
9. Bao, F., Abdulla, W.H.: A new ratio mask representation for CASA-based speech enhancement. IEEE/ACM Trans. Audio Speech Lang. Process. **27**(1), 7–19 (2018)
10. LeCun, Y., Bengio, Y., Hinton, G.: Deep learning. Nature **521**(7553), 436–444 (2015)
11. Schmidhuber, J.: Deep learning in neural networks: an overview. Neural Netw. **61**, 85–117 (2015)
12. Liu, W., Wang, Z., Liu, X., Zeng, N., Liu, Y., Alsaadi, F.E.: A survey of deep neural network architectures and their applications. Neurocomputing **234**, 11–26 (2017)
13. Wang, Y., Wang, D.: Boosting classification based speech separation using temporal dynamics. In: Proceedings of the 13th Annual Conference of the International Speech Communication Association (INTERSPEECH 2012), pp. 1526–1529 (2012)
14. Hershey, J.R., Chen, Z., Le Roux, J., Watanabe, S.: Deep clustering: discriminative embeddings for segmentation and separation. In: 2016 IEEE International Conference on Acoustics, Speech and Signal Processing (ICASSP), pp. 31–35. IEEE (2016)
15. Yu, D., Kolbæk, M., Tan, Z.H., Jensen, J.: Permutation invariant training of deep models for speaker-independent multi-talker speech separation. In: 2017 IEEE International Conference on Acoustics, Speech and Signal Processing (ICASSP), pp. 241–245. IEEE (2017)
16. Xu, C., Rao, W., Chng, E.S., Li, H.: Spex: multi-scale time domain speaker extraction network. IEEE/ACM Trans. Audio Speech Lang. Process. **28**, 1370–1384 (2020)
17. Ge, M., Xu, C., Wang, L., Chng, E.S., Li, H.: Spex+: a complete time domain speaker extraction network. In: Proceedings of the 21st Annual Conference of the International Speech Communication Association (INTERSPEECH 2020), pp. 1406–1410 (2020)
18. Gu, R., Zhang, S.X., Xu, Y., Chen, L., Zou, Y., Yu, D.: Multi-modal multi-channel target speech separation. IEEE J. Select. Top. Signal Process. **14**(3), 530–541 (2020)
19. Gogate, M., Dashtipour, K., Adeel, A., Hussain, A.: CochleaNet: a robust language-independent audio-visual model for real-time speech enhancement. Information Fusion **63**, 273–285 (2020)

20. Yu, J., Zhang, S.X., Wu, J., Ghorbani, S., Wu, B., Kang, S., et al.: Audio-visual recognition of overlapped speech for the lrs2 dataset. In: 2020 IEEE International Conference on Acoustics, Speech and Signal Processing (ICASSP), pp. 6984–6988. IEEE (2020)

21. Hou, J.C., Wang, S.S., Lai, Y.H., Tsao, Y., Chang, H.W., Wang, H.M.: Audio-visual speech enhancement using multimodal deep convolutional neural networks. IEEE Trans. Emerg. Topics Comput. Intell. 2(2), 117–128 (2018)

22. Ephrat, A., Mosseri, I., Lang, O., Dekel, T., Wilson, K., Hassidim, A., et al.: Looking to listen at the cocktail party: a speaker-independent audio-visual model for speech separation. ACM Trans. Graph. 37(4), 1–11 (2018)

23. Xu, Y., Du, J., Dai, L.R., Lee, C.H.: An experimental study on speech enhancement based on deep neural networks. IEEE Signal Process. Lett. 21(1), 65–68 (2013)

24. Isik, Y., Roux, J.L., Chen, Z., Watanabe, S., Hershey, J.R.: Single-channel multi-speaker separation using deep clustering. In: Proceedings of the 17th Annual Conference of the International Speech Communication Association (INTERSPEECH 2016), pp. 545–549 (2016)

25. Wang, Z.Q., Roux, J.L., Wang, D., Hershey, J.R.: End-to-end speech separation with unfolded iterative phase reconstruction. In: Proceedings of the 19th Annual Conference of the International Speech Communication Association (INTERSPEECH 2018), pp. 2708–2712 (2018)

26. Gabbay, A., Ephrat, A., Halperin, T., Peleg, S.: Seeing through noise: visually driven speaker separation and enhancement. In: 2018 IEEE International Conference on Acoustics, Speech and Signal Processing (ICASSP), pp. 3051–3055. IEEE (2018)

27. Luo, Y., Mesgarani, N.: Tasnet: time-domain audio separation network for real-time, single-channel speech separation. In: 2018 IEEE International Conference on Acoustics, Speech and Signal Processing (ICASSP), pp. 696–700. IEEE (2018)

28. Luo, Y., Mesgarani, N.: Conv-tasnet: surpassing ideal time–frequency magnitude masking for speech separation. IEEE/ACM Trans. Audio Speech Lang. Process. 27(8), 1256–1266 (2019)

29. Wu, J., et al.: IEEE automatic speech recognition and understanding workshop (ASRU). IEEE 2019, 667–673 (2019)

30. Pan, Z., Tao, R., Xu, C., Li, H.: Muse: multi-modal target speaker extraction with visual cues. In: ICASSP 2021–2021 IEEE International Conference on Acoustics, Speech and Signal Processing (ICASSP), pp.6678–6682. IEEE (2021)

31. Afouras, T., Chung, J.S., Senior, A., Vinyals, O., Zisserman, A.: Deep audio-visual speech recognition. IEEE Trans. Pattern Anal. Mach. Intell. 44(12), 8717–8727 (2018)

32. Le Roux, J., Wisdom, S., Erdogan, H., Hershey, J.R.: SDR–half-baked or well done? In: 2019 IEEE International Conference on Acoustics, Speech and Signal Processing (ICASSP), pp. 626–630. IEEE (2019)

33. Chung, J.S., Nagrani, A., Zisserman, A.: Voxceleb2: deep speaker recognition. In: Proceedings of the19th Annual Conference of the International Speech Communication Association (INTERSPEECH 2018), pp. 1086–1090. Incheon (2018)

34. Vincent, E., Gribonval, R., Févotte, C.: Performance measurement in blind audio source separation. IEEE Trans. Audio Speech Lang. Process. 14(4), 1462–1469 (2006)

35. Rix, A.W., Beerends, J.G., Hollier, M.P., Hekstra, A.P.: Perceptual evaluation of speech quality (PESQ)-a new method for speech quality assessment of telephone networks and codecs. IEEE Trans. Audio Speech Lang. Process. 2, 749–752 (2001)

36. Taal, C.H., Hendriks, R.C., Heusdens, R., Jensen, J.: A short-time objective intelligibility measure for time-frequency weighted noisy speech. In 2010 IEEE International Conference on Acoustics, Speech and Signal Processing (ICASSP), pp. 4214–4217. IEEE (2010)

NearbyPatchCL: Leveraging Nearby Patches for Self-supervised Patch-Level Multi-class Classification in Whole-Slide Images

Gia-Bao Le[1,2] , Van-Tien Nguyen[1,2] , Trung-Nghia Le[1,2(✉)] ,
and Minh-Triet Tran[1,2]

[1] University of Science, VNU-HCM, Ho Chi Minh City, Vietnam
[2] Vietnam National University, Ho Chi Minh City, Vietnam
lgbao19@apcs.fitus.edu.vn

Abstract. Whole-slide image (WSI) analysis plays a crucial role in cancer diagnosis and treatment. In addressing the demands of this critical task, self-supervised learning (SSL) methods have emerged as a valuable resource, leveraging their efficiency in circumventing the need for a large number of annotations, which can be both costly and time-consuming to deploy supervised methods. Nevertheless, patch-wise representation may exhibit instability in performance, primarily due to class imbalances stemming from patch selection within WSIs. In this paper, we introduce Nearby Patch Contrastive Learning (NearbyPatchCL), a novel self-supervised learning method that leverages nearby patches as positive samples and a decoupled contrastive loss for robust representation learning. Our method demonstrates a tangible enhancement in performance for downstream tasks involving patch-level multi-class classification. Additionally, we curate a new dataset derived from WSIs sourced from the Canine Cutaneous Cancer Histology, thus establishing a benchmark for the rigorous evaluation of patch-level multi-class classification methodologies. Intensive experiments show that our method significantly outperforms the supervised baseline and state-of-the-art SSL methods with top-1 classification accuracy of 87.56%. Our method also achieves comparable results while utilizing a mere 1% of labeled data, a stark contrast to the 100% labeled data requirement of other approaches. Source code: https://github.com/nvtien457/NearbyPatchCL

Keywords: Self-supervised learning · Contrastive learning · Whole-slide image · Representation learning

1 Introduction

Histology image analysis is crucial for understanding biological tissues. As Whole-Slide Images (WSI) become more common, and as cost-effective storage and fast data transfer networks become available, the creation of large

G.-B. Le and V.-T. Nguyen—Contributed equally.

© The Author(s), under exclusive license to Springer Nature Switzerland AG 2024
S. Rudinac et al. (Eds.): MMM 2024, LNCS 14555, pp. 239–252, 2024.
https://doi.org/10.1007/978-3-031-53308-2_18

Fig. 1. In contrastive learning [8], our NearbyPatchCL, and SupCon [16], positive samples are handled differently. Contrastive learning involves augmented pairs from the same image. In SupCon, it includes all images with the same label, while Nearby-PatchCL defines positive samples that encompass all views of the center patch (blue border) and nearby patches (green border). (Color figure online)

databases of digitized tissue sections in hospitals and clinics is on the rise. Currently, the computational pathology community is actively digitizing tissue slides into WSI for automated analysis. There's a growing focus on developing precise algorithms for clinical use, with recent advancements in deep learning making automatic analysis of WSIs, whether supervised or weakly supervised, more popular [4,17,24,28,32].

While there has been notable progress in the automated processing and clinical use of WSIs, challenges persist due to their gigapixel size. These challenges often require the use of tile-level processing and multiple instance learning for predicting clinical endpoints [22,28,32]. Additionally, the large size of WSIs makes the annotation process by human experts cumbersome, requiring annotated data for algorithm development. The shift to deep learning further emphasizes the importance of annotations, but some methods explore the use of pre-trained representations, typically from ImageNet, as an alternative to generating WSI-specific representations [11,18,20].

Self-supervised learning (SSL) methods [1,8,10] are gaining popularity for their ability to acquire competitive, versatile features compared to supervised methods. SSL involves two steps: unsupervised pre-training on unlabeled data and supervised fine-tuning on downstream tasks with limited labeled data. These methods not only require a small amount of labeled data but also enhance model performance across various histology pathology tasks [10,15,18,20,21,23,25,26,29]. Commonly, these methods break WSIs into smaller patches, feed them to an encoder, and extract representation features for downstream tasks. However, imbalanced category annotations within WSIs can lead to unstable performance, especially when random cropping generates image patches. While contrastive learning methods [10,18,20,21,23] have been applied to address this issue, it is important to note that even contrastive learning methods are not immune to imbalanced datasets [14]. Consequently, several SSL methods require further improvements to seamlessly integrate into clinical practice.

In this paper, we propose a simple yet efficient self-supervised learning method called Nearby Patch Contrastive Learning (NearbyPatchCL) that treats adjacent patches as positive samples in a supervised contrastive framework, which makes the training process more robust (See Fig. 1). To grapple with the intricate challenge of imbalanced data, we adopt the decoupled contrastive learning (DCL) loss [30]. The amalgamation of these approaches not only enhances the overall effectiveness and stability of our methodology but also engenders a strong representation that holds substantial promise for clinical integration. Remarkably, our approach yields commendable performance even when furnished with a scant fraction of labeled data, demonstrating its potential utility in real-world applications within the medical domain.

For patch-level multi-class evaluation, we adopt and process WSIs from the public CAnine CuTaneous Cancer Histology (CATCH) [27], resulting in a new benchmark dataset, namely P-CATCH. Intensive experiments on the newly constructed P-CATCH dataset demonstrate the superiority of NearbyPatchCL. Our method achieves the top-1 classification accuracy of 87.56% and significantly outperforms the supervised baseline and existing SSL methods. The proposed method also achieves compatible results when using only 1% labeled data compared to others using 100% labeled data. The source code will be released upon the paper's acceptance. Our contributions are summarized in the following:

- We propose a novel self-supervised method, namely NearbyPatchCL, which incorporates the modified supervised contrastive loss function by leveraging nearby patches as positive samples combined with the decoupled contrastive learning loss for better representation.
- We introduce a new dataset for benchmarking WSI patch-level multi-class classification methods.
- We perform a comprehensive comparison with state-of-the-art SSL methods to demonstrate the superior performance of the proposed method.

2 Related Works

2.1 Self-supervised Learning

SSL methods can be separated into four types based on their learning techniques. **Contrastive learning** algorithms (e.g., SimCLR [8] and MoCo [13]) aim to distinguish individual training data instances from others, creating similar representations for positive pairs and distinct representations for negative pairs. However, these methods require diverse negative pairs, often mitigated by using large batches or memory banks. Yeh et al. [30] introduced decoupled contrastive learning loss, enhancing learning efficiency by removing the positive term from the denominator. This method can achieve competitive performance with reduced sensitivity to sub-optimal hyperparameters, without the need for large batches or extended training epochs. Additionally, **asymmetric networks**, such as BYOL [12] and SimSiam [9], exhibit parallels with contrastive learning methods as they both learn representations of images from various augmented viewpoints. BYOL uses 2 distinct networks to create such representations, while SimSiam utilizes Siamese Networks [3]. A distinguishing factor compared to contrastive methods is that these strategies operate independently of negative pair incorporation, enabling them to function effectively even when dealing with small batch sizes. On the other hand, **clustering-based methods**, including DeepCluster [5] and SwAV [6], pursue the discovery of meaningful and compact representations by leveraging the notion of similarity and dissimilarity between data points. These methods operate under the assumption that similar data points should be closer together in the embedding space, while dissimilar points should be far apart. Meanwhile, **feature decorrelation** methods, exemplified by VICReg [2] and BarlowTwins [31], address redundancy among different dimensions of learned features, preventing collapse or over-reliance on specific dimensions. By reducing redundancy, these methods enable more reliable and comprehensive representations, contributing to advancements in self-supervised learning techniques and achieving results on par with state-of-the-arts on several downstream tasks.

2.2 SSL in Digital Pathology Images Analysis

By leveraging the inherent structure and relationships within the data, SSL techniques can learn rich and meaningful representations without relying on explicit annotations. Ciga et al. [10] highlighted the benefit of amalgamating diverse multi-organ datasets, including variations in staining and resolution, along with an increased number of images during the SSL process for enhanced downstream task performance. They achieved impressive results in histopathology tasks like classification, regression, and segmentation. Srinidhi et al. [23] introduced a domain-specific contrastive learning model tailored for histopathology, which outperforms general-purpose contrastive learning methods on tumor metastasis detection, tissue type classification, and tumor cellularity quantification. They focused on enhancing representations through a pretext task that involves predicting the order of all feasible sequences of resolution generated from the input multi-resolution patches. Wang et al. [25] developed a SSL approach integrating self-attention to learn patch-level embeddings, called semantically-relevant contrastive learning, using convolutional neural network and a multi-scale Swin Transformer architecture as the backbone, which compares relevance between instances to mine more positive pair. The SSL method proposed by Yang et al. [29] comprises two self-supervised stages: cross-stain prediction and contrastive learning, both grounded in domain-specific information. It can merge advantages from generative and discriminative models. In addition, SimTriplet [21] proposed by Liu et al. uses the spatial neighborhood on WSI to provide rich positive pairs (patches with the same tissue types) for triplet representation learning. It maximizes both intra-sample and inter-sample similarities via triplets from positive pairs, without using negative samples. Also, the benchmarking created by Kang et al. [15], which includes MOCO [13], BarlowTwin [31], SwAV [6], and DINO [7], concluded that large-scale domain-aligned pre-training is helpful for pathology, showing its value in scenarios with limited labeled data, longer fine-tuning schedules, and when using larger and more diverse datasets for pre-training.

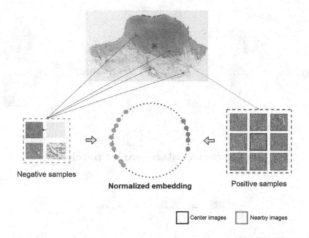

Negative samples

Normalized embedding   Positive samples

☐ Center images ☐ Nearby images

Fig. 2. Overview of the proposed NearbyPatchCL. Normalized embeddings of nearby images are pulled closer to their center while pushing away other images.

Our proposed approach is related to SimTripet [21]. By leveraging a certain number of nearby images in the self-supervised training process, we aim for a robust patch-wise representation for achieving better results on patch-level multi-class datasets.

3 Methodology

3.1 Overview

Inheriting the similar structure from Supervised Contrastive Learning (SupCon) [16], our method aims to learn visual representations by multi-positive samples. Differently from previous contrastive learning methods [8,13] treating one augmented view as a positive sample for the other, we harness label information to add more positives for one sample. However, labels are not provided in the self-supervised setting like SupCon [16]; so we adopt nearby patches as alternative positives as illustrated in Fig. 1.

To this end, we propose a novel Nearby Patch Contrastive Learning (NearbyPatchCL), where we maximize the similarity between not only different views of the same patch image but also adjacent patches (See Fig. 2). Inspired from SimTriplet [21], in our method, neighbor patches share the same tissue (or class) because they are cropped at a small scale of WSI.

3.2 Nearby Patch Contrastive Learning (NearbyPatchCL)

Given a randomly sampled batch of B samples (denoted as "batch"), each image is transformed by two random augmentations to get the training batch containing $2B$ samples (denoted by "multiviewed batch"). Besides, we denote

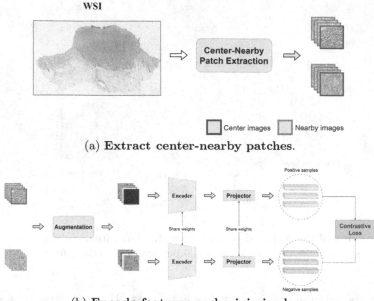

(a) Extract center-nearby patches.

(b) Encode features and minimize loss.

Fig. 3. The architecture of NearbyPatchCL includes two main parts: (a) Extract center patches with 8 corresponding neighbors from WSIs which is a preprocessing step discussed in Sect. 4 and (b) A contrastive loss is used to minimize the distance between learned features for positive samples (center-nearby) while simultaneously maximizing the feature distance from negative samples.

$I \equiv \{1, \ldots, 2B\}$ as the set of indices of all samples in multiviewed batch. For a sample with index i, $P(i)$ and $A(i) \equiv I \backslash P(i)$ are the collection of indices of all positives and negatives related to the sample respectively, and $j(i)$ is the index of the other augmented sample originating from the same source sample in the batch.

Each batch contains C center patches with N corresponding nearby patches ($0 \leq N \leq 8$), so there are $B = C \times (N+1)$ samples, $\{x_k\}_{k=1\ldots B}$, in total. A group of patches including a center and corresponding neighbors belongs to one class; therefore, there are C classes in the batch. By applying two transformations, the multiviewed batch contains $2C(N+1)$ samples, $\{\tilde{x}_l\}_{l=1\ldots 2B}$. For an arbitrary sample with index i, $P(i) \equiv \{j(i)\} \cup \mathcal{N}(i)$ where $\mathcal{N}(i)$ is the set that consists of $2N$ indices of positives (nearby) for the sample and $2(N+1)(C-1)$ remaining instances as negatives. Our naive loss function is as follows:

$$\mathcal{L} = \sum_{i \in I} \frac{-1}{|P(i)|} \sum_{p \in P(i)} \log \frac{\exp\left(z_i \cdot z_p / \tau\right)}{\exp\left(z_i \cdot z_p / \tau\right) + \sum_{a \in A(i)} \exp\left(z_i \cdot z_a / \tau\right)}, \quad (1)$$

where $z_i \equiv g\left(f\left(\tilde{x}_i\right)\right)$ is the output feature of sample $\tilde{x}_i$, the $\cdot$ symbol denotes the inner product, and τ is a scalar temperature parameter.

Algorithm 1. Pseudocode for NearbyPatchCL algorithm.

1: **Input**
      X   Sampled minibatch
      C   Number of center samples in X
      I   Index set of samples in the multiviewed batch
      N   Number of nearby samples per center
      $\mathcal{N}_n(x_i)$ Function return n-th nearby of sample x_i
      $\mathcal{T}$ Distribution of image transformation
      f, g Encoder network and Projection head
      $P(i)$ Set of indices of positive samples for sample with index i.
      $A(i)$ Set of indices of negative samples for sample with index i.
2: **for** sampled minibatch $X = \{x_c\}_{c=1}^{C}$ **do**
3:    $Y = \{1, \ldots, C\}$ # labels
4:    **for** $n \in \{1, \ldots, N\}$ **do**   # Retrieve N nearby samples
5:        **for** $c \in \{1 \ldots C\}$ **do**
6:           $X \leftarrow X + \mathcal{N}_n(x_c)$
7:           $Y \leftarrow Y + c$
8:        **end for**
9:    **end for**
10:   $t \sim \mathcal{T}, t' \sim \mathcal{T}$ # two augmentations
11:   **for** $k \in \{1, \ldots, C(N+1)\}$ **do**
12:      $\tilde{x}_k = t(x_k)$ # first augmentation
13:      $h_k = f(\tilde{x}_k)$
14:      $z_k = g(h_k)$
15:      $\tilde{x}_{k+C(N+1)} = t'(x_k)$ # second augmentation
16:      $h_{k+C(N+1)} = f(\tilde{x}_{k+C(N+1)})$
17:      $z_{k+C(N+1)} = g(h_{k+C(N+1)})$
18:   **end for**
19:   **for** $i \in \{1, \ldots, 2C(N+1)\}$ and $j \in \{1, \ldots, 2C(N+1)\}$ **do**
20:      $s_{i,j} = z_i^T z_j / \|z_i\| \|z_j\|$ # pairwise similarity
21:   **end for**
22:   **for** $i \in I$ **do**
23:      $l(i) = \frac{-1}{|P(i)|} \sum_{p \in P(i)} \log \frac{\exp(z_i \cdot z_p \backslash \tau)}{\sum_{a \in A(i)} \exp(z_i \cdot z_a \backslash \tau)}$
24:   **end for**
25:   $\mathcal{L} = \frac{1}{2C(N+1)} \sum_{i=1}^{2C(N+1)} l(i)$
26:   update networks f and g to minimize $\mathcal{L}$
27: **end for**
28: **Output**   Encoder f

Inspired by Yeh et al. [30], to address the imbalance due to cropping WSIs, we hypothesize that robustness representations can be obtained by removing negative samples in a batch to eliminate the negative pair contrast effect. It means that we do not need a large batch size during the training process to enhance learning efficiency, resulting in stable performance. This hypothesis is explored further in Sect. 5.3. Our naive loss function (Eq. 1) becomes:

$$\mathcal{L} = \sum_{i \in I} \frac{-1}{|P(i)|} \sum_{p \in P(i)} \log \frac{\exp(z_i \cdot z_p / \tau)}{\sum_{a \in A(i)} \exp(z_i \cdot z_a / \tau)}. \tag{2}$$

Table 1. The number of patch images in unlabeled sets, each has 247 WSIs. The number after the dataset name denotes the number of nearby images.

Unlabeled Set	Per WSI		Total	
	Center Images	Nearby Images	Center Images	Nearby Images
P-CATCH-0	1,000	0	247,000	0
P-CATCH-1	500	500	123,500	123,500
P-CATCH-2	333	666	82,251	164,502
P-CATCH-4	200	800	49,400	197,600
P-CATCH-8	111	889	27,417	246,753

Algorithm 1 summarizes our proposed method.

3.3   Implementation Details

Architecture. Figure 3 shows the overall architecture of our NearbyPatchCL, adopted from SupCon's architecture [16]. ResNet-50 backbone is used. The representation from ResNet-50's last fully connected layer remains 2048-dimensional, and it's then reduced to 128 dimensions using a multi-layer perceptron (MLP).

Image Augmentations. We follow the augmentation described in SupCon [16]. Additionally, horizontal flips are randomly applied for rotation invariance in skin tissue images. Images are resized to 128×128 for computational efficiency.

Optimization. We use the SGD optimizer, with a base learning rate of 0.2, momentum of 0.9, and weight decay of 0.0001. To optimize resources, we use Mixed Precision Training (MPT) [19] with a learning rate scheduler. The new learning rate is calculated as $lr \times BatchSize \times (N+1)/256$, where $BatchSize = 512/(N+1)$ is the number of center images and N is the number of nearby patches per center. The scheduler involves 10 warm-up epochs followed by cosine decay. This SSL setup takes about 73 h to train the encoder for 400 epochs using unlabeled data from our newly constructed P-CATCH dataset.

4   Proposed P-CATCH Dataset

4.1   Original CATCH Images

The CATCH dataset [27] consists of 350 WSIs, stored in the pyramidal Aperio file format, having direct access to three resolution levels ($0.25 \frac{\mu m}{px}$, $1 \frac{\mu m}{px}$, $4 \frac{\mu m}{px}$). Pathologists used Slide Runner software to create a database with 12,424 area annotations in total, covering six non-neoplastic tissues (epidermis, dermis, subcutis, bone, cartilage, inflammation/necrosis) and seven tumor classes. Notably, there is a significant imbalance in the distribution of the six non-neoplastic tissue classes within the database.

Table 2. The number of patch images per category in our P-CATCH dataset.

Category	Training Set	Test Set	Unlabeled Set
Dermis	22,020	81,143	-
Epidermis	9,471	16,001	-
Inflamm/Necrosis	19,488	24,612	-
Subcutis	16,566	53,426	-
Tumor	22,341	96,188	-
Background	1,917	4,533	-
Total	91,803	175,903	2,223,000

4.2   Proposed P-CATCH Dataset

Unlabeled Sets. A total of 70% of WSIs contained within the CATCH database, specifically corresponding to 247 WSIs, is used as unlabeled data for training SSL methods. To ensure a balanced representation of the diverse tumor subtypes present, we adopt an equitable distribution strategy, resulting in approximately 35 WSIs per subtype. From each WSI at the $0.25\frac{\mu m}{px}$ resolution level, we randomly extract image patches with size of 512×512 pixels, encompassing center patches along with nearby patches. We create different subsets, denoted by P-CATCH-N, where N indicates the number of nearby samples of a center image. Table 1 shows the statistics of created five subsets, corresponding to 0, 1, 2, 4, and 8 nearby samples. We remark that we try to ensure an equal number of patch images for P-CATCH-N subsets.

Annotated Sets. We partitioned the dataset, with the remaining 97 WSIs reserved for the test set, while 37 WSIs from the unlabeled set were allocated for training purposes. Specifically, we conduct random sampling to extract non-overlapping image patches, each size at 512×512 pixels, from the WSIs at the $0.25\frac{\mu m}{px}$ resolution level. This process yields approximately 92,000 images distributed across six categories for the training set, and around 176,000 images for the test set, spanning the same set of categories. The distribution of images across these categories is detailed in Table 2. This meticulous dataset partitioning ensures a rigorous and comprehensive evaluation of our methodology.

5   Experiments

5.1   Linear Evaluation Protocol

We adhere to the well-established linear evaluation protocol [8,31] which entails training a linear classifier atop a frozen base network (i.e., ResNet-50), and test accuracy is used as a proxy for representation quality. A single transformation is applied, involving resizing to 256×256, subsequent center-cropping to

Table 3. Classification performance, using different amounts of training data. With different numbers of annotated data, we ensure different classes contribute similar numbers of images to address the issue that the annotation is highly imbalanced.

Method	F1 Score				Balanced Accuracy			
	1%	10%	20%	100%	1%	10%	20%	100%
Supervised baseline	66.95	75.92	78.23	81.04	74.79	81.94	83.58	84.24
SimCLR [8]	70.09	72.42	72.43	72.32	77.55	79.45	79.40	79.26
SimSiam [9]	54.24	57.52	57.60	58.43	58.59	60.14	60.69	61.31
SimTriplet [21]	55.18	57.11	56.20	58.67	60.75	62.65	62.78	63.57
BYOL [12]	65.62	77.35	79.24	83.01	74.63	83.41	84.51	85.48
BarlowTwins [31]	76.43	80.88	81.08	82.55	82.73	85.98	86.19	86.36
NearbyPatchCL ($N=4$)	**81.85**	**83.73**	**84.41**	**85.72**	**85.63**	**87.19**	**87.56**	**87.14**

revert to the original resolution of 224×224, followed by normalization. Methods are trained for 15 epochs, utilizing SGD optimizer with a learning rate of 0.2, momentum of 0.9, and weight decay of 0. The batch size is configured at 32 for training on 1% labeled data, while a batch size of 512 is employed for cases involving 10%, 20%, and 100% labeled data. To ensure a robust and comprehensive assessment, we employ 5-fold cross-validation on the training set, resulting in 5 trained models. Notably, MPT is excluded during the linear evaluation. During the evaluation of the test set, the results obtained from these 5 models are averaged to provide a more reliable estimate of the model's performance, thereby effectively mitigating issues such as overfitting and yielding a more robust evaluation outcome.

5.2   Comparison with State-of-the-Arts

We compared our NearbyPatchCL, using $N = 4$ nearby samples, with state-of-the-art SSL methods, including SimCLR [8], SimSiam [9], BarlowTwins [31], SimTriplet [21], and BYOL [12]. For fair evaluation, all methods employed the ResNet-50 backbone pre-trained on the ImageNet dataset. SimTriplet [21] used P-CATCH-1 for self-supervised training phase. Meanwhile, P-CATCH-0 subset was used for training other SSL methods. Training parameters follow their original work.

Comparison results in Table 3 demonstrate that our NearbyPatchCL ($N=4$) significantly outperforms the supervised baseline (i.e., frozen ImageNet pre-trained ResNet-50 with a classifier trained on the training set) and other SSL methods across all data proportions. Furthermore, even with 1% labeled data used for training, NearbyPatchCL still has competitive results compared to BYOL and Barlowtwins using 100% of training data. Leveraging nearby batches as positive samples, our approach achieves better results than state-of-the-art. Additionally, we visualize results on a whole WSI in Fig. 4. The result also indicates that our approach can effectively leverage unlabeled data for improving classification tasks, making them a valuable tool in practical scenarios where annotating WSIs is costly.

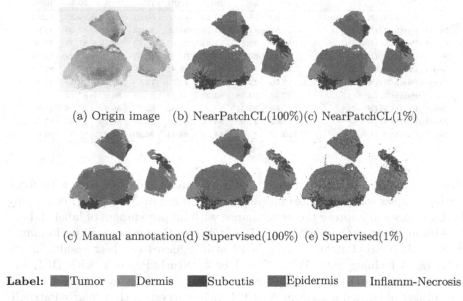

(a) Origin image   (b) NearPatchCL(100%)(c) NearPatchCL(1%)

(c) Manual annotation(d) Supervised(100%) (e) Supervised(1%)

Label: ■Tumor   ■Dermis   ■Subcutis   ■Epidermis   ■Inflamm-Necrosis

Fig. 4. Visualization of classification results on a sample WSI from the test set. Our proposed method shows significantly better results than that of the supervised baseline, even with only 1% of the training data.

5.3   Ablation Study

Number of Nearby Samples. We investigate the influence of the number of nearby samples (N) for each center image on the performance of the proposed method. It is notable that with $N = 0$, NearbyPatchCL becomes SimCLR. The result in Table 4, with DCL loss, shows that the performance of Nearby-PatchCL increases gradually from $N = 0$ to $N = 4$ and achieves the best overall performance at $N = 4$ in using 10%, 20%, and 100% labeled images in the training set. The result in NearbyPatchCL($N = 8$) still has a competitive result with NearbyPatchCL ($N = 4$) and has the second-highest overall performance. It is notable that annotations made by pathologists may not have an accuracy of 100%. Hence, training on a large amount of data can make the model learn some false cases, as we can see the balanced accuracy drop when moving from 20% to 100% in $N = 1, 4, 8$. On the other hand, the performance without DCL loss increasing gradually from $N = 0$ to $N = 8$ also shows the superiority of utilizing nearby patches in the SSL process.

Table 4. Ablation study with different numbers of nearby samples (N) for each image center and using DCL loss.

Method		F1 Score				Balanced Accuracy			
		1%	10%	20%	100%	1%	10%	20%	100%
NearbyPatchCL (N = 0)	DCL	73.83	78.36	79.14	80.69	80.01	83.87	84.35	84.44
NearbyPatchCL (N = 1)		77.97	81.84	82.77	84.40	83.42	86.54	87.00	86.94
NearbyPatchCL (N = 2)		80.27	82.87	83.43	85.07	_85.63_	87.16	87.38	**87.47**
NearbyPatchCL (N = 4)		_81.85_	**83.73**	**84.41**	**85.72**	_85.63_	87.19	**87.56**	_87.14_
NearbyPatchCL (N = 8)		79.66	83.14	83.97	_85.45_	84.60	**87.29**	_87.53_	87.09
NearbyPatchCL (N = 0)	w/o DCL	70.09	72.42	72.43	72.32	77.55	79.45	79.40	79.26
NearbyPatchCL (N = 1)		77.52	82.58	83.31	84.68	83.93	86.92	87.12	86.57
NearbyPatchCL (N = 2)		77.69	82.41	83.25	84.82	82.85	86.42	86.86	86.86
NearbyPatchCL (N = 4)		78.62	82.68	83.62	84.94	83.77	86.90	87.40	87.08
NearbyPatchCL (N = 8)		**82.17**	_83.58_	_84.00_	84.41	**85.86**	_87.22_	87.47	86.67

Effect of DCL Loss. The result shown in Table 4 indicates that with more nearby samples for each center image, as we can see in $N = \{2, 4\}$, employing the DCL loss can improve the performance with all percentages of labeled data.

Although our method, with DCL loss, achieves the best overall performance at $N = 4$, NearbyPatchCL (w/o DCL) still achieves the best results in 1% and 10% of training data. With $N = 1$ or 8, NearbyPatchCL with DCL loss does not have any significant improvement in performance, even worse at some percentages of labeled data as in $N = 1$. Our hypothesis is that NearbyPatchCL gains more benefit from leveraging more nearby patches with DCL loss. However, with $N = 8$, there are only 111 center images that are randomly cropped from each WSI, which are nearly half of that of $N = 4$, this can make the training data more unbalanced due to the differences in area and quantity of each category annotation, leading to a decrease in overall performance from $N = 4$ to $N = 8$.

6   Conclusion and Future Works

In this paper, we have conducted a new benchmark of patch-level multi-class WSI classification using SSL methods. To tackle the scarcity of labeled data and imbalanced datasets issue in digital pathology images, our work has shown that by leveraging nearby patches as positive samples in the SSL phase, the proposed method can have a more robust representation and perform better on downstream tasks. Furthermore, we have shown that using DCL loss can benefit contrastive methods while training on an imbalanced dataset.

In future work, we aim to extend our approach to other medical imaging domains and explore its application in other downstream tasks. We also plan to further investigate methods to enhance the interpretability of learned representations and incorporate domain-specific knowledge to improve the performance of the model in real-world clinical settings.

Acknowledgement. This research was funded by Vingroup and supported by Vingroup Innovation Foundation (VINIF) under project code VINIF.2019.DA19. This project was also supported by the Faculty of Information Technology, University of Science, Vietnam National University - Ho Chi Minh City.

References

1. Azizi, S., et al.: Big self-supervised models advance medical image classification. In: ICCV, pp. 3478–3488 (2021)
2. Bardes, A., Ponce, J., LeCun, Y.: VICReg: variance-invariance-covariance regularization for self-supervised learning. arXiv preprint arXiv:2105.04906 (2021)
3. Bromley, J., Guyon, I., LeCun, Y., Säckinger, E., Shah, R.: Signature verification using a "siamese" time delay neural network. In: NeurIPS, vol. 6 (1993)
4. Campanella, G., et al.: Clinical-grade computational pathology using weakly supervised deep learning on whole slide images. Nat. Med. **25**, 1301–1309 (2019)
5. Caron, M., Bojanowski, P., Joulin, A., Douze, M.: Deep clustering for unsupervised learning of visual features. In: Ferrari, V., Hebert, M., Sminchisescu, C., Weiss, Y. (eds.) Computer Vision – ECCV 2018. LNCS, vol. 11218, pp. 139–156. Springer, Cham (2018). https://doi.org/10.1007/978-3-030-01264-9_9
6. Caron, M., Misra, I., Mairal, J., Goyal, P., Bojanowski, P., Joulin, A.: Unsupervised learning of visual features by contrasting cluster assignments. NeurIPS **33**, 9912–9924 (2020)
7. Caron, M., et al.: Emerging properties in self-supervised vision transformers. In: ICCV, pp. 9650–9660 (2021)
8. Chen, T., Kornblith, S., Norouzi, M., Hinton, G.: A simple framework for contrastive learning of visual representations. In: ICML, pp. 1597–1607 (2020)
9. Chen, X., He, K.: Exploring simple Siamese representation learning. In: CVPR, pp. 15750–15758 (2021)
10. Ciga, O., Xu, T., Martel, A.L.: Self supervised contrastive learning for digital histopathology. Mach. Learn. Appl. **7**, 100198 (2022)
11. Gamper, J., Rajpoot, N.: Multiple instance captioning: learning representations from histopathology textbooks and articles. In: CVPR, pp. 16549–16559 (2021)
12. Grill, J.B., et al.: Bootstrap your own latent-a new approach to self-supervised learning. NeurIPS **33**, 21271–21284 (2020)
13. He, K., Fan, H., Wu, Y., Xie, S., Girshick, R.: Momentum contrast for unsupervised visual representation learning. In: CVPR, pp. 9729–9738 (2020)
14. Jiang, Z., Chen, T., Mortazavi, B.J., Wang, Z.: Self-damaging contrastive learning. In: ICML, pp. 4927–4939 (2021)
15. Kang, M., Song, H., Park, S., Yoo, D., Pereira, S.: Benchmarking self-supervised learning on diverse pathology datasets. In: CVPR, pp. 3344–3354 (2023)
16. Khosla, P., et al.: Supervised contrastive learning. NeurIPS **33**, 18661–18673 (2020)
17. van der Laak, J.A., Litjens, G.J.S., Ciompi, F.: Deep learning in histopathology: the path to the clinic. Nat. Med. **27**, 775–784 (2021)
18. Li, B., Li, Y., Eliceiri, K.W.: Dual-stream multiple instance learning network for whole slide image classification with self-supervised contrastive learning. In: CVPR, pp. 14318–14328 (2021)
19. Li, H., Wang, Y., Hong, Y., Li, F., Ji, X.: Layered mixed-precision training: a new training method for large-scale AI models. J. King Saud Univ. Comput. Inf. Sci. **35**(8), 101656 (2023)

20. Liu, K., et al.: Multiple instance learning via iterative self-paced supervised contrastive learning. In: CVPR, pp. 3355–3365 (2023)
21. Liu, Q., et al.: SimTriplet: simple triplet representation learning with a single GPU. In: de Bruijne, M., et al. (eds.) MICCAI 2021. LNCS, vol. 12902, pp. 102–112. Springer, Cham (2021). https://doi.org/10.1007/978-3-030-87196-3_10
22. Lu, M.Y., Williamson, D.F., Chen, T.Y., Chen, R.J., Barbieri, M., Mahmood, F.: Data-efficient and weakly supervised computational pathology on whole-slide images. Nat. Biomed. Eng. **5**(6), 555–570 (2021)
23. Srinidhi, C.L., Kim, S.W., Chen, F.D., Martel, A.L.: Self-supervised driven consistency training for annotation efficient histopathology image analysis. Med. Image Anal. **75**, 102256 (2022)
24. Wang, D., Khosla, A., Gargeya, R., Irshad, H., Beck, A.H.: Deep learning for identifying metastatic breast cancer. arXiv preprint arXiv:1606.05718 (2016)
25. Wang, X., et al.: TransPath: transformer-based self-supervised learning for histopathological image classification. In: de Bruijne, M., et al. (eds.) MICCAI 2021. LNCS, vol. 12908, pp. 186–195. Springer, Cham (2021). https://doi.org/10.1007/978-3-030-87237-3_18
26. Wang, X., et al.: Transformer-based unsupervised contrastive learning for histopathological image classification. Med. Image Anal. **81**, 102559 (2022)
27. Wilm, F., et al.: Canine cutaneous cancer histology dataset (2022). https://wiki.cancerimagingarchive.net/x/DYITBg
28. Wulczyn, E., et al.: Interpretable survival prediction for colorectal cancer using deep learning. npj Digit. Med. **4**, 71 (2021)
29. Yang, P., et al.: CS-CO: a hybrid self-supervised visual representation learning method for h&e-stained histopathological images. Med. Image Anal. **81**, 102539 (2022)
30. Yeh, C.H., Hong, C.Y., Hsu, Y.C., Liu, T.L., Chen, Y., LeCun, Y.: Decoupled contrastive learning. In: Avidan, S., Brostow, G., Cissé, M., Farinella, G.M., Hassner, T. (eds.) Computer Vision, ECCV 2022. LNCS., vol. 13686, pp. 668–684. Springer, Cham (2022). https://doi.org/10.1007/978-3-031-19809-0_38
31. Zbontar, J., Jing, L., Misra, I., LeCun, Y., Deny, S.: Barlow twins: self-supervised learning via redundancy reduction. In: ICML, pp. 12310–12320 (2021)
32. Zheng, Y., Jiang, B., Shi, J., Zhang, H., Xie, F.: Encoding histopathological WSIs using GNN for scalable diagnostically relevant regions retrieval. In: Shen, D., et al. (eds.) MICCAI 2019. LNCS, vol. 11764, pp. 550–558. Springer, Cham (2019). https://doi.org/10.1007/978-3-030-32239-7_61

Improving Small License Plate Detection with Bidirectional Vehicle-Plate Relation

Songkang Dai[1], Song-Lu Chen[1], Qi Liu[1], Chao Zhu[1(✉)], Yan Liu[2],
Feng Chen[3], and Xu-Cheng Yin[1]

[1] School of Computer and Communication Engineering,
University of Science and Technology Beijing, Beijing, China
{songkangdai,qiliu7}@xs.ustb.edu.cn,
{songluchen,chaozhu,xuchengyin}@ustb.edu.cn
[2] Key Laboratory of Knowledge Automation for Industrial Processes
of Ministry of Education, School of Automation and Electrical Engineering,
University of Science and Technology Beijing, Beijing, China
liuyan@ustb.edu.cn
[3] EEasy Technology Company Ltd., Zhuhai, China
cfeng@eeasytech.com

Abstract. License plate detection is a critical component of license plate recognition systems. A challenge in this domain is detecting small license plates captured at a considerable distance. Previous researchers have proved that pre-detecting the vehicle can enhance small license plate detection. However, this approach only utilizes the one-way relation that the presence of a vehicle can enhance license plate detection, potentially resulting in error accumulation if the vehicle fails to be detected. To address this issue, we propose a unified network that can simultaneously detect the vehicle and the license plate while establishing bidirectional relationships between them. The proposed network can utilize the vehicle to enhance small license plate detection and reduce error accumulation when the vehicle fails to be detected. Extensive experiments on the SSIG-SegPlate, AOLP, and CRPD datasets prove our method achieves state-of-the-art detection performance, achieving an average detection $AP_{0.5}$ of 99.5% on these three datasets, especially for small license plates. When incorporating a license plate recognizer that relies on character detection, we can achieve an average recognition accuracy of 95.9%, surpassing all comparative methods. Moreover, we have manually annotated the vehicles within the CRPD dataset and have made these annotations publicly available at https://github.com/kiki00007/CRPDV.

Keywords: License plate detection · License plate recognition · Small license plate · Bidirectional vehicle-plate relation

1 Introduction

Automatic license plate recognition (ALPR) has recently gained significant popularity in various applications, such as traffic enforcement, theft detection, and

---

S. Dai and S.-L. Chen—Equal contribution.

S. Rudinac et al. (Eds.): MMM 2024, LNCS 14555, pp. 253–266, 2024.
https://doi.org/10.1007/978-3-031-53308-2_19

automatic toll collection. The ALPR system typically consists of three stages: license plate detection, character detection, and character recognition [1]. Among these stages, license plate detection plays a pivotal role in determining the overall accuracy of the ALPR system. Specifically, detecting small license plates presents a significant challenge due to their size.

As shown in Fig. 1(a), many ALPR methods have been proposed to directly detect the license plate from the input image [6,28]. However, detecting the license plate directly can lead to missed detections, primarily due to its small size. To address this issue, Kim et al. [12,14] propose a two-step approach as depicted in Fig. 1(b), where the vehicle is first pre-detected, followed by license plate detection within the vehicle region. These methods reduce the search region and mitigate background noises, enhancing license plate detection. Nevertheless, these methods may encounter error accumulation if the vehicle fails to be detected, resulting in subsequent failures in license plate detection. To minimize error accumulation, Chen et al. [5] propose a fusion approach illustrated in Fig. 1(c), which combines direct license plate detection (Fig. 1(a)) and vehicle pre-detection (Fig. 1(b)), merging both detection branches to obtain the final results. However, this approach is time-consuming due to the involvement in multiple detection branches and the subsequent merge operation.

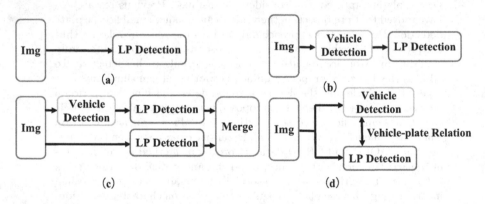

Fig. 1. (a) Direct license plate detection from the input image. (b) License plate detection based on vehicle pre-detection. (c) License plate detection by combining direct detection and vehicle pre-detection. (d) Our proposed method, using bidirectional vehicle-plate relationships to enhance license plate detection.

To address the challenges mentioned earlier, as depicted in Fig. 1(d), we propose simultaneous detection of both the vehicle and the license plate, leveraging their bidirectional relationship to enhance small license plate detection. This approach facilitates mutual reinforcement between vehicles and license plates due to their interdependency. In comparison to direct detection (Fig. 1(a)), our method utilizes the presence of the vehicle to improve license plate detection. Unlike the vehicle pre-detection approach (Fig. 1(b)), our method mitigates error accumulation arising from the one-way relationship between the vehicle and the license plate. Additionally, compared to the fusion approach (Fig. 1(c)), our method enhances inference speed through simultaneous detection and bidirectional relation mining. Extensive experiments on the SSIG-SegPlate [9], AOLP [11], and CRPD [32] datasets validate the effectiveness of our method, achieving an average detection $AP_{0.5}$ of 99.5%, particularly for small license plates. When combined with a YOLO-based character recognizer [15], our method outperforms other state-of-the-art techniques, achieving an average recognition accuracy of 95.9%. Notably, annotations for both vehicles and license plates are available for the SSIG-SegPlate and AOLP datasets within the community. However, for the CRPD dataset, only license plate annotations are provided. To support the community, we manually annotated vehicles in the CRPD dataset and made the annotations publicly available at https://github.com/kiki00007/CRPDV.

2 Related Work

2.1 Object Detection

Object detection is a task that involves locating the bounding box and predicting the category of an object. Previous object detectors can be broadly categorized into two types based on the detection stage: two-stage detectors [17,22] and one-stage detectors [8,18]. Additionally, they can be classified as anchor-based [18,22] or anchor-free [26,30] based on the matching mechanism. However, these methods typically involve complex post-processing and matching procedures. To reduce complexity, DETR-based methods [4,31] utilize the transformer [27] architecture and object queries to directly predict the class and bounding box of an object. However, the aforementioned methods ignore the relationships between different objects, which is suboptimal to small object detection. In this work, besides object detection, we propose to utilize the bidirectional relationships between vehicles and license plates to enhance small license plate detection.

2.2 License Plate Detection

There are two prevailing approaches for license plate detection: direct detection [6,28] and vehicle pre-detection [12,14]. The former involves directly detecting the license plate in the image. However, these methods may not work well for small license plates due to their small size. The latter approach, known as vehicle pre-detection, first detects the vehicle in the image and then locates the license plate within the vehicle region. This approach reduces the search region and mitigates background noises, enhancing small license plate detection. However, these vehicle pre-detection methods are prone to error accumulation because the absence of vehicle detection inevitably leads to the failure of license plate detection. To mitigate error accumulation, Chen et al. [5] introduce a method incorporating two detection branches. One branch focuses on pre-detecting the vehicle, and the other directly detects the license plate. The outputs from these

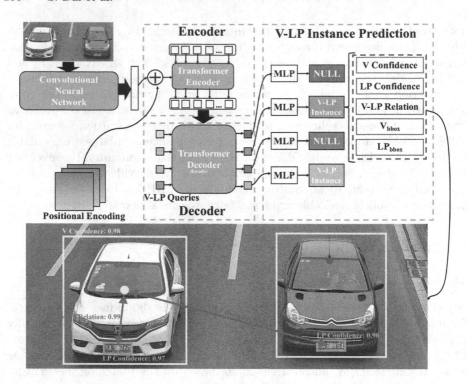

Fig. 2. Overall architecture. The network utilizes an encoder-decoder architecture, taking an image as input and generating predictions for the category, bounding box of vehicles (V) and license plates (LP), and the relationships between them.

branches are then fused to obtain the final results. However, this approach introduces significant computational overhead. In this work, we propose to simultaneously detect vehicles and license plates and leverage bidirectional relationships between them to enhance the effectiveness and efficiency of small license plate detection.

3 Method

As depicted in Fig. 2, our proposed network can simultaneously detect vehicles and license plates and generate their bidirectional relationships. When a license plate subordinates to a vehicle, their relation confidence is higher, and vice versa. This way, it can mutually enhance the detection of vehicles and license plates.

3.1 Network Architecture

The proposed network can be mainly divided into three parts: (I) A CNN backbone to extract visual features from the input image; (II) A transformer encoder-decoder to process visual features and generate global features; (III) A multi-layer perceptron layer (MLP) to generate predictions based on global features.

Backbone: We utilize ResNet-50 [10] to extract visual features from the input image into feature maps. The size of the input image and features maps is $[H_0, W_0, 3]$ and $[H, W, C]$, respectively, s.t., $H = H_0/32$ and $W = W_0/32$. Subsequently, a $1{\times}1$ convolutional layer is utilized to reduce the channel dimension from $C = 2048$ to $d = 256$. Since the subsequent encoder requires a sequence as input, we convert the reduced features into a sequence of length $H{\times}W$, where each step corresponds to a vector of size d. As a result, we obtain a flattened feature map with the dimension of $[H{\times}W, d]$.

Encoder: The encoder follows the vanilla transformer [27], incorporating six identical units. Each unit comprises an eight-head self-attention network and a two-layer feed-forward network (FFN) with the dimension of $d_{ff} = 2048$. The output dimension is set to $d_{model} = 512$. The Query, Key, and Value are all obtained by the sum of positional encodings and visual features from the CNN backbone to generate global features.

Decoder: The decoder also follows the vanilla transformer, incorporating six identical units. Each unit comprises an eight-head cross-attention network, an eight-head self-attention network, and a two-layer feed-forward network. Similar to the encoder, the FFN dimension is $d_{ff} = 2048$, and the output dimension is $d_{model} = 512$. The decoder takes three inputs, i.e., positional encodings, V-LP queries, and global features from the encoder, to generate $N = 100$ embeddings for predictions. In the cross-attention network, the Value is obtained directly from global features. The Key is the sum of global features and positional encodings, and the Query is the sum of positional encodings and V-LP queries.

Vehicle-Plate Instance Prediction: The output embeddings generated by the decoder are converted into vehicle-plate instances using MLPs. We define the vehicle-plate instance as a five-tuple consisting of vehicle confidence, vehicle-plate relation confidence, plate confidence, vehicle box, and plate box. Specifically, two three-layer MLPs are employed to predict the bounding box of the vehicle and the license plate. Additionally, three single-layer MLPs are utilized to estimate the confidence of the vehicle, the plate, and the vehicle-plate relation.

3.2   Training Objective

We treat the prediction of vehicle-plate instances as a problem of set prediction, involving a bipartite matching between the predicted instances and the ground truth. When presented with an input image, our model generates $N = 100$ predicted instances, where N represents the number of V-LP queries. The prediction set is represented as $P = p^i, i = 1, 2, ..., N$. The ground-truth set is represented as $G = g^i, i = 1, 2, ..., M, \phi, ..., \phi$, where ϕ denotes a null value for one-to-one matching between P and G, and M denotes the total number of ground-truth instances, s.t., $M \leq N$. The number of ϕ plus M equals N.

As demonstrated in Eq. (1), we use the Hungarian algorithm [13] to find the best bipartite matching $\hat{\sigma}$ by minimizing the overall matching cost ζ_{cost}, which is composed of the matching cost of all N matching pairs.

$$\hat{\sigma} = argmin\zeta_{cost}, \sigma \in \mathcal{O}_N$$

$$\zeta_{cost} = \sum_i^N \zeta_{match}(g^i, p^{\sigma(i)}) \tag{1}$$

where $\mathcal{O}_N$ represents the one-to-one matching solution space, and σ represents an injective function from the ground-truth set G to the prediction set P. $\zeta_{match}(g^i, p^{\sigma(i)})$ represents the matching cost between the i-th ground truth and $\sigma(i)$-th prediction, where $\sigma(i)$ represents the matching index of the prediction.

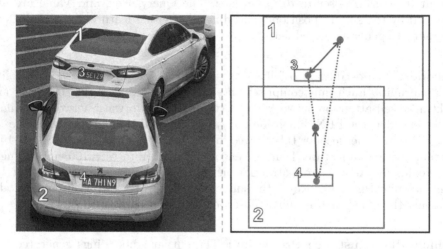

Fig. 3. Ground truth during training. The red and green boxes denote the ground-truth boxes of vehicles and license plates, respectively. The solid purple line represents the ground-truth V-LP relation, i.e., the positive relation sample used during training. The dotted purple line denotes no relation between the vehicle and the license plate, i.e., the negative relation sample, which is not used during training. (Color figure online)

As demonstrated in Eq. (2), the matching cost of each pair contains the classification loss ζ^j_{cls} and bounding box regression loss ζ^k_{box}.

$$\zeta_{match}(g^i, p^{\sigma(i)}) = \beta_1 \sum_{j \in v,p,r} \alpha_j \zeta^j_{cls} + \beta_2 \sum_{k \in v,p} \zeta^k_{box} \tag{2}$$

where v, p, r represents the vehicle, license plate, and vehicle-plate relation, respectively. ζ^j_{cls} is calculated by the softmax cross-entropy loss. ζ^k_{box} is calculated by the weighted sum of L_1 loss and GIoU [23] loss. In this work, we emphasize classification by setting β_1 to 2 and β_2 to 1. Among classification, we emphasize vehicle-plate relation by setting α_r to 2, α_v to 1, and α_p to 1. The ground truth during training is illustrated in Fig. 3.

4   Experiments

4.1   Datasets

We utiliz three publicly available datasets: SSIG-SegPlate [9], AOLP [11], and CRPD [32]. SSIG-SegPlate and AOLP provide the annotations for the vehicle and the license plate, but CRPD only provides the annotations for the license plate. We manually annotated the vehicles in CRPD and made them available at https://github.com/kiki00007/CRPDV.

SSIG-SegPlate comprises 2,000 Brazilian license plates obtained from 101 vehicles. Following the official settings, we use 40% images for training, 20% for validation, and 40% for testing.

AOLP consists of three distinct subsets, each captured using different shooting methods. The AC subset focuses on static vehicles, while the LE subset captures vehicles violating traffic rules via roadside cameras. The RP subset captures images from various viewpoints and distances using cameras mounted on patrol vehicles. In total, the dataset includes 2,049 images containing Taiwanese license plates. When testing on one subset, the other two subsets are used for training and validation.

CRPD has 33,757 Chinese license plates captured by overpasses, which cover various vehicle models, such as cars, trucks, and buses. We follow the official split, i.e., 25,000 images for training, 6,250 for validation, and 2,300 for testing.

4.2   Training Settings

The backbone and transformer are initialized using the pre-trained DETR [4] model. During training, we utilize the Adam optimizer [21] to train the model for 50 epochs with the learning rate of 10^{-4} for the transformer and 10^{-5} for the backbone, weight decay to 10^{-4}, and batch size to 2. Moreover, data augmentation is adopted. First, we apply the image-level augmentation by adjusting the brightness and contrast with a probability of 0.5. Specifically, we randomly select a parameter from the range of [0.8, 1.2] for the brightness and contrast, slightly modifying the original image. Second, we perform scale augmentation by resizing the input image such that the shortest side ranges from 480 to 800 pixels, while the longest side is at most 1333 pixels. The input image is then scaled to the range of [0, 1] and normalized using channel mean and standard deviation. All the experiments are conducted on four NVIDIA 2080Ti GPUs.

4.3   Evaluation Protocols

We use Average Precision (AP) to evaluate license plate detection. Specifically, we utilize the computation method introduced in COCO [19] that calculates AP with different IoU (Intersection over Union) thresholds, i.e., ranging from 0.5 to 0.95 with an interval of 0.05. $AP_{0.5}$ refers to the average precision calculated at the IoU threshold of 0.5. We utilize Accuracy as the evaluation metric for license plate recognition, where all characters must be recognized accurately. We use Frame Per Second (FPS) to calculate the inference speed.

In addition, to verify the effectiveness of small license plate detection, we categorize license plates into three groups based on their height. License plates with a height of 25 pixels or less are categorized as small (S), those exceeding 25 pixels but not exceeding 50 pixels are categorized as medium (M), and license plates taller than 50 pixels are categorized as large (L).

4.4   Ablation Study

Table 1. Ablation study on SSIG-SegPlate. LP: license plate. V: vehicle.

Method	LP	V	Relation	Detection (V)		Detection (LP)		Recognition
				AP	$AP_{0.5}$	AP	$AP_{0.5}$	Accuracy
DETR	√			-	-	45.6%	96.3%	95.4%
	√	√		78.0%	99.2%	50.1%	97.5%	95.6%
Ours	√	√	√	**81.4%**	**100.0%**	**60.6%**	**100.0%**	**96.4%**

Table 2. Ablation study on AOLP. R: relation.

Method	LP	V	R	Detection (V)			Detection (LP)			Recognition		
				AP			AP			Accuracy		
				AC	LE	RP	AC	LE	RP	AC	LE	RP
DETR	√			-	-	-	53.2	52.2	43.6	96.1	94.3	95.3
	√	√		89.2	87.8	83.6	52.2	54.6	40.6	96.2	95.0	94.5
Ours	√	√	√	**93.9**	**90.8**	**91.5**	**65.2**	**60.8**	**58.2**	**98.1**	**98.0**	**97.6**

Table 3. Ablation study on CRPD.

Method	LP	V	Relation	Detection (V)		Detection (LP)		Recognition
				AP	$AP_{0.5}$	AP	$AP_{0.5}$	Accuracy
DETR	√			-	-	53.0%	96.3%	86.0%
	√	√		83.9%	98.5%	54.2%	96.2%	87.5%
Ours	√	√	√	**87.2%**	**98.6%**	**62.8%**	**98.6%**	**89.3%**

As presented in Table 1, Table 2, and Table 3, we investigate the impact of implicit and explicit relationships between vehicles and license plates on the SSIG-SegPlate, AOLP, and CRPD datasets, respectively. We conduct three ablation experiments: (I) direct license plate detection using the vanilla DETR model; (II) simultaneous vehicle and license plate detection using the vanilla DETR model, which implicitly captures the relation between vehicles and license plates;

(III) our proposed method, except for simultaneous vehicle and license plate detection, explicitly incorporating vehicle-plate relationships. After performing license plate detection, we employ the same YOLO-based character recognizer [15] for license plate recognition. Implicit vehicle-plate relationships have minimal impact on license plate detection and recognition. However, when incorporating explicit vehicle-plate relationships, our method substantially improves license plate detection and recognition. Additionally, our method enhances vehicle detection due to the bidirectional relationships between vehicles and license plates.

As shown in Fig. 4, we visualize the attention map of vehicle-plate relationships. The attention map highlights vehicles and their subordinated license plates, which means the relationships are constructed between them. Hence, the detection performance of vehicles and license plates are both enhanced.

Fig. 4. Visualization of vehicle-plate relationships.

4.5   Comparative Experiments

Table 4. Comparative experiments on SSIG-SegPlate.

Method	Detection			Recognition
	AP	$AP_{0.5}$	FPS	Accuracy
RARE [29]	-	-	-	93.7%
Rosetta [3]	-	-	-	94.3%
Direct Detection [4]	45.6%	96.3%	**13.0**	95.4%
Vehicle Pre-detection [6]	52.6%	97.5%	7.7	95.6%
STAR-Net [20]	-	-	-	96.1%
Two Branches [5]	53.8%	98.2%	5.4	96.2%
Ours	**60.6%**	**100.0%**	12.2	**96.4%**

As presented in Table 4, Table 5, and Table 6, we conduct comparative experiments on the SSIG-SegPlate, AOLP, and CRPD datasets, respectively. In all of these datasets, we compare three approaches: direct detection (Fig. 1(a)), vehicle

Table 5. Comparative experiments on AOLP.

Method	Detection						Recognition		
	AC		LE		RP		AC	LE	RP
	AP	$AP_{0.5}$	AP	$AP_{0.5}$	AP	$AP_{0.5}$	Accuracy		
RCLP [16]	-	98.5	-	97.8	-	95.3	94.8	94.2	88.4
DLS [24]	-	92.6	-	93.5	-	92.9	96.2	95.4	95.1
DELP [25]	-	99.3	-	**99.2**	-	99.0	97.8	97.4	96.3
Direct Detection [4]	53.2	98.2	52.2	96.1	43.6	97.8	96.1	94.3	95.3
Vehicle Pre-detection [6]	47.8	98.1	53.8	96.3	44.4	96.9	96.2	95.0	94.5
Two Branches [5]	58.4	96.4	57.8	93.5	48.8	98.2	94.7	92.2	96.2
Ours	**65.2**	**100.0**	**60.8**	99.0	**58.2**	**100.0**	**98.1**	**98.0**	**97.6**

Table 6. Comparative experiments on CRPD.

Method	Detection			Recognition
	AP	$AP_{0.5}$	FPS	Accuracy
SYOLOv4+CRNN [2]	-	-	-	71.0%
RCNN+CRNN [22]	-	-	-	73.7%
UCLP [32]	-	-	-	84.1%
Direct Detection [4]	53.0%	96.3%	**12.8**	86.0%
Vehicle Pre-detection [6]	57.4%	97.8%	7.4	86.2%
Two Branches [5]	58.8%	98.1%	4.8	87.5%
Ours	**62.9%**	**98.3%**	12.5	**89.3%**

pre-detection (Fig. 1(b)), and two branches combining direct detection and vehicle pre-detection (Fig. 1(c)). To ensure a fair comparison, all of these comparative methods utilize the same backbone and transformer as our proposed method. After performing license plate detection, both the comparative methods and our proposed method employ the same YOLO-based character recognizer [15] for license plate recognition. Our proposed method demonstrates superior detection and recognition performance on the SSIG-SegPlate and CRPD datasets while achieving the best performance on most subsets within the AOLP dataset. Concretely, our proposed method achieves an average $AP_{0.5}$ of 99.5% and an average recognition accuracy of 95.9% on the SSIG-SegPlate and CRPD datasets and three subsets of AOLP. However, for the LE subset of AOLP, our proposed method can not effectively handle some low-light images. In future work, we aim to enhance license plate detection under low-light conditions.

Moreover, the direct detection method [4] offers the fastest inference speed but suffers from the lowest detection and recognition performance due to its limited ability to detect small license plates. On the other hand, the vehicle pre-detection method [6] improves license plate detection at the cost of slower

Fig. 5. Visualization examples of license plate detection and recognition.

inference speed. By combining direct detection and vehicle pre-detection, the two branches method [5] further enhances license plate detection and recognition, albeit with the slowest inference speed. In contrast, our proposed method achieves the best detection and recognition performance while maintaining a comparable inference speed to the direct detection method.

Figure 5 demonstrates that our proposed method can accurately detect vehicles and license plates, and the YOLO-based character recognizer [15] can accurately recognize the detected license plates based on character detection.

4.6 Experiments on Multi-scale License Plates

Table 7. Comparative experiments on multi-scale license plates of the CRPD dataset.

Method	Detection (LP)						Recognition		
	S		M		L		S	M	L
	AP	$AP_{0.5}$	AP	$AP_{0.5}$	AP	$AP_{0.5}$	Accuracy		
Direction Detection [4]	45.3	92.4	56.7	96.5	62.6	96.9	82.2	86.1	86.8
Simultaneous Detection [7]	45.0	92.0	56.6	96.8	62.1	96.9	82.0	86.2	86.7
Vehicle Pre-detection [6]	48.7	93.5	59.0	97.3	62.4	98.0	83.5	87.4	87.6
Two Branches [5]	50.5	93.6	60.4	98.1	64.7	98.5	84.0	88.4	88.5
Ours	**55.0**	**95.6**	**62.5**	**98.4**	**67.3**	**99.2**	**85.1**	**89.2**	**89.9**

Table 7 presents comparative experiments involving multi-scale license plates on the CRPD dataset. Notably, we do not conduct multi-scale experiments on the SSIG-SegPlate and AOLP datasets because the size of license plates in these datasets is relatively consistent. In all of these sizes, we compare three approaches: direct detection (Fig. 1(a)), vehicle pre-detection (Fig. 1(b)), and two branches combining direct detection and vehicle pre-detection (Fig. 1(c)). Moreover, the simultaneous detection method denotes detecting vehicles and license plates simultaneously using the vanilla DETR model. To ensure a fair comparison, all of these comparative methods utilize the same backbone and transformer as our proposed method. After performing license plate detection, both the comparative methods and our proposed method employ the same YOLO-based character recognizer [15] for license plate recognition. Our proposed method demonstrates superior performance in both license plate detection and recognition across all sizes, especially for small license plate detection. Concretely, it achieves a 4.5% AP improvement in the detection performance of small license plates compared to the two branches method, with a 2.1% AP improvement for medium license plates and a 2.6% AP improvement for large license plates.

As depicted in Fig. 6, our method can effectively detect small license plates at a considerable distance. Our method can achieve comparative inference speed with the direct detection method, surpassing other comparative methods. Moreover, our method can detect vehicles truncated by image edges due to the bidirectional relationships between vehicles and license plates.

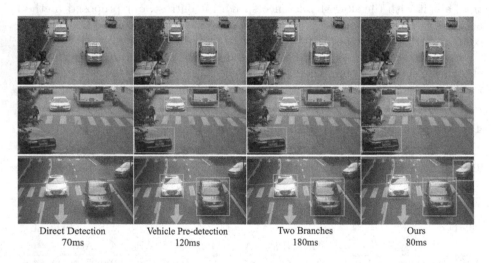

| Direct Detection | Vehicle Pre-detection | Two Branches | Ours |
| 70ms | 120ms | 180ms | 80ms |

Fig. 6. Visualization examples. Under challenging conditions, our proposed method can accurately small license plates at a fast inference speed.

5 Conclusion

We propose to leverage bidirectional relationships between the vehicle and the license plate to enhance small license plate detection. Extensive experiments on the SSIG-SegPlate, AOLP, and CRPD datasets prove our method achieves state-of-the-art detection performance, especially for small license plates. When incorporating a character recognizer, our proposed method can surpass all comparative methods in license plate recognition. In the future, we aim to enhance license plate detection under severe low-light conditions, enabling it to handle more complex scenarios.

Acknowledgements. This work was partly supported by the National Key Research and Development Program of China under Grant 2020AAA0109700 and partly by the National Natural Science Foundation of China under Grant 62076024 and Grant 62006018.

References

1. Batra, P., Hussain, I., Ahad, M.A., Casalino, G., Alam, M.A., Khalique, A., Hassan, S.I.: A novel memory and time-efficient alpr system based on yolov5. Sensors **22**(14), 5283 (2022)
2. Bochkovskiy, A., Wang, C.Y., Liao, H.Y.M.: Yolov4: Optimal speed and accuracy of object detection. arXiv preprint arXiv:2004.10934 (2020)
3. Borisyuk, F., Gordo, A., Sivakumar, V.: Rosetta: large scale system for text detection and recognition in images. In: Proceedings of the 24th ACM SIGKDD International Conference on Knowledge Discovery & Data Mining, pp. 71–79 (2018)
4. Carion, N., Massa, F., Synnaeve, G., Usunier, N., Kirillov, A., Zagoruyko, S.: End-to-end object detection with transformers. In: Vedaldi, A., Bischof, H., Brox, T., Frahm, J.-M. (eds.) ECCV 2020. LNCS, vol. 12346, pp. 213–229. Springer, Cham (2020). https://doi.org/10.1007/978-3-030-58452-8_13
5. Chen, S.L., Liu, Q., Ma, J.W., Yang, C.: Scale-invariant multidirectional license plate detection with the network combining indirect and direct branches. Sensors **21**(4), 1074 (2021)
6. Chen, S.L., et al.: End-to-end trainable network for degraded license plate detection via vehicle-plate relation mining. Neurocomputing **446**, 1–10 (2021)
7. Chen, S.L., Yang, C., Ma, J.W., Chen, F., Yin, X.C.: Simultaneous end-to-end vehicle and license plate detection with multi-branch attention neural network. IEEE Trans. Intell. Transp. Syst. **21**(9), 3686–3695 (2019)
8. Glenn Jocher, Alex Stoken, J.B.: yolov5. https://github.com/ultralytics/yolov5. (2020)
9. Gonçalves, G.R., da Silva, S.P.G., Menotti, D., Schwartz, W.R.: Benchmark for license plate character segmentation. J. Electron. Imaging **25**(5), 053034–053034 (2016)
10. He, K., Zhang, X., Ren, S., Sun, J.: Deep residual learning for image recognition. In: Proceedings of the IEEE Conference on Computer Vision and Pattern Recognition, pp. 770–778 (2016)
11. Hsu, G.S., Chen, J.C., Chung, Y.Z.: Application-oriented license plate recognition. IEEE T-VT **62**(2), 552–561 (2012)
12. Kim, S., Jeon, H., Koo, H.: Deep-learning-based license plate detection method using vehicle region extraction. Electron. Lett. **53**(15), 1034–1036 (2017)
13. Kuhn, H.W.: The Hungarian method for the assignment problem. Naval Res. Logistics Quarterly **2**(1–2), 83–97 (1955)
14. Laroca, R., et al.: A robust real-time automatic license plate recognition based on the yolo detector. In: 2018 International Joint Conference on Neural Networks (ijcnn), pp. 1–10. IEEE (2018)
15. Laroca, R., Zanlorensi, L.A., Gonçalves, G.R., Todt, E., Schwartz, W.R., Menotti, D.: An efficient and layout-independent automatic license plate recognition system based on the yolo detector. IET Intel. Transport Syst. **15**(4), 483–503 (2021)
16. Li, H., Shen, C.: Reading car license plates using deep convolutional neural networks and lstms. arXiv preprint arXiv:1601.05610 (2016)

17. Lin, T.Y., Dollár, P., Girshick, R., He, K., Hariharan, B., Belongie, S.: Feature pyramid networks for object detection. In: Proceedings of the IEEE Conference on Computer Vision and Pattern Recognition, pp. 2117–2125 (2017)
18. Lin, T.Y., Goyal, P., Girshick, R., He, K., Dollár, P.: Focal loss for dense object detection. In: Proceedings of the IEEE International Conference on Computer Vision, pp. 2980–2988 (2017)
19. Lin, T.Y., et al.: Microsoft coco: Common objects in context (2014)
20. Liu, W., Chen, C., Wong, K.Y.K., Su, Z., Han, J.: Star-net: a spatial attention residue network for scene text recognition. In: BMVC, vol. 2, p. 7 (2016)
21. Loshchilov, I., Hutter, F.: Decoupled weight decay regularization. arXiv preprint arXiv:1711.05101 (2017)
22. Ren, S., He, K., Girshick, R., Sun, J.: Faster r-cnn: towards real-time object detection with region proposal networks. Advances in neural information processing systems 28 (2015)
23. Rezatofighi, H., Tsoi, N., Gwak, J., Sadeghian, A., Reid, I., Savarese, S.: Generalized intersection over union: a metric and a loss for bounding box regression. In: Proceedings of the IEEE/CVF Conference on Computer Vision and Pattern Recognition, pp. 658–666 (2019)
24. Selmi, Z., Halima, M.B., Alimi, A.M.: Deep learning system for automatic license plate detection and recognition. In: 2017 14th IAPR International Conference on Document Analysis and Recognition (ICDAR), vol. 1, pp. 1132–1138. IEEE (2017)
25. Selmi, Z., Halima, M.B., Pal, U., Alimi, M.A.: Delp-dar system for license plate detection and recognition. Pattern Recogn. Lett. **129**, 213–223 (2020)
26. Tian, Z., Shen, C., Chen, H., He, T.: Fcos: fully convolutional one-stage object detection. In: Proceedings of the IEEE/CVF International Conference on Computer Vision, pp. 9627–9636 (2019)
27. Vaswani, A., et al.: Attention is all you need. Advances in neural information processing systems 30 (2017)
28. Xu, Z., Yang, W., Meng, A., Lu, N., Huang, H., Ying, C., Huang, L.: Towards end-to-end license plate detection and recognition: a large dataset and baseline. In: Proceedings of the European conference on computer vision (ECCV), pp. 255–271 (2018)
29. Zhang, H., Yao, Q., Yang, M., Xu, Y., Bai, X.: Autostr: efficient backbone search for scene text recognition. In: Computer Vision-ECCV 2020: 16th European Conference, Glasgow, UK, August 23–28, 2020, Proceedings, Part XXIV 16, pp. 751–767. Springer (2020)
30. Zhang, S., Chi, C., Yao, Y., Lei, Z., Li, S.Z.: Bridging the gap between anchor-based and anchor-free detection via adaptive training sample selection. In: Proceedings of the IEEE/CVF Conference on Computer Vision and Pattern Recognition, pp. 9759–9768 (2020)
31. Zhu, X., Su, W., Lu, L., Li, B., Wang, X., Dai, J.: Deformable detr: deformable transformers for end-to-end object detection. arXiv preprint arXiv:2010.04159 (2020)
32. Zou, Y., Zhang, Y., Yan, J., Jiang, X., Huang, T., Fan, H., Cui, Z.: License plate detection and recognition based on yolov3 and ilprnet. SIViP **16**(2), 473–480 (2022)

A Purified Stacking Ensemble Framework for Cytology Classification

Linyi Qian[ID], Qian Huang[✉][ID], Yulin Chen[ID], and Junzhou Chen[ID]

College of Computer Science and Software Engineering,
Hohai University, Nanjing, China
huangqian@hhu.edu.cn

Abstract. Cancer is one of the fatal threats to human beings. However, early detection and diagnosis can significantly reduce death risk, in which cytology classification is indispensable. Researchers have proposed many deep learning-based methods for automated cancer diagnosis. Nevertheless, due to the similarity of pathological features in cytology images and the scarcity of high-quality datasets, neither the limited accuracy of single networks nor the complex architectures of ensemble methods can meet practical application needs. To address the issue, we propose a purified Stacking ensemble framework, which employs three homogeneous convolutional neural networks (CNNs) as base learners and integrates their outputs to generate a new dataset by a k-fold split and concatenation strategy. Then a distance weighted voting technique is applied to purify the dataset, on which a multinomial logistic regression model with a designed loss function is trained as the meta-learner and performs the final predictions. The method is evaluated on the FNAC, Ascites, and SIPaKMeD datasets, achieving accuracies of 99.85%, 99.24%, and 99.75%, respectively. The experimental results outperform the current state-of-the-art (SOTA) methods, demonstrating its potential for reducing screening workload and helping pathologists detect cancer.

Keywords: Cytology classification · Ensemble learning · Stacking

1 Introduction

Cytology is a branch of pathology to study cells under microscopes to analyze the cellular morphology and compositions, usually for cancer screening [1]. Compared with histopathology, cytology focuses on the pathological characteristics of cells instead of tissues, which is a collection of thousands of cells in a specific architecture [2].

Cytology classification plays a vital role in cancer screening and early diagnosis. However, it is a complex and massive undertaking, which requires pathologists to sift through thousands of cells to identify problematic cells. In recent years, many computer-aided diagnostic methods based on deep learning have made significant breakthroughs. However, these models often fail to achieve satisfactory accuracy due to the high similarity between cytology images (e.g., Fig. 1)

© The Author(s), under exclusive license to Springer Nature Switzerland AG 2024
S. Rudinac et al. (Eds.): MMM 2024, LNCS 14555, pp. 267–280, 2024.
https://doi.org/10.1007/978-3-031-53308-2_20

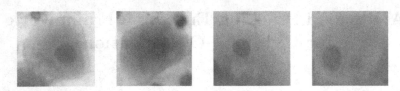

Fig. 1. Examples of two confusable classes in cervical cytology. The two on the left are metaplastic cells, and the two on the right are parabasal cells.

and low quality of datasets (e.g., Imbalanced distribution and limited number). Currently, the commonly used approaches include single networks and ensemble methods. The former makes it easier to misclassify similar image features, causing relatively limited accuracy. The latter have complex architectures, resulting in more parameters and slower inference speeds. The ensemble framework can also propagate errors and noise during the learning process, making it more prone to overfitting.

To address the issues, we propose a purified Stacking ensemble framework for cytology classification. Initially, data preprocessing is performed to increase the size of datasets and enhance image features. Then, we feed them into three homogeneous models (each pre-trained on the ImageNet dataset), which serve as base learners. Those models have similar architecture and can better learn certain image features while reducing the number of parameters. The outputs of the base learners are aggregated to generate a new dataset with a k-fold split and concatenation strategy, which mitigates the problem of overfitting. Next, we use a distance weighted voting strategy to purify the dataset, focusing on preserving confusable image features for relearning. Finally, we apply the purified dataset to train a multinomial logistic regression (MLR) model with a designed adaptive weighted softmax loss function, which can further improve the performance. The trained MLR model is utilized for the final prediction.

The contributions of this paper are as follows:

(1) We propose a novel k-fold split and concatenation (KFSC) strategy, combining k-fold cross-validation with the Stacking method to generate a more diverse dataset and effectively address the overfitting issue.
(2) We design a purification method termed distance weighted voting (DW-Voting) that uses an elaborate voting strategy to filter the newly generated dataset and makes the meta-learner focus on the features of misclassified samples.
(3) We devise an adaptive weighted softmax loss (AW-Softmax) function, which automatically adjusts the weights based on the meta-learner's performance and further enhances the overall framework's robustness.
(4) We conduct experiments on various CNN architectures, and the results demonstrate that the proposed framework significantly improves classification accuracy with fewer parameters and faster inference speed. Furthermore, we evaluate the proposed method on three public cytology datasets using a range of metrics, and the results outperform state-of-the-art (SOTA) methods.

2 Related Work

2.1 Cell-Level Classification

Cell-level classification could be one of the most successful tasks in deep learning-based cytology image analysis [3]. Due to the giga-pixel resolution of collected cytology whole slide images, scholars often crop them into cell patches and use them for training cell classification models [4]. The most common method is to directly feed cell patches into a multi-layer CNN to extract feature maps, then cross the output layer to get the predicted category. Based on this, a series of CNN-based methods have been proposed.

For lung cytology classification, Teramoto *et al.* [5] introduced a deep convolutional neural network (DCNN) to automatize the classification of malignant lung cells from microscopic images, and it reached a performance comparable to that of a cytopathologist. For cervical cytology classification, previous classification methods are only built upon extracting hand-crafted features, such as morphology and texture. Zhang *et al.* [6] designed a CNN called DeepPap to directly classify cervical cells without prior segmentation-based on deep features, which reached a high accuracy when evaluated on both the Pap smear and LBC datasets. In addition, Tripathi *et al.* [7] presented deep learning classification methods applied to the Pap smear dataset to establish a reference point for assessing forthcoming classification techniques. These studies demonstrated the substantial clinical value of classification-assisted cytology image analysis. However, the lack of high-quality datasets and the similarity of cell morphology also pose great challenges to cell-level classification.

2.2 Ensemble Learning

An individual model is limited by its architecture, and there is always an upper bound (i.e., Bayes error) which makes it increasingly difficult to improve the performance currently. Ensemble learning is an alternative solution to the problem, combining multiple models to achieve better predictive performance by taking advantage of the strengths of each model and compensating for their weaknesses [8].

For breast cytology classification, Ghiasi *et al.* [9] proposed a decision tree-based ensemble learning framework. They evaluated it on Wisconsin Breast Cancer Database (WBCD) and achieved satisfactory accuracy. For cervical cytology classification, Manna *et al.* [10] proposed an ensemble scheme that used a fuzzy rank-based fusion of classifiers by considering two non-linear functions on the decision scores generated by base learners. The proposed framework achieved the highest accuracy on the SIPaKMeD and Mendeley datasets. Although these ensemble methods have achieved excellent performance, certain models within the framework may be influenced by image noise, and the ensemble may propagate these errors, resulting in incorrect predictions.

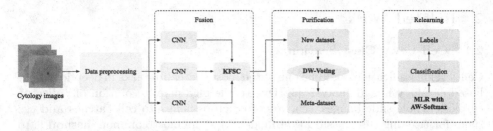

Fig. 2. The overall workflow of the purified Stacking ensemble, where **KFSC** represents k-fold split and concatenation, **DW-Voting** represents distance weighted voting and **MLR with AW-Softmax** represents multinomial logistic regression model with adaptive weighted softmax loss function.

3   Method

The proposed method is based on the Stacking ensemble strategy [11], which trains the base learners on the initial dataset and uses the outputs of these base learners to train the meta-learner. We can divide it into four stages. The first stage is data preprocessing, which includes resizing and data augmentation. The second stage is the fusion of base learners, where three homogeneous models extract features from cellular images. We design a k-fold split and concatenation strategy for aggregating their outputs to generate a new dataset in preparation for the next stage. The third stage focuses on purification, where we use a designed voting filter to sift the newly generated dataset and obtain the meta-dataset. In the last stage, we apply a multinomial logistic regression (MLR) model with a designed loss function to relearn and make the final prediction. An illustration of the complete workflow can be seen in Fig. 2, which will be explained in detail below.

3.1   Data Preprocessing

In the data preprocessing stage, we first uniformly resize the images to fit different inputs of network architectures (e.g., 224×224 pixels for ResNet). Considering the limited number of images in cytology datasets, we employ data augmentation techniques.

Since each cell patch is cropped from a large image slide, resizing and translation operations may result in the loss of image features. To mitigate this, we employ rotation and flipping methods. For each cellular image, we perform one full rotation, rotating it by 20°C each time. Additionally, we apply horizontal flipping and vertical flipping. This augmentation process effectively increases the size of the dataset by a factor of 20.

3.2   Fusion with KFSC

The main task of the second stage is to train multiple base learners and generate a new dataset based on their outputs. Suppose we directly use the initial dataset to construct the target dataset. In this case, there is a risk of overfitting, where the meta-learner becomes too specific to the initial dataset and fails to generalize well to new data.

To avoid the above issues, we propose a novel k-fold split and validation (KFSC) strategy to obtain multiple dataset partitions and generate a more diverse and representative dataset. The pipeline is illustrated in Fig. 3.

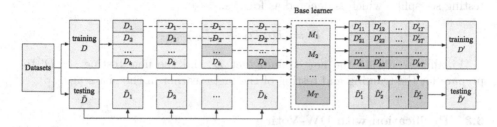

Fig. 3. An overview of k-fold split and concatenation.

Firstly, the dataset is divided into a training set D and a testing set $\tilde{D}$. When implementing k-fold cross-validation, the initial training set is divided into k subsets of similar size, denoted as $D_1, D_2, \cdots, D_k$. Let D_i and $\overline{D}_i = D \backslash D_i$ represent the validation set and the training set for the i-th fold, respectively. We train T base learners $M_1, M_2, \cdots, M_T$ on $\overline{D}_i$ and then validate them on D_i.

During the i-th round of training, each base learner is trained on the training set $\overline{D}_i$ to obtain the corresponding classifier $C_j = M_j(\overline{D}_i), j \in \{1, 2, \cdots, T\}$. The results obtained on the validation set D_i are denoted as follows:

$$D'_{ij} = C_j(D_i), j \in \{1, 2, \cdots, T\} \tag{1}$$

By horizontally concatenating the results, the training set split generated in the i-th round is denoted as follows:

$$D'_i = (D'_{i1}, D'_{i2}, \cdots, D'_{iT}) \tag{2}$$

To obtain the final training set, we can vertically concatenate the training splits from each round, which is denoted as follows:

$$D' = (D'_1; D'_2; \cdots; D'_k) \tag{3}$$

It is evident that the generated training set has the same dimensionality along the x-axis as the original training set, which means the number of generated samples remains the same.

The process of generating the new testing set is similar. During the i-th round, the results of each classifier on the testing set $\tilde{D}$ are defined as follows:

$$\tilde{D}'_{ij} = C_j(\tilde{D}), j \in \{1, 2, \cdots, T\} \tag{4}$$

To maintain consistency in dimensionality, we need to average the testing results of each classifier, so the j-th testing set split is denoted as follows:

$$\tilde{D}'_j = \frac{1}{k} \sum_{i=1}^{k} \tilde{D}'_{ij} \tag{5}$$

The complete testing set can be obtained by horizontally concatenating the testing set splits, which is denoted as follows:

$$\tilde{D}' = (\tilde{D}_1, \tilde{D}_2, \cdots, \tilde{D}_T) \tag{6}$$

Finally, we obtain the new training set D' and the new testing set $\tilde{D}'$ to prepare for the training and testing of the meta-learner in the last stage.

3.3   Purification with DW-Voting

During the third stage, we filter the new dataset to generate the final meta-dataset for the meta-learner (e.g., Fig. 4). Instead of employing the complete data, we sift and retain the misclassified samples. In other words, we introduce a new concept called purity, which refers to the proportion of misclassified samples in a dataset. The purpose of filtering is to enhance the diversity of the dataset and make the meta-learner focus on the confusable features. By excluding correctly classified samples, we can reduce potential interference and improve the final accuracy. Besides, it significantly reduces the size of the dataset, which can accelerate the training and testing of the meta-learner.

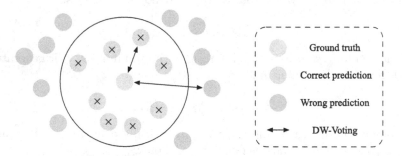

Fig. 4. Visualization of purification. It aims to remove (represented by ×) correct predictions while retaining wrong predictions (namely confusable futures), allowing the meta-learner to relearn. The purifying criterion is based on the distance between prediction and ground truth, measured by the DW-Voting strategy.

Taking the purification of the training set as an example, we define the newly generated training set as $D' = \{(\mathcal{X}_i, y_i)|_{i=1}^m)\}$, and $\mathcal{X}_i$ is defined as follows:

$$\mathcal{X}_i = (P_{i1}, P_{i2}, \cdots, P_{iT}) \tag{7}$$

where P_{ij} represents the probability vector generated by the j-th classifier for the i-th image, and it can be expanded as follows:

$$P_{ij} = (C_j^1(x_i), C_j^2(x_i), \cdots, C_j^c(x_i)), \sum_{k=1}^c C_j^k(x_i) = 1 \tag{8}$$

where $C_j^k(x_i)$ represents the probability corresponding to the k-th class assigned by classifier C_j for the i-th image x_i.

Here we propose a distance weighted voting filter technique. Given the one-hot label encoding T_i of the i-th image, we can calculate the distance between the probability vector P_{ij} and the ground truth T_i for each classifier:

$$T_i = (\cdots, 0, \cdots, 1, \cdots, 0, \cdots), T_{iy_i} = 1$$
$$d_{ij} = \sqrt{(P_{ij} - T_i)^2} \tag{9}$$

The distance indirectly reflects the performance of the classifier. When the distance is smaller, it indicates that the predicted value is closer to the true label. Therefore, in the subsequent voting process, the weight of this classifier should be appropriately increased. We can calculate the proportion of each classifier's distance r_{ij} and then obtain the corresponding weight w_{ij} based on the proportion:

$$r_{ij} = \frac{d_{ij}}{\sum_{k=1}^T d_{ik}} \quad w_{ij} = \frac{1 - r_{ij}}{T - 1}, \sum_{j=1}^T w_{ij} = 1 \tag{10}$$

The final predicted value can be calculated through weighted sum:

$$P_i = \sum_{j=1}^T w_{ij} P_{ij} = (\sum_{j=1}^T w_{ij} C_j^1(x_i), \sum_{j=1}^T w_{ij} C_j^2(x_i), \cdots, \sum_{j=1}^T w_{ij} C_j^c(x_i)) \tag{11}$$
$$\hat{y}_i = \mathbf{argmax}(P_i)$$

After filtering the samples whose predicted label matches with the ground truth, we can obtain a purified meta-training set for the relearning of the meta-learner:

$$D' = \{(\mathcal{X}_i, y_i)|_{i=1}^{m'})\}, \hat{y}_i \neq y_i \tag{12}$$

The purification of the testing set follows the same process as described above. It is important to note that the samples filtered by DW-Voting in the testing set are considered successfully predicted by the ensemble of base learners. Hence, the meta-learner in the testing phase only needs to focus on the misclassified samples.

3.4   Relearning with AW-Softmax

To cope with multi-class classification tasks, we adopt multinomial logistic regression (MLR) as the meta-learner. Besides, we design an adaptive weighted softmax loss (AW-Softmax) function, of which the principle is to dynamically adjust the weights based on the performance of each round's model. The calculation process of the loss function is as shown in Fig. 5 and will be described in detail below:

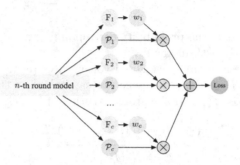

Fig. 5. A process of loss calculation for a single image. Firstly, the F1-score vector F and probability vector $\mathcal{P}$ are obtained based on the model. Then, the weight vector $\mathcal{W}$ is derived. Finally, the loss of the image is calculated through the weighted sum.

For meta-training dataset $D' = \{(\mathcal{X}_i, y_i)|_{i=1}^{m'}\}$, each class $r \in \{1, 2, \cdots, c\}$ has a corresponding weight vector a_r (namely the parameters of the model). The probability of sample $\mathcal{X}_i$ belonging to class r can be calculated as follows:

$$P(r|\mathcal{X}_i) = \text{softmax}(a_r \cdot \mathcal{X}_i) = \frac{\exp(a_r \cdot \mathcal{X}_i)}{\sum\limits_{j=1}^{c} \exp(a_j \cdot \mathcal{X}_i)} \tag{13}$$

So the final predicted probability vector of the meta-learner for sample $\mathcal{X}_i$ can be represented as follows:

$$\mathcal{P}_i = (\mathcal{P}(1|\mathcal{X}_i), \mathcal{P}(2|\mathcal{X}_i), \cdots, \mathcal{P}(c|\mathcal{X}_i)) \tag{14}$$

Based on this, we define the loss function L_j in the the j-th training round as follows:

$$L_j = -\sum_{i=1}^{m'} \log \mathcal{P}_i \cdot \mathcal{W}_j \tag{15}$$

where $\mathcal{W}_j$ represents the weight vector, which can be recursively derived.

Suppose the weight vector for the previous round is defined as $\mathcal{W}_{j-1} = (w_1, w_2, \cdots, w_c)^{\mathrm{T}}$. Based on the predictions of each round, we can calculate the precision P, recall R, and F1-score F for each class $r \in \{1, 2, \cdots, c\}$:

$$\mathrm{P}_r = \frac{\sum_{i=1}^{n}(y_i = r, \hat{y}_i = r)}{\sum_{i=1}^{n}(\hat{y}_i = r)} \qquad \mathrm{R}_r = \frac{\sum_{i=1}^{n}(y_i = r, \hat{y}_i = r)}{\sum_{i=1}^{n}(y_i = r)} \qquad \mathrm{F}_r = \frac{2 \times \mathrm{P}_r \times \mathrm{R}_r}{\mathrm{P}_r + \mathrm{R}_r} \qquad (16)$$

We utilize the F1-score to provide a more comprehensive evaluation of the model performance, and define the weight vector $\mathcal{W}$ with the following formula:

$$\begin{aligned} \mathcal{W}_j &= \mathcal{Z}\text{-score}(\mathcal{W}_{j-1} + 1 - \mathrm{F}) \\ &= \mathcal{Z}\text{-score}([w_1 + 1 - \mathrm{F}_1, w_2 + 1 - \mathrm{F}_2, \cdots, w_c + 1 - \mathrm{F}_c]) \end{aligned} \qquad (17)$$

where $\mathcal{Z}$-score is a standard normalization function to ensure weights sum up to 1 and prevent overflow. A lower F1-score indicates lower precision and recall, signifying poorer performance for specific classes. In such cases, it is appropriate to increase their weights, which shifts the focus of the model towards confusable features in the next training round.

Besides, for the recursive formula, an initial weight needs to be defined. Given the uneven distribution of dataset, we define the initial weight $\mathcal{W}_0$ based on the proportion of each class:

$$\mathcal{W}_0 = \mathcal{Z}\text{-score}(1 - \frac{\sum_{i=1}^{m'}(y_i = 1)}{m'}, 1 - \frac{\sum_{i=1}^{m'}(y_i = 2)}{m'}, \cdots, 1 - \frac{\sum_{i=1}^{m'}(y_i = c)}{m'}) \qquad (18)$$

For classes with fewer samples, we increase their weights appropriately so that the model will not be biased during subsequent training and vice versa. Once the loss function is determined, the weights can be updated using gradient descent during learning. The trained meta-learner will be used for the final predictions (Table 1).

4   Experiments and Analysis

4.1   Datasets

In this paper, we evaluate the proposed method on three publicly available cytology datasets:

1. FNAC Pap Smear dataset for breast cytology classification [12]
2. Ascites Pap Smear dataset for stomach cytology classification [13]
3. SIPaKMeD Pap Smear dataset for cervical cytology classification [14]

Table 1. Detailed description of three public datasets.

	Class	Index	Cell type	Number
FNAC (total: 212)	0	Benign	–	99
	1	Malignant	–	113
Ascites (total: 7880)	0	Benign	Eosinophil granulocyte	30
	1	Benign	Lymphocyte	200
	2	Benign	Mesothelial	800
	3	Benign	Neutrophil granulocyte	150
	4	Malignant	Determined	6000
	5	Malignant	Suspicious	700
SIPaKMeD (total: 4049)	0	Normal	Superficial-intermediate	831
	1	Normal	Parabasal	787
	2	Abnormal	Koilocytotic	825
	3	Abnormal	Dyskeratotic	813
	4	Abnormal	Metaplastic	793

Table 2. The hyperparameters used for experiments.

Hyperparameters	Value/Method
Learning Rate	0.0001
Batch Size	16
Epoch	60
Optimizer	AdamW
Learning Rate Scheduler	ReduceLROnPlateau
Loss	AW-Softmax

4.2 Experimental Configuration

All the experiments are conducted on GeForce RTX 3080 with TensorFlow deep learning framework. The configuration of this study is presented in Table 2. There are two additional points to note: (1) During the training stage of the base learners, we split 20% part of the training set into a validation set to assist in selecting the best-performing models. (2) The stratified sampling strategy is used for all dataset partitioning to address the issue of imbalanced data distributions.

4.3 Experimental Results on CNN Architectures

Table 3. Experimental results on different CNN architectures.

Model	FNAC(%)	Ascites(%)	SIPaKMeD(%)
VGG13	92.16	90.21	91.24
VGG16	93.17	91.14	92.28
VGG19	93.24	91.86	94.33
Ours with VGG-Ensemble	**95.32**	**93.45**	**95.12**
ResNet50	94.24	91.67	92.45
ResNet101	95.16	92.13	93.24
ResNet152	95.85	92.97	93.65
Ours with ResNet-Ensemble	**98.25**	**94.16**	**95.19**
EfficientNetV2S	95.38	92.64	93.41
EfficientNetV2M	96.01	93.14	94.32
EfficientNetV2L	96.36	93.54	95.23
Ours with EfficientNet-Ensemble	**98.84**	**95.54**	**97.21**
ConvNeXtSmall	95.64	94.75	94.32
ConvNeXtBase	96.32	95.64	94.75
ConvNeXtLarge	97.35	96.19	95.18
Ours with ConvNeXt-Ensemble	**99.52**	**98.87**	**98.75**
Xception	97.45	95.34	96.99
InceptionV3	96.25	95.25	94.86
InceptionResNetV2	97.12	96.32	96.25
Ours with Inception-Ensemble	**99.85**	**99.24**	**99.75**

We conduct a series of experiments using several popular CNN architectures as base learners, including VGG [15], ResNet [16], Inception [17], EfficientNet [18], and ConvNeXt [19]. We evaluate the performance of individual models and ensemble in our proposed framework on three public datasets with the mean accuracy, and the results are shown in Table 3.

It can be observed that all architectures achieve significant performance improvements when combined with our framework. In particular, the Inception-family models exhibit the best performance, achieving accuracies of 99.85%, 99.24%, and 99.75% on the FNAC, Ascites, and SIPaKMeD datasets, respectively. Therefore, we will use the three Inception models as our base learners in subsequent experiments.

4.4 Comparison with Other Methods

In the comparative experiments, we compare our method with some other ensemble methods. For a more comprehensive comparison, in addition to the four commonly used classification metrics - accuracy, recall, precision, and F1-score, we also compare complexities, including the number of parameters (corresponding to spatial complexity) and inference speed (corresponding to time complexity).

Table 4. Comparison with other methods, where P represents the number of parameters, S represents inference speed, Acc represents accuracy, Pre represents precision and Rec represents recall.

Method	P(M)	S(ms)	FNAC 2-class				Ascites 6-class				SIPaKMeD 5-class			
			Acc(%)	Pre(%)	Rec(%)	F1(%)	Acc(%)	Pre(%)	Rec(%)	F1(%)	Acc(%)	Pre(%)	Rec(%)	F1(%)
DTE [9]	100	20	97.03	97.12	97.08	97.10	96.32	96.28	96.32	96.30	95.74	95.69	95.72	95.70
FRE [10]	105	30	97.42	97.41	97.43	97.42	96.64	96.52	96.75	96.63	96.02	95.87	96.13	96.00
FDE [20]	105	29	97.67	97.45	97.74	97.59	96.82	96.76	96.98	96.87	96.96	96.92	96.97	96.91
EHDLF [22]	118	25	98.12	98.16	98.12	98.14	97.02	97.09	97.05	97.07	97.26	97.27	97.28	97.28
PCA-GWO [23]	112	27	98.21	98.24	98.21	98.22	97.18	97.14	97.16	97.15	97.87	98.56	99.12	98.89
Ours	**96**	**18**	**99.85**	**99.86**	**99.86**	**99.86**	**99.24**	**99.12**	**99.36**	**99.24**	**99.75**	**99.75**	**99.76**	**99.75**

The results are shown in Table 4. DTE [9], as an ensemble of machine learning methods, has fewer parameters and consequently faster inference speed. FRE [10] and FDE [20] represent ensembles of heterogeneous CNNs, which sacrifice inference speed for improved accuracy. EHDLF [22] and PCA-GWO [23] both involve feature fusion, which enhances accuracy at the expense of parameter count. Our method employs an ensemble of homogeneous CNNs, which has the fewest parameters compared to other methods. Additionally, we utilize a designed voting strategy to filter the dataset, significantly reducing inference times. It can be observed that our method achieved the best results across all metrics on the three datasets, demonstrating the superiority of the framework.

4.5   Ablation Study

We conduct ablation experiments to evaluate the importance of each component in the framework, and the results are shown in Table 5, from which several conclusions can be drawn:

Table 5. Ablation study on the four most important parts of the framework, where ✓ represents selected.

Number of base learners	KFSC	DW-Voting	AW-Softmax	Accuracy(%)		
				FNAC	Ascites	SIPaKMeD
1	✓	✓	✓	97.24	96.15	96.28
2	✓	✓	✓	97.35	96.51	96.57
3		✓	✓	97.64	96.82	97.02
3	✓		✓	98.25	97.63	98.05
3	✓	✓		99.06	98.53	99.02
3	✓	✓	✓	**99.85**	**99.24**	**99.75**

(1) More base learners means better classification accuracy. Utilizing more classifiers enables deeper learning of features, reducing classification error and improving accuracy (as shown in rows 1, 2, and 6, with an average increase of 3.06% in accuracy). It is worth noting that when we increase the number of base learners to 4, the parameter count of the overall framework increases significantly while the improvement in accuracy is minimal. Therefore, we have limited the number to 3. (2) Using the KFSC strategy to generate a new dataset can significantly enhance the generalization ability, reducing the risk of overfitting and improving accuracy (as shown in rows 3 and 6, with an average increase of 2.45% in accuracy). (3) Applying the DW-Voting strategy to filter the dataset increases the purity of datasets, making the meta-learner focus more on the features of confusable images, further enhancing classification accuracy (as shown in rows 4 and 6, with an average increase of 1.64% in accuracy). (4) The AW-Softmax loss function considers the distribution of datasets and the model's performance in each round, improving the framework's robustness and increasing classification accuracy (as shown in rows 5 and 6, with an average increase of 0.74% in accuracy).

5   Conclusion

This paper proposes a purified Stacking ensemble strategy for cytology classification, mainly consisting of four stages. The first stage is data preprocessing, which includes resizing and data augmentation. The second stage trains three homogeneous networks and uses their outputs to generate a new dataset using a KFSC strategy. The third stage is the implementation of purification, in which the new dataset is filtered by a DW-Voting method. The last stage focuses on relearning using an MLR model with AW-Softmax loss function. We evaluate the proposed method on three benchmark datasets: FNAC (99.85%), Ascites (99.24%), and

SIPaKMeD (99.75%), and achieve better performance in accuracy, recall, precision, and F1-score than the current SOTA methods. The experimental results demonstrate that our method can effectively improve the accuracy of cytology classification and has promising prospects for future applications in computer-aided diagnostic systems.

Acknowledgements. This paper was supported by the Key Research and Development Program of Yunnan Province under grant no. 202203AA080009, the 14th Five-Year Plan for Educational Science of Jiangsu Province under grant no. D/2021/01/39, and the Jiangsu Higher Education Reform Research Project "Research on the Evaluation of Practical Teaching Reform in Information Majors based on Student Practical Ability Model" under grant no. 2021JSJG143.

References

1. Alberts, B., et al.: Essential cell biology. Garland Science (2015)
2. Morrison, W., DeNicola, D.: Advantages and disadvantages of cytology and histopathology for the diagnosis of cancer. In: Seminars in veterinary medicine and surgery (small animal), vol. 8, pp. 222–227 (1993)
3. Jantzen, J., Norup, J., Dounias, G., Bjerregaard, B.: Pap-smear benchmark data for pattern classification. Nature inspired Smart Information Systems (NiSIS 2005), pp. 1–9 (2005)
4. Zhang, C., Liu, D., Wang, L., Li, Y., Chen, X., Luo, R., Che, S., Liang, H., Li, Y., Liu, S., Tu, D., Qi, G., Luo, P., Luo, J.: DCCL: a benchmark for cervical cytology analysis. In: Suk, H.-I., Liu, M., Yan, P., Lian, C. (eds.) MLMI 2019. LNCS, vol. 11861, pp. 63–72. Springer, Cham (2019). https://doi.org/10.1007/978-3-030-32692-0_8
5. Teramoto, A., et al.: Automated classification of benign and malignant cells from lung cytological images using deep convolutional neural network. Inf. Med. Unlocked **16**, 100205 (2019)
6. Zhang, L., Lu, L., Nogues, I., Summers, R.M., Liu, S., Yao, J.: Deeppap: deep convolutional networks for cervical cell classification. IEEE J. Biomed. Health Inform. **21**(6), 1633–1643 (2017)
7. Tripathi, A., Arora, A., Bhan, A.: Classification of cervical cancer using deep learning algorithm. In: 2021 5th International Conference on Intelligent Computing and Control Systems (ICICCS), pp. 1210–1218. IEEE (2021)
8. Sagi, O., Rokach, L.: Ensemble learning: a survey. Wiley Interdisciplinary Rev. Data Mining Knowl. Discovery **8**(4), e1249 (2018)
9. Ghiasi, M.M., Zendehboudi, S.: Application of decision tree-based ensemble learning in the classification of breast cancer. Comput. Biol. Med. **128**, 104089 (2021)
10. Manna, A., Kundu, R., Kaplun, D., Sinitca, A., Sarkar, R.: A fuzzy rank-based ensemble of cnn models for classification of cervical cytology. Sci. Rep. **11**(1), 14538 (2021)
11. Wolpert, D.H.: Stacked generalization. Neural Netw. **5**(2), 241–259 (1992)
12. Saikia, A.R., Bora, K., Mahanta, L.B., Das, A.K.: Comparative assessment of CNN architectures for classification of breast FNAC images. Tissue Cell **57**, 8–14 (2019)
13. Su, F., et al.: Development and validation of a deep learning system for ascites cytopathology interpretation. Gastric Cancer **23**, 1041–1050 (2020)

14. Plissiti, M.E., Dimitrakopoulos, P., Sfikas, G., Nikou, C., Krikoni, O., Charchanti, A.: Sipakmed: a new dataset for feature and image based classification of normal and pathological cervical cells in pap smear images. In: 2018 25th IEEE International Conference on Image Processing (ICIP), pp. 3144–3148. IEEE (2018)

15. Simonyan, K., Zisserman, A.: Very deep convolutional networks for large-scale image recognition. arXiv preprint arXiv:1409.1556 (2014)

16. He, K., Zhang, X., Ren, S., Sun, J.: Deep residual learning for image recognition. In: Proceedings of the IEEE Conference on Computer Vision and Pattern Recognition, pp. 770–778 (2016)

17. Szegedy, C., et al.: Going deeper with convolutions. In: Proceedings of the IEEE Conference on Computer Vision and Pattern Recognition, pp. 1–9 (2015)

18. Tan, M., Le, Q.: Efficientnetv2: smaller models and faster training. In: International Conference on Machine Learning, pp. 10096–10106. PMLR (2021)

19. Liu, Z., Mao, H., Wu, C.Y., Feichtenhofer, C., Darrell, T., Xie, S.: A convnet for the 2020s. In: Proceedings of the IEEE/CVF Conference on Computer Vision and Pattern Recognition, pp. 11976–11986 (2022)

20. Dey, S., Das, S., Ghosh, S., Mitra, S., Chakrabarty, S., Das, N.: SynCGAN: using learnable class specific priors to generate synthetic data for improving classifier performance on cytological images. In: Babu, R.V., Prasanna, M., Namboodiri, V.P. (eds.) NCVPRIPG 2019. CCIS, vol. 1249, pp. 32–42. Springer, Singapore (2020). https://doi.org/10.1007/978-981-15-8697-2_3

21. Pramanik, R., Biswas, M., Sen, S., de Souza Júnior, L.A., Papa, J.P., Sarkar, R.: A fuzzy distance-based ensemble of deep models for cervical cancer detection. Comput. Methods Programs Biomed. **219**, 106776 (2022)

22. Nanni, L., Ghidoni, S., Brahnam, S., Liu, S., Zhang, L.: Ensemble of handcrafted and deep learned features for cervical cell classification. Deep Learners and Deep Learner Descriptors for Medical Applications, pp. 117–135 (2020)

23. Basak, H., Kundu, R., Chakraborty, S., Das, N.: Cervical cytology classification using pca and gwo enhanced deep features selection. SN Comput. Sci. **2**(5), 369 (2021)

SEAS-Net: Segment Exchange Augmentation for Semi-supervised Brain Tumor Segmentation

Jing Zhang and Wei Wu[✉]

Inner Mongolia University, 235 University West Road, Huhhot,
Inner Mongolia, China
32109194@mail.imu.edu.cn, cswuwei@imu.edu.cn

Abstract. Accurate segmentation of brain tumors is crucial for cancer diagnosis, treatment planning, and evaluation. However, semi-supervised brain tumor image segmentation methods often face the challenge of mismatched distribution between labeled and unlabeled data. This study proposes a Segment Exchange Augmentation for Semi-supervised Segmentation Network (SEAS-Net) based on a teacher-student model that exchanges labeled and unlabeled data segments. This approach fosters unlabeled data to grasp general semantics from labeled data, and the consistent learning for both types notably reduces the empirical distribution gap. The SEAS-Net consists of two stages: supervised pre-training and semi-supervised segmentation. In the pre-training stage, segment exchange strategies optimize labeled images. During the semi-supervised stage, the teacher network generates pseudo-labels for unlabeled images. Subsequently, labeled and unlabeled images through segment exchange and input to the student network for generating predictive segmentation templates. Pseudo-labels and real mixed supervised signals supervise these predictions. Additionally, the discriminator after the student network enhances the reliability of prediction results. This methodology excels on the BraTS2019 and BraTS2021 datasets, effectively mitigating data distribution disparities and substantially enhancing brain tumor segmentation accuracy.

Keywords: Semi-supervised learning · Segment exchange augmentation · Medical image processing

1 Introduction

Brain tumors are among the malignancies with exceedingly high incidence and mortality rates, accounting for 2.4% of the overall cancer incidence. Early diagnosis of brain tumors is instrumental in devising appropriate treatment plans, thereby enhancing patient survival prospects [22]. In this context, the accurate segmentation of brain tumors is paramount. However, this task is technically challenging due to the complexity of brain structures and the heterogeneity of MRI data [6].

While deep learning techniques such as FCN [14], U-Net [16], and 3D U-Net [7] have made substantial strides in medical image analysis, these advancements still need to fully address the crux of Acquiring large-scale, high-quality

S. Rudinac et al. (Eds.): MMM 2024, LNCS 14555, pp. 281–295, 2024.
https://doi.org/10.1007/978-3-031-53308-2_21

annotated data in the medical domain is time-consuming and labor-intensive. The need for such labeled data limits the applicability of these methods. Consequently, Supervised and semi-supervised medical image segmentation techniques are gaining increasing attention. Although these methods perform well in scenarios with limited labeled data, they still face the distributional mismatch between labeled and unlabeled data, further impacting the model's generalization capabilities.

This study designed a Segment Exchange Augmentation for Semi-supervised Segmentation Network (SEAS-Net) to address the distributional imbalance between labeled and unlabeled data. The network employs a teacher-student architecture and incorporates Generative Adversarial Training. A key innovation is the implementation of the SEA strategy. During the pre-training phase, the SEA technique optimizes the labeled images. In the semi-supervised segmentation phase, segment swaps between labeled and unlabeled images to produce mixed images into the student network for segmentation prediction. Both pseudo-labels and actual segmentation templates, enhanced by SEA, serve as supervisory signals for guiding the predictive templates. SEA facilitates comprehensive learning for the unlabeled data from the labeled data and achieves distribution alignment. Moreover, an efficient discriminator is integrated into the system to elevate the prediction accuracy.

Experiments were conducted on the BraTS2019 and BraTS2021 datasets. Results show that the synergistic application of the SEA strategy and the discriminator not only resolves the issue of data distribution misalignment but also significantly enhances the accuracy of brain tumor segmentation. Ablation studies further substantiate the efficacy of each component in the model.

2   Related Work

Semi-supervised medical image segmentation uses contrastive learning [20], consistency regularization [5,13], and self-training [2,17]. Additionally, other works have expanded the MeanTeacher framework in various ways. SASSNet [13] applies shape constraints, and DTC [15] uses dual-task consistency. However, these methods need more effective knowledge transfer between labeled and unlabeled data.

The GANs framework improves accuracy of medical image segmentation through adversarial learning, especially in complex cancer imaging scenarios [1]. Discriminators evaluate mask-image similarity, aiding in handling diverse cancer datasets [12]. The models also tackle specific challenges like distinguishing tumors and identifying post-treatment responses [4]. However, data augmentation's limited variability can affect accuracy, and training GANs in data-scarce fields like cancer imaging is challenging.

Segment exchange is a prevalent data augmentation method in tasks like instance and semantic segmentation and object detection [8,11,19]. GuidedMix-Net [19] uses mixup for better pseudo-labels, while InstaBoost [10] and Context Copy-Paste [8] focus on contextual foreground placement. Prior approaches lacked a unified strategy, causing distribution issues [9]. They either copied unlabeled data or used labeled crops, ignoring the need for a consistent learning strategy and leading to distribution gaps.

Our Segment Exchange Augmentation for Semi-supervised Segmentation Network (SEAS-Net) integrates the teacher-student model and generative adversarial training, facilitating knowledge transfer between labeled and unlabeled data. Through the Segment Exchange Augmentation(SEA) strategy, we have reduced the dependency on a large amount of labeled data and enriched sample diversity. Additionally, we designed a consistent learning strategy and an efficient discriminator to mitigate data distribution disparities, thereby enhancing the model's generalizability and the accuracy of brain tumor segmentation.

3   Method

We define the volume of 3D medical images as $\mathbf{X} \in \mathbb{R}^{W \times H \times L}$. In the context of semi-supervised medical image segmentation, the objective is to accurately identify the background and target locations in $\mathbf{X}$, by predicting labels for each pixel $\hat{\mathbf{Y}} \in \{0, 1, ..., K-1\}^{W \times H \times L}$.

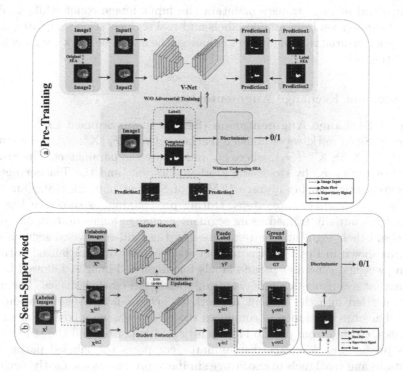

Fig. 1. Overview of the Segment Exchange Augmentation for Semi-supervised Segmentation Network (SEAS-Net). (a) pre-training: optimization of labeled images.(b) semi-supervised segmentation: generation of hybrid images by fragment exchange and supervision using pseudo-labels generated by the teacher network. A discriminator is integrated to optimize the model.

K represents different class labels used to describe various structures or tissues within medical images. Our training dataset $\mathcal{D}$, consists of N labeled images and M unlabeled images $(N \ll M)$, which can be represented as two subsets: $\mathcal{D} = \mathcal{D}^l + \mathcal{D}^u$, where $\mathcal{D}^l = \left\{ \left(\mathbf{X}_i^l, \mathbf{Y}_i^l \right) \right\}_{i=1}^N$ and $\mathcal{D}^u = \left\{ \left(\mathbf{X}_i^u \right) \right\}_{i=N+1}^{M+N}$.

We employ a generative adversarial Teacher-Student model for semi-supervised brain tumor segmentation. Specifically, an unlabeled image $\mathbf{X}^u$ and a labeled image $\mathbf{X}^l$ are randomly selected initially, and $\mathbf{X}^u$ is fed into the teacher model to generate pseudo-labels $\mathbf{Y}^p$. We then perform segment exchange between $\mathbf{X}^u$ and $\mathbf{X}^l$ to create mixed images $\mathbf{X}^{in1}$ and $\mathbf{X}^{in2}$, where $\mathbf{X}^u$ and $\mathbf{X}^l$ are dominant, respectively. These mixed images are then input into the student network to generate predicted segmentation templates $\mathbf{Y}^{in1}$ and $\mathbf{Y}^{in2}$. Using the same segment exchange principle labels $\mathbf{Y}^{out1}$ and $\mathbf{Y}^{out2}$ are generated by swapping segments of the teacher model's pseudo-label $\mathbf{Y}^p$ for $\mathbf{X}^u$ and the real label $\mathbf{Y}^l$ for $\mathbf{X}^l$. The student network also functions as the generator in a Generative Adversarial Network (GAN). By continuously applying segment exchange, the labeled portions in the students' predicted segmentation become more consistent with their true labels. These predictions and true labels are then input into a discriminator for feature extraction. This approach encourages the student network's predictions to align with actual segmentation templates closely. Our segment exchange and mixing strategy maintain the input image count while facilitating joint supervision from pseudo-labels and real segmentation templates. Moreover, the adversarial training further enhances the neural network's segmentation capabilities.

3.1   Segment Exchange Augmentation

Segment Exchange Augmentation (SEA). In our Segment Exchange Augmentation(SEA) work, we establish a teacher network $\mathcal{F}_t \left(\mathbf{X}^u; \Theta_t \right)$ and a student network $\mathcal{F}_s \left(\mathbf{X}^{in1}; \mathbf{X}^{in2}; \Theta_s \right)$, where Θ_s and Θ_t are the parameters. The Student network is optimized by stochastic gradient descent, and the Teacher network is by exponential moving average (EMA) of Student network [18]. Our training strategy consists of three steps. During pre-training, as shown in Fig. 1(a), the model is initially trained using segment exchange augmentation on labeled images and their segmentation templates. In the semi-supervised segmentation phase, as shown in Fig. 1(b), the teacher-student network, initialized with pre-training weights, generates pseudo-labels for unlabeled images, followed by segment exchange augmentation between unlabeled and labeled images. The mixed images are fed into the student network to get predicted segmentation templates. The objective function computes the error between mixed segmentation templates, created from pseudo-labels and real templates, and the student network's predictions. Also, the discriminator compares the labeled parts of the predictions and real labels to encourage similar representations. Lastly, each iteration involves optimizing the student network parameters, Θ_s, using stochastic gradient descent, followed by updating the teacher network parameters, Θ_t, are updated using the exponential moving average (EMA) of the student parameters.

Pre-training of the Backbone Network and Discriminator. Inspired by previous related work [3], we employ fragment exchange augmentation to train labeled data to develop supervised models. The supervised model generates pseudo-labels for unlabeled data during semi-supervised segmentation. This strategy is effective in improving segmentation performance. In order to avoid the initialization of the generator parameters in the subsequent stages from being too powerful and causing the discriminator to fail to function properly, we use the real labels and the model-predicted segmentation templates to train the discriminator to be able to initially distinguish between the real segmentation labels and the predicted segmentation templates. Note that there is no process of generating confrontation between the segmentation model and the discriminator; they are independent. More detailed information will be described in a subsequent ablation study.

Inter-image Segment Exchange Augmentation. To facilitate segment exchange between images, we begin by creating a one-centered mask, denoted as $\mathcal{M} \in \{0,1\}^{W \times H \times L}$, which differentiates voxels originating from the original image (1) and those from other images (0). The size of the one-value region is $\beta H \times \beta W \times \beta L$, where $\beta \in (0,1)$. Subsequently, we employ the technique of segment exchange augmentation to process both labeled and unlabeled images as follows:

$$\mathbf{X}^{in1} = \mathbf{X}^l \odot \mathcal{M} + \mathbf{X}^u \odot (1 - \mathcal{M}) \tag{1}$$

$$\mathbf{X}^{in2} = \mathbf{X}^u \odot \mathcal{M} + \mathbf{X}^l \odot (1 - \mathcal{M}) \tag{2}$$

where $\mathbf{X}^l \in \mathcal{D}^l$, $\mathbf{X}^u \in \mathcal{D}^u$, $1 \in \{1\}^{W \times H \times L}$, and $\odot$ means element-wise multiplication. It is important to note that incorporating random segment exchanges between labeled and unlabeled images enhances the diversity of the input data.

Supervision Signals Segment Exchange Augmentation. To train the student network, we utilize the SEA approach to generate supervision signals. Specifically, we feed the unlabeled image $\mathbf{X}^u$ into the teacher network and calculate the probability map:

$$\mathbf{P}^u = \mathcal{F}_t (\mathbf{X}^u; \Theta_t) \tag{3}$$

In binary segmentation, initial pseudo-labels $\mathbf{Y}^u$ is set using a 0.5 threshold on $\mathbf{P}^u$. For multi-class tasks, initial pseudo-label $\mathbf{Y}^u$ is determined via argmax on $\mathbf{P}^u$. The final $\mathbf{Y}^p$ is obtained by selecting its largest connected component, removing outlier voxels. We recommend using segment exchange between $\mathbf{Y}^p$ and ground-truth labels following the same methodology outlined in Eq. 1 and Eq. 2 to obtain the supervisory signal.

$$\mathbf{Y}^{out1} = \mathbf{Y}^t \odot \mathcal{M} + \mathbf{Y}^p \odot (1 - \mathcal{M}) \tag{4}$$

$$\mathbf{Y}^{out2} = \mathbf{Y}^p \odot \mathcal{M} + \mathbf{Y}^t \odot (1 - \mathcal{M}) \tag{5}$$

The student network's predictions are supervised by $\mathbf{Y}^{out1}$ and $\mathbf{Y}^{out2}$, guiding $\mathbf{X}^{in1}$ and $\mathbf{X}^{in2}$.

Generative Adversarial Training. In addition to evaluating the error between the predicted segmentation template and the corresponding label, we use adversarial training with a pre-trained discriminator (discussed in Section Pre-training of the Backbone Network and Discriminator) to improve segmentation performance. The generator G (student network) and the discriminator D are involved in a min-max optimization problem, which can be formulated as follows:

$$\min_{\theta_G} \max_{\theta_D} \mathcal{L}\left(\theta_G, \theta_D\right) = \mathbb{E}_{x \sim P_{data}}\left[\log D\left(x\right)\right] + \mathbb{E}_{z \sim P_z} \log\left[\left(1 - D\left(G\left(z\right)\right)\right)\right] \quad (6)$$

In this objective function, x represents the actual image, z is the segmentation prediction template $\mathbf{Y}^l$ of the labeled image $\mathbf{X}^l$ formed by $\mathbf{Y}^{in1}$ and $\mathbf{Y}^{in2}$ again after segment exchange enhancement. Both are fed into the discriminator to distinguish whether the input is a segmentation template predicted by the generator or an actual label, classifying them as 0 and 1, respectively. During the early stages of discriminator training, the aim is to have the output for actual labels approximate 1 and the output for generator-predicted results approximate 0. This endows the discriminator with a preliminary ability to discern between real and fake. When training the generator, the objective is to have its predicted output, when passed through the discriminator, approximate 1. This not only interferes with the discriminator's judgment, embodying the concept of adversarial training but also encourages the generator to produce segmentation templates similar to actual labels. As the iterative training of the discriminator and generator progresses, the generator increasingly maps the mixed input images to actual labels, reducing the error. This deceives the discriminator into making incorrect judgments, thereby increasing its error over time.

3.2 Loss Function

Each input image of the student network is composed of components from both labeled and unlabeled images. Intuitively, the ground truth mask of labeled images is usually more accurate than the pseudo-labels of unlabeled images. To control the contribution of unlabeled image pixels in the loss function, we introduce a parameter α. The loss functions for $\mathbf{X}^{in1}$ and $\mathbf{X}^{in2}$ can be respectively expressed as:

$$\mathcal{L}^{in1} = \mathcal{L}_{seg}(\mathbf{Y}^{out1}, \mathbf{Y}^{in1}) \odot \mathcal{M} + \alpha \mathcal{L}_{seg}(\mathbf{Y}^{out1}, \mathbf{Y}^{in1}) \odot (1 - \mathcal{M}) \quad (7)$$

$$\mathcal{L}^{in2} = \mathcal{L}_{seg}(\mathbf{Y}^{out2}, \mathbf{Y}^{in2}) \odot (1 - \mathcal{M}) + \alpha \mathcal{L}_{seg}(\mathbf{Y}^{out1}, \mathbf{Y}^{in1}) \odot \mathcal{M} \quad (8)$$

The combination of Dice loss and Cross-entropy loss is represented by $\mathcal{L}_{seg}$. The calculation of $\mathbf{Y}^{out1}$ and $\mathbf{Y}^{out2}$ is given by the following equations:

$$\mathbf{Y}^{out1} = \mathcal{F}_x\left(\mathbf{X}^{in1}; \Theta_s\right), \mathbf{Y}^{out2} = \mathcal{F}_x\left(\mathbf{X}^{in2}; \Theta_s\right) \quad (9)$$

The loss function of the student network includes an adversarial loss with the discriminator, represented as:

$$\mathcal{L}_{adv} = \mathcal{L}_{mse}\left(D\left(\mathbf{Y}^l\right) - 1\right) \quad (10)$$

$\mathcal{L}_{mse}$ denotes the widely used root mean square error, $\mathbf{Y}^l$ represents the completed segmentation prediction template of $\mathbf{X}^l$ obtained through the segment exchange augmentation on $\mathbf{Y}^{in1}$ and $\mathbf{Y}^{in2}$. The parameters Θ_s in the student network are updated at each iteration using stochastic gradient descent with the loss function:

$$\mathcal{L}_{all} = \mathcal{L}^{in1} + \mathcal{L}^{in2} + \mathcal{L}_{adv} \tag{11}$$

Then, the parameters $\Theta_t^{(k+1)}$ of the teacher network are updated for the $(k+1)$th iteration:

$$\Theta_t^{(k+1)} = \lambda\Theta_t^{(k)} + (1 - \lambda)\Theta_s^{(k)} \tag{12}$$

λ serves as the smoothing coefficient parameter.

3.3 Testing Phase

During testing, for the test image $\mathbf{X}_{test}$, we utilize the trained parameters $\hat{\Theta}_s$ of the student network to compute the probability map $\mathbf{Q}_{test}$ using the equation $\mathbf{Q}_{test} = \mathcal{F}\left(\mathbf{X}_{test}; \hat{\Theta}_s\right)$. Subsequently, we derive the final label map based on the obtained probability map $\mathbf{Q}_{test}$.

4 Experiments

4.1 Experimental Setting

Dataset. We introduce two multimodal brain tumor segmentation challenge datasets for evaluation, namely the BraTS2019 dataset and the BraTS2021 dataset, which are popular datasets in the field of tumor region segmentation. BraTS2019 includes 335 cases, and BraTS2021 includes 2040 cases, and both consist of MRI images in four modalities, namely T1, T2, Flair, and T1CE. To validate the effectiveness of our proposed algorithm, we focused on segmenting the complete tumor section using the Flair mode, which includes 97 cases in BraTS2019 and 110 cases in BraTS2021, respectively. The data were divided into fixed sets of 70, 9, and 18 patients versus 80, 10, and 20 patients for training, validation, and testing, respectively. We use four metrics to evaluate the segmentation performance of the proposed model: Dice score, Jaccard score, 95% Hausdorff distance (95HD), and Average Surface Distance (ASD) in voxels.

Implementation Details. After extensive tuning, we followed the following settings in our experiments: $\alpha = 0.5$ and $\beta = 2/3$. All experiments used a fixed random seed on an NVIDIA RTX 4600 GPU. Data augmentation with rotations and flips was employed to enhance the model's generalization ability. The proposed model uses the SGD optimizer with an initial learning rate of 0.01, decaying by 10% every 2.5K iterations. The 3D VNet was used as the backbone network, as shown in Fig. 2, and the discriminator consists of 5 layers (Conv3d-BatchNorm3d-LeakyReLU), as shown in Fig. 3, where the kernel size

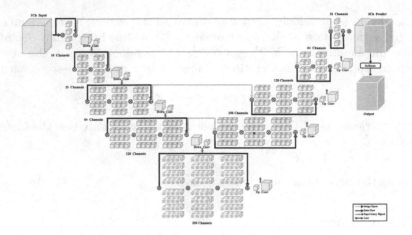

Fig. 2. V-net Network

is 4, the stride is 2, and padding is 1. Subsequently, a fully connected layer transforms the intermediate features into a scalar value ranging between 0 and 1. During training, to comply with the downsampling rules of the 3D VNet, the first and last slices of the input volume were discarded, resulting in a fixed size of 240×240×114. The size of the region of interest with a value of 1 was set to 160×160×76 ($\beta = 2/3$). The batch size was set to 8, consisting of 4 labeled images and 4 unlabeled images. The pre-training and self-training iterations were set to 2k and 15k, respectively.

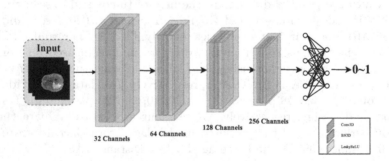

Fig. 3. Discriminator Network.

4.2   Comparison with Sate-of-the-Art Methods

Our SEAS-Net outperformed state-of-the-art methods SASSNet [13], DTC [15], SS-Net [20], and UA-MT [21] on different scales of the BraTS2019 and BraTS2021 brain tumor dataset. As shown in Table 1, our method achieved the best performance with an average Dice score of 86.96%, surpassing SASSNet [13] by 1.55%. Although our average 95HD metric for tumor segmentation was 5.6218 (slightly lower than the top result of 5.4132), our network demonstrated superior segmentation performance. The segment-exchanging augmentation strategy enabled the network to capture boundary regions and voxel variations, yielding excellent shape-related performance without explicit constraints. Our method successfully captured fine details as shown in Fig. 4.

Table 1. Comparison with state-of-the-art semi-supervised segmentation methods on different scales of the BraTS2019 and BraTS2021 datasets, *highlighted* improvements over the second-best result.

Method	Brats2019						Brats2021					
	Scans used		Metrics				Scans used		Metrics			
	Labeled	Unlabeled	Dice↑	Jaccard↑	95HD↓	ASD↓	Labeled	Unlabeled	Dice↑	Jaccard↑	95HD↓	ASD↓
SASSNet [13]	3(5%)	67(95%)	82.49	68.83	6.57	2.74	4(5%)	76(95%)	83.38	70.25	6.48	2.49
UA-MT [21]			78.24	67.04	7.91	3.48			79.92	68.72	7.34	3.81
DTC [15]			80.06	67.95	7.23	3.15			81.17	70.21	6.72	2.64
SS-Net [20]			79.32	67.57	7.72	3.29			80.53	69.40	6.81	3.35
SEAS-Net			**83.77**	**71.32**	**6.14**	**2.36**			**84.75**	**73.54**	**6.02**	**2.17**
SASSNet [13]	7(10%)	63(90%)	83.67	70.63	5.84	2.43	8(10%)	72(90%)	85.41	74.53	**5.41**	2.55
UA-MT [21]			81.65	69.15	6.79	3.26			82.37	70.54	7.05	3.18
DTC [15]			82.43	70.24	6.03	2.81			83.22	73.46	6.97	2.39
SS-Net [20]			82.58	69.87	6.26	3.05			82.79	71.72	6.51	2.60
SEAS-Net			**85.16**	**71.80**	**5.61**	**2.20**			**86.96**	**75.86**	5.62	**2.06**

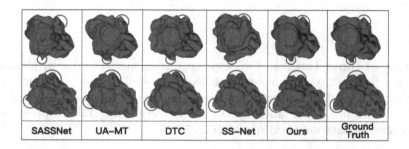

Fig. 4. Visualization of several semi-supervised segmentation methods using 10% labeled data and ground truth on the BraTS2021 dataset (best viewed by zooming in on the screen). Our method can accurately segment fine details, particularly edges that are prone to misidentification or omission, *highlighted* with blue circles. (Color figure online)

4.3   Ablation Studies

Segment Exchange Augmentation Direction. We devised three experiments to investigate the impact of different segment exchange orientations outlined in Table. 2 presents the impact on segmentation performance. Exchange inwards and outwards (referred to as In and Out in the Table) implies the use of either $\mathbf{X}^l \odot \mathcal{M} + \mathbf{X}^u \odot (1 - \mathcal{M})$ or $\mathbf{X}^u \odot \mathcal{M} + \mathbf{X}^l \odot (1 - \mathcal{M})$ for training the network. We also conducted within-set segment exchange (SE in the Table), segment exchanging labeled data on another labeled data and segment exchanging unlabeled data on other unlabeled data.

Compared to our approach, all these variations exhibit inferior performance due to a lack of uniform treatment of labeled and unlabeled data during training or a deficiency in general semantic transfer between labeled and unlabeled data.

Table 2. Ablation study of the segment exchange directions. In: inward segment exchange (foreground: unlabeled, background: labeled). Out: outward segment exchange (foreground: labeled, background: unlabeled). SE: direct segment exchange (background & foreground: labeled & labeled and unlabeled & unlabeled).

Method	Brats2019						Brats2021					
	Scans used		Metrics				Scans used		Metrics			
	Labeled	Unlabeled	Dice↑	Jaccard↑	95HD↓	ASD↓	Labeled	Unlabeled	Dice↑	Jaccard↑	95HD↓	ASD↓
In	3(5%)	67(95%)	83.10	70.27	6.99	2.78	4(5%)	76(95%)	83.42	72.62	7.57	2.35
Out			83.07	70.23	7.32	2.37			83.36	72.60	8.31	2.21
SE			76.32	61.45	11.58	3.51			79.10	71.89	13.46	3.84
Ours			**83.77**	**71.32**	**6.14**	**2.36**			**84.75**	**73.54**	**6.02**	**2.17**
In	7(10%)	63(90%)	84.55	70.87	6.88	2.42	8(10%)	72(90%)	84.12	73.44	7.26	2.27
Out			83.91	69.15	7.73	4.06			83.85	72.93	9.42	3.96
SE			82.74	68.02	7.24	3.81			83.07	73.15	7.31	3.18
Ours			**85.16**	**71.80**	**5.61**	**2.20**			**86.96**	**75.86**	**5.92**	**2.06**

Quantity of Input Images. Two segment mixing methods: one with changed input image quantity and merged labeled/unlabeled segments, reducing the image count by half. The other maintained the original input image count with separate mixed images for labeled and unlabeled segments. Table 3 presents the impact on segmentation performance (Fig. 5).

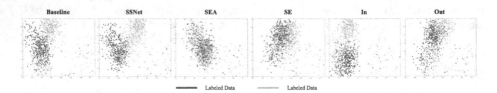

Fig. 5. Kernel dense estimations of different methods, trained on the 10% labeled BraTS2021 dataset. Baseline: training utilized only labeled data. SE, In, and Out as defined in Table 2. Our SEA offers better feature alignment and highlights the necessity of segment exchanging.

Table 3. Revealing decreased performance when reducing input image quantity. This suggests a preference for larger batch sizes in neural network iterations when the sample size is limited.

Method	Scans used		Metrics			
	Labeled	Unlabeled	Dice↑	Jaccard↑	95HD↓	ASD↓
Mixup	4(5%)	76(95%)	75.28	63.81	21.52	6.41
Ours			**84.75**	**73.54**	**6.02**	**2.17**
Mixup	8(10%)	72(90%)	82.55	68.65	10.33	3.61
Ours			**86.96**	**75.86**	**5.92**	**2.06**

The Volume of Segment Exchange. We examined the impact of segment volume on the semi-supervised segmentation performance. Setting the optimal ratio at $\beta = 2/3$, we conducted comparative experiments with varying ratios of $\beta = \{\frac{2}{5}, \frac{1}{2}, \frac{2}{3}, \frac{3}{4}, \frac{5}{6}\}$. Table 4 showed a positive correlation between increasing β and improved metrics, but a decline in performance beyond $\beta = 2/3$. Extremely small or large segment volumes compromised the effectiveness of segment exchanging augmentation, reducing the transfer of semantic information between labeled and unlabeled data.

Role of Discriminator. We compared conventional training with adversarial training on semi-supervised segmentation performance. Table 5 showed that adversarial training outperformed conventional training across all metrics. Including a discriminator encouraged the generator network to generate more accurate representations like real segmentation masks.

Table 4. Ablation study of β on the Brats2021.

β	Scans used		Metrics			
	Labeled	Unlabeled	Dice↑	Jaccard↑	95HD↓	ASD↓
2/5	4(5%)	76(95%)	82.99	70.93	6.14	2.23
1/2			83.68	72.20	6.32	2.35
2/3			**84.75**	**73.54**	**6.02**	**2.17**
3/4			84.16	72.65	6.18	2.28
5/6			84.39	72.98	6.44	2.31
2/5	8(10%)	72(90%)	84.40	73.02	6.32	2.15
1/2			84.09	72.54	6.00	2.35
2/3			**86.96**	**75.86**	**5.92**	**2.06**
3/4			85.12	73.74	6.19	2.46
5/6			85.31	74.02	6.87	2.96

Table 5. Ablation study on generative adversarial training on the Brats2021.

Mode	Scans used		Metrics			
	Labeled	Unlabeled	Dice↑	Jaccard↑	95HD↓	ASD↓
w/o Discriminator	4(5%)	76(95%)	82.14	70.48	8.72	2.58
Ours			**84.61**	**73.54**	**6.02**	**2.17**
w/o Discriminator	8(10%)	72(90%)	83.87	71.54	7.46	2.29
Ours			**86.96**	**75.86**	**5.92**	**2.06**

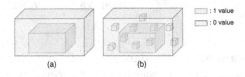

Fig. 6. Different segment strategies. **(a)**one-centered mask;**(b)**random mask.

Shapes of Segment Exchange. We analyze how shape distribution in segment exchanging affects semi-supervised segmentation models. To ensure a fair comparison, we maintained equal volumes for all segments, with the same number of voxels set to 1. The difference lies in the presence or absence of zero-value cavities within the segments, as shown in Fig. 6. Segments with equal volumes and voxel values set to 1 compared based on the presence of zero-filled holes. Randomly selecting nine cubes of size $\beta H \times \beta W \times \beta L$ from all-ones segments and moving them outside creates nine zero-filled holes. Table 6 shows the effects of the two training methods on the four categories of indicators.

Weight in Loss Function. Based on their reliability, we assign higher weight coefficients to predicted results with real labels and lower weight coefficients to predicted results with pseudo labels. The default value is $\alpha = 0.5$. Varying $\alpha = \{0.5, 1.5, 2.5\}$, we observe the impact on four evaluation metrics in Table 7. Results show that performance remains stable for α values from 0.5 to 1.5 but

Table 6. Ablation study on shape distribution in segment exchange on the Brats2021.

Mode	Scans used		Metrics			
	Labeled	Unlabeled	Dice↑	Jaccard↑	95HD↓	ASD↓
Random	4(5%)	76(95%)	82.78	70.85	8.51	2.64
Ours			**84.75**	**73.54**	**6.02**	**2.17**
Random	8(10%)	72(90%)	82.84	71.73	8.64	2.51
Ours			**86.96**	**75.86**	**5.92**	**2.06**

Table 7. Ablation study on the weight α in the loss function on the BraTS2019 and BraTS2021.

α	Brats2019						Brats2021					
	Scans used		Metrics				Scans used		Metrics			
	Labeled	Unlabeled	Dice↑	Jaccard↑	95HD↓	ASD↓	Labeled	Unlabeled	Dice↑	Jaccard↑	95HD↓	ASD↓
0.5	3(5%)	67(95%)	**83.77**	**71.32**	**6.14**	**2.36**	4(5%)	76(95%)	**84.75**	**73.54**	**6.02**	**2.17**
1.5			82.56	70.85	7.03	2.59			82.92	72.31	6.79	2.44
2.5			81.94	70.48	8.25	3.14			82.07	71.28	7.94	2.79
0.5	7(10%)	63(90%)	**85.16**	**71.80**	**5.61**	**2.20**	8(10%)	72(90%)	**86.96**	**75.86**	**5.92**	**2.06**
1.5			83.28	71.16	6.82	2.47			86.61	75.32	6.57	2.39
2.5			82.47	70.73	7.56	2.98			85.95	74.22	6.94	2.51

decreases significantly at $\alpha = 2.5$. A weight coefficient of 0.001 for the adversarial loss to avoid model instability.

5 Conclusion

This study proposes a Segment Exchange Augmentation for a Semi-supervised Segmentation Network (SEAS-Net). We successfully employ the teacher-student model to utilize limited labeled data for standard segmentation. By blending labeled and unlabeled images and training the student network with pseudo-labels and mixed supervision signals, we enhance the reliability of prediction results by including a discriminator. However, limitations include focusing on single-modal images, necessitating future exploration of multi-modal applications, and the potential for improving backbone network selection. The strategy will also apply to tumor segmentation in other organs or anatomical regions.

References

1. Alshehhi, R., Alshehhi, A.: Quantification of uncertainty in brain tumor segmentation using generative network and bayesian active learning. In: VISIGRAPP (4: VISAPP), pp. 701–709 (2021). https://doi.org/10.5220/0010341007010709
2. Bai, W., Oktay, O., Sinclair, M., Suzuki, H., Rajchl, M., Tarroni, G., Glocker, B., King, A., Matthews, P.M., Rueckert, D.: Semi-supervised learning for network-based cardiac MR image segmentation. In: Descoteaux, M., Maier-Hein, L., Franz, A., Jannin, P., Collins, D.L., Duchesne, S. (eds.) MICCAI 2017. LNCS, vol. 10434, pp. 253–260. Springer, Cham (2017). https://doi.org/10.1007/978-3-319-66185-8_29
3. Bai, Y., Chen, D., Li, Q., Shen, W., Wang, Y.: Bidirectional copy-paste for semi-supervised medical image segmentation. In: Proceedings of the IEEE/CVF Conference on Computer Vision and Pattern Recognition, pp. 11514–11524 (2023). https://doi.org/10.1109/cvpr52729.2023.01108
4. Bi, W.L., Hosny, A., Schabath, M.B., Giger, M.L., Birkbak, N.J., Mehrtash, A., Allison, I.F., others: Artificial intelligence in cancer imaging: clinical challenges and applications 69(2), 127–157 (2019). https://doi.org/10.3322/caac.21552, publisher: Wiley Online Library

5. Bortsova, G., Dubost, F., Hogeweg, L., Katramados, I., de Bruijne, M.: Semi-supervised medical image segmentation via learning consistency under transformations. In: Shen, D., Liu, T., Peters, T.M., Staib, L.H., Essert, C., Zhou, S., Yap, P.-T., Khan, A. (eds.) MICCAI 2019. LNCS, vol. 11769, pp. 810–818. Springer, Cham (2019). https://doi.org/10.1007/978-3-030-32226-7_90

6. Bray, F., Ferlay, J., Soerjomataram, I., Siegel: Global cancer statistics 2018: GLOBOCAN estimates of incidence and mortality worldwide for 36 cancers in 185 countries **68**(6), 394–424 (2018). https://doi.org/10.3322/caac.21492, publisher: Wiley Online Library

7. Çiçek, Ö., Abdulkadir, A., Lienkamp, S.S., Brox, T., Ronneberger, O.: 3D U-Net: learning dense volumetric segmentation from sparse annotation. In: Ourselin, S., Joskowicz, L., Sabuncu, M.R., Unal, G., Wells, W. (eds.) MICCAI 2016. LNCS, vol. 9901, pp. 424–432. Springer, Cham (2016). https://doi.org/10.1007/978-3-319-46723-8_49

8. Dvornik, N., Mairal, J., Schmid, C.: Modeling visual context is key to augmenting object detection datasets. In: ECCV, pp. 364–380 (2018). https://doi.org/10.1007/978-3-030-01258-8_23

9. Fan, J., Gao, B., Jin, H., Jiang, L.: Ucc: Uncertainty guided cross-head co-training for semi-supervised semantic segmentation (2023). https://doi.org/10.48550/arXiv.2205.10334

10. Fang, H.S., Sun, J., Wang, R., Gou, M., Li, Y.L., Lu, C.: Instaboost: boosting instance segmentation via probability map guided copy-pasting. In: Proceedings of the IEEE/CVF International Conference on Computer Vision, pp. 682–691 (2019). https://doi.org/10.1109/iccv.2019.00077

11. Ghiasi, G., Cui, Y., Srinivas, A., Qian, R., Lin, T.Y., Cubuk, E.D., Le: Simple copy-paste is a strong data augmentation method for instance segmentation. In: Proceedings of the IEEE/CVF Conference on Computer Vision and Pattern Recognition, pp. 2918–2928 (2021). https://doi.org/10.1109/cvpr46437.2021.00294

12. Kohl, S., et al.: Adversarial networks for the detection of aggressive prostate cancer. arXiv preprint arXiv:1702.08014 (2017). https://doi.org/10.48550/arXiv.1702.08014

13. Li, S., Zhang, C., He, X.: Shape-aware semi-supervised 3d semantic segmentation for medical images. In: Martel, A.L., Abolmaesumi, P., Stoyanov, D., Mateus, D., Zuluaga, M.A., Zhou, S.K., Racoceanu, D., Joskowicz, L. (eds.) MICCAI 2020. LNCS, vol. 12261, pp. 552–561. Springer, Cham (2020). https://doi.org/10.1007/978-3-030-59710-8_54

14. Long, J., Shelhamer, E., Darrell, T.: Fully convolutional networks for semantic segmentation. In: Proceedings of the IEEE Conference on Computer Vision and Pattern Recognition, pp. 3431–3440 (2015). https://doi.org/10.1109/cvpr.2015.7298965

15. Luo, X., Chen, J., Song, T., Wang, G.: Semi-supervised medical image segmentation through dual-task consistency. In: Proceedings of the AAAI Conference on Artificial Intelligence, vol. 35, pp. 8801–8809 (2021). https://doi.org/10.1609/aaai.v35i10.17066

16. Ronneberger, O.: Invited talk: u-net convolutional networks for biomedical image segmentation. In: Bildverarbeitung für die Medizin 2017. I, pp. 3–3. Springer, Heidelberg (2017). https://doi.org/10.1007/978-3-662-54345-0_3

17. Shi, Y., Zhang, J., Ling, T., Lu, J., Zheng, Y., Yu, Q., Qi, L., Gao, Y.: Inconsistency-aware uncertainty estimation for semi-supervised medical image segmentation **41**(3), 608–620 (2021). https://doi.org/10.1109/tmi.2021.3117888, publisher: IEEE

18. Tarvainen, A., Valpola, H.: Mean teachers are better role models: Weight-averaged consistency targets improve semi-supervised deep learning results. Advances in neural information processing systems 30 (2017). https://doi.org/10.5555/3294771. 3294885

19. Tu, P., Huang, Y., Zheng, F., He, Z., Cao, L., Shao, L.: GuidedMix-net: Semi-supervised semantic segmentation by using labeled images as reference. In: Proceedings of the AAAI Conference on Artificial Intelligence, vol. 36, pp. 2379–2387 (2022). https://doi.org/10.1609/aaai.v36i2.20137, issue: 2

20. Wu, Y., Wu, Z., Wu, Q., Ge, Z., Cai: exploring smoothness and class-separation for semi-supervised medical image segmentation. In: MICCAI 2022: 25th International Conference, Singapore, September 18–22, 2022, Proceedings, Part V, pp. 34–43. Springer (2022). https://doi.org/10.1007/978-3-031-16443-9_4

21. Yu, L., Wang, S., Li, X., Fu, C.-W., Heng, P.-A.: Uncertainty-aware self-ensembling model for semi-supervised 3D left atrium segmentation. In: Shen, D., Liu, T., Peters, T.M., Staib, L.H., Essert, C., Zhou, S., Yap, P.-T., Khan, A. (eds.) MICCAI 2019. LNCS, vol. 11765, pp. 605–613. Springer, Cham (2019). https://doi.org/10. 1007/978-3-030-32245-8_67

22. Zülch, K.J.: Brain tumors: their biology and pathology. Springer (2013). https:// doi.org/10.1007/978-3-642-68178-3

Super-Resolution-Assisted Feature Refined Extraction for Small Objects in Remote Sensing Images

Lihua Du[1], Wei Wu[1]([✉]), and Chen Li[2]

[1] Inner Mongolia University, Hohhot, China
32109051@mail.imu.edu.cn, cswuwei@imu.edu.cn
[2] Dalian Maritime University, Dalian, China
lichen217@dlmu.edu.cn

Abstract. Despite achieving impressive results in object detection in natural scenes, the task of object detection in remote sensing images is still full of challenges due to the large number of small objects in remote sensing images caused by the dense object distribution, complex backgrounds, and diverse scale variations. We propose a Super-Resolution-Assisted Feature Refined Extraction (SRRE) approach to address the difficulties of detecting small objects. Firstly, we employ a deeper level of feature fusion to effectively harness deep semantic information and shallow detailed information. Secondly, in the feature extraction process, a Feature Refined Extraction Module (FREM) is introduced to capture a wider range of contextual information, enhancing the global perceptual capability of features. Lastly, we introduce Super-Resolution (SR) branches at various feature layers to better integrate local textures and contextual information. We compared our method against commonly used approaches in remote sensing image object detection, including state-of-the-art (SOTA) methods. Our approach outperforms these methods and achieves superior results on the DOTA-v1.0, DIOR, and SODA-A datasets.

Keywords: Remote sensing · Object detection · Super resolution

1 Introduction

Remote sensing image object detection aims to acquire both the categorical and positional information about valuable objects within remote sensing images [22]. This field holds significant potential applications in urban management planning, natural disaster forecasting, resource management, military surveillance, and various other domains. Although object detection in natural images has achieved notable advancements, there are still many difficulties for remote sensing images. Remote sensing images are characterized by their large dimensions, scene diversity, varying imaging altitudes, and dense distribution of objects. Consequently, they encompass a considerable number of objects with extremely limited sizes. For instance, in Fig. 1, the airplane and the small vehicle appear at the same

S. Rudinac et al. (Eds.): MMM 2024, LNCS 14555, pp. 296–309, 2024.
https://doi.org/10.1007/978-3-031-53308-2_22

time. Owing to the size discrepancy, small vehicles comprise only a few dozen pixels. These size-constrained targets present numerous challenges for object detection in remote sensing imagery [4]. For such small object detection studies, in addition to some inherent problems beyond the generalized detection tasks, such as intra-class differences and inter-class similarities, data characterization of small targets in remotely sensed images makes detection more challenging. [2] [9]Existing optical remote sensing datasets tend to use area definitions, such as the SODA-A dataset which treats areas smaller than 1200 pixel instances as small targets.

Deep learning-based object detection methods often rely on backbone networks to capture high-dimensional features, reduce redundancy by decreasing resolution, and enhance channel dimensionality for effective feature representation. However, significant limitations arise for small objects, as shown in Fig. 1, where the vehicle occupies a small area and contains limited information. After the downsampling process in the backbone network, the spatial information will be lost in a large number, resulting in a significant reduction of the responding features. Meanwhile, the small vehicles in Fig. 1 are easily submerged in the complex background, so foreground-background confusion and missed detection are also major problem faced.

To address the above problems, the challenge of small object detection lies in how to maintain a robust feature representation while avoiding the detrimental effects caused by deep networks. To solve the problem of information loss during downsampling, multi-scale feature fusion is usually adopted. However, this method still loses some useful information, especially when the features at different scales have conflicts or inconsistencies. To resolve the conflict, the design of the fusion method needs to balance the retention and removal of information, which may lead to performance degradation. For targets with large-scale variations, because targets at different scales may have different shapes, textures, and appearances, it may not be possible to adequately capture their variation patterns. The detection field may cover a large area of the background, which can lead to an increase in background interference. As a result, multi-scale fea-

Fig. 1. Example of a smaller target with a larger size

ture fusion methods may not be able to effectively distinguish between target and background, which leads to an increase in the false detection rate.

In this paper, we propose an object detection network for small objects, referred to as the Super-Resolution-Assisted Feature Refined Extraction (SRRE) network. This network preserves the texture information of the low-level features, provides better access to contextual information, and better combines low-level features with high-level features. Specifically, we adopt a deeper level of feature fusion, which allows for more effective integration of low-level and high-level features. The Feature Refined Extraction Module(FREM) is introduced in the extraction process, which introduces an attention module for which we designed EBAM, allows better access to semantic information. Additionally, a Super-resolution auxiliary branch is introduced, utilizing an Encoder-Decoder structure to connect our layer 3 and layer 5 features. This module enables high-level features to capture the texture and other low-level details from the underlying features. Our contributions are mainly as follows:

- A Super-Resolution-Assisted Feature Refined Extraction (SRRE) network is proposed, which can obtain more underlying and high-level information about small objects and make them better balanced semantically and spatially.
- A deeper feature fusion approach is adopted and a Feature Refined Extraction Module(FREM) is designed to obtain more semantic and contextual information about small objects in the extracted feature layer.
- Super-Resolution-Assisted branching is introduced into the detector, which can distinguish the small objects from the vast and complex background, and the lost local texture information and high-level semantic information are fused.

2   Related Work

Thanks to the rapid development of general object detection tasks, small object detection for optical remote sensing images has seen several excellent works in recent years. These methods often use the detection algorithms that have made a big splash in the field of general-purpose target detection, such as Faster RCNN [16], RetinaNet [13], FPN [12], and YOLO [15], as the base model. The detection of small targets in remote sensing images is accomplished through targeted optimization of the input image processing or structural design. To address the issue of low detection accuracy of small objects in remote sensing images, numerous innovations have emerged. In recent years, with the continuous advancement of deep learning techniques, there has been a proliferation of methods aimed at enhancing the precision of small object detection within remote sensing imagery. These methods can be primarily categorized into approaches based on feature fusion, and feature super-resolution.

2.1   Feature Fusion-Based Methods

Inspired by the successful performance of Feature Pyramid Networks (FPN [12]) in natural image object detection tasks, it is introduced into the field of optical remote sensing image object detection to improve the accuracy of small object detection. Cao et al. [1] proposed the MARNet, which introduces a global attention mechanism to FPN. This mechanism uses high-level feature maps to obtain the attention of low-level features in the channel dimension and then fuses it with the low-level feature maps weighted by that attention to highlight the information in the small object region. Hong et al. [7] proposed the SSPNet, which employs a combination of a scale enhancement module and a scale selection module to improve the representation of small objects in the FPN. Wu et al. [20] proposed the FSANet, which incorporates a feature-sensitive alignment module to align features of differing resolutions before fusing adjacent layer feature maps. This approach has yielded notable accuracy improvements in small object detection tasks within optical remote sensing images. We adopt a deeper feature fusion approach and design a Feature Refined Extraction Module(FREM) to obtain more contextual and semantic information.

2.2 Super-Resolution-Based Methods

Focusing on the idea of designing a network to enhance small object regions through super-resolution for achieving detection capabilities comparable to those of larger and medium-sized objects, Zhang et al. [28] proposed Deconv-RCNN to improve the detection accuracy by recovering the loss of small target information brought about in the pooling layer through the deconvolution operation on the feature map output from the backbone network. Kaiming He et al. [11] proposed the EDSR structure, which represents a highly effective method for image super-resolution. This technique excels in transforming low-resolution images into high-resolution counterparts while preserving intricate image details. Courtrai et al. [3] performed resolution enhancement of the input image by adding GAN-based super-resolution branches to the detection network to get a clearer image for detection. We propose a super-resolution-assisted branching of the Encoder-decoder structure that can better integrate texture information and semantic information.

3 Proposed Methods

We propose a Super-Resolution-Assisted Feature Refined Extraction(SRRE) Network to enhance the performance of small object detection in remote sensing images. As illustrated in Fig. 2, in the feature extraction stage, we introduce a Feature Refined Extraction Module (FREM), which incorporates a new EBAM attention module.

As the high-level feature maps are continuously downsampled, the details of the object features are lost layer by layer. Low-level feature maps have clear contours with high resolution, contain relatively more primitive information, are closer to pixel-level image information, and are very helpful for detecting small-scale objects, but lack sufficient semantic information. To better use the high-level semantic information and the low-level detail information in the feature fusion stage, we propose a 4-layer pyramid structure in the Neck section. Bottom-up and top-down fusion of the feature map is performed to enhance the receptive field of small object features. Finally, the prediction is performed in the bottom three layers, which can effectively improve the accuracy of detecting small objects.

Finally, during the training phase, we introduce a Super-Resolution Auxiliary Branch (SR) that reconstructs high-level features. The details are as follows.

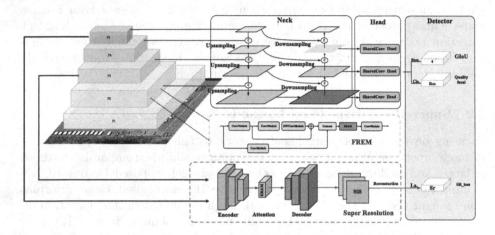

Fig. 2. Overall of our proposed method

3.1 Feature Refined Extraction Module

The purpose of the FREM is to obtain contextual information in a wider range and enhance the global perceptual capability of features. Because small objects are prone to receive interference from background noise and other salient objects, as shown in Fig. 3, our proposed module first splits the feature map into two branches. One branch is processed by a ConvModule(Convolution, Normalization, SiLu). The other branch is passed to deeper convolutional layers for feature extraction, including residual-connected ConvModule and DWConvModule(Depthwise Convolutions, Normalization, SiLu) with a large kernel, which increases the Receptive Field and helps to obtain high-level semantic information, and then the two branches are connected together in the channel dimension to form a new feature representation.

The feature of small objects is then specially enhanced by an EBAM attention module to suppress background noise interference. The structure of the module is illustrated in Fig. 4, the input feature map undergoes global average pooling (GAP), and an adaptive selection function is used to ascertain the appropriate convolutional kernel size. Feature extraction is carried out via a 1D convolution to enable cross-channel interaction, resulting in channel weights. The normalized weights are multiplied channel-wise with the original input feature map to generate the final feature map. The output feature map of the channel attention mechanism undergoes both Maxpooling and Avgpooling across the channel dimension at each feature point. After that, these two feature results are stacked, and a 7×7 convolutional kernel is employed to fuse channel information. Finally, the spatial weights of the feature map are normalized through the sigmoid function, and the input feature map is multiplied by the weights to yield the outcome. Through the Efficient Channel Attention Module(ECM), potent semantic information is harnessed, augmenting the discriminative capacity of the features. Moreover, the Spatial Attention Module(SAM) is instrumental in expanding the network's receptive field, facilitating the acquisition of a more extensive context, and enhancing the global perceptual capability of the features.

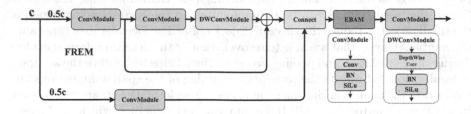

Fig. 3. The structure of FREM

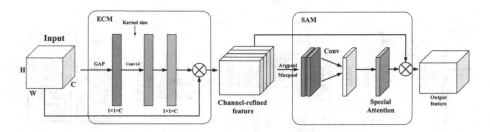

Fig. 4. The structure of EBAM

3.2   Super-Resolution Module

The size of the features retained in the backbone network for multiscale detection is much smaller than the size of the original input image. Existing methods [12] [16] [24] perform an upsampling operation to recover the feature size. However, the effectiveness of this method is limited due to the loss of information in the texture and patterns. To solve this problem. As shown in Fig. 5, we propose an auxiliary SR branch.

The SR branch exists only in the training phase and does not participate in the inference phase to reduce the inference speed. Specifically, the SR structure can be viewed as a simple Encoder-decoder model. We opt to fuse local textures and patterns, along with semantic information, using low-level and high-level features from the backbone network. As depicted in Fig. 2, we select the feature maps from the 3rd and 5th layers as low-level and high-level features, respectively. The Encoder combines the low-level features and high-level features generated by the backbone network.

As shown in Fig. 5, in the Encoder, the high-level features are first processed with CBR modules (Convolution, Normalization, and ReLU). For the high-level features, we use an upsampling operation to match the spatial dimensions of the low-level features, and then use a concatenation operation and two CBR modules to merge the low-level and high-level features. For the decoder, up-sampling is performed in the high-level features, where the output size of the SR module is twice as large as the input image. Then through the EBAM module, as shown in Fig. 4, the fused features can be better focused on the corresponding region of the small object to reduce the effect of object noise. The Decoder uses three anti-convolutional layers that are able to recover some of the detailed information lost during the convolution and pooling process, which helps to improve the accuracy of localization. SR directs the correlation learning of the spatial dimensions and passes it to the main branch, which improves the performance of target detection. In addition, we introduce EDSR [11] as our Encoder structure, which can be used in the generation process to utilize its learned feature extraction capabilities, resulting in high-quality super-resolution reconstruction. The whole SR structure can better combine local texture information and contextual information.

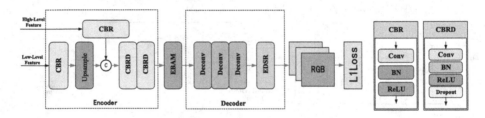

Fig. 5. Super Resolution Structure

3.3   Loss Function

The overall loss of our network comprises two components: the detection loss $L_{quality}$ [14] and the SR structure loss L_s, which can be represented as Eq. (1):

$$L_{total} = cL_s + (1 - c)L_{quality} \tag{1}$$

where c is a parameter set to 0.3, used to balance the training of the object detection and Super-Resolution tasks. The Super-Resolution reconstruction loss is employed to assess the network's performance in the super-resolution task. An L_1 loss [29] is adopted to compute the loss between the input image X and the Super-Resolution result S, represented as Eq. (2):

$$L_s = \| S - X \|_1 \tag{2}$$

where $\| S - X \|_1$ represents the L_1 norm, which is used to compute the difference between the two. The L_1 loss is widely used in image generation tasks because it produces sharper images while preserving image details. The $L_{quality}$ loss encompasses the initial three losses: the localization prior loss (L_{center}) [14], the sample regression loss (L_{reg}) [14], and the sample classification loss (L_{cls}) [14]. The details are shown in Eq. (3):

$$L_{quality} = \lambda_1 L_{cls} + \lambda_2 L_{reg} + \lambda_3 L_{center} \tag{3}$$

where $\lambda_1=1$, $\lambda_2=3$, and $\lambda_3=1$ are the weights assigned to these three types of losses.

4   Experiments

4.1   Datasets

SODA-A [2] dataset is the latest dataset specifically proposed for the small target challenge, in which the images have high resolution, with most of the images having a resolution greater than 4700×2700. 2513 high-resolution images of aerial scenes are included, of which 872,069 instances are annotated with oriented rectangular box annotations on 9 categories. The dataset has 9 categories: Airplane (PL), Helicopter (HL), Small-vehicle (SV), Largevehicle (LV), Ship (SH), Container (CN), Storage-tank (ST), Swimming-pool (SP) and Windmill (WD). The dataset was partitioned into training, validation, and test sets with a ratio of 5:2:3.

DOTA-v1.0 [21] is a large-scale dataset designed for evaluating directed object detection performance in aerial images. It comprises 2806 images with a total of 188,282 instances distributed across 15 distinct categories characterized by varying orientations, scales, and shapes. The dataset encompasses instances across 15 categories: Plane (PL), Baseball Diamond (BD), Bridge (BR), Ground Track Field (GTF), Small Vehicle (SV), Large Vehicle (LV), Ship (SH), Tennis Court (TC), Basketball Court (BC), Storage Tank (ST), Soccer-ball Field (SBF), Roundabout (RA), Harbor (HA), Swimming Pool (SP), and Helicopter (HC). The DOTA benchmark dataset comes complete with comprehensive annotations, incorporating 188,282 instances, each delineated by arbitrary quadrilateral markings. Notably, the DOTA dataset features small objects primarily concentrated in specific categories, such as Small Vehicles and Ships.

DIOR [9] dataset is a large-scale benchmark dataset designed for optical remote sensing image object detection. This dataset comprises 23,463 images encompassing a total of 192,472 instances, distributed across 20 distinct object classes. Notably, several classes within the dataset, such as aircraft, storage tanks, overpasses, vehicles, and ships, are categorized as small objects.

4.2   Implementation Details

The backbone network of the proposed architectural framework is pre-trained on ImageNet and is denoted as CSPNet [17]. RTMDet [14] stands as the SOTA approach in remote sensing image object detection tasks, and it is the baseline method adopted in our paper. Images from all three datasets were cropped into blocks of 1024×1024 pixels with an overlap of 256 pixels. Image augmentation was applied using both single-image techniques (RandomResize, RandomCrop, HSVRandomAug, RandomFlip) and mixed-class data augmentation techniques (Mosaic, MixUp). We use AdamW as our optimizer in training, with an initial learning rate of 0.00025 and a weight decay of 0.05, using a fixed learning rate for the first half and a Flat Cosine for the second half.

4.3   Results Comparison

We compared our method with other target detection methods on SODA-A and DOTA-v1.0 datasets and our method achieved the best results. All methods take the same image processing approach and all achieve the best results for this model. As shown in Table 1, on the DOTA-v1.0 dataset, our approach showcases a higher overall accuracy compared to other methods, and it particularly excels in categories like Small Vehicles (SV) and ship (SH), surpassing other methods by more than 3%. As shown in Table 2, on the SODA-A dataset tailored for small object detection, our approach also attains a remarkable 75.8% accuracy across these nine specific categories, surpassing other methods by more than 5%. It proves that our proposed FREM and SR-assisted branching can better fuse the low-level texture information with the high-level semantic information, which has a superior performance on the detection of small objects.

4.4   Ablation Study

We conduct comparative experiments on the DOTA-v1.0, SODA-A, and DIOR datasets to test the effectiveness of each component in our proposed framework as well as the chosen parameter configurations.

Comparison of Module Effectiveness. The three modules we proposed for small targets are compared one by one, as shown in Table 3, deeper feature fusion can obtain deeper information, and it can be seen that the enhancement is 0.9%, 0.8%, and 0.8% above the three datasets of SODA-A, DOTA-v1.0, and DIOR respectively. Then we add the FREM module to the extracted feature maps to enhance the feature reinforcement of small objects and suppress the background noise interference to obtain more contextual and global information, which improves 1.8%, 1.5%, and 1.7% on the three datasets, respectively, compared with Baseline. Finally, our SR branch is introduced to better fuse the local texture information and contextual information, and the effect is improved by 3.8%, 2.5%, and 2.0% compared with Baseline. Thus, it can be proved that each of our proposed methods improves the detection of small targets in remote-sensing images.

Table 1. Comparison of our method and with other state-of-the-art methods for some small objects and categories that are highly affected by background noise and overall detection accuracy on the DOTA-v1.0 dataset

Methods	PL	SV	LV	SH	ST	RA	HA	HC	mAP/%
Rotated-Faster-Rcnn [16](2022)	90.9	77.0	85.4	88.0	71.6	74.9	78.0	90.1	82.8
Sasm [8](2022)	90.4	72.5	78.1	79.9	77.2	68.6	72.4	70.8	73.9
Rotated-Atss [27](2020)	90.5	82.6	85.9	88.0	79.0	80.0	77.4	89.1	82.3
Oriented-RepPoints [10](2022)	90.3	75.0	81.2	88.1	78.0	64.8	70.1	77.8	75.9
RoI-Trans [4](2019)	90.8	78.9	88.9	81.2	72.3	87.4	88.4	90.2	86.1
RoI-KFIoU [26](2023)	90.7	76.6	87.1	80.6	71.4	72.2	86.1	87.2	81.7
R^3Det [25](2021)	90.8	76.1	84.4	80.0	77.2	83.6	78.1	75.0	81.5
Oriented R-CNN [23](2021)	90.7	77.9	88.3	80.9	72.0	86.5	85.9	89.4	84.8
Rotated Retainet [13](2017)	90.0	74.1	79.4	78.3	70.5	86.1	73.2	79.8	80.7
Gliding Vertex [24](2020)	90.7	77.9	84.0	80.6	71.8	83.5	78.7	81.2	82.2
ReDet [6](2021)	90.5	75.0	84.9	80.4	68.5	66.9	77.5	86.1	76.5
S^2A-Net [5](2021)	90.8	79.9	87.1	80.7	80.8	88.7	79.3	88.7	85.2
SRRE(Ours)	**90.8**	**87.3**	**90.0**	**90.1**	**88.8**	**89.2**	**89.6**	**90.9**	**89.3**

Table 2. Comparison of detection accuracy on SODA-A dataset with other state-of-the-art methods

Methods	PL	HP	SV	LV	SH	CN	ST	SP	WD	mAP/%
Rotated-Faster-Rcnn [16](2022)	89.5	55.1	57.4	57.7	80.1	52.0	69.3	87.6	67.4	67.3
Sasm [8](2022)	88.6	47.8	62.4	41.4	76.9	49.9	75.6	86.2	71.9	66.7
Rotated-Atss [27](2020)	89.2	54.3	66.0	47.0	86.2	56.3	76.2	87.3	72.7	70.6
Oriented-RepPoints [10](2022)	88.4	48.6	63.1	23.9	68.1	35.5	74.6	85.1	72.3	62.2
RoI-Trans [4](2019)	89.5	52.3	59.0	54.8	80.7	53.1	70.2	87.9	67.7	68.3
RoI-KFIoU [26](2023)	88.9	54.0	58.4	52.9	80.2	52.7	69.4	87.3	73.4	68.6
R^3Det [25](2021)	88.8	51.1	62.3	48.4	77.5	49.5	75.5	87.0	66.8	66.5
Oriented R-CNN [23](2021)	89.4	55.1	58.3	54.3	80.4	52.8	68.8	87.5	67.3	68.2
Rotated Retainet [13](2017)	88.6	50.3	53.2	27.5	73.5	48.6	69.1	84.2	73.6	63.2
Gliding Vertex [24](2020)	88.5	53.0	53.7	40.3	78.1	49.0	68.3	84.5	74.1	65.5
ReDet [6](2021)	88.4	49.1	64.0	32.3	68.7	37.2	76.1	84.9	76.1	64.1
S^2A-Net [5](2021)	89.3	52.9	65.5	48.8	84.2	54.4	77.4	87.4	67.4	69.7
SRRE(Ours)	**89.7**	**69.6**	**79.3**	**54.5**	**88.5**	**58.7**	**79.8**	**87.3**	**74.9**	**75.8**

Table 3. Comparison of the detection accuracy of the three improved modules on the SODA-A, DOTA, and DIOR datasets

Method	SODA-A	DOTA-v1.0	DIOR
Baseline	72.0	86.8	63.1
Baseline+4L	72.9(**+0.9**)	87.6(**+0.8**)	63.9(**+0.8**)
Baseline+4L+FREM	73.8(**+1.8**)	88.3(**+1.5**)	64.8(**+1.7**)
Baseline+4L+FREM+SR	75.8(**+3.8**)	89.3(**+2.5**)	65.9(**+2.0**)

Comparison of Attention Module Methods. For our EBAM attention module introduced by the FRERM module, we compared it with other attention methods. As shown in Table 4, we compare CA, ECA [18], CAM [30], and CBAM [19] with our method on both datasets and we can find that better results are achieved in terms of accuracy with little difference in Params and FLOPS.

Table 4. Comparison of different attentional modules for the SODA-A and DOTA datasets, where the bolded data indicate the best results for each indicator comparing several approaches

Method	Params(M)	FLOPS(G)	SODA-A	DOTA-v1.0
CA	14.57	**139.48**	79.2	87.6
ECA [18]	**14.38**	139.48	72.7	87.7
CAM [30]	14.4	**139.48**	73.3	87.9
CBAM [19]	14.43	139.5	73.5	88.0
EBAM	**14.38**	139.49	**73.8**	**88.3**

Comparison of SR Loss Parameters. For the choice of parameter c in front of the Ls loss function, a comparison was made. As shown in Table 5, it can be seen that the best results are achieved at c = 0.3.

Table 5. Comparison of different parameters c in front of the L_s loss on the SODA-A dataset

c	0.1	0.2	0.3	0.4	0.5	0.6
mAP/%	70.9	73.2	**75.8**	75.2	70.3	68.2

Comparison of Layer Selection for SR. We compared the selection of layers for low-level features and high-level features in SR by selecting four layers and three layers at different feature layers, respectively. As shown in Table 6, at three layers, the channel numbers 256 and 1024 can better fuse the texture information of the low layer and the semantic information of the high layer.

Table 6. Comparison of Layer Selection for SR on the SODA-A Dataset

Layers	Low Feature	High Feature	mAP/%
3	64	256	70.3
	192	768	71.4
	256	**1024**	**75.8**
4	64	512	72.1
	128	1024	73.3

4.5   Visualization

As shown in Fig. 6, our method is compared with several methods with better accuracy in terms of detection effect. The first row shows that the first two methods can only detect airplanes, and the selected baseline can detect carts, but there are still multiple misses, our method has a better detection effect and can detect carts between two airplanes, which shows that our method can still get better results under the influence of large size difference and background noise. As shown in the second row, the background of dense targets contains a lot of cluttered information, the boundary-blurring of small objects, and the similarity of color and background affect the detection. The detection effect of the proposed network is closer to the Ground Truth than the Baseline and other methods compared to the detection effect.

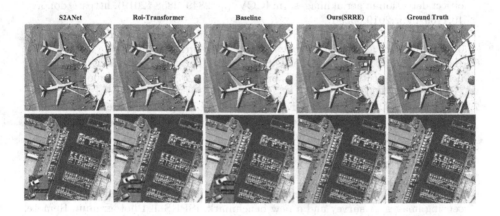

Fig. 6. Comparison of our method with other best methods such as S^2A-Net [5], Roi-Trans [4] and Baseline [14]. The green boxes represent examples of missed detections compared to Ground Truth

5   Conclusion

We propose a Super-Resolution-Assisted Feature Refined Extraction (SRRE) to enhance the performance of small object detection in remote sensing images. In the feature extraction stage, a Feature Refined Extraction Module (FREM) is introduced, which incorporates a newly designed EBAM attention module to obtain more intricate features. In the feature fusion stage, a deeper level of fusion is used. Finally, during the training phase, a Super-Resolution (SR) Assisted Branch is incorporated, enabling the better integration of low-level texture details and high-level semantic information. Experimental results substantiate that our approach yields superior performance in the detection of small objects within remote sensing images.

Acknowledgements. This work is supported by the project of the Engineering Research Center of Ecological Big Data, Ministry of Education, China).

References

1. Cao, L., Zhang, X., Wang, Z., Ding, G.: Multi angle rotation object detection for remote sensing image based on modified feature pyramid networks. Int. J. Remote Sens. **42**(14), 5253–5276 (2021). https://doi.org/10.1080/01431161.2021.1910371
2. Cheng, G., et al.: Towards large-scale small object detection: survey and benchmarks. IEEE TPAMI (2023). https://doi.org/10.1109/tpami.2023.3290594
3. Courtrai, L., Pham, M.T., Lefèvre, S.: Small object detection in remote sensing images based on super-resolution with auxiliary generative adversarial networks. Remote Sens. **12**(19), 3152 (2020). https://doi.org/10.3390/rs12193152
4. Ding, J., Xue, N., Long, Y., Xia, G.S., Lu, Q.: Learning roi transformer for oriented object detection in aerial images. In: ICCV, pp. 2849–2858 (2019). https://doi.org/10.1109/cvpr.2019.00296
5. Han, J., Ding, J., Li, J., Xia, G.S.: Align deep features for oriented object detection. TGRS **60**, 1–11 (2021). https://doi.org/10.1109/tgrs.2021.3062048
6. Han, J., Ding, J., Xue, N., Xia, G.S.: Redet: a rotation-equivariant detector for aerial object detection. In: CVPR, pp. 2786–2795 (2021). https://doi.org/10.1109/cvpr46437.2021.00281
7. Hong, M., Li, S., Yang, Y., Zhu, F., Zhao, Q., Lu, L.: Sspnet: scale selection pyramid network for tiny person detection from UAV images. IEEE Geosci. Remote Sens. Lett. **19**, 1–5 (2021). https://doi.org/10.1109/lgrs.2021.3103069
8. Hou, L., Lu, K., Xue, J., Li, Y.: Shape-adaptive selection and measurement for oriented object detection. In: AAAI. vol. 36, pp. 923–932 (2022). https://doi.org/10.1609/aaai.v36i1.19975
9. Li, K., Wan, G., Cheng, G., Meng, L., Han, J.: Object detection in optical remote sensing images: A survey and a new benchmark. ISPRS J. Photogramm. Remote. Sens. **159**, 296–307 (2020). https://doi.org/10.1016/j.isprsjprs.2019.11.023
10. Li, W., Chen, Y., Hu, K., Zhu, J.: Oriented reppoints for aerial object detection. In: CVPR, pp. 1829–1838 (2022). https://doi.org/10.1109/cvpr52688.2022.00187
11. Lim, B., Son, S., Kim, H., Nah, S., Mu Lee, K.: Enhanced deep residual networks for single image super-resolution. In: CVPR Workshops, pp. 136–144 (2017). https://doi.org/10.1109/cvprw.2017.151
12. Lin, T.Y., Dollár, P., Girshick, R., He, K., Hariharan, B., Belongie, S.: Feature pyramid networks for object detection. In: CVPR, pp. 2117–2125 (2017). https://doi.org/10.1109/cvpr.2017.106
13. Lin, T.Y., Goyal, P., Girshick, R., He, K., Dollár, P.: Focal loss for dense object detection. In: ICCV, pp. 2980–2988 (2017). https://doi.org/10.1109/iccv.2017.324
14. Lyu, C, et al.: Rtmdet: an empirical study of designing real-time object detectors. arXiv preprint arXiv:2212.07784 (2022). https://doi.org/10.48550/arXiv.2212.07784
15. Redmon, J., Divvala, S., Girshick, R., Farhadi, A.: You only look once: unified, real-time object detection. In: Proceedings of the IEEE Conference on Computer Vision and Pattern Recognition, pp. 779–788 (2016). https://doi.org/10.1109/cvpr.2016.91
16. Ren, S., He, K., Girshick, R., Sun, J.: Faster r-cnn: towards real-time object detection with region proposal networks. In: Advances in Neural Information Processing Systems, vol. 28 (2015). https://doi.org/10.1109/tpami.2016.2577031

17. Wang, C.Y., Liao, H.Y.M., Wu, Y.H., Chen, P.Y., Hsieh, J.W., Yeh, I.H.: Cspnet: a new backbone that can enhance learning capability of cnn. In: CVPR, pp. 390–391 (2020). https://doi.org/10.1109/cvprw50498.2020.00203

18. Wang, Q., Wu, B., Zhu, P., Li, P., Zuo, W., Hu, Q.: Eca-net: efficient channel attention for deep convolutional neural networks. In: CVPR, pp. 11534–11542 (2020). https://doi.org/10.1109/cvpr42600.2020.01155

19. Woo, S., Park, J., Lee, J.Y., Kweon, I.S.: Cbam: Convolutional block attention module. In: ECCV, pp. 3–19 (2018). https://doi.org/10.1007/978-3-030-01234-2_1

20. Wu, J., Pan, Z., Lei, B., Hu, Y.: Fsanet: feature-and-spatial-aligned network for tiny object detection in remote sensing images. TGRS 60, 1–17 (2022). https://doi.org/10.1109/tgrs.2022.3205052

21. Xia, G.S., et al.: Dota: a large-scale dataset for object detection in aerial images. In: CVPR, pp. 3974–3983 (2018). https://doi.org/10.1109/cvpr.2018.00418

22. Xiaolin, F., et al.: Small object detection in remote sensing images based on super-resolution. Pattern Recogn. Lett. 153, 107–112 (2022). https://doi.org/10.3390/rs13091854

23. Xie, X., Cheng, G., Wang, J., Yao, X., Han, J.: Oriented r-cnn for object detection. In: ICCV, pp. 3520–3529 (2021). https://doi.org/10.1109/iccv48922.2021.00350

24. Xu, Y., et al.: Gliding vertex on the horizontal bounding box for multi-oriented object detection. TPAMI 43(4), 1452–1459 (2020). https://doi.org/10.1109/tpami.2020.2974745

25. Yang, X., Yan, J., Feng, Z., He, T.: R3det: refined single-stage detector with feature refinement for rotating object. In: AAAI. vol. 35, pp. 3163–3171 (2021). https://doi.org/10.1609/aaai.v35i4.16426

26. Yang, X., et al.: The kfiou loss for rotated object detection. arXiv preprint arXiv:2201.12558 (2022). https://doi.org/10.48550/arXiv.2201.12558

27. Zhang, S., Chi, C., Yao, Y., Lei, Z., Li, S.Z.: Bridging the gap between anchor-based and anchor-free detection via adaptive training sample selection. In: CVPR, pp. 9759–9768 (2020). https://doi.org/10.1109/cvpr42600.2020.00978

28. Zhang, W., Wang, S., Thachan, S., Chen, J., Qian, Y.: Deconv r-cnn for small object detection on remote sensing images. In: IGARSS, pp. 2483–2486. IEEE (2018). https://doi.org/10.1109/igarss.2018.8517436

29. Zhao, H., Gallo, O., Frosio, I., Kautz, J.: Loss functions for image restoration with neural networks. IEEE Trans. Comput. Imaging 3(1), 47–57 (2016). https://doi.org/10.1109/tci.2016.2644865

30. Zhou, B., Khosla, A., Lapedriza, A., Oliva, A., Torralba, A.: Learning deep features for discriminative localization. In: CVPR, pp. 2921–2929 (2016). https://doi.org/10.1109/cvpr.2016.319

Lightweight Image Captioning Model Based on Knowledge Distillation

Zhenlei Cui[1], Zhenhua Tang[1,2(✉)], Jianze Li[1], and Kai Chen[1]

[1] School of Computer and Electronics Information, Guangxi University,
Nanning 530004, China
tangedward@126.com

[2] Guangxi Key Laboratory of Multimedia Communication and Network Technology,
Nanning 530004, China

Abstract. The performance of image captioning models based on deep learning has been significantly improved compared with traditional algorithms. However, due to the complex network structure and huge parameters, these image captioning models are hard to apply to resource-constrained appliances such as mobile embedded systems. To address this issue, we propose two lightweight image captioning models based on two pre-training models, i.e. M^2 Transformer and CaMEL. To improve the performances of the proposed lightweight models, we also develop an optimization training strategy based on knowledge distillation. Specifically, we design knowledge distillation loss functions in the encoder, decoder, and response modules, which can use the multi-modal feature mapping of the original models to transfer prior knowledge, improving the feature learning ability and generalization ability of the lightweight models. Experimental results show that the proposed lightweight models are superior to the original pre-trained models in performance when reducing the parameters greatly. Besides, the proposed lightweight models also have a good knowledge transfer ability.

Keywords: Lightweight model · Image captioning · Knowledge distillation

1 Introduction

Deep learning-based image captioning algorithms, which can realize the multimodal transformation from images to text, have received increasing research attention. It can be widely applied to many fields such as image analysis and intelligent human-computer interaction [13,18]. The performance of image captioning models using transformer [30] with a multi-head attention mechanism has been proven to be excellent, but they may also bring high computational costs. Due to a large number of parameters and high computational complexity, the model training and inference time become long, making it hard to apply to resource-constrained devices such as mobile embedded systems [6]. Therefore, it is necessary to study lightweight image captioning models.

© The Author(s), under exclusive license to Springer Nature Switzerland AG 2024
S. Rudinac et al. (Eds.): MMM 2024, LNCS 14555, pp. 310–324, 2024.
https://doi.org/10.1007/978-3-031-53308-2_23

At present, only several pieces of research work focus on lightweight image captioning models [3, 27–29]. Although the models mentioned above help reduce the number of parameters and computational complexity, their performances are degraded. In addition, problems such as structural clutter, reduced generalization ability, and decreased robustness may also occur in the compressed model. Knowledge distillation is an effective technique to improve the performance of compression models, but less work has adopted knowledge distillation techniques to improve the performances of lightweight image captioning models.

To solve the above problems, we construct two lightweight image captioning models based on two pre-trained models M^2 Transformer [7] and CaMEL [5], respectively. Furthermore, we employ knowledge distillation technology to train them to improve the performance of the lightweight models. Specifically, in the cross-entropy training stage, we use the minimized KL divergence to measure the difference in the probability distribution of each word between the lightweight model and the teacher model. We also utilize the output features of the encoder and the last layer of the decoder in the teacher model as source information and use the mean square error (MSE) to measure the feature of the intermediate layer. This makes the lightweight models contain the low-level and high-level characteristics of the image and the context semantic information. In the reinforcement learning stage, we employ the probability distribution of the caption sequence generated by the original model in beam search as supervision information. Experimental results show that compared with two teacher models M^2 Transformer and CaMEL, the model parameters of our proposed lightweight models are reduced by 50.8% and 23.5%, respectively, while the CIDEr scores are 132.4 and 142.3, which are 0.9% and 1.2% higher than the teacher model, respectively. The main contributions of this paper are as follows:

- We propose two lightweight image captioning models. Based on the two pre-trained large models, M^2 Transformer and CaMEL, we construct two single-layer lightweight models with the same intra-layer structure as the original ones. The proposed lightweight models can reduce the parameters significantly and lower the demand for computing resources and memory.
- We design a distillation loss function for training the lightweight models according to the output features of the encoder and the last layer of the decoder in the original model, which can effectively exploit the advantages of the multi-modal feature mapping of the original model, improving the performance of the lightweight model.
- We take the probability distribution information of the caption sequence generated by the original model in beam search as the supervision information in the reinforcement learning optimization stage, which alleviates the performance degradation caused by the improper allocation of caption rewards in the lightweight model.

2   Related Work

2.1   Image Captioning

In the field of image captioning, mainstream model algorithms widely rely on a multi-layer encoder-decoder framework based on the transformer model [30]. Researchers primarily focus on designing better visual feature extraction modules [2], attempting to find a more effective way to utilize the rich visual features in images [7,11,23,35], which are decoded by the language model to generate a caption of the image scene, thus improving the overall performance of the model. Simply extracting the region feature from images will lack context information and local detail information. To address this, visual features from image grids and regions are combined to realize the complementary advantages of these two visual features [21]. With the support of these feature processing modules and attention modules [14], the image captioning pre-training model has achieved efficient performance. However, they all have a common drawback, which is that these algorithms neglect the existence of redundant parameters in the model and lack effective lightweight processing steps for the model.

2.2   Knowledge Distillation in Image Captioning

Hinton et al. [12] proposed knowledge distillation, whose core idea is to train by matching students' predictions with teachers' predictions through knowledge transfer. The research on knowledge distillation algorithms mainly focuses on classification problems, and further optimization and improvement are needed for the application of complex tasks such as image captioning. To our knowledge, only a few research works have introduced knowledge distillation algorithms in image captioning tasks. Barraco et al. [5] employed the interaction of two mutually related language models for online distillation. Yang et al. [33] exploited the knowledge distillation strategy to enhance the inductive bias transfer from the pure language domain to the vision-language domain. Dong et al. [9] applied standard word-level knowledge distillation to image captioning models. Huang et al. [15] trained a teacher model by incorporating additional image attribute data into the baseline model, acting as a bridge between ground-truth captions and the caption model. The above research work mainly uses knowledge distillation as a means to improve the performance of the original model, without considering the lightweight of the model.

3   Lightweight Image Captioning Model Based on Knowledge Distillation

We constructed two lightweight image captioning models based on knowledge distillation on the basis of two pre-trained models M^2 Transformer [7] and CaMEL [5]. Specifically, we reduced the redundancy of the model parameters significantly by reducing the depth of the network model. Please see Sect. 3.1

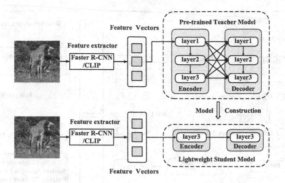

Fig. 1. Construction of lightweight image captioning model.

for details. To enhance the performance of the lightweight models, besides the traditional response loss function, we designed the distillation loss function of the encoder and decoder, which can improve the multi-modal reasoning ability of the lightweight models. In addition, we adopted different distillation loss functions as training strategies according to the features of the lightweight image captioning models in the cross-entropy (XE) and reinforcement learning (RL) stages. Please refer to Sect. 3.2 for details.

3.1 Lightweight Model Construction

Image captioning tasks usually include visual feature extraction and encoding/decoding of the language model, as shown in Fig. 1. Specifically, visual features of the input image are first extracted to generate feature vectors. Then, the language model encodes and decodes the feature vectors and outputs the text that describes the image scene. In the stage of visual feature extraction, M^2 Transformer [7] uses Faster R-CNN with ResNet-101 to extract region features from the image. CaMEL [5] employs the contrastive language-image pre-training (CLIP) algorithm [25] to extract the features of grids on a large-scale image-text dataset. The language model adopts an encoder-decoder architecture with a multi-layer based on Transformer [30]. Each layer of encoder is composed of multi-head self-attention and feed-forward modules. And each layer of decoder has three modules: masked multi-head self-attention, cross-attention, and feed-forward network.

To improve the total performance, both M^2 Transformer and CaMEL adopt a stacking multi-layer structure to increase the depth of the model. The experimental results on the MSCOCO dataset show that when applying the encoder of three layers and the decoder of three layers, the model performs the best [7]. In these models, the input of the current layer is the output of the previous layer for the encoder, while the input of the current layer of the decoder contains not only the output of the previous decoding layer but also the output of the last encoding layer. In addition, the M^2 Transformer and CaMEL establish a mesh-like between different encoding and decoding layers and use persistent memory

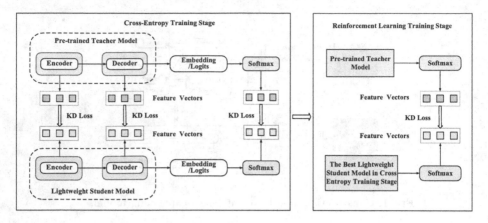

Fig. 2. Overview of the performance optimization framework of lightweight image Captioning model based on knowledge distillation.

vectors to learn and encode prior knowledge, which can transfer the low-level and high-level features of the image to lightweight models as supervision information when using knowledge distillation.

Although neural networks can improve their performance by increasing the depth and the number of neurons, the operations would bring redundancies in the depth and dimension of the model [8]. Transformer-based language models such as M^2 Transformer [7] and CaMEL [5] use a multi-head attention mechanism to project image feature information into different spaces, so they can extract more interaction information between various vectors and have powerful representation ability of input data. Therefore, a well-trained single-layer encoder-decoder architecture can also have strong performance. Based on the above analysis, we construct a single-layer lightweight image captioning model by reducing the depth of the network model on the basis of M^2 Transformer and CaMEL models, as shown in Fig. 1. Since the proposed lightweight model is a single-layer codec structure, here we denote it by layer 3.

3.2   Model Performance Optimization

We employ the knowledge distillation technology to improve the performance of the lightweight model by using M^2 Transformer and CaMEL as teacher models and the corresponding lightweight model as student models. Because of the complexity of the model, the lightweight model can learn the knowledge of the teacher model from different aspects, so choosing an appropriate distillation strategy is critical [34]. According to the features of the image captioning model, we design the knowledge distillation loss function in the encoder, decoder, and response module. Specifically, the response-based knowledge distillation method can convert the word embedding vector of the teacher model into the form of soft prediction, whose prediction distribution carries more abundant information. To

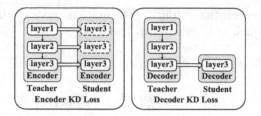

Fig. 3. Knowledge distillation loss function mapping for Encoder and Decoder.

make full use of the prior knowledge of the teacher network, we also designed a distillation loss function in the encoder and decoder respectively, aiming to map the prior knowledge of the teacher network to the student network.

The lightweight model training strategy is shown in Fig. 2. In the cross-entropy training stage, we combine the distillation loss functions of encoder, decoder, and response to take advantage of multi-modal feature mapping between different modules of the teacher model. On the other hand, we alleviate the impact of improper caption reward allocation during reinforcement learning.

Loss Function Design of Encoder and Decoder. In Natural language processing tasks, Sun et al. [26] proposed a Patient-KD method, which adopts two different distillation training strategies of PKD-Last and PKD-Skip. Jiao et al. [16] introduced TinyBERT, which utilized a layer-wise distillation strategy to effectively transfer a large amount of intermediate-layer knowledge from the teacher model to the student model. Wang et al. [32] proposed deep self-attention distillation, which extracted the self-attention module from the last transformer layer of the teacher model to train the student model.

Unlike the aforementioned models, the encoder for image captioning needs to encode features of the image region or grid in a multi-level manner. These low-level and high-level relationships are crucial for generating high-quality captions in the model's language decoder. Therefore, considering this characteristic of image captioning models, we have designed an intermediate layer distillation loss function for both the model encoder and decoder, as shown in Fig. 3. In order to enable the single-layer student model encoder to have low-level and high-level visual features of the image, we did not adopt a hierarchical distillation training strategy in the teacher model encoder. Instead, we aligned the encoder of the teacher model by concatenating the encoding layer of the single-layer student model, which is then inputted into the decoding layer. Using the feature outputs of all layers of the teacher model encoder and the last layer of the decoder as information sources effectively map knowledge to the encoder and decoder of the student network.

MSE denotes the Mean squared error loss function, which is used to measure the difference between the real value and the predicted value. The intermediate layer feature distillation loss transmits priori knowledge by penalizing the distance difference between the output representation space of the student model and the teacher model, as shown in Eq. 1 and Eq. 2.

$$L_e(\theta) = MSE(F_\theta^S(e_3, e_3, e_3), F_\theta^T(e_1, e_2, e_3)) \tag{1}$$

$$L_d(\theta) = MSE(F_\theta^S(d_3), F_\theta^T(d_3)) \tag{2}$$

$$L_{kd-f}^{XE}(\theta) = L_e(\theta) + L_d(\theta) \tag{3}$$

F_θ^S and F_θ^T represent the feature output functions of the student network and the teacher network, respectively. The objective is to convert the network outputs into feature vector representations, which can be defined as the output of any layer in the network, abstracting the current stage's output information of the network. Where e_i and d_i represent the encoding layer and decoding layer of the model, respectively, while θ represents the model training parameters.

Loss Function Design of Response. Some researchers [10] focus on the output features of the penultimate layer of the model. They believe that the spatial distribution of word vectors approximates the soft target probability in knowledge distillation and multimodal information can be extracted from the word embedding space. However, we normalize the network logits to generate word probability distributions, which helps to better convey knowledge to student models. As shown in the following equation:

$$p_\theta^S = softmax(fc_\theta^S/\tau) \tag{4}$$

$$p_\theta^T = softmax(fc_\theta^T/\tau) \tag{5}$$

where fc_θ^S and fc_θ^T are student and teacher network output logits respectively, and τ refers to the distillation temperature, which we set to 1 here. p_θ^S and p_θ^T are denoted as the predicted word probability distributions for the processed student and teacher models.

The distillation loss function of the response in the cross-entropy stage quantifies the difference in the probability distribution of each word position between the student network and the teacher network by minimizing the Kullback-Leibler (KL) divergence. As shown in Eq. 6:

$$L_{kd-r}^{XE}(\theta) = KL(logp_\theta^S, p_\theta^T) \tag{6}$$

During the optimization of CIDEr in the reinforcement learning stage, Huang et al. [15] discovered an issue of improper reward allocation during training, where inaccurate words in the generated captions were also rewarded for optimization. Through experimental data, we have also observed the occurrence of improper reward allocation, which was reflected in the decline of some evaluation metrics (such as BLEU). So we designed the distillation loss function shown in Eq. 7. In beam search, the predicted sequence probability distribution of the output caption of the teacher model decoder is used as supervisory information, which is mapped to the student network caption probability distribution through the mean square error (MSE) function.

$$L_{kd-r}^{RL}(\theta) = \quad MSE(\log p_\theta^S(y_{1:T}^i), \quad \log p_\theta^T(y_{1:T}^i)) \tag{7}$$

where $p_\theta^S(y_{1:T}^i)$ and $p_\theta^T(y_{1:T}^i)$ represent the probability distribution of the i-th caption sequence produced by the student model and the teacher model in the beam search.

Objective of Cross Entropy Training Stage. Given the input image I and the target ground truth sentence $y_{1:T}^*$, where $y_{1:T}^*$ is a sequence containing T words, the input data is encoded as a feature vector. The training objective of the image captioning model is to find a set of parameters and optimize the model by minimizing the loss of cross-entropy (XE) as the objective function, as shown in Eq. 8. In the cross-entropy training stage of the lightweight image captioning model, we fuse the encoder, decoder, and the response distillation loss function, as shown in Eq. 9.

$$L_{XE}(\theta) \quad = \quad -\sum_{t=1}^{T} \log(p_\theta(y_t^*|y_{1:(t-1)}^*)) \tag{8}$$

$$L_{kd}^{XE}(\theta) = L_{XE}(\theta) + \alpha \cdot L_{kd-f}^{XE}(\theta) + \beta \cdot L_{kd-r}^{XE}(\theta) \tag{9}$$

where hyperparameters α and β are tuning coefficients that control the amount of knowledge from the different modules of the teacher network. Note that due to the same internal structure of the teacher network and student network, the probability distribution generated is uniform and similar, and the response loss function value in cross-entropy training is small. By training the dataset, observe the initial value changes of this distillation loss term. The experimental shows that the hyperparameter β order of magnitude is set to 10^5 for the best effect.

Objective of Reinforcement Learning Training Stage. To address the issue of inconsistent training objectives and evaluation metrics as well as exposure bias, it is common to use Self-Critical Sequence Training (SCST) [20] for fine-tuning. Specifically, CIDEr is used as the reward, and the expectation of the reward is used as the baseline to maximize the expected CIDEr score of the generated captions. The gradient of the sample is approximated as Eq. 11 :

$$L_{RL}(\theta) = -E_{y_{1:T} \sim p_\theta(y_{1:T})}[r(y_{1:T})] \tag{10}$$

$$\nabla L_{RL}(\theta) \quad \approx \quad -\frac{1}{k} \sum_{i=1}^{k} (r(y_{1:T}^i) - b) \nabla_\theta \log p_\theta(y_{1:T}^i)) \tag{11}$$

where $r(\cdot)$ is the reward function; b is the reward baseline, and k is the number of samples in a batch; $y^i_{1:T}$ is the i-th sampled caption in beam search.

In the reinforcement learning training stage of the lightweight student model, design a distillation loss function such as Eq. 12, which alleviates the improper distribution of student network caption rewards during the training process.

$$L^{RL}_{kd}(\theta) = L_{RL}(\theta) + \lambda \cdot L^{RL}_{kd-r} \tag{12}$$

where hyperparameter λ is the adjustment factor that controls the amount of knowledge from the teacher model.

4 Experimental Results and Discussion

4.1 Dataset and Evaluation Metrics

Since the MSCOCO dataset which consists of more than 120000 images is widely used for image captioning, we employ this dataset as a benchmark for offline training and evaluation of models. To train and test our lightweight models more effectively, we used the "Karpathy" split paradigm [17] to divide the dataset. The training set consists of 113287 images, while the validation and testing set contains 5000 images, respectively. Each image is accompanied by five descriptive sentences. We adopt the following assessment metrics, including BLEU [24], METEOR [4], ROUGE [19], CIDEr [31], and SPICE [1] to evaluate the performances of proposed models.

4.2 Experimental Settings

To make a fair comparison, the proposed lightweight models $M^2(S_{base})$ and CaMEL(S_{base}) follow the implementation setting of the original model M^2 Transformer [10] and CaMEL [11] during training. In the cross-entropy training stage, following the learning rate scheduling strategy [30] with a warmup equal to 10000 iterations and the batch size is set to 50. During the reinforcement learning training phase, a fixed learning rate of 5×10^{-6} is used, and the $M^2(S_{base})$ and CaMEL(S_{base}) batch sizes are 50 and 30, respectively. In addition, we introduce three additional hyperparameters. In Eq. 9, we set the values of hyperparameter α as 0.2, 0.5, and 1, and the range of values for hyperparameter β as 1 to 4 with an interval of 0.5 for experimental testing. In Eq. 12, the hyperparameter λ value ranges from 0.01 to 0.03, with an interval of 0.005 for experimental testing.

4.3 Performance Evaluation

The parameter numbers of lightweight models are reduced by 50.8% and 23.5% compared with the M^2 Transformer and CaMEL model, respectively. We conducted experiments on the MSCOCO dataset and the results are as shown in Table 1. We can see that the performances of the proposed two lightweight models are superior to those of the corresponding pre-trained teacher models. For the

Table 1. Performance comparison on the MSCOCO Karpathy test split.

Model	BLEU-1	BLEU-4	METEOR	ROUGE-L	CIDEr	SPICE
ORT [11]	80.5	38.6	28.7	58.4	128.3	22.6
AoANet [14]	80.2	38.9	29.2	58.8	129.8	22.4
TCTS [15]	81.2	40.1	29.5	59.3	132.3	23.5
X-Transformer [23]	80.9	39.7	29.5	59.1	132.8	23.4
DLCT [21]	81.4	39.8	29.5	59.1	133.8	23.0
RSTNet [35]	81.1	39.3	29.4	58.8	133.3	23.0
LSTNet [22]	81.5	40.3	29.6	59.4	134.8	23.1
M^2Transformer [7]	80.8	39.1	29.2	58.6	131.2	22.6
CaMEL [5]	82.8	41.3	30.2	60.1	140.6	23.9
Ours $M^2(S_{base})$	**82.1**	**40.4**	29.1	**59.0**	**132.4**	22.6
Ours CaMEL(S_{base})	**83.6**	**42.2**	**30.4**	**60.6**	**142.3**	**24.1**

Table 2. Performance results of two-stage training for lightweight models without introducing knowledge distillation.

	BLEU-1	BLEU-2	BLEU-3	BLEU-4	METEOR	ROUGE-L	CIDEr	SPICE
	Cross-Entropy Training Stage							
$M^2(S_{base})$	75.6	59.7	46.1	35.6	27.6	56.1	113.3	20.6
CaMEL(S_{base})	78.3	62.8	49.3	38.6	29.2	58.4	124.9	22.1
	Reinforcement Learning Training Stage							
$M^2(S_{base})$	79.8	64.3	49.8	38.0	28.9	58.0	128.6	22.5
CaMEL(S_{base})	82.4	67.4	53.0	40.7	30.0	59.8	137.8	24.1

results of the lightweight model $M^2(S_{base})$, compared with the teacher model, the metric values of BLEU-4, ROUGE, and CIDEr increase by 3.3%, 0.7%, and 0.9%, respectively. Only the value of METEOR slightly decreased. For the results of the lightweight model CaMEL(S_{base}), the proposed lightweight student model outperforms the teacher model by 2.2%, 0.7%, 0.8%, 1.2%, and 0.8% on BLEU-4, METEOR, ROUGE, CIDEr, and SPICE, respectively. These results show that although the number of parameters of the two lightweight models is significantly reduced, their performances are not degraded but increased compared to their teacher networks.

In addition, the proposed lightweight models outperform the SOTA models in performance, as shown in Table 1. The main reason is that we utilize knowledge distillation technology to train the lightweight model by using the supervision information provided by the large model, which can improve the performance of lightweight models. Besides, we design distillation loss functions in different modules such as encoder and decoder, which makes the knowledge transfer of the teacher model more comprehensive.

4.4   Ablation Study

To verify the advantage of the proposed knowledge distillation strategy, we perform an ablation study on the training of the proposed lightweight model.

Experiment 1: The lightweight models are trained in two stages without using knowledge distillation algorithms. Table 2 shows the experimental results. From the results, we can see that the metric scores of the lightweight models are much lower than those of the corresponding teacher models, which indicates that the lightweight models cannot capture the features of the image well without additional information sources, leading to lower expression ability and performance. Therefore, introducing the knowledge distillation method to improve the lightweight model is necessary.

Table 3. The results of ablation experiments on introducing knowledge distillation loss terms during the two-stage training of $M^2(S_{base})$.

		BLEU-1	BLEU-2	BLEU-3	BLEU-4	METEOR	ROUGE-L	CIDEr	SPICE
Teacher		80.8	65.3	50.9	39.1	29.2	58.6	131.2	22.6
L_{kd-r}^{XE}	L_{kd-f}^{XE}	Cross-Entropy Training Stage							
✓		80.7	65.5	51.1	39.1	29.3	58.8	131.6	22.6
✓	✓	81.0	65.8	51.4	39.5	29.2	59.0	131.9	22.7
	L_{kd-r}^{RL}	Reinforcement Learning Training Stage							
		81.2	65.8	51.2	39.2	29.2	58.7	132.0	22.8
	✓	**82.1**	**66.9**	**52.4**	**40.4**	29.1	**59.0**	**132.4**	22.6

Table 4. The results of ablation experiments on introducing knowledge distillation loss terms during the two-stage training of CaMEL(S_{base}).

		BLEU-1	BLEU-2	BLEU-3	BLEU-4	METEOR	ROUGE-L	CIDEr	SPICE
Teacher		82.8	68.0	53.5	41.3	30.2	60.1	140.6	23.9
L_{kd-r}^{XE}	L_{kd-f}^{XE}	Cross-Entropy Training Stage							
✓		83.2	68.6	54.3	42.2	30.4	60.5	140.4	24.2
✓	✓	83.5	68.9	54.6	42.4	30.4	60.7	141.1	24.4
	L_{kd-r}^{RL}	Reinforcement Learning Training Stage							
		83.2	68.3	53.8	41.5	30.3	60.3	141.8	24.1
	✓	**83.6**	**68.8**	**54.4**	**42.2**	**30.4**	**60.6**	**142.3**	24.1

Experiment 2: We conducted experiments to evaluate the performances of the lightweight models under the following two cases: (1) introducing distillation loss function in the training stages of cross entropy; and (2) adding distillation loss function in the reinforcement learning stage. The results are shown in Table 3 and Table 4. The experimental results show that in the cross-entropy training stage, the lightweight models' performances have been superior to the teacher

models, which indicates that the proposed strategy is efficient. In the reinforcement learning stage, if the CIDEr score is used as a reward to further optimize the lightweight model obtained by cross-entropy training, it leads to a decrease in BLEU and ROUGE scores. But adding the distillation loss function of knowledge in the reinforcement learning stage will make the lightweight model achieve a more balanced performance. The results show that our method can help alleviate the effect of the reward misallocation in the process of training, improving the performances of the lightweight models.

4.5   Performance of Tiny Models

A lightweight model is suitable for some resource-constrained applications, but it might cause performance degradation. Therefore, it is necessary to keep a balance between model performance and network parameters in practical applications. To verify the knowledge transfer ability of the proposed lightweight models, we further construct smaller models from the proposed lightweight models, namely $M^2(S_{tiny})$ and CaMEL(S_{tiny}), whose structure configuration is shown in Table 5. We use the lightweight models as the teacher models to train the small model S_{tiny}, and their performance results are demonstrated in Table 6. Experimental results show that compared with the lightweight models, the performance of the small models $M^2(S_{tiny})$ and CaMEL(S_{tiny}) show a slight decrease when the parameters are reduced by 79.0% and 74.5%, respectively. Especially, their CIDEr scores are 129.4 and 140.5, which are only 2.3% and 1.3% lower than those of the corresponding lightweight models, respectively. This indicates that the proposed lightweight models have a strong knowledge transfer ability.

Table 5. Model structure configuration, including layers of encoder and decoder, dimensions of each layer in the model(dim-1), feed-forward dimensions(dim-2), vocabulary size, Embedding parameters, and overall model parameters.

Model	layer	dim-1	dim-2	vcab size	Emb Params	Total Params
M^2 Transformer	3	512	2048	10201	10.45M	38.44M
$M^2(S_{base})$	1	512	2048	10201	10.45M	18.92M
$M^2(S_{tiny})$	1	128	512	10201	2.61M	3.98M
CaMEL	3	512	2048	49408	50.59M	80.64M
CaMEL(S_{base})	1	512	2048	49408	50.59M	61.65M
CaMEL(S_{tiny})	1	128	512	49408	12.65M	15.72M

Table 6. The single-layer lightweight model serves as a teacher network, and the final performance of the tiny model after Reinforcement Learning training.

	BLEU-1	BLEU-2	BLEU-3	BLEU-4	METEOR	ROUGE-L	CIDEr	SPICE
$M^2(S_{base})$	82.1	66.9	52.4	40.4	29.1	59.0	132.4	22.6
$M^2(S_{tiny})$	81.5	66.2	51.7	39.7	28.6	58.4	129.4	22.1
CaMEL(S_{base})	83.6	68.8	54.4	42.2	30.4	60.6	142.3	24.1
CaMEL(S_{tiny})	83.3	68.7	54.4	42.1	30.2	60.4	140.5	23.9

5   Conclusion

In this paper, we propose two lightweight image captioning models based on the pre-training models, i.e. M^2 Transformer and CaMEL. We also develop an optimization training strategy based on knowledge distillation to improve the performance of the proposed lightweight models. Experimental results show that the proposed lightweight models are superior to the original pre-trained models in performance. In future work, we will consider combining the knowledge distillation technique with other model compression techniques to optimize the proposed lightweight models.

Acknowledgement. This work was supported in part by the Natural Science Foundation of Guangxi, China (No. 2021GXNSFAA220058).

References

1. Anderson, P., Fernando, B., Johnson, M., Gould, S.: SPICE: semantic propositional image caption evaluation. In: Leibe, B., Matas, J., Sebe, N., Welling, M. (eds.) ECCV 2016. LNCS, vol. 9909, pp. 382–398. Springer, Cham (2016). https://doi.org/10.1007/978-3-319-46454-1_24
2. Anderson, P., et al.: Bottom-up and top-down attention for image captioning and visual question answering. In: Proceedings of the IEEE Conference on Computer Vision and Pattern Recognition, pp. 6077–6086 (2018)
3. Atliha, V., Šešok, D.: Image-captioning model compression. Appl. Sci. **12**(3), 1638 (2022)
4. Banerjee, S., Lavie, A.: METEOR: an automatic metric for MT evaluation with improved correlation with human judgments. In: Proceedings of the ACL Workshop on Intrinsic and Extrinsic Evaluation Measures for Machine Translation and/or Summarization, pp. 65–72 (2005)
5. Barraco, M., Stefanini, M., Cornia, M., Cascianelli, S., Baraldi, L., Cucchiara, R.: CaMEL: mean teacher learning for image captioning. In: 2022 26th International Conference on Pattern Recognition (ICPR), pp. 4087–4094. IEEE (2022)
6. Cornia, M., Baraldi, L., Cucchiara, R.: SMArT: training shallow memory-aware transformers for robotic explainability. In: 2020 IEEE International Conference on Robotics and Automation (ICRA), pp. 1128–1134. IEEE (2020)
7. Cornia, M., Stefanini, M., Baraldi, L., Cucchiara, R.: Meshed-memory transformer for image captioning. In: Proceedings of the IEEE/CVF Conference on Computer Vision and Pattern Recognition, pp. 10578–10587 (2020)

8. Denil, M., Shakibi, B., Dinh, L., Ranzato, M., De Freitas, N.: Predicting parameters in deep learning. In: Advances in Neural Information Processing Systems, vol. 26 (2013)

9. Dong, J., Hu, Z., Zhou, Y.: Revisiting knowledge distillation for image captioning. In: Fang, L., Chen, Y., Zhai, G., Wang, J., Wang, R., Dong, W. (eds.) Artificial Intelligence CICAI 2021. Lecture Notes in Computer Science, vol. 13069, pp. 613–625. Springer, Cham (2021)

10. Hahn, S., Choi, H.: Self-knowledge distillation in natural language processing. arXiv preprint arXiv:1908.01851 (2019)

11. Herdade, S., Kappeler, A., Boakye, K., Soares, J.: Image captioning: transforming objects into words. In: Advances in Neural Information Processing Systems, vol. 32 (2019)

12. Hinton, G., Vinyals, O., Dean, J.: Distilling the knowledge in a neural network. arXiv preprint arXiv:1503.02531 (2015)

13. Hsieh, H.Y., Huang, S.A., Leu, J.S.: Implementing a real-time image captioning service for scene identification using embedded system. Multimed. Tools Appl. **80**, 12525–12537 (2021)

14. Huang, L., Wang, W., Chen, J., Wei, X.Y.: Attention on attention for image captioning. In: Proceedings of the IEEE/CVF International Conference on Computer Vision, pp. 4634–4643 (2019)

15. Huang, Y., Chen, J.: Teacher-critical training strategies for image captioning. arXiv preprint arXiv:2009.14405 (2020)

16. Jiao, X., et al.: TinyBERT: Distilling BERT for natural language understanding. arXiv preprint arXiv:1909.10351 (2019)

17. Karpathy, A., Fei-Fei, L.: Deep visual-semantic alignments for generating image descriptions. In: Proceedings of the IEEE Conference on Computer Vision and Pattern Recognition, pp. 3128–3137 (2015)

18. Li, X., Guo, D., Liu, H., Sun, F.: Robotic indoor scene captioning from streaming video. In: 2021 IEEE International Conference on Robotics and Automation (ICRA), pp. 6109–6115. IEEE (2021)

19. Lin, C.Y.: Rouge: A package for automatic evaluation of summaries. In: Text summarization branches out, pp. 74–81 (2004)

20. Luo, R.: A better variant of self-critical sequence training. arXiv preprint arXiv:2003.09971 (2020)

21. Luo, Y., et al.: Dual-level collaborative transformer for image captioning. In: Proceedings of the AAAI Conference on Artificial Intelligence, vol. 35, pp. 2286–2293 (2021)

22. Ma, Y., Ji, J., Sun, X., Zhou, Y., Ji, R.: Towards local visual modeling for image captioning. Pattern Recogn. **138**, 109420 (2023)

23. Pan, Y., Yao, T., Li, Y., Mei, T.: X-linear attention networks for image captioning. In: Proceedings of the IEEE/CVF Conference on Computer Vision and Pattern Recognition, pp. 10971–10980 (2020)

24. Papineni, K., Roukos, S., Ward, T., Zhu, W.J.: BLEU: a method for automatic evaluation of machine translation. In: Proceedings of the 40th Annual Meeting of the Association for Computational Linguistics, pp. 311–318 (2002)

25. Radford, A., et al.: Learning transferable visual models from natural language supervision. In: International Conference on Machine Learning, pp. 8748–8763. PMLR (2021)

26. Sun, S., Cheng, Y., Gan, Z., Liu, J.: Patient knowledge distillation for BERT model compression. arXiv preprint arXiv:1908.09355 (2019)

27. Tan, J.H., Chan, C.S., Chuah, J.H.: COMIC: toward a compact image captioning model with attention. IEEE Trans. Multimed. **21**(10), 2686–2696 (2019)

28. Tan, J.H., Chan, C.S., Chuah, J.H.: End-to-end supermask pruning: learning to prune image captioning models. Pattern Recogn. **122**, 108366 (2022)

29. Tan, J.H., Tan, Y.H., Chan, C.S., Chuah, J.H.: ACORT: a compact object relation transformer for parameter efficient image captioning. Neurocomputing **482**, 60–72 (2022)

30. Vaswani, A., et al.: Attention is all you need. In: Advances in Neural Information Processing Systems, vol. 30 (2017)

31. Vedantam, R., Zitnick, C.L., Parikh, D.: CIDEr: consensus-based image description evaluation. In: Proceedings of the IEEE Conference on Computer Vision and Pattern Recognition, pp. 4566–4575 (2015)

32. Wang, W., Wei, F., Dong, L., Bao, H., Yang, N., Zhou, M.: MINILM: deep self-attention distillation for task-agnostic compression of pre-trained transformers. In: Advances in Neural Information Processing Systems, vol. 33, pp. 5776–5788 (2020)

33. Yang, X., Zhang, H., Cai, J.: Auto-encoding and distilling scene graphs for image captioning. IEEE Trans. Pattern Anal. Mach. Intell. **44**(5), 2313–2327 (2020)

34. Zhang, Q., Cheng, X., Chen, Y., Rao, Z.: Quantifying the knowledge in a DNN to explain knowledge distillation for classification. IEEE Trans. Pattern Anal. Mach. Intell. **45**(4), 5099–5113 (2022)

35. Zhang, X., et al.: RSTNet: captioning with adaptive attention on visual and non-visual words. In: Proceedings of the IEEE/CVF Conference on Computer Vision and Pattern Recognition, pp. 15465–15474 (2021)

Irregular License Plate Recognition via Global Information Integration

Yuan-Yuan Liu[1], Qi Liu[1], Song-Lu Chen[1], Feng Chen[2],
and Xu-Cheng Yin[1(✉)]

[1] School of Computer Science and Technology, University of Science and Technology
Beijing, Beijing, China
{yyliu,qiliu7}@xs.ustb.edu.cn, {songluchen,xuchengyin}@ustb.edu.cn
[2] EEasy Technology Company Ltd., Beijing, China
cfeng@eeasytech.com

Abstract. Irregular license plate recognition remains challenging due
to the irregular layouts of characters, such as multi-line and perspective-
distorted layouts. Many previous methods are based on different attention mechanisms, which generate attention weights and aggregate the
features to obtain character features for recognition. However, we found
that attention-based methods suffer from attention deviation and character misidentification of similar glyphs. We infer that the lack of global
perception is the main reason for these problems. Hence, we propose
to integrate sufficient global information into the network to improve
irregular license plate recognition. Firstly, we propose the deformable
spatial attention module to integrate global layout information into
attention calculations, thus generating more accurate attention. Secondly, we propose the global perception module that integrates global
visual information into feature extraction to enhance the completeness
of character features, making them more representative, and thereby
alleviating character misidentification. Experiments demonstrate that
our method achieves state-of-the-art results, with a significant improvement of 6.7% on irregular license plates. Our codes are available at
https://github.com/MMM2024/GP_LPR.

Keywords: Irregular license plate recognition · Deformable spatial
attention · Global perception

1 Introduction

License plate recognition (LPR) is an essential part of intelligent transportation systems and is widely used for parking management and law enforcement.
Many existing methods [4,14,24,35] have achieved promising performance on
horizontal single-line license plates. However, license plates are usually in irregular layouts, such as severe perspective distortion and multi-line character layouts,
making it challenging to recognize irregular license plates accurately.

Y.-Y. Liu and Q. Liu—These authors contributed equally to this work.

S. Rudinac et al. (Eds.): MMM 2024, LNCS 14555, pp. 325–339, 2024.
https://doi.org/10.1007/978-3-031-53308-2_24

N(H)NF9415   M(P)PY6301   O(J)JM1920   6(9)511DZ

Fig. 1. Two types of wrong predictions. Left: attention deviation. Right: character misidentification of similar glyphs.

Currently, irregular license plate recognition methods can be roughly classified into rectification-based and attention-based methods. The former propose to rectify distorted license plate images (LPs) to horizontal before recognition, with the help of STN [16,23] or affine transformation [19,28,30]. However, the rectification process consumes a large number of computational resources and is ineffective for multi-line license plates. Therefore, the latter [15,30,33] leverage attention modules to select the representative features of each character from two-dimensional (2D) visual features, thus recognizing irregular license plates accurately and robustly. However, as shown in Fig. 1, the attention-based method has two problems: one is that the attention of misidentified characters deviates from the corresponding location, causing the wrong prediction. For this problem, we infer that the global layouts of the characters may not be fully captured due to the limited receptive field, which leads to inaccurate attention. The other problem is that characters may be misidentified as characters with partially similar glyphs, even if the attention is correct. For example, 'J' is incorrectly recognized as 'O', as both 'J' and 'O' have a similar arc. This is due to that the recognition models tend to overfit local details of characters and lack the complete perception of the characters. To summarize, it is necessary to integrate sufficient global information for irregular license plate recognition.

To solve the above problems, we propose to integrate two types of global information into attention computation. **Firstly**, we propose an attention module with layout perception ability, which can globally perceive the irregular layout of characters, generating more accurate attention. **Secondly**, we propose a global perception module to improve the completeness of character features, making them more representative, thus alleviating character misidentification. **Based on this**, we propose a fast and effective license plate recognition network with global perception ability. Extensive experiments show that our proposed method performs well on both irregular and regular license plates. Specifically, for irregular license plates, it achieves performance improvements of 1.3% on AOLP, 3.0% on CCPDv2, and 6.7% on RodoSol-ALPR. Additionally, our method achieves a maximum performance improvement of 2.3% on regular license plates.

2   Related Work

2.1   Regular License Plate Recognition

With the development of artificial neural networks, deep learning-based license plate recognition (LPR) methods are gaining attention. Existing LPR methods can be classified into two categories: segment-based methods and segment-free methods. The segment-based methods [23,37] firstly segment or detect license plate characters and recognize each character. However, this kind of method requires character-level annotations, which is labor-intensive. The segment-free methods consider LPR as a sequential recognition task, avoiding character segmentation. Spanhel et al. [24] propose to use multiple classifiers to recognize license plates, each of which corresponds to a character of fixed reading order. There are some methods [17,28] following this idea to use multiple classifiers for LPR. Besides, many LPR methods [14,35] leverage CTC loss [6] to align visual features with each character to avoid segmentation. Moreover, with the wide application of attention mechanism in the field of text recognition, some LPR methods based on attention mechanism have emerged [4]. However, the above methods treat license plates as 1D signals, suitable only for regular license plates. In contrast, we propose a novel method for irregular license plate recognition.

2.2 Irregular License Plate Recognition

Since the segmentation and detection of characters are not affected by their layouts, some methods [10,11] propose directly detecting characters to recognize irregular license plates. However, the acquisition of character-level annotations is very costly, leading to a focus on segmentation-free methods. Considering the layout of characters, irregular license plates include perspective-distorted and multi-line license plates. To address perspective-distorted license plate recognition, some methods [23,28] use rectification modules to rectify distorted license plates to horizontal ones. However, the rectification module requires additional computational resources and is difficult to train. Since the characters on the irregular license plates are distributed in a 2D space, many methods propose [15,33,34] 2D attention modules for recognition. However, we experimentally found that the attention computation lacks perception of global information, resulting in recognition errors. Consequently, we propose integrating two types of global information into the attention computation.

3   Preliminary

In this section, we first describe the vanilla attention mechanism in a generic form to further clarify the proposed method. As shown in Fig. 2 (a), the attention modules can be considered as follows: given a query and a key-value set, it first calculates the similarity of the query (Q) and key (K) to obtain attention weights, and then aggregates the value (V) according by the weights. The key-value set is usually derived from visual features extracted by CNNs. The query differs depending on the attention algorithm, and it can be the learnable weights of the network or the hidden states of the network, among other possibilities. In our method, the query is the learnable weights of the network. Specifically, for the license plate recognition task, the V represents the visual features of the whole license plate, the K represents the corresponding indexes of the features,

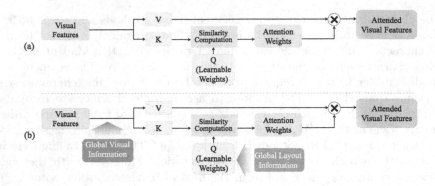

Fig. 2. (a) The schematic diagram of the vanilla attention mechanism in a generic form. (b) The schematic diagram of our proposed method in a generic form. We propose to integrate global visual information and global layout information into the attention computational process.

the Q represents the query vector used to find the features of each character, and the attention weights calculated by Q and K represents the locations of characters. To address the issues of attention deviation and character misidentification caused by the lack of global information, we propose two key enhancements: One is to integrate global layout information into Q to perceive the locations of characters, and the other is to integrate global visual information into the K-V set to improve the completeness of character features, making them more representative (refer to Fig. 2 (b)). This way, accurate attention will be generated and character misidentification will be alleviated.

4 Proposed Method

As shown in Fig. 3, the proposed method consists of four components, i.e., the encoder, the global perception module (GPM), the deformable spatial attention module (DSAM), and the predictor. Each module is described in detail below.

4.1 Encoder

We adopt a lightweight CNN network as the backbone to extract visual features. Specifically, we down-sample the feature maps by alternating convolutional and pooling layer structures. The encoder is a network consisting of 6 convolutional layers and 2 pooling layers, with a down-sampling rate of 4. The input size of the encoder (W, H) is fixed to (96, 32).

4.2 Global Perception Module

Based on the aforementioned analysis in the introduction, we propose a novel component known as the global perception module (GPM). The primary objective of the GPM is to facilitate global interactions among features, enabling the effective integration of global visual information into the key-value set.

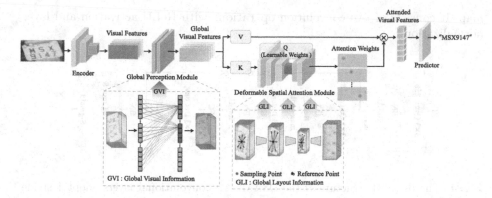

Fig. 3. The overall architecture of the proposed network. First, the encoder extracts the visual features of license plate images. Then, the global perception module integrates global visual information into the extracted features by the global interaction of features. Next, the deformable spatial attention module (DASM) calculates the similarity of the key and query (i.e. learnable weights of DASM) in the form of convolutional operations to generate the attention weights. During this process, DASM adaptively adjusts convolution kernels to adapt to the irregular layout of characters and perceive the locations of characters, so the global layout information (GLI) is effectively integrated into the query. Finally, the character features are aggregated by attention weights and recognized by the predictor.

One of the core advantages of GPM is its ability to improve the completeness of character features. This improvement proves crucial in distinguishing characters from others that share partially similar glyphs, a common challenge in recognition tasks. It helps in mitigating the overfitting of recognition models to the similar local details of distinct characters, making the character features more representative. GPM empowers our model to make finer distinctions between characters, thereby enhancing recognition accuracy. The functional schematic diagram of GPM is illustrated in Fig. 3.

Inspired by Vaswani et al. [25], the proposed GPM consists of 2 global perception layers, which have two sub-layers per layer. The first sub-layer is a multi-headed self-attention layer (SA). In this sub-layer, the visual features F are first mapped to F_q, F_k and F_v by Eq. 1. The self-attention output matrices are calculated as in Eq. 2,

$$F_{q/k/v} = F \times W_{q/k/v} \tag{1}$$

$$SA\left(F_q, F_k, F_v\right) = softmax\left(\frac{F_q F_k^T}{\sqrt{d_k}}\right) F_v \tag{2}$$

where d_k is the dimension of the query and key, $W_{q/k/v}$ are trainable parameters and F_k^T is the transposed matrix of F_k. After that, the second sub-layer is a fully connected feed-forward network applied to each position on the feature map. It consists of two convolution operations with ReLU activation and layer normalization.

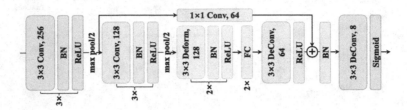

Fig. 4. The detailed structure of the DSAM. All convolutional layers adopt 1 stride and 1 padding.

4.3   Deformable Spatial Attention Module

The attention module aims to extract the most relevant visual features of each character on the license plate. The vanilla spatial attention module calculates attention weights by CNNs and aggregates the visual features based on the weights. Then all attended features are concatenated and sent to the predictor for parallel character prediction.

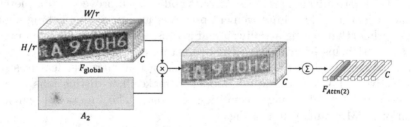

Fig. 5. The computation process of F_{Attn} corresponding to character position 2. $\textcircled{\Sigma}$: all elements within the same channel are summed at the spatial dimension. "r": the down-sampling rate of the encoder.

For irregular license plate images, the spatial layouts of characters are diverse. However, CNNs with fixed kernel shapes learn queries with limited receptive fields, which may potentially lead to information loss or blurriness. Thus the ability of CNNs to perceive the layout of characters is very constrained. To enhance the perception of character layout (i.e., to integrate global layout information), we employ deformable convolutional networks [36] to generate the query of the spatial attention module. Deformable convolution allows the network to dynamically adjust the shape and position of convolution kernels during the convolution process, based on the shape of the target. In our method, the query of attention is the learnable weights of the deformable convolutional layers, and the similarity of the key and query is calculated by deformable convolutional operations to obtain the attention weights. In this way, the network learns an offset to adaptively adjust the position and shape of the convolution kernels, based on the layout of characters. This enables better capture of crucial features in the image, expanding the receptive field of query. The query of the attention module has a larger receptive field, allowing for the utilization of more relevant global information to calculate attention weights. This enhancement improves the module's understanding of contextual information in the image and facilitates better adaptation to the multi-line or perspective-distorted layout of characters. As a result, the convolutional layers sample region more closely fit the layout of characters, resulting in the global irregular character layout being accurately sampled. Therefore, the generated attention is more accurately focused on the location of each character, avoiding the extraction of redundant or even incorrect features. The functional schematic diagram of the DSAM is shown in Fig. 3.

The detailed structure of DSAM is shown in Fig. 4. The generated attention weights are denoted as $A \in \mathbb{R}^{T \times H/r \times W/r}$, where r is the down-sampling rate of the encoder and T is set to the maximum decoding length of the dataset to accommodate different character lengths. The t-th channel of A corresponds to the character at the t-th order, represented as $A_t \in \mathbb{R}^{H/r \times W/r}$. As illustrated in Fig. 5, the attention features $(F_{Attn(t)} \in \mathbb{R}^{C \times 1 \times 1})$ of the t-th character are generated from the weighted summation of visual features (F_{global}) based on attention weights (A_t). The computation process of $F_{Attn(t)}$ as Eq. 3,

$$F_{Attn(t,j)} = \sum_{m,n=0}^{m=H/r, n=W/r} A_{t,m,n} \times F_{global_{j,m,n}} , \; j \in [0, C-1] \qquad (3)$$

where $F_{global} \in \mathbb{R}^{C \times H/r \times W/r}$ are the output features of the GPM. To compute the attention features of all characters in parallel, we first flatten A and F_{global} into one dimension in the spatial dimension. Then the attention features of all characters are computed in parallel by the matrix multiplication as Eq. 4,

$$F_{Attn} = A' \left(F_{global}' \right)^T \qquad (4)$$

where T denotes the transpose of F_{global}'. $A \in \mathbb{R}^{T \times H/r \times W/r}$ is reshaped to $A' \in \mathbb{R}^{T \times (H/r \times W/r)}$ and $F_{global} \in \mathbb{R}^{C \times H/r \times W/r}$ is reshaped to $F_{global}' \in \mathbb{R}^{C \times (H/r \times W/r)}$.

4.4   Predictor

The predictor predicts each character in parallel. Considering characters in license plate images have almost no semantic information, we simply adopt an FC layer as the predictor. We adopt cross-entropy loss as the loss function during the training phase, as shown in Eq. 5.

$$loss = \frac{1}{T} \sum_{t=1}^{T} -log\, p_t \left(y_t^{GT} | X \right) \tag{5}$$

where $p_t \left(y_t^{GT} \right)$ represents the predicted probability of the output being y_t^{GT} at decoding step t, y_t^{GT} represents the t-th character of the ground truth (GT), and X represents the input image. For parallel training, multiple blank characters are padded at the end of the ground truth during training to match the maximum decoding length T.

5   Experiment

5.1   Datasets

To demonstrate the effectiveness of our proposed method on irregular license plate images, we conducted evaluations on three public datasets as follows.

CCPD. [31] is the largest public license plate dataset with over 250k Chinese urban license plate images. It encompasses subsets with diverse challenging conditions like blur, tilt, rainy, snowy, etc. There are two versions, CCPDv1 and CCPDv2, and we conducted separate experiments on both versions.

RodoSol-ALPR. [10] consists of 20k images captured by static cameras at toll collection points. It includes four subsets of LP images with two types of vehicles, cars (horizontal single-line LPs) and motorcycles (multi-line LPs).

AOLP. [7] comprises 2049 images of Taiwan license plates, divided into three subsets: access control (AC), traffic law enforcement (LE), and road patrol (RP). In the RP subset, LP images are mostly with severe perspective distortion and blur. We follow the dataset partitioning protocol from previous works [13,14,29], using any two of these subsets for training, and the remaining one for testing.

5.2   Implement Details

In our experiments, the maximum decoding length T is set to 8. The network is trained with the ADAM optimizer [9]. To avoid oscillations during training, we adopt the StepLR strategy to adjust the learning rate. The initial learning rate is set to 1e-3 and multiplied by 0.8 every 50 epochs. We use 96×32 pixels fixed-size images as inputs with a batch size of 128 on a single GPU. We adopt the same training settings as ours to reproduce the previous methods for comparison.

For the above three datasets, we apply the same data processing approach. We crop the whole image according to the ground truth to get the license plate image. Besides, perspective transformation and pixel transformation (e.g., random noise, contrast transformation, brightness transformation, equalization, sharpening, etc.) are used for data augmentation.

Table 1. License plate recognition accuracy on AOLP. Our method achieves optimal performance across all three subsets, particularly on RP, which contains a large number of perspective-distorted LPs.

Model	AC	LE	RP
Li et al. (2016) [13]	94.9%	94.2%	88.4%
Wu et al. (2018) [29]	96.6%	97.8%	91.0%
Li et al. (2019) [14]	95.3%	96.6%	83.7%
Björklund et al. (2019) [2]	94.6%	97.8%	96.9%
Selmi et al. (2020) [21]	97.8%	97.4%	96.3%
Zhang et al. (2021) [33]	97.3%	98.3%	91.9%
Fan et al. (2022) [5]	79.0%	46.9%	88.1%
Ke et al. (2023) [8]	97.7%	98.2%	87.9%
Ours	**98.2%**	**99.2%**	**98.2%**

5.3 Experiments on Benchmarks

Performances on AOLP: Table 1 shows the results of LP recognition methods on three subsets of AOLP. Our proposed method achieves the best accuracy among state-of-the-art LP recognition methods. Particularly on the RP subset, where LP images often exhibit severe perspective distortion and blur, our method demonstrates a 1.3% improvement in accuracy compared to Björklund et al.'s method. [2]. Additionally, our proposed method achieves performance gains of 0.4% and 0.9% on AC and LE, respectively. The results of our aforementioned experiments robustly demonstrate the effectiveness of our method in addressing the challenges posed by license plates with severe perspective distortions.

Table 2. License plate recognition accuracy on CCPDv2. Our method excels on most subsets, particularly showing marked improvement on Rotate and Tilt subsets featuring perspective-distorted license plates.

Model	Avg	DB	Blur	FN	Rotate	Tilt	Cha.
Xu et al. (2018) [31]	43.4%	34.5%	25.8%	45.2%	52.8%	52.0%	44.6%
Zherzdev et al. (2018) [35]	69.0%	63.0%	59.2%	71.9%	73.1%	62.8%	74.6%
Luo et al. (2019) [18]	64.2%	55.5%	57.6%	66.1%	70.1%	55.9%	69.3%
Wang et al. (2020) [26]	46.8%	43.9%	39.9%	52.6%	50.4%	35.5%	53.4%
Li et al. (2022) [12]	67.8%	67.3%	**75.3%**	67.0%	63.3%	47.5%	78.4%
Wang et al. (2022) [27]	42.1%	38.2%	36.1%	51.0%	36.5%	36.6%	46.2%
Ke et al. (2023) [8]	74.8%	68.3%	69.2%	77.5%	77.9%	72.1%	78.3%
Ours	**76.7%**	**70.6%**	71.6%	**78.6%**	**80.9%**	**73.9%**	**80.1%**

Table 3. License plate recognition accuracy on CCPDv1. Our method achieves state-of-the-art performance on most CCPDv1 subsets, primarily consisting of regular LPs.

Model	Avg	Base	DB	FN	Rotate	Tilt	Wea.	Cha.
Zhang et al. (2016) [32]	93.0%	99.1%	96.3%	97.3%	95.1%	96.4%	97.1%	83.2%
Joseph et al. (2017) [20]	93.7%	98.1%	96.0%	88.2%	84.5%	88.5%	87.0%	80.5%
Zherzdev et al. (2018) [35]	93.0%	97.8%	92.2%	91.9%	79.4%	85.8%	92.0%	69.8%
Xu et al. (2018) [31]	95.5%	98.5%	96.9%	94.3%	90.8%	92.5%	87.9%	85.1%
Li et al. (2019) [14]	94.4%	97.8%	94.8%	94.5%	87.9%	92.1%	86.8%	81.2%
Luo et al. (2019) [18]	98.3%	99.5%	98.1%	98.6%	98.1%	98.6%	97.6%	86.5%
Wang et al. (2020) [26]	96.6%	98.9%	96.1%	96.4%	91.9%	93.7%	95.4%	83.1%
Zhang et al. (2021) [33]	98.5%	99.6%	98.8%	98.8%	96.4%	97.6%	98.5%	88.9%
Liu et al. (2021) [15]	98.7%	**99.7%**	99.0%	99.2%	97.7%	98.3%	98.8%	89.2%
Zou et al. (2022) [38]	97.5%	99.2%	98.1%	98.5%	90.3%	95.2%	97.8%	86.2%
Chen et al. (2023) [3]	98.8%	**99.7%**	99.1%	98.8%	98.4%	98.5%	98.8%	90.0%
Ours	**99.1%**	99.1%	**99.3%**	**99.5%**	**98.8%**	**99.0%**	**99.1%**	**90.7%**

Performances on CCPD: Compared to CCPDv1, CCPDv2 contains more irregular license plates. Table 2 shows that our method achieves remarkable state-of-the-art performance compared to previous methods on almost all subsets. Notably, our method outperforms previous methods by at least 3.0% and 1.8% on the Rotate and Tilt subsets (irregular license plates), respectively. As for the commonly used CCPDv1, our method achieved an overall accuracy of 99.1%, which is 0.3% better than the previous state-of-the-art method [3], as shown in Table 3. These results demonstrate that our method is equally effective for regular and irregular LP images.

Performances on RodoSol-ALPR: As shown in Table 4, our method outperforms the previous state-of-the-art methods by 4.9% and 6.7% on the multi-line subsets motorcycles-br and motorcycles-me, respectively. On the single-line subsets (cars-br and cars-me), our method also achieves competitive performance. The significant performance degradation of the previous methods on the multi-line subsets is attributed to the fact that they are ineffective for multi-line license plates. In contrast, our method performs remarkably well on both multi-line and single-line license plates.

5.4   Ablation Study

In this section, we divided AOLP and RodoSol-ALPR into regular (horizontal single-line LPs) and irregular (perspective-distorted LPs and multi-line LPs) datasets and conducted ablation studies. Specifically, in AOLP, we combine the AC and LE subsets to form the regular dataset, while the RP subset (perspective-distorted LPs) serves as the irregular dataset. In RodoSol-ALPR, we create the regular dataset by merging the cars subsets and the irregular dataset by merging the motorcycles subsets (multi-line LPs).

Table 4. License plate recognition accuracy on RodoSol-ALPR. Our method exhibits significant performance improvements on multi-line subsets (moto-br and moto-me) while achieving competitive performance on single-line subsets (cars-br and cars-me).

Model	Avg	cars-br	cars-me	moto-br	moto-me
Shi et al. (2017) [22]	45.4%	73.8%	78.5%	9.4%	19.8%
Spanhel et al. (2017) [24]	78.5%	83.6%	91.2%	69.6%	69.7%
Zherzdev et al. (2018) [35]	51.3%	94.6%	**99.1%**	7.1%	5.2%
Luo et al. (2019) [18]	77.7%	87.9%	88.4%	69.2%	65.3%
Baek et al. (2019) [1]	84.7%	95.4%	98.3%	60.1%	84.9%
Wang et al. (2020) [26]	88.9%	91.5%	97.0%	82.2%	85.2%
Li et al. (2022) [12]	74.8%	89.4%	97.3%	53.7%	58.9%
Fan et al. (2022) [5]	91.3%	92.7%	96.1%	87.6%	88.7%
Ke et al. (2023) [8]	74.6%	94.1%	98.1%	49.6%	56.6%
Ours	**95.7%**	**96.0%**	99.0%	**92.5%**	**95.4%**

Effectiveness of each module: We first experimentally validate the effectiveness of the proposed module. In the rows where the DSAM column is not checked, the model adopts the baseline 2D spatial attention, which is composed of multiple vanilla convolutional layers. As shown in Table 5, the addition of GPM results in 2.13%~4.42% and 1.82%~4.25% performance improvement to the model on the two irregular subsets, respectively. DSAM enhances model performance by 0.33%~2.62% and 1.5%~3.93% on the two irregular subsets. Meanwhile, GPM and DSAM also remarkably improve the performance of both regular subsets, demonstrating the effectiveness of our method. Besides, we also conducted ablation studies on another dataset containing irregular LPs, i.e., CCPDv2. The results demonstrate that our method has a 3% advantage over the baseline on irregular LPs (Rotated and Tilted subsets).

Table 5. Ablation study on the effectiveness of the GPM and DSAM. DSAM and GPM each significantly boost model performance. AOLP Regular: AC ∪ LE. AOLP Irregular: RP. RodoSol-ALPR Regular: cars-br ∪ cars-me. RodoSol-ALPR Irregular: motorcycles-br ∪ motorcycles-me.

GPM	DSAM	AOLP		RodoSol-ALPR	
		Regular	Irregular	Regular	Irregular
		96.52%	93.45%	95.40%	88.18%
✓		98.40%	97.87%	96.08%	90.00%
	✓	97.70%	96.07%	95.88%	89.68%
✓	✓	**98.75%**	**98.20%**	**97.48%**	**93.93%**

Table 6. Ablation study on the structure of the DSAM. DSAM achieves the best performance with a comparable number of parameters. "C": convolutional layer. "D": deformable convolutional layer. "C×6+D×2": our proposed method.

Structure	AOLP		RodoSol-ALPR		Model size
	Regular	Irregular	Regular	Irregular	(MB)
C×8	97.71%	95.25%	96.63%	90.63%	11.5
C×6+D×2	**98.75%**	**98.20%**	97.48%	**93.93%**	**10.0**
C×10	98.12%	94.60%	**97.90%**	91.93%	14.3

The superiority of the DSAM: DSAM can flexibly expand the receptive field. To further demonstrate the superiority of DSAM, we replace it with structures with the same or even more vanilla convolutional layers for comparison. As shown in Table 6, whether the attention module has 8 or 10 vanilla convolutional layers, the proposed DSAM (middle row of Table 6) performs optimally on the two irregular subsets. Compared with the 8-convolution method, the proposed method achieves gains of 2.95% and 3.30% on the two irregular subsets. Compared with the 10-convolution method, the proposed method outperforms it by a large margin on all irregular subsets, i.e., 3.60% and 2.00%. The performance of the proposed DSAM is slightly lower than that of the 10-convolution method on the regular subset of RodoSol-ALPR. The reason might be that the correction effect of DSAM on attention deviation is insignificant on the license plates with regular character layouts. Besides, we also conducted experiments on the structure of fully deformable convolutions but found it to be less effective compared to our current design. We believe that excessive use of deformable convolutions in the network can pose training challenges and may lead to deviations in the offsets of deformable convolutions, resulting in failure.

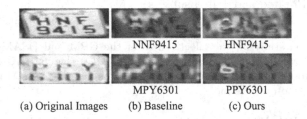

(a) Original Images   (b) Baseline   (c) Ours

Fig. 6. Examples of attention visualization on misidentification characters. Compared to the baseline, our proposed method can generate more accurate attention.

5.5   Attention Visualization

In this section, we delve into the visual analysis of the attention maps generated by our proposed attention module and compare them to those obtained using a baseline method. The baseline method is a network that removes the GPM and adopts baseline 2D spatial attention.

As shown in Fig. 6, the stark contrast between the attention maps produced by the two methods becomes evident. The attention maps generated by the baseline method exhibit a notable issue-they are often severely distracted and tend to deviate significantly from the intended target character. Consequently, this distraction and deviation of attention have a detrimental impact on the accuracy of the recognition process, leading to incorrect outputs. In sharp contrast, our method demonstrates a remarkable ability to generate attention maps that accurately and precisely focus on the target character. These attention maps deviate less from the corresponding character, effectively correcting the recognition results. Based on these observations, we draw a significant conclusion: the judicious integration of global information contributes to learning attention. The strategy of integrating global information not only contributes to improving attention learning but also serves as a catalyst for enhancing the overall performance of attention-based irregular license plate recognition models.

6 Conclusion

In this paper, we introduce an innovative method to enhance irregular license plate recognition through the integration of global information. Our method relies on the utilization of two key components: the deformable spatial attention module and the global perception module. These components collaboratively enable our model to attain a comprehensive understanding of the license plate image, surpassing the limitations of local information. Extensive experimental results demonstrate the superiority of our proposed method when compared to existing methods. This not only underscores the significance of integrating global information but also establishes our method as a robust solution for tackling irregular license plate recognition challenges.

Acknowledgement. This work was partly supported by the National Key Research and Development Program of China under Grant 2020AAA0109700 and partly by the National Natural Science Foundation of China under Grant 62076024 and Grant 62006018.

References

1. Baek, J., et al.: What is wrong with scene text recognition model comparisons? Dataset and model analysis. In: ICCV, pp. 4714–4722 (2019)
2. Björklund, T., et al.: Robust license plate recognition using neural networks trained on synthetic images. PR **93**, 134–146 (2019)
3. Chen, S.L., et al.: End-to-end multi-line license plate recognition with cascaded perception. In: ICDAR (2023)
4. Duan, S., et al.: Attention enhanced convnet-RNN for Chinese vehicle license plate recognition. In: PRCV, pp. 417–428 (2018)

5. Fan, X., Zhao, W.: Improving robustness of license plates automatic recognition in natural scenes. TITS **23**(10), 18845–18854 (2022)
6. Graves, A., et al.: Connectionist temporal classification: labelling unsegmented sequence data with recurrent neural networks. In: ICML, pp. 369–376 (2006)
7. Hsu, G., Chen, J., Chung, Y.: Application-oriented license plate recognition. IEEE Trans. Veh. Technol. **62**(2), 552–561 (2013)
8. Ke, X., Zeng, G., Guo, W.: An ultra-fast automatic license plate recognition approach for unconstrained scenarios. TITS **24**(5), 5172–5185 (2023)
9. Kingma, D.P., Ba, J.: Adam: a method for stochastic optimization. In: ICLR (2015)
10. Laroca, R., et al.: On the cross-dataset generalization in license plate recognition. In: VISIGRAPP, pp. 166–178 (2022)
11. Laroca, R., et al.: An efficient and layout-independent automatic license plate recognition system based on the YOLO detector. CoRR abs/1909.01754 (2019)
12. Li, C., et al.: PP-OCRv3: more attempts for the improvement of ultra lightweight OCR system. CoRR abs/2206.03001 (2022)
13. Li, H., Shen, C.: Reading car license plates using deep convolutional neural networks and LSTMs. CoRR abs/1601.05610 (2016)
14. Li, H., Wang, P., Shen, C.: Toward end-to-end car license plate detection and recognition with deep neural networks. TITS **20**(3), 1126–1136 (2019)
15. Liu, Q., et al.: Fast recognition for multidirectional and multi-type license plates with 2D spatial attention. In: ICDAR, pp. 125–139 (2021)
16. Lu, N., et al.: Automatic recognition for arbitrarily tilted license plate. In: ICVIP, pp. 23–28 (2018)
17. Lu, Q., Liu, Y., et al.: License plate detection and recognition using hierarchical feature layers from CNN. Multim. Tools Appl. **78**(11), 15665–15680 (2019)
18. Luo, C., Jin, L., Sun, Z.: MORAN: a multi-object rectified attention network for scene text recognition. PR **90**, 109–118 (2019)
19. Qin, S., Liu, S.: Towards end-to-end car license plate location and recognition in unconstrained scenarios. Neural Comput. Appl. **34**(24), 21551–21566 (2022)
20. Redmon, J., Farhadi, A.: YOLO9000: better, faster, stronger. In: CVPR, pp. 6517–6525 (2017)
21. Selmi, Z., Halima, M.B., Pal, U., et al.: DELP-DAR system for license plate detection and recognition. Pattern Recogn. Lett. **129**, 213–223 (2020)
22. Shi, B., Bai, X., Yao, C.: An end-to-end trainable neural network for image-based sequence recognition and its application to scene text recognition. TPAMI **39**(11), 2298–2304 (2017)
23. Silva, S.M., Jung, C.R.: License plate detection and recognition in unconstrained scenarios. In: ECCV, pp. 593–609 (2018)
24. Spanhel, J., et al.: Holistic recognition of low quality license plates by CNN using track annotated data. In: AVSS, pp. 1–6 (2017)
25. Vaswani, A., et al.: Attention is all you need. In: NeurIPS, pp. 5998–6008 (2017)
26. Wang, T., et al.: Decoupled attention network for text recognition. In: AAAI, pp. 12216–12224 (2020)
27. Wang, T., et al.: Efficient license plate recognition via parallel position-aware attention. In: PRCV, vol. 13536, pp. 346–360 (2022)
28. Wang, Y., Bian, Z., Zhou, Y., et al.: Rethinking and designing a high-performing automatic license plate recognition approach. TITS **23**(7), 8868–8880 (2022)
29. Wu, C., et al.: How many labeled license plates are needed? In: PRCV, pp. 334–346 (2018)

30. Xu, H., Zhou, X., Li, Z., et al.: EILPR: toward end-to-end irregular license plate recognition based on automatic perspective alignment. TITS **23**(3), 2586–2595 (2022)
31. Xu, Z., et al.: Towards end-to-end license plate detection and recognition: a large dataset and baseline. In: ECCV, pp. 261–277 (2018)
32. Zhang, K., Zhang, Z., Li, Z., et al.: Joint face detection and alignment using multitask cascaded convolutional networks. IEEE Signal Process. Lett. **23**(10), 1499–1503 (2016)
33. Zhang, L., Wang, P., Li, H., et al.: A robust attentional framework for license plate recognition in the wild. TITS **22**(11), 6967–6976 (2021)
34. Zhang, Y., Wang, Z., Zhuang, J.: Efficient license plate recognition via holistic position attention. In: AAAI, pp. 3438–3446 (2021)
35. Zherzdev, S., Gruzdev, A.: LPRNet: license plate recognition via deep neural networks. CoRR abs/1806.10447 (2018)
36. Zhu, X., et al.: Deformable convnets V2: more deformable, better results. In: CVPR, pp. 9308–9316 (2019)
37. Zhuang, J., et al.: Towards human-level license plate recognition. In: ECCV, pp. 314–329 (2018)
38. Zou, Y., et al.: License plate detection and recognition based on YOLOv3 and ILPRNET. Sign. Image Video Process. **16**(2), 473–480 (2022)

TNT-Net: Point Cloud Completion by Transformer in Transformer

Xiaohai Zhang, Jinming Zhang$^{(\boxtimes)}$, Jianliang Li, and Ming Chen

School of Computer Science and Technology, Xinjiang University, Urumqi, China
zhjm@xju.edu.cn, hi_klaus@foxmail.com

Abstract. Estimating the overall structure of a point cloud from a partial 3D point cloud input is a crucial task in computer vision. However, existing point cloud completion methods often overlook object detail information and the local correlation within the incomplete point cloud. To address this challenge, we propose an enhanced point cloud completion approach called TNT-Net, which leverages a transformer in transformer architecture for accurate and refined point cloud completion. TNT-Net incorporates a local feature extraction module to capture long-range correlations within the input point cloud. Moreover, we introduce stacked feature extractors to simplify subsequent calculations and gather more comprehensive feature information on the spatial distribution of the point cloud. Additionally, we present an efficient method that integrates the kNN-transformer into the existing point transformer to address the deficiency of local detail information in previous works. This method enables TNT-Net to capture fine-grained object details and local correlations more effectively. Extensive experiments conducted on synthetic datasets PCN, ShapeNet55/34, and real-world dataset KITTI demonstrate the superior quantitative and qualitative performance of TNT-Net compared to state-of-the-art methods.

Keywords: point cloud completion · deep learning · 3D vision

1 Introduction

Point clouds provide a new way to help us reconstruct and store 3D objects. Normally, LiDAR scanners obtain these points by emitting laser points onto objects to capture their geometric shape. However, due to the limitation of sensor resolution or object self-occlusion, the point cloud obtained in this way is usually incomplete, so the point cloud completion is very meaningful.

Since the release of PointNet [1] and PointNet++ [2], many pioneering works based on them have achieved remarkable results in the point cloud completion task [3–5]. However, limited by the size of the convolutional receptive field, learning features layer by layer through convolution will lead to the loss of some important details.

With transformers' amazing achievements in Natural Language Processing (NLP) [6] and Computer Vision (CV) [7], many pioneers have introduced transformers to the task of point cloud completion [8,9]. Using transformer in point

© The Author(s), under exclusive license to Springer Nature Switzerland AG 2024
S. Rudinac et al. (Eds.): MMM 2024, LNCS 14555, pp. 340–352, 2024.
https://doi.org/10.1007/978-3-031-53308-2_25

cloud tasks, we can learn spatially distant but closely related structural information. Compared with the previous only use of multi-layer perceptron (MLP) to extract features, transformers have significant advantages in extracting global features.

In order to solve the problem that existing point cloud completion methods cannot fully utilize known structural information, inspired by some recent works [8,10], we propose a model named TNT-Net. We design a local feature extractor to fully capture the spatial structure features of point cloud. Relying on the local-transformer, our local feature extractor module can obtain patterns with far geometric distance but close semantic distance. And it is convenient to reduce the excessive computational burden in the subsequent transformer. Transformer is good at dealing with long-range sequence information, but for extracting local information, there is no inductive bias to space like the convolution kernel. We design a skeleton-aware transformer block that can learn geometric information more widely to avoid the loss of significant detail information caused by premature max-pooling. Once the information is lost in the encoding phase, it can hardly be recovered in the decoding phase.

The main contributions of this work are:

1) Based on the encoder-decoder structure, the novel TNT-Net is designed for point cloud completion tasks. Our model can effectively retain local geometric details, avoid the loss of important information, and improve the performance of point cloud completion.
2) In order to reduce the impact of early pooling on feature information loss, we design a novel (k-nearest neighbors) kNN-transformer block for extracting local information from k-nearest neighbors and fuse the local features of its output with the central feature, which allows the global transformer to master more features to restore the point cloud.
3) We explore the robustness of our model for the task of point cloud reconstruction at different resolutions. Experiments and results show that our method outperforms other existing methods.

2 Related Work

In the early stage, researchers used voxel-based point cloud completion methods, which usually do not directly use point clouds but convert them into 3D voxels, are important for early research on the surface reconstruction of 3D objects [11–16]. There have been a series of achievements in the methods of processing voxels in 3D point cloud data. However, the computational overhead and resource consumption of these methods are often huge when generating high-resolution data, and it is difficult to deal with detailed information.

Since the PointNet [1] was proposed, researchers have used the multi-layer perceptron(MLP) to directly process the raw point cloud data in an end-to-end way for 3D tasks. PointNet++ [2] proposes a point cloud grouping algorithm farthest point sampling(FPS) to obtain the structural features of point clouds in the same group. Spatial features of point cloud are learned step-by-step in

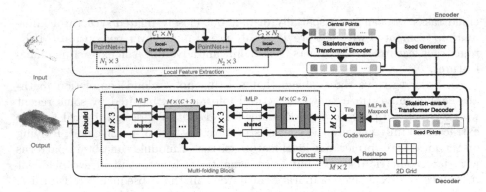

Fig. 1. Overview of TNT-Net structure. We use two stacked LFEM modules to perform feature extraction on the input point cloud and downsample the number of input point clouds to reduce the computation of the transformer.

a hierarchical way and obtained to a certain extent by mapping one point to multiple points. AtlasNet [3] proposes a local approximate target surface found by mapping a set of 2D squares to the surface of a 3D object. FoldingNet [4] put a new point cloud reconstruction method that concatenates features with a 2D grid. PCN [5] combines the methods of PointNet and FoldingNet to complete the point cloud in a coarse-to-fine manner. DGCNN [17] proposes edge convolution operation, which can preserve permutation invariance while capturing the structural characteristics of the local geometry of the point cloud. TopNet [18] proposes to build a point cloud into a rooted tree topology, and use this to reconstruct point clouds with arbitrary structures. PFNet [19] proposes a pyramid decoder for generating point clouds, which utilizes a multi-stage loss to supervise the generation of key points, thus reducing the missing geometric structure. PMP-Net [20] proposes to generate a complete point cloud by moving the points in the partial point cloud in multi-steps. Based on the Vision Transformer [7], PoinTr [8] proposes to use a set-to-set approach to process point clouds. The point cloud is transformed into a series of point proxies, and the geometrically aware transformer block is used to extract and reconstruct the feature of the point cloud. SnowflakeNet [9] and SeedFormer [21] explored the decoder for point cloud completion, SnowflakeNet proposed SPD deconvolution to recover the point cloud shape gradually, and SeedFormer proposed upsampling transformer layer to gradually recover the point cloud.

However, the existing methods lack the necessary inductive bias when dealing with local detail features, which leads to the unsatisfactory effect of point cloud completion.

3   Method

Inspired by [10] in the image field, the 2D image data is divided into small patches and input into transformer to obtain better local information. This has a similar situation to our point cloud completion. Our TNT-Net is based on an end-to-end encoder-decoder architecture and adopts a coarse-to-fine approach to generate point clouds. The pipeline of our method is shown in Fig. 1. In the encoder, we design the local feature extraction module(LFEM) and the skeleton-aware feature encoder to fully learn the spatial feature distribution of the point cloud. The decoder is composed of the skeleton-aware feature decoder and the Multi-folding module based on FoldingNet [4] to generate the complete point cloud.

3.1 Problem Statement

We assume that the set composed of partial point clouds is $X = \{x_i \mid i = 1, 2, ..., N\}$, where N is the number of input point clouds. The set composed of ground truth point clouds is Y_{gt}, the two point clouds output by the model is coarse point cloud $\tilde{Y}_{coarse}$ and fine point cloud $\tilde{Y}_{fine}$.

The task of point cloud completion is to compare the predicted results $\tilde{Y}_{fine}$ with the ground truth Y_{gt}, making them as close as possible. In the following sections, we will introduce several commonly used metrics to measure point cloud similarity and how to calculate them.

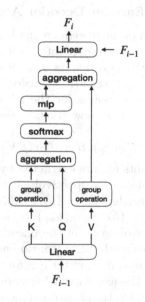

Fig. 2. Local-transformer for extracting point features.

3.2 Local Feature Extraction Module

Considering that directly processing each point of input point cloud in the skeleton-aware transformer module will bring much more computational complexity, we first need to reduce the number of input points while preserving enough structure information of point cloud. Our LFEM applies set abstraction layers to downsample the point cloud, where the center points $X_p \in \mathbb{R}^{N_p \times 3}$ are obtained by FPS and then k-NN is employed to aggregate the features of neighbor points corresponding to the current center point, the local feature is marked $F_p \in \mathbb{R}^{N_p \times C_p}$. Then, X_p and F_p are fed into the local-transformer to fuse the semantic connection of local feature information through attention. local-transformer is shown in Fig. 2.

Local-transformer takes center points extracted by PointNet++ and the features of the neighboring points of the center point as input. Firstly, the center point set is embedded as the position, and the grouping operation is performed with key and value respectively. Then the key and query are correlated to obtain attention and then input into the normalization function softmax. Finally, the output features of the previous layer are embedded and added to obtain the final output of the layer. By stacking two LFEM layers, we downsample the center points to appropriate amount and obtain sufficient local features, then feed them into the skeleton-aware transformer. In the subsequent experiments, it can be proved that our LFEM module has better performance in obtaining the structural information of the point cloud.

3.3   Transformer-Based Encoder-Decoder Architecture

From the previous section, in order to reduce the length of the sequence input into the transformer, TNT-Net converts it into a set of feature vectors of center points, where each center point represents its own local region. The center points can be regarded as the input sequence of the transformer, and these center points will be used to generate the seed points for the missing part. The transformer encoder-decoder architecture can be viewed as:

$$V = \mathcal{T}_E(F_p), H = \mathcal{T}_D(Q, V), \tag{1}$$

where F_p is the center points feature extracted by LFEM, $\mathcal{T}_E, \mathcal{T}_D$ represents the transformer encoder and decoder respectively, $V = \{v_1, v_2, ..., v_N\}$ represents the output feature of the encoder, $Q = \{q_1, q_2, ..., q_M\}$ represents the query feature of the seed point, $H = \{h_1, h_2, ..., h_M\}$ is the seed point for predicting the missing part. The number of layers in the encoder-decoder is L_E layer and L_D layer. In the encoder, by using the self-attention mechanism, the center point can learn more rich long-range and short-term information to update the point features and then input into the feed-forward network(FFN) to further update the point features. In the decoder, the skeleton structure information is obtained based on self-attention and cross-attention mechanisms. The input seed points are enriched with local feature information by the self-attention mechanism, and cross-attention is used to find the relationship between the query and the output.

3.4   Skeleton-Aware Transformer Block

Applying the excellent ability of the transformer to process sequence-to-sequence to the task of point cloud completion is a recent hot topic. We try to endow the transformer with more feature information in order to achieve better point cloud completion performance. Transformer is good at dealing with the correlation information of long sequences, but it lacks inductive bias like convolution operations [7]. To solve this problem, we design kNN-transformer, so as to learn more comprehensive feature distribution information. We design the skeleton-aware transformer block to embed the kNN-transformer with neighbor semantic information into the existing transformer structure. The kNN-transformer is based on point transformer [22]. Skeleton-aware transformer block is shown in Fig. 3. By embedding kNN information into the existing transformer, the model learns more inductive bias.

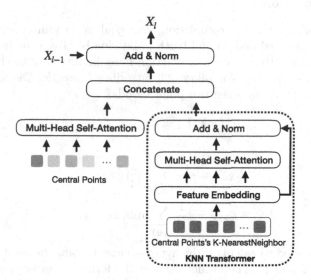

Fig. 3. The proposed Skeleton-aware transformer block.

3.5   Seed Generator

This module is used to output the rough structure of the predicted geometric object, which is used as the query for the subsequent transformer. Our seed generator dynamically generates seed points based on the output of the transformer encoder. Similar to PCN [5], we enlarge the query features obtained by the generator to a higher dimension through MLP and then use point-wise max-pool to obtain codewords. Then, the codewords are linearly mapped to $M \times 3$ dimensional point features, and the residual connection with the previous global features are entered into the MLP as the seed feature for subsequent completion.

3.6   Multi-Folding Moudle

Our TNT-Net model is used to predict the missing part of the incomplete point cloud, and the skeleton-aware transformer decoder outputs the predicted seed point cloud. Here, we propose multi-layer folding modules based on Fold-ingNet [4] in order to reuse these seed points to generate missing point clouds at different resolutions.

$$N_i = f(H_i) + c_i, i = 1, 2, ..., M \tag{2}$$

where N_i is a series of neighbor points centered at c_i and M is the number of seed points. After generating the missing part, we concatenate it with the input partial point cloud to obtain the final complete point cloud.

3.7   Loss Function

In the task of point cloud completion, our goal is to minimize the distance between the generated and ground truth point clouds. Given the inherent disor-der of point clouds, the chosen loss function must be invariant to changes in the order of the input points. We follow [23] introduce Charmfer Distance (CD) for computing the metric between two point clouds.

$$CD(X, Y) = \mathcal{L}_{X,Y} + \mathcal{L}_{Y,X} \tag{3}$$

, where

$$
\begin{aligned}
\mathcal{L}_{X,Y} &= \frac{1}{|X|} \sum_{x \in X} \min_{y \in Y} \|x - y\|_p, \\
\mathcal{L}_{Y,X} &= \frac{1}{|Y|} \sum_{y \in Y} \min_{x \in X} \|y - x\|_p
\end{aligned}
\tag{4}
$$

The CD is used to calculate the average closest point distance between two point clouds X and Y. The former term of the formula can be seen as in order to constrain the predicted points to be near the true points, and the latter term can be seen as the true points being uniformly distributed around the predicted points. Note that there are two variants for CD, which are $p = 1$ and $p = 2$. The difference between CD-ℓ_1 and CD-ℓ_2 is the square root of the Euclidean distance.

Our training loss is given by:

$$
\begin{aligned}
\mathcal{L}(Y_{coarse}, Y_{fine}, Y_{gt}) = CD(Y_{coarse}, Y_{gt}) + \\
CD(Y_{fine}, Y_{gt})
\end{aligned}
\tag{5}
$$

Table 1. Completion results on ShapeNet-55 dataset. We use CD-S, CD-M, and CD-H to represent the CD-$\ell_2 \times 10^3$ (lower is better) results under the Simple, Moderate, and Hard settings, and the F-Score@1% (higher is better).

Methods	Table	Chair	Plane	Car	Sofa	CD-S	CD-M	CD-H	CD-Avg	F1
FoldingNet [4]	2.53	2.81	1.43	1.98	2.48	2.67	2.66	4.05	3.12	0.082
PCN [5]	2.13	2.29	1.02	1.85	2.06	1.94	1.96	4.08	2.66	0.133
TopNet [18]	2.21	2.53	1.14	2.18	2.36	2.26	2.16	4.30	2.91	0.126
PFNet [19]	3.95	4.24	1.81	2.53	3.34	3.83	3.87	7.97	5.22	0.339
GRNet [24]	1.63	1.88	1.02	1.64	1.72	1.35	1.71	2.85	1.97	0.238
PoinTr [8]	0.81	0.95	0.44	0.91	0.79	0.58	0.88	1.79	1.09	0.464
SeedFormer [21]	0.72	0.81	0.40	0.89	0.71	0.50	0.77	1.49	0.92	0.472
Ours	**0.64**	**0.74**	**0.37**	**0.75**	**0.62**	**0.45**	**0.69**	**1.38**	**0.84**	**0.544**

Table 2. Completion results on ShapeNet-34 dataset. We use CD-S, CD-M, and CD-H to represent the CD-$\ell_2 \times 10^3$ (lower is better) results under the Simple, Moderate, and Hard settings, and the F-Score@1% (higher is better).

CD-ℓ_2(×1000)	34 seen categories					21 unseen categories				
	CD-S	CD-M	CD-H	Avg	F1	CD-S	CD-M	CD-H	Avg	F1
FoldingNet [4]	1.86	1.81	3.38	2.35	0.139	2.76	2.74	5.36	3.62	0.095
PCN [5]	1.87	1.81	2.97	2.22	0.154	3.17	3.08	5.29	3.85	0.101
TopNet [18]	1.77	1.61	3.54	2.31	0.171	2.62	2.43	5.44	3.50	0.121
PFNet [19]	3.16	3.19	7.71	4.68	0.347	5.29	5.87	13.33	9.16	0.322
GRNet [24]	1.26	1.39	2.57	1.74	0.251	1.85	2.25	4.87	2.99	0.216
PoinTr [8]	0.76	1.05	1.88	1.23	0.421	1.04	1.67	3.44	2.05	0.384
SeedFormer [21]	0.48	0.70	1.30	0.83	0.452	0.61	1.07	2.35	1.34	0.402
Ours	**0.43**	**0.61**	**1.12**	**0.72**	**0.527**	**0.56**	**0.97**	**2.22**	**1.25**	**0.489**

4 Experiments

In this section, we start by introducing a selection of widely utilized point cloud completion datasets. Subsequently, we adhere to the benchmark established by [8] and conduct an extensive series of experiments, including ablation experiments, to evaluate our proposed approach. We compare our TNT-Net with other SOTA models, including FoldingNet [4], PCN [5], TopNet [18], GRNet [24], PFNet [19] ,PoinTr [8], SnowflakeNet [9] and SeedFormer [21], on the ShapeNet55/34, KITTI and PCN datasets.

4.1 Experiments on ShapeNet55/34

We use ShapeNet-55/34 benchmark [8] to evaluate our model. This dataset contains many complex object shapes and incomplete patterns, which can comprehensively evaluate the performance of point cloud completion models. There are

55 types of objects in the ShapeNet55, which can fully show the performance of our model when dealing with different objects. We use 41,952 shapes for training and 10,518 shapes for testing.

The ShapeNet34 is divided from the ShapeNet55, and 34 categories are used for training, and the remaining 21 unseen categories are used for testing, which can fully demonstrate the ability of the model in dealing with unseen objects. In the seen categories, 100 object from different categories is randomly obtained for testing, the remaining 46,765 object shapes are used for training, and 2,305 shapes of 21 unseen categories are used for evaluation.

In both datasets, we sample 2048 points from the object as input and 8192 points as ground truth. We follow the benchmark [8] and fix 8 viewpoints and the model input points n is set to 2048(25%), 4096(50%), 6144(75%) of the complete point cloud, It corresponds to three kinds of difficulty levels simple, moderate, and hard.

Experiments show that TNT-Net has better generalization performance in unseen model categories. Results of the experiment can be seen in Table 1 and Table 2, some of the visualizations are in Fig. 4.

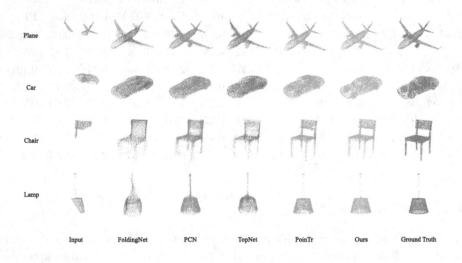

Fig. 4. The visualization results of point cloud completion in ShapeNet55/34, both in hard setting. The completion effect of our method is better than others.

4.2   Experiments on KITTI

The KITTI dataset [25] comprises LiDAR scans captured from real-world scenes, making it an ideal benchmark for assessing our model's real-world performance. We trained our model on the ShapeNetCars [5] dataset and evaluated its effectiveness on the KITTI dataset. Two metrics were employed for evaluation:(1)

Fidelity: This metric quantifies the average distance between each input point and its nearest neighbor in the model's output. It serves as a measure of how well the input points are preserved during the model's processing.(2) Minimum matching distance (MMD): This metric calculates the Chamfer Distance (CD) between the output and the car point cloud in ShapeNet, offering a measure of similarity between the output and a typical car.

The data results are presented in Table 3, while Fig. 5 showcases a visual representation of a subset of cars. TNT-Net exhibits impressive performance, demonstrating both qualitative and quantitative excellence.

Table 3. Result on KITTI dataset use Fidelity Distance and Minimal Matching Distance(MMD), both lower is better.

CD-ℓ_2(×1000)	FoldingNet [4]	PCN [5]	TopNet [18]	GRNet [24]	PoinTr [8]	SeedFormer [21]	Ours
Fidelity	7.467	2.235	5.354	0.816	**0.000**	0.151	**0.000**
MMD	0.537	1.366	0.636	0.568	0.526	0.516	**0.374**

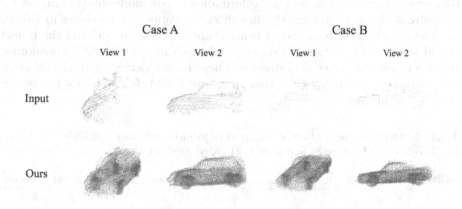

Fig. 5. Visualization results on KITTI dataset. We provide two different perspectives for each case.

4.3   Experiments on PCN

The PCN dataset [5] is the most widely used dataset for point cloud completion tasks, it has 30,974 shapes from 8 categories: airplane, cabinet, cat, chair, lamp, sofa, table, and vessel. All of these shapes are generated from the ShapeNet dataset by sampling 16384 points uniformly on the mesh surfaces and the partial point clouds are generated by back-projecting 2.5D depth images into 3D.

On the PCN benchmark, we compare it with the previous state-of-the-art methods, the excellent performance of our TNT-Net can be seen in Table 4.

Table 4. The performance of our methods and other methods on PCN benchmark, uses the CD-$\ell_1 \times 10^3$ results.

CD-ℓ_1($\times$1000)	Avg	Air	Cab	Car	Cha	Lam	Sof	Tab	Wat
FoldingNet [4]	14.31	9.49	15.80	12.61	15.55	16.41	15.97	13.65	14.99
AtlasNet [3]	10.85	6.37	11.94	10.10	12.06	12.37	12.99	10.33	10.61
TopNet [18]	12.15	7.61	13.31	10.90	13.82	14.44	14.78	11.22	11.12
PCN [5]	9.64	5.50	22.70	10.63	8.70	11.00	11.34	11.68	8.59
GRNet [24]	8.83	6.45	10.37	9.45	9.41	7.96	10.51	8.44	8.04
PMP-Net [20]	8.73	5.65	11.24	9.64	9.51	6.95	10.83	8.72	7.25
PoinTr [8]	7.92	4.67	10.07	8.70	8.67	6.98	10.06	7.27	6.96
SnowflakeNet [9]	7.21	4.29	**9.16**	8.08	7.89	**6.07**	**9.23**	6.55	**6.40**
Ours	**7.16**	**4.05**	9.31	**7.97**	**7.65**	6.12	9.27	**6.47**	6.43

4.4   Ablation Study

In order to verify the effectiveness of our proposed module, we conduct ablation experiments to evaluate the performance of our model by training on the ShapeNet34 dataset and using the unseen 21 categories test. As shown in Table 5, compared with the branch model using PointNet++ alone (a) and the branch model with local-transformer (b) and the branch model with kNN-transformer (c), TNT-Net has better performance. These results demonstrate the effectiveness of the two modules local-transformer and kNN-transformer for the performance of TNT-Net.

Table 5. Ablation study. Effect of each part of local-transformer and kNN-transformer in TNT-Net on ShapeNet34 and unseen 21 categories, use the CD-$\ell_2 \times 10^3$ results.

Methords	PointNet++	DGCNN	local-transformer	kNN-transformer	ShapeNet34	Unseen21
A	✓	✓			1.230	2.050
B	✓		✓		0.744	1.262
C	✓	✓		✓	0.811	1.387
TNT-Net	✓		✓	✓	**0.720**	**1.251**

5   Conclusion

In this paper, we present a novel encoder-decoder network TNT-Net for point cloud completion by transformer in transformer. In the construction of an incomplete point cloud geometry relationship, we extract rich original input information based on PointNet++ and point transformer. We propose embedding the kNN-transformer into the vanilla transformer to provide sufficient neighboring information. Our TNT-Net learns these features to progressively generate geometric objects with detailed information. Comprehensive experiments on

ShapeNet55, ShapeNet34, KITTI, and PCN benchmarks show that TNT-Net is superior to existing point cloud completion networks.

Acknowledgement. This work was sponsored by Natural Science Foundation of Xinjiang Uygur Autonomous Region under Grant 2022D01C690.

References

1. Charles, R.Q., Su, H., Kaichun, M., Guibas, L.J.: PointNet: deep learning on point sets for 3D classification and segmentation. In: 2017 IEEE Conference on Computer Vision and Pattern Recognition (CVPR), pp. 77–85 (2017). https://doi.org/10.1109/CVPR.2017.16
2. Qi, C.R., Yi, L., Su, H., Guibas, L.J.: PointNet++: deep hierarchical feature learning on point sets in a metric space. In: Advances in Neural Information Processing Systems, vol. 30, pp. 5099–5108 (2017)
3. Groueix, T., Fisher, M., Kim, V.G., Russell, B.C., Aubry, M.: A papier-Mache approach to learning 3D surface generation. In: 2018 IEEE/CVF Conference on Computer Vision and Pattern Recognition, pp. 216–224 (2018). https://doi.org/10.1109/CVPR.2018.00030
4. Yang, Y., Feng, C., Shen, Y., Tian, D.: FoldingNet: point cloud auto-encoder via deep grid deformation. In: 2018 IEEE/CVF Conference on Computer Vision and Pattern Recognition, pp. 206–215 (2018). https://doi.org/10.1109/CVPR.2018.00029
5. Yuan, W., Khot, T., Held, D., Mertz, C., Hebert, M.: PCN: point completion network. In: 2018 International Conference on 3D Vision (3DV), pp. 728–737 (2018). https://doi.org/10.1109/3DV.2018.00088
6. Vaswani, A., et al.: Attention is all you need. In: Advances in Neural Information Processing Systems, vol. 30, pp. 5998–6008 (2017)
7. Dosovitskiy, A., et al.: An image is worth 16×16 words: transformers for image recognition at scale. In: Proceedings of the International Conference on Learning Representations (2021)
8. Yu, X., Rao, Y., Wang, Z., Liu, Z., Lu, J., Zhou, J.: PoinTr: diverse point cloud completion with geometry-aware transformers. In: 2021 IEEE/CVF International Conference on Computer Vision (ICCV), pp. 12478–12487 (2021). https://doi.org/10.1109/ICCV48922.2021.01227
9. Xiang, P., et al.: SnowflakeNet: point cloud completion by snowflake point deconvolution with skip-transformer. In: 2021 IEEE/CVF International Conference on Computer Vision (ICCV), pp. 5479–5489 (2021). https://doi.org/10.1109/ICCV48922.2021.00545
10. Han, K., Xiao, A., Wu, E., Guo, J., Xu, C., Wang, Y.: Transformer in transformer. In: Advances in Neural Information Processing Systems, vol. 34, pp. 15908–15919 (2021)
11. Choy, C.B., Xu, D., Gwak, J.Y., Chen, K., Savarese, S.: 3D-R2N2: a unified approach for single and multi-view 3D object reconstruction. In: Leibe, B., Matas, J., Sebe, N., Welling, M. (eds.) ECCV 2016. LNCS, vol. 9912, pp. 628–644. Springer, Cham (2016). https://doi.org/10.1007/978-3-319-46484-8_38
12. Girdhar, R., Fouhey, D.F., Rodriguez, M., Gupta, A.: Learning a predictable and generative vector representation for objects. In: Leibe, B., Matas, J., Sebe, N., Welling, M. (eds.) ECCV 2016. LNCS, vol. 9910, pp. 484–499. Springer, Cham (2016). https://doi.org/10.1007/978-3-319-46466-4_29

13. Wang, P., Liu, Y., Guo, Y., Sun, C., Tong, X.: O-CNN: octree-based convolutional neural networks for 3D shape analysis. ACM Trans. Graph. **36**(4), 72:1–72:11 (2017)

14. Maturana, D., Scherer, S.: VoxNet: A 3D convolutional neural network for real-time object recognition. In: 2015 IEEE/RSJ International Conference on Intelligent Robots and Systems (IROS), pp. 922–928 (2015). https://doi.org/10.1109/IROS.2015.7353481

15. Dai, A., Qi, C.R., Nießner, M.: Shape completion using 3D-encoder-predictor CNNs and shape synthesis. In: 2017 IEEE Conference on Computer Vision and Pattern Recognition (CVPR), pp. 6545–6554 (2017). https://doi.org/10.1109/CVPR.2017.693

16. Han, X., Li, Z., Huang, H., Kalogerakis, E., Yu, Y.: High-resolution shape completion using deep neural networks for global structure and local geometry inference. In: 2017 IEEE International Conference on Computer Vision (ICCV), pp. 85–93 (2017). https://doi.org/10.1109/ICCV.2017.19

17. Wang, Y., Sun, Y., Liu, Z., Sarma, S.E., Bronstein, M.M., Solomon, J.M.: Dynamic graph CNN for learning on point clouds. ACM Trans. Graph. **38**(5), 146:1–146:12 (2019)

18. Tchapmi, L.P., Kosaraju, V., Rezatofighi, H., Reid, I., Savarese, S.: TopNet: structural point cloud decoder. In: 2019 IEEE/CVF Conference on Computer Vision and Pattern Recognition (CVPR), pp. 383–392 (2019). https://doi.org/10.1109/CVPR.2019.00047

19. Huang, Z., Yu, Y., Xu, J., Ni, F., Le, X.: PF-Net: point fractal network for 3D point cloud completion. In: 2020 IEEE/CVF Conference on Computer Vision and Pattern Recognition (CVPR), pp. 7659–7667 (2020). https://doi.org/10.1109/CVPR42600.2020.00768

20. Wen, X., et al.: PMP-Net: point cloud completion by learning multi-step point moving paths. In: 2021 IEEE/CVF Conference on Computer Vision and Pattern Recognition (CVPR), pp. 7439–7448 (2021). https://doi.org/10.1109/CVPR46437.2021.00736

21. Zhou, H., et al.: SeedFormer: patch seeds based point cloud completion with upsample transformer. In: Avidan, S., Brostow, G., Cissé, M., Farinella, G.M., Hassner, T. (eds.) ECCV 2022. LNCS, vol. 13663, pp. 416–432. Springer, Cham (2022). https://doi.org/10.1007/978-3-031-20062-5_24

22. Zhao, H., Jiang, L., Jia, J., Torr, P.H.S., Koltun, V.: Point transformer. In: Proceedings of the IEEE/CVF International Conference on Computer Vision, pp. 16239–16248 (2021)

23. Fan, H., Su, H., Guibas, L.: A point set generation network for 3D object reconstruction from a single image. In: 2017 IEEE Conference on Computer Vision and Pattern Recognition (CVPR), pp. 2463–2471 (2017). https://doi.org/10.1109/CVPR.2017.264

24. Xie, H., Yao, H., Zhou, S., Mao, J., Zhang, S., Sun, W.: GRNet: gridding residual network for dense point cloud completion. In: Vedaldi, A., Bischof, H., Brox, T., Frahm, J.-M. (eds.) ECCV 2020. LNCS, vol. 12354, pp. 365–381. Springer, Cham (2020). https://doi.org/10.1007/978-3-030-58545-7_21

25. Geiger, A., Lenz, P., Stiller, C., Urtasun, R.: Vision meets robotics: the KITTI dataset. Int. J. Robotics Res. **32**(11), 1231–1237 (2013)

Fourier Transformer for Joint Super-Resolution and Reconstruction of MR Image

Jiacheng Chen(iD), Fei Wu(iD), Wanliang Wang$^{(\boxtimes)}$(iD), and Haoxin Sheng(iD)

The College of Computer Science and Technology, Zhejiang University of Technology, Hangzhou, China
15968812143@163.com, zjutwwl@zjut.edu.cn

Abstract. Super resolution (SR) and fast magnetic resonance (MR) imaging play a paramount role in medical diagnosis community. Though significant progress has been achieved, existing methods are mostly trapped at the local feature dependencies. Transformer holds the potential to tackle this issue, yet it is notoriously known with heavy computational burden. As a remedy, we propose a Fourier Transformer Network (FTN), which leverages 2D Fast Fourier Transform (FFT) to harvest the global relationships. Specifically, multiple Fourier Transformer Blocks (FTBs) are stacked as the backbone, comprehensively extracting the sematic representations. Due to the appealing efficiency of FFT, the overall model complexity is linear to tokens, as opposed to the quadratic complexity of previous transformer-based models. Besides, an edge-enhancement branch is incorporated into FTB in a flexible manner, which further sharpens the contour information of the organs and tissues. Extensive experiments on IXI dataset validate the superiority of FTN over most state-of-the-art methods.

Keywords: MRI · Super Resolution · Reconstruction · Transformer · Fourier transform

1 Introduction

Magnetic Resonance (MR) images are frequently employed in disease diagnosis and pathology experiments due to the appealing properties of non-radiation, non-invasion, and high contrast [14] [17]. However, the resolution and imaging speed of MRs are often constrained by the physical limitations of practical devices [2]. Generally, the prolonged acquisition time and the requirement of breath-holding usually cause patient discomfort and motion artifacts [25]. To facilitate an accurate medical diagnosis, there is an urgent need to speed up acquisition process and enhance image resolution [5] [27].

Among the current solutions, deep learning-based methods enjoy high reputation with promising results, which use large amounts of external data to learn the intrinsic properties hidden in the images. For example, Schlemper et al. [19]

Supported by organization x.

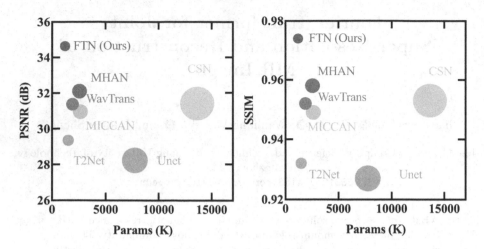

Fig. 1. Comparisons of PSNR(SSIM)/Params among various MRI reconstruction and super-resolution methods. Y-axis and x-axis respectively indicate the metric of PSNR (SSIM) and Params. The circle radius also represents the metrics of Params. Our FTN outperforms previous methods while enjoying high-efficiency assurance.

developed a deep cascade of CNN for dynamic MR image reconstruction, which outperforms the traditional compressed sensing approaches in terms of both speed and reconstruction error. MICCAN [11] recovered high-quality MR image details from a channel-wise attention network, which can attend to prominent features while filtering out the irrelevant ones. Hu et al. [10] proposed a pyramid orthogonal attention network (POAN), using both pixel-level and patch-level similarity to fully exploit the inherent structure within the MR images. Besides, a wide weighted attention multi-scale network (W2AMSN) [21] was proposed to ensure the generation of full-scale features. Furthermore, channel splitting network [26] integrated residual branch and dense branch in a uniform manner, efficiently dealing with the hierarchical characteristics on distinct channels. However, above approaches tend to ignore the intrinsic relationship between reconstruction task and super-resolution task, resulting in degraded network performance. By using two separate branches, Feng et al. [6] proposed a task transformer network (T2Net) to jointly conduct MRI reconstruction and super-resolution. However, the multi-branch treatment inevitably suffers from heavy model burden. Moreover, the generalization is also limited due to the individual learning of task-concerned features. As shown in Fig. 1, we can observe that our proposed FTN based on single-branch treatment outperforms previous methods while enjoying high-efficiency assurance.

With the aid of Fourier transform, in this work, we propose a novel Fourier Transformer Network (FTN) to address the issues. To the best of our knowledge, it is the first attempt to integrate FFT with transformer architecture. Moreover, the proposal is applied for jointly MR image SR and Reconstruction. First, a Fourier Transformer Block (FTB) is elaborately designed as the backbone, which could simultaneously capture the local and global features. On that basis, an edge-enhanced module is equipped, allowing FTB to adaptively absorb meaningful regions (e.g., pathology information and anatomical structure) while eliminating valueless characteristics (e.g., background and smooth areas).

2   Related Work

2.1   MRI Reconstruction

Magnetic resonance imaging (MRI) is a non-invasive medical imaging technique that has become a critical tool in diagnosing diseases and monitoring treatment outcomes. However, the reconstruction process of MRI images can be computationally complex, resulting in slow imaging speeds, which can hinder its widespread use in clinical practice. To address this issue, researchers have developed various acceleration techniques, among which compressed sensing and deep learning are the most commonly employed. Compressed sensing [20] techniques aim to reduce the amount of data acquired during MRI scans, which in turn improves the image reconstruction speed. Various compressed sensing techniques, including sparse regularization [24], edge-based methods [12], and dictionary learning [16], have been proposed to accelerate MRI image reconstruction. However, these hand-designed methods often require long optimization times and have poor generalization capabilities due to the need for strong priori information [9]. Recently, deep learning methods have also shown promising results in improving MRI image reconstruction speeds. Convolutional neural networks (CNNs) have been widely used to reconstruct MRI images from highly undersampled k-space data, achieving high-quality results at fast speeds. Schlemper et al. [19] develop a deep cascade of CNN for dynamic MR image reconstruction, which outperforms compressed sensing approaches in terms of speed and reconstruction error. MICCAN [11] recovered high-quality MR image details form a cascaded channel-wise attention network which can attend to prominent features while filtering out irrelevant information. Unlike these methods mentioned above, Wang et al. [22] design a network that performs directly on the complex values, named DeepcomplexMRI, which is able to preserve the intrinsic relationship between the real and imaginary parts. Then, Feng et al. [8] introduce OctConv [3] to address the shortcoming that CNN cannot compute complex numbers. Since artifacts in the image domain are often non-local and structure, some researchers have started to introduce hybrid domain learning paradigm that exploits both image space and k-space domain. Zhou et al. [27] propose a dual domain recurrent network (DuDoRNet) to simultaneously recover k-space data and images. It is experimentally proven to outperform single domain reconstruction methods.

2.2   MRI Supre-Resolution

Note again MRI is a radiation-free and non-invasive technique that highlights deep details of tissues through multiple sequences of images. Nevertheless, MRI also suffers form low resolution and artifacts due to long scan time and unavoidable patient motions. Deep learning-based methods [14] are widely used for MR image SR. Hu et al. [10] proposed a pyramid orthogonal attention network (POAN), using both pixel-level and patch-level similarity to full exploit the inherent self-similarity within the MR images. Furthermore, a wide weighted attention multi-scale network (W2AMSN) [21] was proposed, which could extracted multi-scale features effectively. Technically, W2AMSN focused the network on graph information-rich regions by using a non-reduction attention mechanism to distinguish the foreground from background. Besides, channel splitting network (CSN) was developed by Zhao et al. [26], which integrated residual branch and dense branch in a uniform manner, dealing with hierarchical characteristics efficiently along distinct channels. However, all the above approaches for MRI-SR depend solely on a single contrast image(e.g., T1WI, T2WI or PDWI), neglecting the complicated relations among different modalities. To that end, MINet [4] was proposed , which leveraged another modality to assist the reconstruction of existing modality. Based on MINet, Feng et al. [7] further designed a separation attention to make the network focused on the relevant information. Unfortunately, obtaining aligned multi-modal data is extraordinary in practice. Therefore, the concentration of our study remains on the restoration of single-modality MR images. Note again that most current methods tend to ignore the relationship between reconstruction and super-resolution, which greatly increases the difficulty of training and reduce the effectiveness of recovery. Therefore, we design a network that is capable of performing both reconstruction and super-resolution.

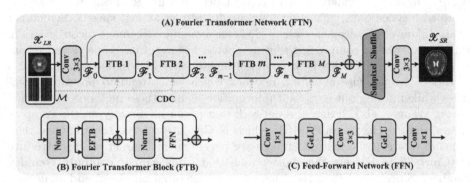

Fig. 2. (A) The overall architecture of the proposed Fourier Transformer Network (FTN). (B) Fourier Transformer Block (FTB) (C). Feed-Forward Network (FFN).

3   Methodology

3.1   Overview Framework

Given a Low Resolution (LR) input $X_{LR} \in R^{H \times W \times 1}$, FTN aims at predicting a full-sampling MR image $X_{SR} \in R^{sH \times sW \times 1}$($H, W$, and s are the height, width and upscale factor, respectively) that is as close as possible to the ground truth $X_{HR} \in R^{sH \times sW \times 1}$. Generally, as shown in Fig. 2 (A), we first leverage one convolution layer $\mathcal{H}_{3 \times 3}$ to extract the shallow features $F_0 \in R^{H \times W \times C}$ from X_{LR}:

$$F_0 = \mathcal{H}_{3 \times 3}(X_{LR}) \tag{1}$$

where C denotes the channel size. Afterwards, M fourier transformer blocks (FTBs) are stacked to learn the mapping from F_0 to deeper features $F_M \in R^{H \times W \times C}$. The formulation can be written as:

$$F_m = \mathcal{H}_{FTB}^m(...\mathcal{H}_{FTB}^1(F_0)) \tag{2}$$

where $\mathcal{H}_{FTB}^m(\cdot)$ and F_m respectively denote the function of m-th FTB and the learned features, $1 \leq m \leq M$. Note that each output of FTB serves immediately as the input for the following block, without any modifications and extra manipulations required. On that basis, both deep feature F_M and shallow feature F_0 are transferred to an up-sampling layer for the reconstruction of high resolution (HR) MR image:

$$F_{up} = \mathcal{H}_{up}(F_M + F_0) \tag{3}$$

where $\mathcal{H}_{up}(\cdot)$ is the up-sampling operation consisting of a convolutional layer and a sub-pixel layer. Within this operation, a large amount of low frequency information is directly transmitted to the upsampling module through a residual connection. As a result, the pending mission mainly lies in recovering the missing high-frequency information, which minimizes the training difficulty and speeds up the convergence of the network. Finally, the reconstruction can be attained by a single convolutional layer that transforms the information from the feature domain to the image domain, such as:

$$X_{SR} = \mathcal{H}_{3 \times 3}(F_{up}) \tag{4}$$

where $\mathcal{H}_{3 \times 3}(\cdot)$ represents the convolution operation with kernel size 3.

L_1 loss is employed to train the network, which can be formulated as:

$$\mathcal{L}(\theta) = \frac{1}{N} \sum_{i=1}^{N} \|\mathcal{H}_{FTN}(X_{LR}^i; \theta) - X_{HR}^i\|_1 \tag{5}$$

where $\mathcal{H}_{FTN}(\cdot)$ is the mapping function given learned parameter θ. Practically, FTN is trained using data pairs $\{X_{LR}^i, X_{HR}^i\}_{i=1}^N$, which contain N low-quality MR images and their corresponding high-quality images.

3.2 Fourier Transformer Block (FTB)

In this subsection, the main components of our network, i.e., FTB, will be elaborated, which taps the potential of capturing global features with linear complexity. Particularly, as shown in Fig. 2 (B), FTB is comprised of two subassemblies: a normalization layer followed by an Edge-enhanced Fourier Transform Block

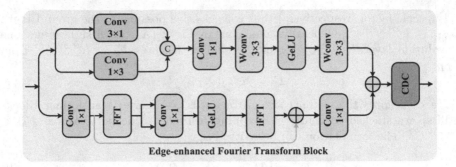

Fig. 3. The architecture of our Edge-enhanced Fourier Transform Block (EFTB).

(EFTB) and a normalization layer followed by a Feed-Forward Network (FFN). The practical computation of the m-th FTB can be represented as:

$$F'_m = \mathcal{H}_{EFTB}(\mathcal{H}_{LN}(F_m)) + F_m \tag{6}$$

$$F_{m+1} = \mathcal{H}_{FFN}(\mathcal{H}_{LN}(F'_m)) + F'_m \tag{7}$$

where F_m and F_{m+1} are the outputs of the EFTB module and FFN module, respectively. $\mathcal{H}_{LN}(\cdot)$ denotes the layer normalization [1] operation.

Edge-enhanced Fourier Transform Block (EFTB). To explore both the global information (e.g., tissue structure and pathological information) and local information (e.g., organ edges and texture), as shown in Fig. 3, EFTB mainly contains two branches, i.e., Fourier Transform Branch (ETB) and Edge-enhanced Branch (EEB). Note the complexity of the vanilla Transformer architecture grows quadratically with the spatial resolution of MR image. In this work, we attempt to relieve the computational burden yet hold the capability of capturing the long-range dependencies. Fast Fourier Transform is the natural choice to behave as the substitute for the multi-head attention mechanism, which enjoys linear complexity and whose frequency features cover the spatial information globally. Given feature map $Y \in R^{H \times W \times C}$, we first leverage a 1×1 Conv to reduce the gap between image domain and k-space domain. Then, 2D Fast Fourier Transform is employed to convert the output Y_1 to k-space features:

$$Y_1 = \mathcal{H}_{1 \times 1}(Y), Y' = \mathcal{H}_{2DFFT}(Y_1) \tag{8}$$

where Y' represents the complex value k-space features. Since the network cannot handle complex numbers directly, we concatenate the real and imaginary parts of Y' in the channel dimension. A 1×1 Conv is then utilized to recover the global information, which can be formulated as:

$$Y'' = \sigma(\mathcal{H}_{1 \times 1}([\mathcal{R}(Y'), \mathcal{I}(Y')])) \tag{9}$$

where Y'' indicates the refined global features. $\sigma(\cdot)$ and $[\cdot]$ represent the GELU function and concat operation, respectively. Moreover, 2D Inverse FFT is appended, converting Y'' back to the image domain, which can be written as:

$$Y_{FFT} = \mathcal{H}_{1 \times 1}(\mathcal{H}_{2DIFFT}(Y'') + Y_1) \tag{10}$$

where $\mathcal{H}_{2DIFFT}(\cdot)$ and $\mathcal{H}_{1\times1}(\cdot)$ denote 2D Inverse FFT and 1×1 Conv, respectively. To facilitate a better training, residual connection is introduced in this process.

EEB is used to harvest detailed information on tissue texture and organ edge, which are known as crucial factors for the subsequent MRI treatment. Specifically, an asymmetric convolution layer is firstly leveraged to extract edge characteristics.

$$Y_{1\_3} = \mathcal{H}_{1\_3}(Y), Y_{3\_1} = \mathcal{H}_{3\_1}(Y) \tag{11}$$

where $\mathcal{H}_{1\_3}(\cdot)$ and $\mathcal{H}_{3\_1}(\cdot)$ denote 1×3 Conv and 3×1 Conv, respectively. Note the two convolution kernels 1×3 and 3×1 own distinct shapes compared with the conventional one, which holds the potential of successfully matching diverse regional patterns. To further enhance the representational power, a wide convolution layer is further employed, such as:

$$Y_{EDGE} = \mathcal{H}_W(\sigma(\mathcal{H}_W(\mathcal{H}_{1\times1}([Y_{1\_3}, Y_{3\_1}])))) \tag{12}$$

By using wide Conv $\mathcal{H}_W$, more information can be made available through the GELU activation (σ). Afterwards, Y_{EDGE} and Y_{FFT} are fed to the channel-wise data consistency (CDC) layer to avoid corruption of sampled data.

Feed-Forward Network (FFN). As depicted in Fig. 2 (C), FFN consists of a 1×1 Conv layer followed by a GELU activation function, a Depth-wise Conv with 3×3 kernel size with a GELU activation function, and another 1×1 Conv layer, such as:

$$F_m = \mathcal{H}_{1\times1}(\sigma(\mathcal{H}_{DW3\times3}(\sigma(\mathcal{H}_{LN}(F_m'))))) + F_m' \tag{13}$$

where F_m' and F_m are the input and output of m-th FFN, respectively.

4   Experiments

4.1   Dataset and Implementation Details

Datasets: The dataset used in this paper is derived initially from IXI, which can be download from http://brain-development.org/ixi-dataset/ and consists of three types of MR images (578 PD volumes, 581 T1 volumes, and 578 T2 volumes). We split dataset patient-wisely into a ratio of 7:1:2 for training/validation/testing. In our research, two distinct k-space undersampling masks, i.e., random mask and equispaced fraction mask, are explored. For numerical presentation, Peak Signal-to-Noise Ratio (PSNR) and Structure Similarity Index Measurement (SSIM) are used.

Implementation Details: With acceleration rates 4/8 and scale factors 2/4, we train all the competing models to reconstruct zero-filled low-resolution images. The used optimizer is Adam [13]. Epoch and batch size are respectively set as 50 and 8. The learning rate is set to 1×10^{-4} in the beginning and is halved every 10 epochs during the training procedure. In addition, we randomly augment the image data during training phase by flipping horizontally or vertically and rotating at different angles. In our FTN, all convolution layers contain 48 filters except the 1×1 Conv layers. Finally, 15 FTBs, i.e., $M = 15$, are contained in our network.

4.2 Ablation Study

Table 1. Ablation study of different components in FTN.

Case Index	Model 0	Model 1	Model 2	Model 3	Model 4
EEB	✗	✔	✗	✔	✔
FTB	✗	✗	✔	✔	✔
CDC	✗	✗	✗	✗	✔
PSNR↑	25.972	27.805	27.586	28.680	**29.074**
SSIM↑	0.884	0.917	0.911	0.928	**0.934**

In this subsection, we show the ablation study in Table 1 to investigate the roles of different components in FTN. All variants (Models 0–4) are trained with acceleration rate 4 and upscale scale factor 4. The PSNR and SSIM values are evaluated on T2 modality for equispaced fraction mask. Model 0 represents the baseline network without any of the developed modules equipped. From Table 1, we can see that EEB brings 1.833 dB (PSNR) and 0.033 (SSIM) gains. This can be attributed to the fact that EEB enables model to learn clearer anatomical structure and edge information. Additionally, the PSNR value raises by 1.614 dB and the SSIM rises from 0.884 to 0.911 when FTB is added to the baseline. This demonstrates the importance of FTB and reveals that the performance of SR and reconstruction can be truly promoted via global information. Moreover, Model 3 achieves 2.708 dB (PSNR) and 0.044 (SSIM) improvement with the help of EEB and FTB together. Note the variant with CDC further improves restoration results in terms of PSNR and SSIM. The CDC is designed to prevent feature biases incurred by integrating irrelevant information representation during the updating stage.

4.3 Comparison with State-of-the-Arts

To evaluate the superiority of our proposal, several state-of-the art methods on MR images are compared both quantitatively and qualitatively, including Unet [18], T2Net [6], MICCAN [11], CSN [26], WavTrans [15] and MHAN [23].

Quantitative Results: As shown in Table 2 and 3, all the quantitative results for different acceleration rates (×4 and ×8) and upscale factors (×2 and ×4) on IXI T1 and T2 modality are respectively reported. Compared with other

Table 2. Quantitative results of T1 modality. The best and second results are highlighted and underlined, respectively.

scale	accelerate	4		8	
	mask	Random PSNR/SSIM	Equispaced PSNR/SSIM	Random PSNR/SSIM	Equispaced PSNR/SSIM
2	UNet	25.728/0.861	25.534/0.860	23.218/0.786	23.531/0.801
	T2Net	26.657/0.867	26.326/0.863	23.302/0.782	23.560/0.793
	MICCAN	28.335/0.900	28.136/0.898	24.503/0.799	25.506/0.836
	CSN	27.977/0.897	27.798/0.895	24.816/0.827	25.092/0.835
	WavTrans	27.914/0.895	27.716/0.894	24.709/0.824	25.021/0.834
	MHAN	28.606/0.907	28.476/0.907	25.162/0.833	25.573/0.846
	FTN (ours)	**29.991/0.928**	**30.029/0.930**	**27.060/0.844**	**27.112/0.846**
4	UNet	22.378/0.741	23.300/0.781	21.040/0.687	22.156/0.733
	T2Net	23.405/0.770	24.641/0.811	21.651/0.711	23.039/0.759
	MICCAN	23.711/0.763	24.801/0.805	22.046/0.702	23.378/0.754
	CSN	24.265/0.799	25.524/0.836	22.297/0.736	23.837/0.786
	WavTrans	24.131/0.796	25.446/0.837	22.162/0.735	23.730/0.785
	MHAN	24.180/0.796	25.531/0.837	22.420/0.739	23.835/0.785
	FTN (ours)	**24.533/0.800**	**26.292/0.856**	**23.502/0.751**	**24.121/0.799**

Table 3. Quantitative results of T2 modality. The best and second results are highlighted and underlined, respectively.

scale	accelerate	4		8	
	mask	Random PSNR/SSIM	Equispaced PSNR/SSIM	Random PSNR/SSIM	Equispaced PSNR/SSIM
2	UNet	28.230/0.927	28.136/0.927	25.692/0.889	26.203/0.898
	T2Net	29.347/0.932	28.934/0.929	25.825/0.887	26.187/0.893
	MICCAN	30.955/0.949	30.669/0.947	27.548/0.908	27.678/0.907
	CSN	31.402/0.953	31.000/0.951	27.713/0.917	28.288/0.924
	WavTrans	31.372/0.952	30.669/0.947	27.548/0.908	27.678/0.907
	MHAN	32.106/0.958	31.618/0.956	27.962/0.921	28.571/0.928
	FTN (ours)	**34.641/0.974**	**34.851/0.972**	**30.060/0.9440**	**30.445/0.946**
4	UNet	24.295/0.848	25.075/0.866	23.079/0.817	24.093/0.844
	T2Net	25.226/0.869	26.319/0.891	23.483/0.832	24.805/0.861
	MICCAN	25.199/0.858	26.154/0.881	23.870/0.829	24.858/0.853
	CSN	26.206/0.890	27.603/0.912	24.355/0.852	25.750/0.882
	WavTrans	26.258/0.892	27.865/0.917	24.295/0.854	25.703/0.883
	MHAN	26.437/0.898	27.874/0.917	24.389/0.869	25.816/0.889
	FTN (ours)	**26.778/0.901**	**28.872/0.931**	**24.509/0.858**	**26.201/0.892**

advanced methods, it can be seen that, on all scaling factors and acceleration rates, the proposed FTN shows superiority and outperforms other state-of-the-art approaches by a large margin. Particularly, under ×2 upscale factor with ×4 Cartesian acceleration for random sampling mask at T2 modality, FTN exceeds the second-best method (CSN) by over 3.2 dB (PSNR) and 0.021 (SSIM), while enjoying mild model parameters (as shown in Table 3). Moreover, FTN also obtains the best PSNR of 30.445, SSIM of 0.946 with ×2 upscale factor and ×8 equispaced sampling mask. In the more challenging case with ×4 upscale factor, FTN again achieves the best results.

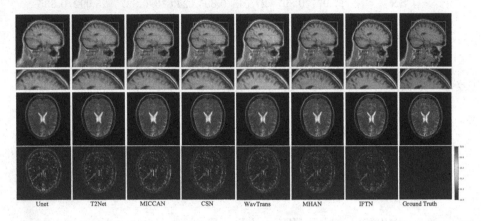

Fig. 4. Visual comparisons on IXI T2 modality with acceleration rate 4 and upscale factor 4.

Visual Comparisons On an example MR image of IXI, Fig. 4 visualizes the reconstructed results by different methods. As can be observed, the image reconstructed by FTN enjoys more precise features and sharper edges when compared to other competitors, most of those results are evidently blurred. These findings cab also be ensured from the error maps. All the other competitors suffer severely from some noisy points or abnormal lines in the central position. Comparatively, the difference between FTN and GT is negligible.

5   Conclusions

In this paper, we propose a Fourier Transformer Network for joint reconstruction and super resolution of MR images. Specifically, Fast Fourier Transform is merged with the transformer architecture to improve the global feature exploration capability. Since we utilize FFT to transfer the features from image domain to k-space, global characteristics can be naturally captured, thereby allowing FTN to hold richer and more valuable information for promoting discrimination ability of the network. In addition, an edge-enhanced branch is integrated into FTB to further refine edges and textures. Experimental results show the superiority of our proposed FTN against most currently leading MR image restoration approaches.

References

1. Ba, J.L., Kiros, J.R., Hinton, G.E.: Layer normalization. arXiv preprint arXiv:1607.06450 (2016)
2. Carmi, E., Liu, S., Alon, N., Fiat, A., Fiat, D.: Resolution enhancement in MRI. Magn. Reson. Imag. **24**(2), 133–154 (2006)
3. Chen, Y., et al.: Drop an octave: Reducing spatial redundancy in convolutional neural networks with octave convolution. In: Proceedings of the IEEE/CVF International Conference on Computer Vision, pp. 3435–3444 (2019)
4. Feng, C.-M., Fu, H., Yuan, S., Xu, Y.: Multi-contrast MRI super-resolution via a multi-stage integration network. In: de Bruijne, M., et al. (eds.) MICCAI 2021. LNCS, vol. 12906, pp. 140–149. Springer, Cham (2021). https://doi.org/10.1007/978-3-030-87231-1_14
5. Feng, C.M., Yan, Y., Chen, G., Xu, Y., Hu, Y., Shao, L., Fu, H.: Multi-modal transformer for accelerated mr imaging. IEEE Transactions on Medical Imaging (2022)
6. Feng, C.-M., Yan, Y., Fu, H., Chen, L., Xu, Y.: Task transformer network for joint MRI reconstruction and super-resolution. In: Bruijne de, M., et al. (eds.) MICCAI 2021. LNCS, vol. 12906, pp. 307–317. Springer, Cham (2021). https://doi.org/10.1007/978-3-030-87231-1_30
7. Feng, C.M., Yan, Y., Yu, K., Xu, Y., Shao, L., Fu, H.: Exploring separable attention for multi-contrast mr image super-resolution. arXiv preprint arXiv:2109.01664 (2021)
8. Feng, C.M., Yang, Z., Fu, H., Xu, Y., Yang, J., Shao, L.: Donet: dual-octave network for fast MR image reconstruction. IEEE Transactions on Neural Networks and Learning Systems (2021)
9. Hammernik, K., et al.: Learning a variational network for reconstruction of accelerated MRI data. Magn. Reson. Med. **79**(6), 3055–3071 (2018)
10. Hu, X., Wang, H., Cai, Y., Zhao, X., Zhang, Y.: Pyramid orthogonal attention network based on dual self-similarity for accurate mr image super-resolution. In: 2021 IEEE International Conference on Multimedia and Expo (ICME), pp. 1–6. IEEE (2021)
11. Huang, Q., Yang, D., Wu, P., Qu, H., Yi, J., Metaxas, D.: MRI reconstruction via cascaded channel-wise attention network. In: 2019 IEEE 16th International Symposium on Biomedical Imaging (ISBI 2019), pp. 1622–1626. IEEE (2019)
12. Jiang, M., Zhai, F., Kong, J.: A novel deep learning model ddu-net using edge features to enhance brain tumor segmentation on MR images. Artif. Intell. Med. **121**, 102180 (2021)
13. Kingma, D.P., Ba, J.: Adam: a method for stochastic optimization. arXiv preprint arXiv:1412.6980 (2014)
14. Li, G., et al.: Transformer-empowered multi-scale contextual matching and aggregation for multi-contrast mri super-resolution. In: Proceedings of the IEEE/CVF Conference on Computer Vision and Pattern Recognition, pp. 20636–20645 (2022)
15. Li, G., Lyu, J., Wang, C., Dou, Q., Qin, J.: Wavtrans: synergizing wavelet and cross-attention transformer for multi-contrast mri super-resolution. In: International Conference on Medical Image Computing and Computer-Assisted Intervention, pp. 463–473. Springer (2022). https://doi.org/10.1007/978-3-031-16446-0_44

16. Liu, Q., Yang, Q., Cheng, H., Wang, S., Zhang, M., Liang, D.: Highly undersampled magnetic resonance imaging reconstruction using autoencoding priors. Magn. Reson. Med. **83**(1), 322–336 (2020)
17. Lyu, Q., et al.: Multi-contrast super-resolution MRI through a progressive network. IEEE Trans. Med. Imaging **39**(9), 2738–2749 (2020)
18. Ronneberger, O., Fischer, P., Brox, T.: U-Net: convolutional networks for biomedical image segmentation. In: Navab, N., Hornegger, J., Wells, W.M., Frangi, A.F. (eds.) MICCAI 2015. LNCS, vol. 9351, pp. 234–241. Springer, Cham (2015). https://doi.org/10.1007/978-3-319-24574-4_28
19. Schlemper, J., Caballero, J., Hajnal, J.V., Price, A.N., Rueckert, D.: A deep cascade of convolutional neural networks for dynamic MR image reconstruction. IEEE Trans. Med. Imaging **37**(2), 491–503 (2017)
20. Sun, J., Li, H., Xu, Z., et al.: Deep admm-net for compressive sensing MRI. In: Advances in Neural Information Processing Systems 29 (2016)
21. Wang, H., Hu, X., Zhao, X., Zhang, Y.: Wide weighted attention multi-scale network for accurate MR image super-resolution. IEEE Trans. Circuits Syst. Video Technol. **32**(3), 962–975 (2021)
22. Wang, S., et al.: Deepcomplexmri: exploiting deep residual network for fast parallel MR imaging with complex convolution. Magn. Reson. Imaging **68**, 136–147 (2020)
23. Wang, W., Shen, H., Chen, J., Xing, F.: Mhan: multi-stage hybrid attention network for MRI reconstruction and super-resolution. Computers in Biology and Medicine, p. 107181 (2023)
24. Zhang, M., Zhang, M., Zhang, F., Chaddad, A., Evans, A.: Robust brain MR image compressive sensing via re-weighted total variation and sparse regression. Magn. Reson. Imag. **85**, 271–286 (2022)
25. Zhao, C., et al.: A deep learning based anti-aliasing self super-resolution algorithm for MRI. In: Frangi, A.F., Schnabel, J.A., Davatzikos, C., Alberola-López, C., Fichtinger, G. (eds.) MICCAI 2018. LNCS, vol. 11070, pp. 100–108. Springer, Cham (2018). https://doi.org/10.1007/978-3-030-00928-1_12
26. Zhao, X., Zhang, Y., Zhang, T., Zou, X.: Channel splitting network for single MR image super-resolution. IEEE Trans. Image Process. **28**(11), 5649–5662 (2019)
27. Zhou, B., Zhou, S.K.: Dudornet: learning a dual-domain recurrent network for fast mri reconstruction with deep t1 prior. In: Proceedings of the IEEE/CVF Conference on Computer Vision and Pattern Recognition, pp. 4273–4282 (2020)

MVD-NeRF: Resolving Shape-Radiance Ambiguity via Mitigating View Dependency

Yangjie Cao[1], Bo Wang[1], Zhenqiang Li[1], and Jie Li[2(✉)]

[1] Zhengzhou University, Zhengzhou, China
caoyj@zzu.edu.cn, {wangbo_3,lizhenqiang}@gs.zzu.edu.cn
[2] Shanghai Jiao Tong University, Shanghai, China
lijiecs@sjtu.edu.cn

Abstract. We propose MVD-NeRF, a method to recover high-fidelity mesh from neural radiance fields(NeRFs). The phenomenon of shape radiance ambiguity, where the radiance of a point changes significantly when viewed from different angles, leads to incorrect geometry. We propose directed view appearance encoding to achieve the desired viewpoint invariance by optimizing the appearance embedding for each input image, enabling the network to learn the shared appearance representation of the entire image set. Moreover, we utilize skip connections and layer scale module in the network architecture to capture complex and multi-faceted scene information. The introduction of layer scale module allows the shallow information of the network to be transmitted to the deep layer more accurately, maintaining the consistency of features. Extensive experiments demonstrate that, despite the inevitability of shape-radiance ambiguity, the proposed method can effectively minimize its impact on geometry which essentially mitigates the impact of view-dependent variations. The extracted geometry and textures can be deployed in any traditional graphics engine for downstream tasks.

Keywords: neural radiance fields · mesh recover · shape-radiance ambiguity · appearance encoding · skip connection · view-dependent

1 Introduction

Multi-view reconstruction of three-dimensional scenes has been a persistent focal point in the realms of both computer vision and computer graphics. Its applications encompass a range of fields including robotics, 3D modeling, and virtual reality. While conventional multi-view stereo methods have achieved certain levels of success in three-dimensional reconstruction, they still yield unsatisfactory outcomes when addressing specific details such as non-Lambertian surfaces and thin structures. Recently, 3D reconstruction methods founded on neural implicit representations have gradually emerged as promising alternatives to classical reconstruction methodologies [1,9,25]. Concurrently, the domain of novel view synthesis (NVS), as a significant offshoot within this field, has made substantial

S. Rudinac et al. (Eds.): MMM 2024, LNCS 14555, pp. 365–378, 2024.
https://doi.org/10.1007/978-3-031-53308-2_27

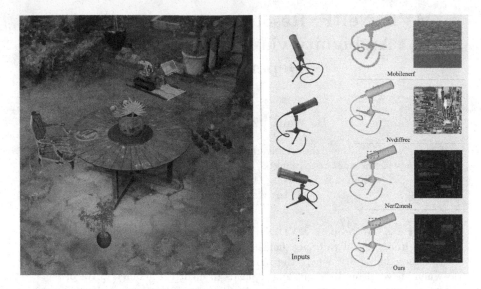

Fig. 1. We aim for quality geometry with NeRF reconstruction under inconsistency when viewing the same 3D point. In contrast to similar methodologies, our approach excels in producing finer details and delivering more precise reconstruction. The comparison is shown on the right side of the picture. This proficiency seamlessly extends from object-level datasets to scene-level datasets. The resultant textured mesh can be effortlessly integrated into mainstream graphics hardware and software. The left side of the image shows the reconstructed object displayed in the Unreal Engine.

strides over the past few years. Neural Radiance Fields (NeRFs) [17], characterized by their representation of scenes as volumetric functions parameterized by multi-layer perceptrons (MLPs), have successfully generated photorealistic renderings. Although present efforts primarily concentrate on enhancing image generation quality [2], training speed [18], rendering velocity [12], and robustness [15,20], NeRF's primary objective remains novel view synthesis, as opposed to authentic geometric reconstruction. This methodology encapsulates both geometric shapes and material attributes as elements within neural networks, leading to incompatibility with conventional 3D software and the inability to meet real-time rendering demands. Although some recent methods [7,8,11,27,30] aim to enhance geometric reconstruction within the context of volume rendering, these improvements have not explicitly explored the impact of view inconsistency on geometric reconstruction. Within the NeRF framework, multi-layer perceptrons (MLPs) accept 3D positions and 2D direction vectors, subsequently outputting corresponding point densities and colors from specific viewpoints. If the viewpoint of the observer(2D direction vectors) changes and the radiance of the same point changes accordingly, the multi-layer Perceptrons (MLPs) must perform well enough to reasonably reconstruct the correct geometry [34]. However, the same multi-layer perceptrons (MLPs) cannot represent multiple scenes well, resulting in shape-radiance ambiguity.

This paper presents a new framework called MVD-NeRF for improved mesh recovery from NeRF by mitigating view dependency, as illustrated in Fig. 1. We introduce the concept of directed view appearance encoding to achieve the desired viewpoint invariance. The network learns a shared appearance representation invariant to viewpoints across the entire image set by optimizing the appearance embedding for each input image. This augmentation allows for a more flexible interpretation of photometric and environmental variations between images. Simultaneously, we effectively capture fine-grained details within the images by incorporating ideas from residual networks and introducing layer scale module to the network responsible for learning scene appearances. This approach is equally applicable to geometric reasoning tasks in unbounded scenarios. In summary, we make the following contributions:

1. We propose a method that leverages directed view appearance encoding to mitigate shape-radiance ambiguity.
2. We present a combination of skip connections and layer scale module to improve the network's ability to capture complex image details during fitting.
3. Our approach effectively enhances the quality of the surface mesh and results in a reduction in mesh size by approximately 30%.

2   Related Work

Implicit Representation. The emergence of NeRF [17] has showcased the potential of utilizing neural networks to represent implicit surfaces, opening new avenues for efficient geometric representation. This methodology has found applications in fields such as human body reconstruction [23,24], view synthesis [14,26], and differentiable rendering [13,21]. Leveraging volumetric rendering techniques, NeRF represents a scene as a continuous volume function parameterized by multi-layer perceptrons (MLPs), thereby emulating the scene's volume density and brightness information related to the observing viewpoint. Consequently, it achieves high-quality view synthesis. Subsequent research [3,22,33] has further improved the NeRF to enhance its performance and applicability.

Surface Mesh. Some methods in the volumetric rendering domain adopt a strategy of decomposing density and color to achieve more accurate geometric shapes and appearances. These methods [7,8,11,28,31,32] often utilize signed distance functions (SDF) to better model shape-guided volumetric rendering weights. NVdiffrec [19] integrates differentiable rendering with trainable tetrahedral grids to directly optimize surface meshes for the recovery of material and lighting information, albeit constrained by mesh resolution limitations.

Recently, the method proposed by nerf2mesh [27] addresses the challenge of thin structure reconstruction. This approach encompasses training a NeRF model, extracting a mesh from it, refining the derived mesh to obtain geometric details, and extending its utility to boundary-less scene reconstruction. However,

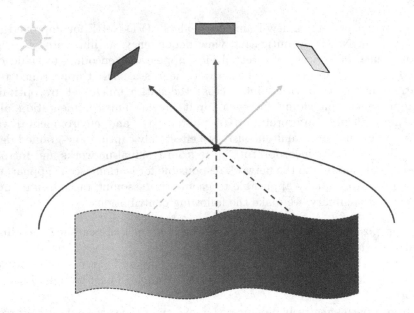

Fig. 2. Illustration of the shape-radiance ambiguity. A single point on a false surface (solid) can satisfy the incoming radiance to the cameras through view-dependent colors, even though these rays correspond to three distinct points on the real surface (dotted).

it does not explicitly consider the issue of shape-radiance ambiguity. Nerfingmvs [30] introduced a strategy of learning sparse structures from monocular depth networks, and then enhancing the sampling process of volumetric rendering using adaptive depth priors. VDN-NeRF [34] employs view normalization to mitigate the impact on geometric reconstruction by optimizing the alignment of orientation function capacities for each scene under shape-radiance ambiguity. These two methods were proposed to address the challenge of non-lambertian surfaces. Nevertheless, their applicability is limited to single objects and cannot be readily extended to boundaryless scenarios.

Shape-Radiance Ambiguity. In Nerf++ [9], the authors proposed a method that suggests introducing view-dependent color information into the MLP structure of NeRF. The integration of this approach, coupled with the constraints of finite resolution imposed by the size of the view-dependent MLP, serves to alleviate the issue of shape-radiance ambiguity in the context of novel view synthesis. Nevertheless, even with this approach, the presence of artifacts may still be observed in the recovery of the reconstructed geometry. By endowing the view-dependent color with sufficient angular resolution, nearly any shape could satisfy the incident radiance conditions of a given camera. In Fig. 2, a single (incorrect) point that satisfies the incident radiance conditions for three different views corresponds to three distinct colored points in reality. Nonetheless, given a sufficiently large capacity for the view-dependent color function, specifically the capacity of the multilayer perceptrons, object reconstruction remains feasible even when confronted with geometric inaccuracies.

3   Method

Our endeavor aims to elevate the quality of the transformation from neural radiance fields to geometric shapes by mitigating view inconsistency. In this section, we introduce the concept of directed view appearance encoding [4] to achieve the desired viewpoint invariance. Subsequently, incorporating layer scale module and drawing inspiration from residual networks, which are responsible for learning scene appearances, effectively captures intricate details within the images. Then, accurate geometric and material estimations are achieved in scenarios with sets of photos exhibiting view inconsistency through a joint iterative mesh refinement algorithm. Our method is implemented on the basis of nerf2mesh [27], and the overall framework is shown in Fig. 3.

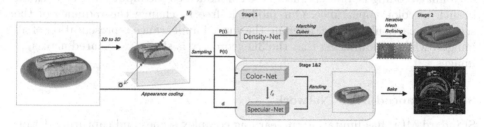

Fig. 3. MVD-NeRF framework, our approach begins by employing a density network to learn the initial geometric shape, which is subsequently extracted into a coarse mesh. Then, we optimize this coarse mesh to achieve a more precise surface with an adaptive distribution of face densities. Concurrently, we add appearance coding that uses a color network to learn appearance information and decompose it into diffuse and specular components.

3.1   Appearance Coding

This method involves modeling the appearance variations for each image in a learned low-dimensional latent space, encompassing factors such as exposure, lighting, and post-processing effects. By optimizing the appearance embedding for each input image, the network is capable of learning a shared appearance representation across the entire image set. This augmentation enables our method to more flexibly interpret photometric and environmental changes between images, thereby enhancing viewpoint invariance.

To begin, consider a set of images $x_1, ..., x_m$, where each image $x_i \in X$ has dimensions of $3 \times w \times h$. In continuation, initialize a set of d-dimensional random vectors $z_1, ..., z_m$, where $z_i \in Z \in R^d$ for all $i = 1, ..., m$. Thereafter, match the image dataset with the random vectors to create the dataset $\{(x_1, z_1), ..., (x_m, z_m)\}$. This is effectively assigning a corresponding real-valued appearance embedding vector $z_i^{(a)}$ of length n^a to each image x_i. We replace the image-independent radiance $c(t)$ with the image-dependent radiance $c_i(t)$, introducing a dependency on the image index i and the approximate pixel color C_i. Ultimately, our volumetric rendering equation takes the form of Eq. 1.

$$\hat{C}_i(o, v) = \int_0^{+\infty} \omega(t) c_i \left(p(t), v, z_i^{(a)} \right) dt \qquad (1)$$

where $\hat{C}_i(o, v)$ is the output color for this pixel, $\omega(t)$ a weight for the point $p(t)$, and $c_i \left(p(t), v, z_i^{(a)} \right)$ the color at the point p along the viewing direction v, and $z_i^{(a)}$ is the appearance embedding vector.

We emphasize the dependency of the color on each image and introduce a pixel-by-pixel dependency on the final rendered color. Because the appearance of each image coding is different, use these embedded as the input into the network branch of color appearance. Our model is free to change the radiance of the scene in a particular image, learning the shared appearance representation of all images, while guaranteeing that the 3D geometry is static and shared across all images.

3.2 Enhanced MLP Network

Standard MLP has limitations in learning complex scenes and capturing details in images, as deep networks tend to learn low-frequency functions, exacerbating the shape-radiance ambiguity issue. Nerf2mesh [27] draws inspiration from Instant-NGP [18] by employing hash-coded leveraging CUDA programming, using two separate MLP networks to fit appearance and density. However, this approach fails to effectively capture the specular surface and does not address the influence of direction-dependent views on the neural radiance field's geometric reconstruction. Nerf2mesh employs two separate shallow MLP feature networks to represent objects in 3D space, which provides valuable guidance for mitigating shape-radiance ambiguity.

We continue to use the structure from nerf2mesh for the network representing density. We propose an enhanced MLP network to represent appearance. The network architecture is shown in fig 4.

$$x = cat \left(E \left(p(t), A(z_i^a) \right) \right) \qquad (2)$$

$$Aff_{\alpha, \beta}(x) = Diag(\alpha) \cdot x + \beta \qquad (3)$$

Initially, we enhance the sampling points of input by employing multi-resolution hash encoding of feature vectors and then concatenate them with the appearance encoded values to obtain x. After affine transformation, which has the effect of layer normalization. α and β are learnable weight vectors. The affine layer, which applies linear transformations and translation operations to the input data, incurs no additional inference cost compared to other normalization operations and can be absorbed by neighboring linear layers.

$$F = Linear\left(Sigmoid\left(Linear\left(Aff\left(x\right)\right)\right)\right) \tag{4}$$

$$d, f_s = Linear\left(ReLU\left(Aff\left(F\right)\right)\right) \tag{5}$$

The notation Aff(x) denotes the independent application of an affine operation to each column of the matrix x. To maintain the consistency of the shallow features and deep features of the neural network, we introduce the residual connection of x to the last layer of the network. To better capture the high-frequency information, the layer scale module is introduced. Some experiments [6,10] have shown that the layer scale module significantly enhances the standard MLP's ability to capture high-frequency components. Following the approach proposed in MobileNeRF [5], we merge the weights of the Specular-Net network into the fragment shader to achieve the viewpoint-dependent specular reflection effect. d represents the diffuse color, and f_s signifies the accumulated gradient of the specular component.

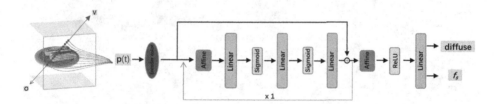

Fig. 4. Enhanced MLP Network Architecture. The coordinates of the ray sampling points, denoted as $p(t)$, are mapped through a hash encoding before being fed into the color prediction network. The affine transform layer is used to accelerate the convergence speed. To enhance the network's ability to capture high-frequency information, techniques such as Layerscale and Skip Connection are employed.

$$Aff_{\alpha,\beta}\left(f_s, d\right) = Diag\left(\alpha\right) \cdot cat\left(f_s, d\right) + \beta \tag{6}$$

$$F = Linear\left(Sigmoid\left(Linear\left(Aff\left(\left(f_s, d\right)\right)\right)\right)\right) \tag{7}$$

$$S = linear\left(ReLU\left(Aff\left(F\right)\right)\right) \tag{8}$$

After obtaining f_s, we concatenate it with the view vector d and feed it into the network to obtain the specular highlight S. We still use Enhanced MLP Network Architecture to get.

3.3 Iterative Mesh Refining

The iterative mesh refinement process [27] aims to improve initial coarse meshes, which often have inaccuracies and dense faces when extracted using Marching Cubes [29]. We want to make them look more delicate like human-made ones by refining vertex positions and face density. The main idea is to adjust face density based on previous training errors. During training, we project 2D pixel-wise rendering errors onto mesh faces and accumulate face-wise errors. After a certain number of iterations, we sort face errors (E_{face}) and set thresholds:

$$e_{subdivide} = percentile\,(E_{face}, 95) \tag{9}$$

$$e_{decimate} = percentile\,(E_{face}, 70) \tag{10}$$

Faces with errors above $e_{subdivide}$ are subdivided to increase density. Faces with errors below $e_{decimate}$ are decimated and remeshed to reduce density. We then reset vertex offsets and face errors and continue training. This iterative process repeats until stage 2 is completed. Through a large number of experiments, we have verified that the optimal effect is achieved with the above two thresholds.

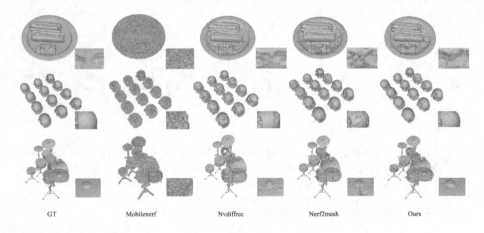

Fig. 5. Surface reconstruction quality based on the NeRF synthetic dataset. Compared to previous methods, our approach achieves superior mesh reconstruction quality, particularly on surfaces with specular highlights and reflections.

4   Experimental

4.1   Datasets

We conducted experiments on three datasets to validate the effectiveness and generalization capability of our method:1)NeRF-Synthetic [17]: This dataset comprises 8 synthetic scenes. 2)LLFF [16]: This dataset includes 8 real-world forward-facing scenes. 3)Mip-NeRF 360 [2]: This dataset consists of 3 publicly available real unbounded outdoor scenes. Our method demonstrates strong generalization across different types of datasets, even excelling in accurately reconstructing meshes for challenging unbounded scenes. All experiments were conducted on a single NVIDIA 3090Ti GPU. Two stages of training and the grid export of each scenario takes about 90 min.

4.2   Comparisons

Mesh Quality. Measuring surface reconstruction quality is challenging for real-world scenes due to the lack of ground truth meshes. Therefore, we mainly compare the results on synthetic datasets. We qualitatively evaluate the extracted meshes from different methods, as shown in Fig. 5. Specifically, we focus on the reconstruction of specular reflection areas. Our method successfully reconstructs the structure of these regions with high fidelity, whereas other methods fail to accurately reconstruct them.

Table 1. Chamfer Distance ↓ (Unit is 10^{-3}) on the NeRF-synthetic dataset compared to the ground truth meshes.

	Chair	Drums	Ficus	Hotdog	Lego	Materials	Mic	ship	Mean
Neus [28]	3.95	6.68	2.84	8.36	6.62	4.10	**2.99**	9.54	5.64
NVdiffrec [19]	**4.13**	8.27	5.47	7.31	**5.78**	4.98	3.38	25.89	8.15
Nerf2mesh [27]	4.36	6.24	**2.42**	5.81	5.81	4.38	3.55	8.29	5.11
Ours	4.24	**5.85**	2.44	**4.71**	5.80	**3.87**	3.21	**7.90**	**4.75**

Table 2. Number of Vertices and Faces ↓ (Unit is 10^3). Our method uses relatively fewer vertices and faces, resulting in the most obvious decrease in NeRF-synthetic, approximately 30%.

	NeRF-synthetic		LLFF		Mip-NeRF360	
	#V	#F	#V	#F	#V	#F
Ground Truth	631	873	–	–	–	–
NeuS [28]	1020	2039	–	–	–	–
NVdiffrec [19]	75	80	–	–	–	–
MobileNeRF [5]	494	224	830	339	1436	609
Nerf2mesh [27]	89	176	132	255	124	239
Ours	63	126	121	218	113	222

Table 3. Rendering Quality Comparison. We present PSNR, SSIM, and LPIPS metrics for various datasets and compare our method against different categories of approaches. Our method demonstrates comparable performance, achieving results in line with the state-of-the-art methods across these metrics.

	NeRF-synthetic			LLFF			Mip-NeRF360		
	PSNR↑	SSIM↑	LPIPS↓	PSNR↑	SSIM↑	LPIPS↓	PSNR↑	SSIM↑	LPIPS↓
MobileNeRF [5]	30.90	0.947	0.0062	25.91	0.825	0.183	23.06	0.527	0.434
NVdiffrec [19]	29.05	0.939	0.081	–	–	–	–	–	–
Nerf2mesh [27]	29.90	0.937	0.071	24.40	0.757	0.280	23.93	0.637	0.397
Ours	30.03	0.947	0.054	24.46	0.731	0.204	23.38	0.606	0.263

To quantify the surface reconstruction quality, we employ the Chamfer Distance (CD) metric. However, as the ground truth meshes may not be surface meshes, we adopted the approach from nerf2mesh [27]. We cast rays from the test cameras and sample 2.5M points from these ray-surface intersections per scene. In Table 1, we present the average CD for all scenes, indicating that our method achieves the best results. We notice that our method performs excellently in scenes with complex topologies (such as ficus, ship, and lego) as well as in scenes with a large number of non-Lambertian surfaces (such as materials and Drums).

We also evaluated the practical applicability by comparing the number of vertices and faces in the exported meshes, as shown in Table 2. To ensure fair comparison, the mesh file format was uncompressed OBJ, textures were in PNG format, and other metadata was stored in JSON format. Our exported meshes have fewer vertices and faces. Because the iterative mesh refinement process can increase the vertex count to enhance surface details, while reducing the face count to control mesh size.

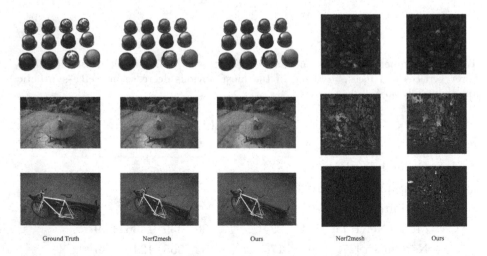

Fig. 6. Rendering quality and visualization of texture maps. We achieved comparable on different data sets of rendering quality, due to the improvement of surface quality, our texture more complete and intuitive.

Render Quality. We present our rendering quality comparison results in Table 3. We demonstrate that our method achieves improved geometric quality while maintaining high-quality rendering compared to nerf2mesh. NVdiffrec is only suitable for single-object reconstruction, and MobileNeRF's exported meshes lack smoothness and struggle to align with object surfaces well, relying heavily on textures for surface detailing. Clearly, our method outperforms both. Figure 6 shows the rendering result visualization and comparison with other methods. We demonstrate that our rendered images and extracted texture images are more complete and intuitive than those obtained by other methods.

4.3   Ablation Studies

We conducted an ablation study on both Appearance Encoding and Enhanced MLP Network. Specifically, we compared the complete model with variants excluding either our proposed enhanced MLP network or the appearance encoding. The results in Fig. 7 demonstrate the following: 1)The generated surface mesh exhibits improved geometric quality and a more complete reconstruction of specular regions when the enhanced MLP network is removed. However, the overall result appears smoother and loses some detail in terms of reconstruction. 2) Excluding appearance coding verifies that our network is capable of effectively reconstructing surface details of objects. Nevertheless, the reconstruction of specular regions remains less satisfactory with some loss. In contrast, our model (MVD-NeRF) excels in recovering specular surfaces and fine regional details.

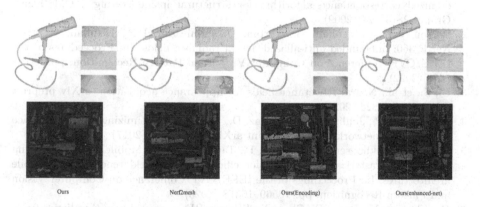

Ours   Nerf2mesh   Ours(Encoding)   Ours(enhanced-net)

Fig. 7. Ablation experiment. We compared the geometry and texture results under different methods. "Ours(app)" represents the result with the enhanced MLP network excluded, "Ours(color-net)" represents the result with appearance encoding excluded, compared with the outcomes of the complete model "Ours", and the Nerf2mesh method.

5   Conclusion

Although our method has shown promising results, it still has several limitations. Due to the difficulty of estimating unknown lighting conditions from images without compromising reconstruction quality, we have chosen to bake illumination into textures, which consequently restricts our ability to perform relighting. In line with other mesh-based methods, we utilize single-pass rasterization and are unable to handle transparency. As we advance our research, we aim to address these limitations by harnessing enhanced techniques and cutting-edge technologies.

In conclusion, we have introduced an effective framework for reconstructing high-quality textured surface meshes from multi-view RGB images. Our method mitigates the impact of radiance on geometry by employing directed view appearance encoding and utilizes an enhanced MLP network to capture scene details. The reconstructed meshes demonstrate improved surface quality and are well-suited for manipulation and editing in downstream applications.

Acknowledgements. The authors would like to thank the Zhengzhou University, the Zhengzhou City Collaborative Innovation Major Project, and the GPU support provided by the 3DCV lab.

References

1. Barnes, C., Shechtman, E., Finkelstein, A., Goldman, D.B.: PatchMatch: a randomized correspondence algorithm for structural image editing. ACM Trans. Graph. **28**(3), 24 (2009)
2. Barron, J.T., Mildenhall, B., Verbin, D., Srinivasan, P.P., Hedman, P.: Mip-NeRF 360: unbounded anti-aliased neural radiance fields. In: Proceedings of the IEEE/CVF Conference on Computer Vision and Pattern Recognition, pp. 5470–5479 (2022)
3. Bi, S., et al.: Neural reflectance fields for appearance acquisition. arXiv preprint arXiv:2008.03824 (2020)
4. Bojanowski, P., Joulin, A., Lopez-Paz, D., Szlam, A.: Optimizing the latent space of generative networks. arXiv preprint arXiv:1707.05776 (2017)
5. Chen, Z., Funkhouser, T., Hedman, P., Tagliasacchi, A.: MobileNeRF: exploiting the polygon rasterization pipeline for efficient neural field rendering on mobile architectures. In: Proceedings of the IEEE/CVF Conference on Computer Vision and Pattern Recognition, pp. 16569–16578 (2023)
6. Dai, S., Cao, Y., Duan, P., Chen, X.: SRes-NeRF: improved neural radiance fields for realism and accuracy of specular reflections. In: International Conference on Multimedia Modeling, pp. 306–317. Springer, Cham (2023). https://doi.org/10.1007/978-3-031-27077-2_24
7. Darmon, F., Bascle, B., Devaux, J.C., Monasse, P., Aubry, M.: Improving neural implicit surfaces geometry with patch warping. In: Proceedings of the IEEE/CVF Conference on Computer Vision and Pattern Recognition, pp. 6260–6269 (2022)
8. Fu, Q., Xu, Q., Ong, Y.S., Tao, W.: Geo-Neus: geometry-consistent neural implicit surfaces learning for multi-view reconstruction. Adv. Neural. Inf. Process. Syst. **35**, 3403–3416 (2022)

9. Furukawa, Y., Ponce, J.: Accurate, Dense, and Robust multiview stereopsis. IEEE Trans. Pattern Anal. Mach. Intell. **32**(8), 1362–1376 (2009)

10. Jacot, A., Gabriel, F., Hongler, C.: Neural tangent kernel: convergence and generalization in neural networks. In: Advances in Neural Information Processing Systems, vol. 31 (2018)

11. Jiang, Y., Ji, D., Han, Z., Zwicker, M.: SDFDiff: differentiable rendering of signed distance fields for 3D shape optimization. In: Proceedings of the IEEE/CVF Conference on Computer Vision and Pattern Recognition, pp. 1251–1261 (2020)

12. Lin, C.H., Ma, W.C., Torralba, A., Lucey, S.: BARF: bundle-adjusting neural radiance fields. In: Proceedings of the IEEE/CVF International Conference on Computer Vision, pp. 5741–5751 (2021)

13. Liu, S., Zhang, Y., Peng, S., Shi, B., Pollefeys, M., Cui, Z.: DIST: rendering deep implicit signed distance function with differentiable sphere tracing. In: Proceedings of the IEEE/CVF Conference on Computer Vision and Pattern Recognition, pp. 2019–2028 (2020)

14. Lombardi, S., Simon, T., Saragih, J., Schwartz, G., Lehrmann, A., Sheikh, Y.: Neural volumes: learning dynamic renderable volumes from images. arXiv preprint arXiv:1906.07751 (2019)

15. Martin-Brualla, R., Radwan, N., Sajjadi, M.S., Barron, J.T., Dosovitskiy, A., Duckworth, D.: Nerf in the wild: neural radiance fields for unconstrained photo collections. In: Proceedings of the IEEE/CVF Conference on Computer Vision and Pattern Recognition, pp. 7210–7219 (2021)

16. Mildenhall, B., et al.: Local light field fusion: practical view synthesis with prescriptive sampling guidelines. ACM Trans. Graph. (TOG) **38**(4), 1–14 (2019)

17. Mildenhall, B., Srinivasan, P.P., Tancik, M., Barron, J.T., Ramamoorthi, R., Ng, R.: NeRF: representing scenes as neural radiance fields for view synthesis. Commun. ACM **65**(1), 99–106 (2021)

18. Müller, T., Evans, A., Schied, C., Keller, A.: Instant neural graphics primitives with a multiresolution hash encoding. ACM Trans. Graph. (ToG) **41**(4), 1–15 (2022)

19. Munkberg, J., et al.: Extracting triangular 3D models, materials, and lighting from images. In: Proceedings of the IEEE/CVF Conference on Computer Vision and Pattern Recognition, pp. 8280–8290 (2022)

20. Niemeyer, M., Barron, J.T., Mildenhall, B., Sajjadi, M.S., Geiger, A., Radwan, N.: RegNeRF: regularizing neural radiance fields for view synthesis from sparse inputs. In: Proceedings of the IEEE/CVF Conference on Computer Vision and Pattern Recognition, pp. 5480–5490 (2022)

21. Niemeyer, M., Mescheder, L., Oechsle, M., Geiger, A.: Differentiable volumetric rendering: learning implicit 3D representations without 3D supervision. In: Proceedings of the IEEE/CVF Conference on Computer Vision and Pattern Recognition, pp. 3504–3515 (2020)

22. Rebain, D., Jiang, W., Yazdani, S., Li, K., Yi, K.M., Tagliasacchi, A.: DeRF: decomposed radiance fields. In: Proceedings of the IEEE/CVF Conference on Computer Vision and Pattern Recognition, pp. 14153–14161 (2021)

23. Saito, S., Huang, Z., Natsume, R., Morishima, S., Kanazawa, A., Li, H.: PIFu: pixel-aligned implicit function for high-resolution clothed human digitization. In: Proceedings of the IEEE/CVF International Conference on Computer Vision, pp. 2304–2314 (2019)

24. Saito, S., Simon, T., Saragih, J., Joo, H.: PIFuHD: multi-level pixel-aligned implicit function for high-resolution 3D human digitization. In: Proceedings of the IEEE/CVF Conference on Computer Vision and Pattern Recognition, pp. 84–93 (2020)

25. Schönberger, J.L., Zheng, E., Frahm, J.-M., Pollefeys, M.: Pixelwise view selection for unstructured multi-view stereo. In: Leibe, B., Matas, J., Sebe, N., Welling, M. (eds.) ECCV 2016. LNCS, vol. 9907, pp. 501–518. Springer, Cham (2016). https://doi.org/10.1007/978-3-319-46487-9_31
26. Sitzmann, V., Zollhöfer, M., Wetzstein, G.: Scene representation networks: continuous 3D-structure-aware neural scene representations. In: Advances in Neural Information Processing Systems, vol. 32 (2019)
27. Tang, J., et al.: Delicate textured mesh recovery from nerf via adaptive surface refinement. arXiv preprint arXiv:2303.02091 (2023)
28. Wang, P., Liu, L., Liu, Y., Theobalt, C., Komura, T., Wang, W.: NeuS: learning neural implicit surfaces by volume rendering for multi-view reconstruction. arXiv preprint arXiv:2106.10689 (2021)
29. We, L.: Marching Cubes: a high resolution 3D surface construction algorithm. Comput. Graph. **21**, 163–169 (1987)
30. Wei, Y., Liu, S., Rao, Y., Zhao, W., Lu, J., Zhou, J.: NerfingMVS: guided optimization of neural radiance fields for indoor multi-view stereo. In: Proceedings of the IEEE/CVF International Conference on Computer Vision, pp. 5610–5619 (2021)
31. Yariv, L., Gu, J., Kasten, Y., Lipman, Y.: Volume rendering of neural implicit surfaces. Adv. Neural. Inf. Process. Syst. **34**, 4805–4815 (2021)
32. Yariv, L., et al.: Multiview neural surface reconstruction by disentangling geometry and appearance. Adv. Neural. Inf. Process. Syst. **33**, 2492–2502 (2020)
33. Zhang, K., Riegler, G., Snavely, N., Koltun, V.: NeRF++: analyzing and improving neural radiance fields. arXiv preprint arXiv:2010.07492 (2020)
34. Zhu, B., Yang, Y., Wang, X., Zheng, Y., Guibas, L.: VDN-NeRF: resolving shape-radiance ambiguity via view-dependence normalization. In: Proceedings of the IEEE/CVF Conference on Computer Vision and Pattern Recognition, pp. 35–45 (2023)

DPM-Det: Diffusion Model Object Detection Based on DPM-Solver++ Guided Sampling

Jingzhi Zhang, Xudong Li, Linghui Sun, and Chengjie Bai[✉]

School of Physics and Electronics, Shandong Normal University, Jinan 250358,
Shandong, China
sdnu_baichengjie@163.com

Abstract. DiffusionDet is the first neural network model to apply the
diffusion model to object detection, which is a new object detection
paradigm that achieves good results compared to other mature object
detection models. However, the diffusion model is an iterative model
by nature, which requires setting numerous sampling steps to get good
results. DiffusionDet has difficulty obtaining stable samples with fewer
sampling steps, and calling the detection head once per iteration will
cause the inference time to keep increasing. To alleviate this problem,
the DPM-Det algorithm is proposed, which only changes the sampling
process from noise boxes to bounding boxes in progressive refinement
during inference, while keeping the structure of the DiffusionDet model
unchanged. It replaces the original denoising diffusion implicit models in
DiffusionDet by introducing DPM-Solver++. On the MS-COCO dataset,
with ResNet-50 with FPN as the backbone and the training and evaluation boxes fixed at 500, DPM-Det achieved an AP of 47.0 with 5 steps
and an AP of 47.3 with 8 steps. Compared with DiffusionDet, DPM-
Det has a certain degree of improvement in both detection accuracy and
speed, which can achieve a better balance of accuracy and speed.

Keywords: Object Detection · Diffusion Model · DPM-Solver++

1 Introduction

With the advancement of deep learning techniques, the field of computer vision
has been developed unprecedentedly, and researchers have tried to tackle the
object detection task from different perspectives, resulting in numerous sophisticated modern object detectors with good performance.

Diffusion models [1,2] belong to a class of generative models, which in computer vision are a class of models capable of generating synthetic images. The
diffusion model can be divided into diffusion and inverse diffusion processes.
The diffusion process gradually adds noise to the image, and the inverse diffusion process derives backward from the noise and gradually eliminates the noise
to reverse the generated image. Once the diffusion model is trained, generating
images by randomly sampling Gaussian noise is possible. With its various excellent features, the diffusion model has been rapidly developed and has obtained
remarkable results in recent years.

© The Author(s), under exclusive license to Springer Nature Switzerland AG 2024
S. Rudinac et al. (Eds.): MMM 2024, LNCS 14555, pp. 379–393, 2024.
https://doi.org/10.1007/978-3-031-53308-2_28

DiffusionDet [3] is the first neural network model that applies the diffusion model to object detection, a new object detection paradigm that achieves good results compared to other established object detection models, this work is highly creative and has attracted a lot of attention. However, the diffusion model is by nature iterative, and the drawbacks of applying it to the object detection task are obvious. The diffusion model requires setting more sampling steps to get good results, which results in a slower generation of samples. With a few iterations, the diffusion model cannot generate a stable sample, which will have some impact on the accuracy. Therefore, reducing the number of iterations and obtaining stable and high-quality samples to improve detection accuracy while speeding up detection is a question that deserves in-depth consideration when diffusion models are applied to object detection.

We proposed the DPM-Det algorithm based on DiffusionDet for model improvement and obtained good results. We keep the structure of the Diffusion-Det model unchanged, but only change the sampling process from the noise box to the bounding boxes in progressive refinement during inference, and replace the original DDIM in DiffusionDet by introducing DPM-Solver++ [4], which can obtain high-quality samples with fewer sampling steps and effectively improve the AP value.

2   Related Work

2.1   Object Detection-Related Work

With the rise of deep learning, many high-performing object detection algorithms have emerged. Depending on the object candidates, deep learning-based object detection methods can be broadly classified into six categories, namely sliding window, regional proposals, anchor boxes, reference points, object queries, and random box methods.

Sliding window class method. To detect different object types at different observation distances, windows of different sizes and aspect ratios are used, sliding the window from left to right and from top to bottom on the image to identify targets using classification, as in the Overfeat [5]. Regional proposals class method. The candidate boxes are first extracted from the images using algorithms such as selective search or region suggestion networks, and then the detection results are obtained by secondary correction of the candidate box targets, such as Fast R-CNN [6]. Anchor boxes class method. A set of fixed reference boxes covering almost all positions and scales, and positive and negative samples are obtained by calculating the intersection and ratio of each reference box to the true value box, and some anchor boxes with higher scores are selected for position regression, and the final results are obtained after NMS processing, as in the SSD [7]. Reference points class method. It can be further divided into the corner and center point-based methods. Corner-based methods predict bounding boxes by merging corner points learned from feature maps, such as the YOLO-MS [8]. Center-based methods predict the probability of being the object center at each position of the feature map and directly regress the height and width without any anchor point prior, as in the FSAF [9]. Object queries class methods. Objects are examined using a fixed set of learnable queries, such as the DETR [10] and Sparse DETR [11]. Random box class method. A set of randomly generated boxes is progressively refined into output results by an incremental approach, as in the DiffusionDet [3] and Diff3Det [12].

2.2 Diffusion Model Improvement-Related Work

The literature [13] states that the improved algorithms for existing diffusion models can be classified into four categories, depending on the improvement method. Among them, training-free sampling is an important category. The goal of training-free sampling is to propose an efficient sampling algorithm to obtain knowledge from pre-trained models in fewer steps and with higher accuracy, which in turn contains four categories: analytical methods, implicit samplers, dynamic programming adjustments, and differential equation solver samplers.

2022, analytical methods were first proposed by Bao et al. [14] for solving covariance optimization problems to dynamically perform sampling steps, such as Analytic-DPM [14] and Extend analytic-DPM [15]. Implicit samplers are a class of jump step samplers that do not require diffusion model retraining, such as DDIM [2] and FastDPM [16]. Dynamic programming adjustments is achieved by memory tricks to traverse all choices to find an optimized solution in less time, e.g., Efficient Sampling [17] and DDSS [18].

Compared to traditional discrete-step diffusion methods, the numerical formulation of differential equations is more efficiently sampled by higher-order solvers, which are centered on efficient sampling with advanced differential equation solvers that minimize the approximation error with fewer sampling steps. The existing methods can be divided into two categories based on ordinary differential equation (ODE) and stochastic differential equation (SDE), such as INDM [19] and DPM-solver++ [4].

3 Model Design

3.1 DPM-Det Design Idea

The Image Encoder and Detection Decoder of DPM-Det remain the same as DiffusionDet, and only the progressive optimization process of bounding boxes is changed in the inference stage. In the asymptotic optimization process of bounding boxes, for problems such as difficulty in obtaining high-quality samples with few sampling steps of the original DDIM, DPM-Solver++ [4], a training-free fast diffusion ODE solver for bootstrap sampling, is introduced to accelerate the sampling, and higher quality samples can be obtained by using fewer sampling steps.

During DPM-Det training, noise is added to the data sample z_0 to transform z_0 into a potentially noisy sample z_{t_0} with end time t_0. A neural network $f_\theta(z_{t_0}, t)$ is trained to predict z_0 from z_{t_0} by minimizing the training objective with ℓ_2 loss [3].

In DPM-Det inference, the potentially noisy samples z_{t_0} are reconstructed z_0 by f_θ trained and update rules iterative [4], and the corresponding category labels are generated. The process of reconstructing z_0 can be expressed as $z_{t_0} \rightarrow z_{t-\Delta} \rightarrow \cdots \rightarrow z_{t_{M+1}}$, where is a set of bounding boxes for reconstruction recovery, $z_{t_{M+1}} \in R^{N \times 4}$ and with N boxes.

DPM-Det's Image Encoder is the same as DiffusionDet and is also run only once to extract the depth features in the image. The Detection Decoder uses this depth feature as a condition to generate prediction boxes $z_{t_{M+1}}$ from noise boxes z_{t_0} in a stepwise optimized manner. Since the diffusion model used is an iterative model, it is necessary to run the model f_θ several times until a stable sampling sample is generated.

3.2   DPM-Solver++ in DPM-Det

Diffusion probabilistic model (DPM) sampling can also be seen as solving the corresponding diffusion ordinary differential equations (ODE), and the commonly used DDIM is a fast ordinary differential equation (ODE) sampling solver for the first-order diffusion equation. However, DDIM usually requires 100 to 250 steps to obtain high-quality samples, while DPM-Solver++ [4], a higher-order solver for DPM bootstrap sampling, can generate high-quality samples in only 15 to 20 steps.

DPM-Solver++ proposes a variety of solvers of different orders, and can obtain higher quality samples in a short period stable, the 2nd and 3rd order DPM-Solver++ are mainly used in DPM-Det, namely DPM-Solver++2 and DPM-Solver++3, with the number after them corresponding to the order of the solver, see [4] for details. We used the higher-order DPM-Solver++ for the sampling process from noise boxes to bounding boxes of DPM-Det and obtained higher-quality bounding boxes.

The DPM-Det uses the same detection head as the DiffusionDet. Compared to DiffusionDet, which calls the detection head once in each sampling step, DPM-Det only needs to call the detection head once in each higher-order DPM-Solver++, not in every step. Using this paradigm for detection allows skipping the generation of lower-quality samples and directly detecting high-quality samples, reducing the use of detection heads while enhancing model performance and efficiency.

M DPM-Solver++3 with a single DPM-Solver++2 is used in the DPM-Det sampling step for the solution. In the progressive refinement of DPM-Det, the first M DPM-Solver++3 sampling solutions are performed first, and the M value is determined by setting the number of diffusion steps. Then, the final result is obtained by a DPM-Solver++2. The sampling process of the DPM-Det progressive refinement is shown in Fig. 1.

We select during the solution of each solver to make predictions and perform box renewal [3], and then move to the next iteration. We use the detection head once in each higher-order DPM-Solver++, optimize it using a progressive refinement strategy, and finally output the result.

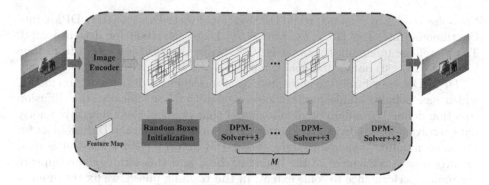

Fig. 1. Sampling process of DPM-Det progressive refinement.

Given a decreasing sequence $\{t_i\}_{i=0}^{M+1}$ from $t_0 = T$ to $z_{t_{M+1}} = 0$ and an initial value $z_0 \sim N(0 \mid \sigma^2 I)$, where z_{t_0} is the noise frame conforming to the Gaussian distribution, σ is referred to as the noise schedules of a DPM. The solver aims to iteratively compute a sequence $\{z_{t_i}\}_{i=0}^{M+1}$ to approximate the exact solution at each time t_i. The final $z_{t_{M+1}}$ is a sample of the diffusion ODE approximation [4], i.e., a set of bounding boxes. To obtain stable sampling quickly in DPM-Det, it is necessary to use DPM-Solver++3 as much as possible. In summary, we set the number of sampling steps of DPM-Det to *3M+2* time steps($\{t_i\}_{i=0}^{M+1}$ and $\{t_s\}_{i=0}^{M+1}$), and M is the number of DPM-Solver++3 that satisfies $t_0 > s_1 > t_1 > \cdots > t_{M-1} > s_M > t_M > s_{M+1} > t_{M+1}$.

Algorithm 1 provides the basic steps of DPM-Solver++2 in DPM-Det sampling.

3.3 DPM-Det Training

In the training phase, the strategy used by DPM-Det and DiffusionDet is the same, i.e., constructing a forward diffusion process from the ground truth to the noisy box and training the model to invert this process. First, the original ground truth is filled with random boxes so that the total number is a fixed number N_{train} and the noise is added as a noise box. We adjusted the noise addition strategy to have $t_0 = 1$ for DPM-Det during noise addition compared to DiffusionDet adding noise with an integer end time $t \in [1, 1000]$. Gaussian noise is scheduled by the variance schedule [20]. Then, taking the noise boxes as input and predicting the classification as well as the boxes, we apply the ensemble prediction loss to the set of N_{train} predictions. We use the same loss function as DiffusionDet, see [3] for more details. Algorithm 2 provides the pseudocode of the DPM-Det training procedure.

3.4 DPM-Det Sampling

The inference process of DPM-Det is a denoising sampling process from the noise box to the bounding box. DiffusionDet is used in Discrete-time DPMs [2] to train the noise model in N steps, while DPM-Det converts Discrete-time DPMs into Continuous-time, $t \in [T/N, T]$, where, $T=1$ and $N=1000$ for details see [4]. Therefore, the minimum time unit is 10^{-3}. Solve the diffusion ODEs from time $t=1$ to $t=10^{-3}$ to obtain the final samples.

DPM-Det transforms the noise prediction model into a data prediction model, which can achieve stable performance acceleration. Specifically, the diffusion equation defined by a data prediction model that predicts clean data given noisy data is solved. More efficient and better model performance than DiffusionDet for the same number of sampling steps. The DPM-Solver++ for DPM-Det is used for progressive refinement of the bounding box, and this strategy can improve the model performance to some extent. In the training phase, we fix the number of noise boxes N_{train}, and in the inference phase, we also have the flexibility to vary the number of noise boxes N_{eval} used for evaluation of inference. Algorithm 3 provides the pseudocode for DPM-Det sampling inference.

Algorithm 1. DPM-Solver++2 in DPM-Det Sampling.

```
1:  def dpm_solver2(z_m, t_m, t_m+1, s_m+1)
2:      # z_m: the output after M DPM-Solver++3 samples
3:      # t_m, t_m+1 and s_m+1: time steps
4:      # model_fnn: outputs data prediction model
5:      # model_fn: predict output coord and class
6:      # lambda: obtained through the transformation of σ
7:      h=transform1 (lambda, t_m+1, t_m)
8:      r=transform2 (lambda, h, t_m+1, t_m)
9:      x_1=model_fnn (z_m, t_m)
10:     # Solve for the intermediate value of the time step
11:     u=median (x_1, r, h, s_m+1)
12:     x_2=model_fnn (u, s_m+1)
13:     d=solve (x_1, x_2, r, t_m+1, s_m+1)
14:     cls, box=model_fn (d, s_m+1)
15:     # Solving the final data sample with intermediate values
16:     z_m+1=fin_data (z_m, d)
17:     # Process the obtained class and boxes
18:     box=box_renewal (box)
19:     box, score, label=infer (cls, box)
20: return z_m+1, box, score, label
```

Algorithm 2. DPM-Det Training.

```
1: def train(images, gt_boxes)
2:     # images: [B, H, W, 3]; gt_boxes: [B, *, 4]; B: batch
3:     # N: number of proposal boxes; t: add the end time of the noise
4:     # Extract feature
5:     feats=encoder (images)
6:     # gt_boxes: [B, *, 4] → pb:[B, N, 4]
7:     pb=pad_boxes (gt_boxes)
8:     # Add noise
9:     ratio=1; t=ratio_to_time (ratio); noise=randn (B, N, 4)
10:    diff_b=add_noise (pb, t, noise)
11:    # Predict
12:    predict=head (features, diff_b, t)
13:    # Train loss
14:    Loss=set_loss (predict, gt_boxes)
15: return Loss
```

Algorithm 3. DPM-Det Sampling.

```
1: def infer(images, steps, order)
2:     # images: [B, H, W, 3]; betas: variance with a value in the range 0 to 1
3:     # steps: number of sample steps; order: the order of DPM-Det
4:     # t_0: the starting time of the sampling
5:     # t_T: the ending time of the sampling
6:     # Extract feature
7:     feats=encoder (images)
8:     # noisy boxes: [B, N 4]
9:     img=randn (mean=0, std=1)
10:    # Create wrapper function for the data prediction model
11:    model=model_wrapper
12:    # Create a wrapper class for the SDE
13:    vp=NoiseScheduleVP (betas)
14:    # Construct a DPM-Solver++
15:    dpm=DPM_Solver++ (model, image, vp)
16:    # DPM-Det progressive optimization sampling process
17:    res=DPM-Det_dpm (img, orders, steps, images, feats)
18: return res
```

4 Experiment Validation

4.1 Experimental Environmental Conditions

We used an Nvidia GeForce RTX 3090 for the hardware environment with 24 GB of video memory. For the software environment, our program was performed under the Ubuntu 18.04 operating system and built under Python 3.8 and PyTorch 1.10 via Detectron2 [21].

4.2   Data Sets and Evaluation Metrics

We selected the MS-COCO [22] dataset. Specifically, the COCO 2017 train set has about 118K training images, the COCO 2017 val set has 5K validation images, and the COCO 2017 test-dev set has about 20K unlabeled test images, for a total of 80 categories. We use AP, AP_{50}, AP_{75}, AP_s, AP_m, AP_l, FPS as the evaluation metrics of the model.

4.3   Experimental Details

Referring to the training strategy of DiffusionDet, the Encoder ResNet [23] with FPN [24] as backbone is initialized on ImageNet-1K [25] using pre-trained weights, and the Decoder is initiated by Xavier init [26]. We train the diffusion detector using the AdamW [27] optimizer with an initial learning rate of 2.5×10^{-5} and a weight decay of 10^{-4}. The DPM-Det training time for the ResNet-50 with FPN as the backbone is 450K iterations, and the learning rate is divided by 10 at 350K and 420K iterations to train in small batches of 8. The DPM-Det training time for the ResNet-101 with FPN as the backbone is 600K iterations, and the learning rate is divided by 10 at 470K and 560K iterations to train in small batches of 6. Data enhancement strategies include horizontal flipping, scaling the input image size, dithering, and random cropping expansion.

4.4   Ablation Study

We conducted ablation experiments on the COCO 2017 val set to study the details of DPM-Det. All experiments use ResNet-50 with FPN [24] as the backbone trained and inferred.

The Impact of Steps. The number of sampling steps directly affects the use of the higher-order DPM-Solver++ in DPM-Det and can influence the performance of the final object detection. We used 500 boxes for inference, as shown in Table 1, and a steady performance gain is obtained as the steps increase. The DPM-Det with 2 steps achieved an AP of 46.0 and the DPM-Det with 5 steps achieved an AP of 47.0, a rise of 1.0. Further, by increasing the number of sampling steps, the DPM-Det with 8 steps improved to 47.3 AP.

The Impact of Proposal Numbers. Proposal numbers will largely influence the DPM-Det. We started with 300 proposals for training and later increased to 500 proposals and obtained better performance. We also investigated the degree of influence of proposal numbers on the DPM-Det, and the experimental results when $N_{train}=N_{eval}$ are shown in Table 2. From 100 proposals to 500 proposals, the performance continues to get stronger, and DPM-Det is performing better and better for object detection tasks. However, increasing proposal numbers led to an increase in the training time of the model, so we selected 300 proposals and 500 proposals as the main configurations.

Table 1. Effect of steps.

steps	AP	AP$_{50}$	AP$_{75}$
2	46.0	65.3	49.6
5	47.0	66.6	50.8
8	47.3	67.0	51.4

Table 2. Effect of proposal numbers.

proposals	AP	AP$_{50}$	AP$_{75}$
100	43.5	61.6	46.9
300	46.5	63.7	50.4
500	47.0	66.6	50.8

Table 3. Matching between N_{train} with N_{eval}.

eval \ train	100	300	500
100	43.5	44.4	43.5
300	44.5	46.5	46.7
500	44.7	46.4	47.0

Matching Between N_{train} with N_{eval}. When $N_{train} \neq N_{eval}$, the performance of the model fluctuates. DPM-Det can use different numbers of boxes for learning in the training phase, and can also be evaluated by different numbers of boxes in the inference phase. We train the model using $N_{train} = \{100, 300, 500\}$ separately and then evaluate the model by $N_{eval} = \{100, 300, 500\}$ separately, and the results are shown in Table 3. It is obtained experimentally that although the number of boxes used for training is important, the model performance is also substantially improved by using more boxes for evaluation in the inference phase. For example, training the DPM-Det with 500 boxes and evaluating it with 300 boxes yields 46.5 AP. The reasoning phase boosts 300 boxes to 500 boxes, and AP can reach 47.0.

4.5   Experimental Results and Analysis

Accuracy and Speed. The experiments were conducted in the COCO 2017 val set. Compared with DiffusionDet with ResNet-50 with FPN as the backbone and using 300 boxes and 500 boxes, DPM-Det achieved 46.5 AP with 17.4 FPS using only 300 boxes, outperforming different kinds of DiffusionDet in terms of accuracy and speed. Using 500 boxes, the DPM-Det with a step of 5 not only obtained an AP of 47.0 but also had an FPS of 15.9. The specific results are shown in Table 4.

Table 4. Comparison of accuracy and speed between DPM-Det and DiffusionDet.

Method	boxes	steps	AP	AP$_{50}$	AP$_{75}$	FPS
DiffusionDet [3]	300	4	45.8	65.7	49.2	12.4
DiffusionDet [3]	500	4	46.1	66.0	49.2	7.4
DiffusionDet [3]	500	5	46.2	66.4	49.9	5.9
DiffusionDet [3]	500	8	46.2	66.4	49.5	4.1
DPM-Det	300	5	46.5	65.9	50.4	**17.4**
DPM-Det	500	5	**47.0**	**66.6**	**50.8**	15.9

Comparison of Results. We compare DPM-Det with previous detectors on the MS-COCO dataset, and DPM-Det uses 500 boxes for training and inference.

At the COCO 2017 val set, DPM-Det obtained 47.0 AP when steps were 5 using ResNet-50 with FPN as the backbone, and further improved to 47.3 AP when steps were 8, surpassing DiffusionDet's 46.2 when steps were 8 when using ResNet-50 with FPN as the backbone the highest AP. It also surpasses previously established detectors such as DETR [10] and Deformable DETR [28] and Sparse R-CNN [29]. DPM-Det obtained 47.5 AP with 5 steps and 47.8 AP with 8 steps when using ResNet-101 with FPN as the backbone. The specific results are shown in Table 5.

Table 5. Comparisons on the COCO 2017 val set.

Method	AP	AP_{50}	AP_{75}	AP_s	AP_m	AP_l
ResNet-50						
DETR [10]	42.0	62.4	44.2	20.5	45.8	61.1
Deformable DETR [28]	43.8	62.6	47.7	26.4	47.1	58.0
Sparse DETR [11]	46.3	66.0	50.1	29.0	49.5	60.8
Sparse R-CNN [29]	45.0	63.4	48.2	26.9	47.2	59.5
DiffusionDet (1 step) [3]	45.5	65.1	48.7	27.5	48.1	61.2
DiffusionDet (4 step) [3]	46.1	66.0	49.2	28.6	48.5	61.3
DiffusionDet (8 step) [3]	46.2	66.4	49.5	28.7	48.5	61.5
DPM-Det (5 steps)	47.0	66.6	50.8	29.7	49.6	62.2
DPM-Det (8 steps)	**47.3**	**67.0**	**51.4**	**30.3**	**49.9**	**62.4**
ResNet-101						
DETR [10]	43.5	63.8	46.4	21.9	48.0	61.8
Sparse R-CNN [29]	46.4	64.6	49.5	28.3	48.3	61.6
RecursiveDet (DiffusionDet) [30]	46.9	66.7	50.3	29.5	49.8	62.7
DiffusionDet (1 step) [3]	46.6	66.3	50.0	30.0	49.3	62.8
DiffusionDet (4 step) [3]	46.9	66.8	50.4	30.6	49.5	62.6
DiffusionDet (8 step) [3]	47.1	67.1	50.6	30.2	49.8	62.7
DPM-Det (5 steps)	47.5	67.1	51.3	**31.0**	50.4	**63.2**
DPM-Det (8 steps)	**47.8**	**67.4**	**51.6**	**31.0**	**51.0**	63.1

The DPM-Det also performed well on the COCO 2017 test-dev set. DPM-Det used ResNet-50 with FPN as the backbone and obtained 46.6 AP with 5 steps. When the steps were 8, the AP was further increased to 46.9, achieving better results. When using ResNet-101 with FPN as the backbone, DPM-Det obtained 47.8 AP with 5 steps and 48.0 AP with 8 steps, surpassing DW [31], VFL [32], and MuSu [33], which are more effective and mature detectors. The specific results are shown in Table 6.

In the same test environment, both ResNet-50 and ResNet-101 with FPN as the backbones were used with training and evaluation boxes of 500, respectively, and tested using the COCO 2017 val set, and the results are shown in Fig. 2. For the trade-off between model accuracy and speed, DPM-Det can do better than DiffusionDet. When the backbone is ResNet-50 with FPN, the best accuracy of DiffusionDet with multiple steps has 46.2 and FPS is 4.1, and the best accuracy of DPM-Det can reach 47.3 and FPS reaches 11.4, which is more than 2 times faster. When the backbone is ResNet-101 with FPN, DPM-Det also achieves better results. The best accuracy of DiffusionDet has 47 and FPS is 5.4, and the best accuracy of DPM-Det can reach 47.8 and FPS reaches 10.1. DPM-Det is almost 2 times faster with improved accuracy.

Table 6. Comparisons on the COCO 2017 test-dev set.

Method	AP	AP_{50}	AP_{75}	AP_s	AP_m	AP_l
ResNet-50						
DPM-Det (5 steps)	46.6	66.8	50.1	28.0	48.6	59.6
DPM-Det (8 steps)	**46.9**	**67.2**	**50.4**	**28.2**	**49.0**	**59.7**
ResNet-101						
MuSu [33]	44.8	63.2	49.1	26.2	47.9	56.4
VFL [32]	44.9	64.1	48.9	27.1	48.7	55.1
DW [31]	46.2	64.8	50.0	27.1	49.4	58.5
DPM-Det (5 steps)	47.8	**67.4**	51.8	28.2	50.3	61.8
DPM-Det (8 steps)	**48.0**	**67.4**	**52.3**	**28.4**	**50.4**	**62.0**

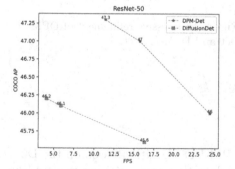
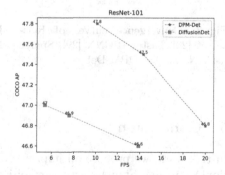

Fig. 2. Accuracy and speed of DPM-Det and DiffusionDet on the COCO 2017 val set with ResNet-50 and ResNet-101 with FPN as the backbone.

Convergence Rate. As an example, DPM-Det with ResNet-50 with FPN as the backbone with 5 steps, using 500 boxes for training and inference, has achieved a relatively good performance on the COCO 2017 val set at 30 epochs. Compared with different previous mature object detectors, DPM-Det has a faster convergence speed and can achieve higher accuracy in the same training time, as shown in Fig. 3.

Progressive Refinement. In contrast to DiffusionDet's progressive refinement strategy, which leads to improved model performance, DPM-Det's progressive refinement strategy also leads to improved model performance. The performance of DPM-Det and DiffusionDet on the COCO 2017 val set with ResNet-50 with FPN as the backbone, using 500 boxes for training and inference, is shown in Fig. 4. DPM-Det can improve the model performance more substantially by improving the progressive refinement of AP to a small extent compared to DiffusionDet. The AP value of DPM-Det with 5 steps is 1.0 higher than that of DPM-Det with 2 steps, which is a relatively large improvement. The DPM-Det with 8 sampling steps has an AP of 47.3, and the DPM-Det with 5 steps also has a higher FPS compared to the DiffusionDet with 5 steps.

 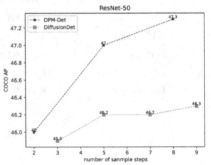

Fig. 3. Convergence curves of RetinaNet [29], Faster R-CNN [29], Sparse R-CNN [29] and DPM-Det.

Fig. 4. Progressive refinement of DPM-Det and DiffusionDet.

5   Conclusion

In this work, we propose an improved DPM-Det algorithm based on the DiffusionDet model. Similarly, we view object detection as a denoising process from noise boxes to bounding boxes, changing the inference process from noise boxes to bounding boxes in a gradual refinement process. By introducing DPM-Solver++ to replace the original DDIM in DiffusionDet, it is possible to obtain high-quality samples using fewer sampling steps. Based on the DiffusionDet training algorithm and sampling algorithm, the training algorithm and sampling algorithm of DPM-Det adapted to DPM-Solver++ are proposed, while inheriting the features of DiffusionDet progressive refinement and decoupling of the training box and evaluation box. Experiments on the MS-COCO dataset show that the DPM-Det with ResNet-50 and ResNet-101 with FPN as the backbone, respectively, can use fewer sampling steps to obtain better performance and have higher FPS than DiffusionDet with the same number of sampling steps, achieving a better balance of accuracy and speed. Compared to the previously proven detectors, DPM-Det has obtained good performance.

References

1. Ho, J., Jain, A., Abbeel, P.: Denoising diffusion probabilistic models. Adv. Neural. Inf. Process. Syst. **33**, 6840–6851 (2020)
2. Song, J., Meng, C., Ermon, S.: Denoising diffusion implicit models. arXiv preprint arXiv:2010.02502 (2020)
3. Chen, S., Sun, P., Song, Y., Luo, P.: DiffusionDet: diffusion model for object detection. arXiv preprint arXiv:2211.09788 (2022)
4. Lu, C., Zhou, Y., Bao, F., Chen, J., Li, C., Zhu, J.: DPM-Solver++: fast solver for guided sampling of diffusion probabilistic models. arXiv preprint arXiv:2211.01095 (2022)
5. Sermanet, P., Eigen, D., Zhang, X., Mathieu, M., Fergus, R., LeCun, Y.: OverFeat: integrated recognition, localization and detection using convolutional networks. arXiv preprint arXiv:1312.6229 (2013)
6. Girshick, R.: Fast R-CNN. In: Proceedings of the IEEE International Conference on Computer Vision, pp. 1440–1448 (2015). https://doi.org/10.1109/ICCV.2015.169
7. Liu, W., et al.: SSD: single shot multibox detector. In: Leibe, B., Matas, J., Sebe, N., Welling, M. (eds.) ECCV 2016. LNCS, vol. 9905, pp. 21–37. Springer, Cham (2016). https://doi.org/10.1007/978-3-319-46448-0_2
8. Chen, Y., Yuan, X., Wu, R., Wang, J., Hou, Q., Cheng, M.M.: YOLO-MS: rethinking multi-scale representation learning for real-time object detection. arXiv preprint arXiv:2308.05480 (2023)
9. Zhu, C., He, Y., Savvides, M.: Feature selective anchor-free module for single-shot object detection. In: Proceedings of the IEEE/CVF Conference on Computer Vision and Pattern Recognition, pp. 840–849 (2019). https://doi.org/10.1109/CVPR.2019.00093
10. Carion, N., Massa, F., Synnaeve, G., Usunier, N., Kirillov, A., Zagoruyko, S.: End-to-end object detection with transformers. In: Vedaldi, A., Bischof, H., Brox, T., Frahm, J.-M. (eds.) ECCV 2020. LNCS, vol. 12346, pp. 213–229. Springer, Cham (2020). https://doi.org/10.1007/978-3-030-58452-8_13
11. Roh, B., Shin, J., Shin, W., Kim, S.: Sparse DETR: efficient end-to-end object detection with learnable sparsity. arXiv preprint arXiv:2111.14330 (2021)
12. Zhou, X., Hou, J., Yao, T., Liang, D., Liu, Z., Zou, Z., et al.: Diffusion-based 3D object detection with random boxes. arXiv preprint arXiv:2309.02049 (2023)
13. Cao, H., Tan, C., Gao, Z., Chen, G., Heng, P.A., Li, S.Z.: A survey on generative diffusion model. arXiv preprint arXiv:2111.14330 (2022)
14. Bao, F., Li, C., Zhu, J., Zhang, B.: Analytic-DPM: an analytic estimate of the optimal reverse variance in diffusion probabilistic models. arXiv preprint arXiv:2201.06503 (2022)

15. Bao, F., Li, C., Sun, J., Zhu, J., Zhang, B.: Estimating the optimal covariance with imperfect mean in diffusion probabilistic models.arXiv preprint arXiv:2206.07309 (2022)
16. Kong, Z., Ping, W.: On fast sampling of diffusion probabilistic models. arXiv preprint arXiv:2106.00132 (2021)
17. Jolicoeur-Martineau, A., Li, K., Piché-Taillefer, R., Kachman, T., Mitliagkas, I.: Gotta go fast when generating data with score-based models. arXiv preprint arXiv:2105.14080 (2021)
18. Watson, D., Chan, W., Ho, J., Norouzi, M.: Learning fast samplers for diffusion models by differentiating through sample quality. In: International Conference on Learning Representations (2021)
19. Kim, D., Na, B., Kwon, S.J., Lee, D., Kang, W., Moon, I.C.: Maximum likelihood training of implicit nonlinear diffusion model. Adv. Neural Inf. Process. Syst. **35**, 32270–32284 (2022). https://doi.org/10.48550/arXiv.2205.13699
20. Nichol, A.Q., Dhariwal, P.: Improved denoising diffusion probabilistic models. In: International Conference on Machine Learning, pp. 8162–8171. PMLR (2021)
21. Wu, Y., Kirillov, A., Massa, F., Lo, W.Y., Girshick, R.: Detectron2. https://github.com/facebookresearch/detectron2 (2019)
22. Lin, T.-Y., et al.: Microsoft COCO: common objects in context. In: Fleet, D., Pajdla, T., Schiele, B., Tuytelaars, T. (eds.) ECCV 2014. LNCS, vol. 8693, pp. 740–755. Springer, Cham (2014). https://doi.org/10.1007/978-3-319-10602-1_48
23. He, K., Zhang, X., Ren, S., Sun, J.: Deep residual learning for image recognition. In: Proceedings of the IEEE Conference on Computer Vision and Pattern Recognition, pp. 770–778 (2016). https://doi.org/10.1109/CVPR.2016.90
24. Lin, T.Y., Dollár, P., Girshick, R., He, K., Hariharan, B., Belongie, S.: Feature pyramid networks for object detection. In: Proceedings of the IEEE Conference on Computer Vision and Pattern Recognition, pp. 2117–2125 (2017). https://doi.org/10.1109/CVPR.2017.106
25. Deng, J., Dong, W., Socher, R., Li, L.J., Li, K., Fei-Fei, L.: ImageNet: a large-scale hierarchical image database. In: 2009 IEEE Conference on Computer Vision and Pattern Recognition, pp. 248–255. IEEE (2009). https://doi.org/10.1109/CVPR.2009.5206848
26. Glorot, X., Bengio, Y.: Understanding the difficulty of training deep feedforward neural networks. In: Proceedings of the Thirteenth International Conference on Artificial Intelligence and Statistics, pp. 249–256. JMLR Workshop and Conference Proceedings (2010)
27. Loshchilov, I., Hutter, F.: Decoupled weight decay regularization. arXiv preprint arXiv:1711.05101 (2017)
28. Zhu, X., Su, W., Lu, L., Li, B., Wang, X., Dai, J.: Deformable DETR: deformable transformers for end-to-end object detection. arXiv preprint arXiv:2010.04159 (2020)
29. Sun, P., Zhang, R., Jiang, Y., Kong, T., Xu, C., Zhan, W., et al.: Sparse R-CNN: end-to-end object detection with learnable proposals. In: Proceedings of the IEEE/CVF Conference on Computer Vision and Pattern Recognition, pp. 14454–14463 (2021). https://doi.org/10.1109/CVPR46437.2021.01422
30. Zhao, J., Sun, L., Li, Q.: RecursiveDet: end-to-end region-based recursive object detection. arXiv preprint arXiv:2307.13619 (2023)
31. Li, S., He, C., Li, R., Zhang, L.: A dual weighting label assignment scheme for object detection. In: Proceedings of the IEEE/CVF Conference on Computer Vision and Pattern Recognition, pp. 9387–9396 (2022). https://doi.org/10.1109/CVPR52688.2022.00917

32. Zhang, H., Wang, Y., Dayoub, F., Sunderhauf, N.: VarifocalNet: an IoU-aware dense object detector. In: Proceedings of the IEEE/CVF Conference on Computer Vision and Pattern Recognition, pp. 8514–8523 (2021). https://doi.org/10.1109/CVPR46437.2021.00841

33. Gao, Z., Wang, L., Wu, G.: Mutual supervision for dense object detection. In: Proceedings of the IEEE/CVF International Conference on Computer Vision, pp. 3641–3650 (2021). https://doi.org/10.1109/ICCV48922.2021.00362

CT-MVSNet: Efficient Multi-view Stereo with Cross-Scale Transformer

Sicheng Wang, Hao Jiang$^{(\boxtimes)}$, and Lei Xiang

Nanjing University of Information Science and Technology, Nanjing, China
{scwang,jianghao}@nuist.edu.cn, xl1294487391@gmail.com

Abstract. Recent deep multi-view stereo (MVS) methods have widely incorporated transformers into cascade network for high-resolution depth estimation, achieving impressive results. However, existing transformer-based methods are constrained by their computational costs, preventing their extension to finer stages. In this paper, we propose a novel cross-scale transformer (CT) that processes feature representations at different stages without additional computation. Specifically, we introduce an adaptive matching-aware transformer (AMT) that employs different interactive attention combinations at multiple scales. This combined strategy enables our network to capture intra-image context information and enhance inter-image feature relationships. Besides, we present a dual-feature guided aggregation (DFGA) that embeds the coarse global semantic information into the finer cost volume construction to further strengthen global and local feature awareness. Meanwhile, we design a feature metric loss (FM Loss) that evaluates the feature bias before and after transformation to reduce the impact of feature mismatch on depth estimation. Extensive experiments on DTU dataset and Tanks and Temples (T&T) benchmark demonstrate that our method achieves state-of-the-art results. Code is available at https://github.com/wscstrive/CT-MVSNet.

Keywords: Multi-view stereo · Feature matching · Transformer · 3D reconstruction

1 Introduction

Multi-View Stereo (MVS) is a fundamental task in computer vision that aims to reconstruct the dense geometry of the scene from a series of calibrated images, and has been extensively studied by traditional methods [2,6,18,19] for years. Compared to traditional techniques, learning-based MVS methods [7,30,31] use convolutional neural networks for dense depth prediction, which achieve superior advancements in challenging scenarios (e.g., illumination changes, non-Lambertian surfaces, and textureless areas).

Recently, transformer [21] has shown great potential in various tasks [4,10, 17,20,23,25], especially in MVS. By treating MVS as a feature matching task between different images, i.e., finding the pixel with the smallest matching cost

© The Author(s), under exclusive license to Springer Nature Switzerland AG 2024
S. Rudinac et al. (Eds.): MMM 2024, LNCS 14555, pp. 394–408, 2024.
https://doi.org/10.1007/978-3-031-53308-2_29

under differentiable homography. The transformer-based method [4] can capture long-range context information within and across different images to improve reconstruction quality. Additionally, some networks [10,25,29] use dimensionality reduction or epipolar constraints to decrease the parameters of the attention mechanism, resulting in reducing the computational burden. However, an interesting phenomenon was observed: these networks prefer to add attention modules to the coarse stage, while the finer stage is just a simple baseline. It makes the whole network appear top-heavy and quite unstable. Thus, how to allocate attention mechanisms in different stages of the feature pyramid network (FPN) [11] without excessive burdens is particularly important.

To address this problem, we revisit the hierarchical design of the FPN and discover that this design has unique characteristics at each stage. Consequently, customizing attention modules based on these individual characteristics becomes a more logical design strategy. In this paper, we propose a novel cross-scale transformer-based multi-view stereo network termed CT-MVSNet. Specifically, we introduce an adaptive matching-aware transformer (AMT) that rearranges and combines interleaved intra-attention and inter-attention blocks to capture intra-image context information and inter-image feature relationships at each stage. For the coarser stage characterized by inherently lower image resolutions, we emphasize using more inter-attention blocks to discern local details from diverse perspectives. As the network progresses to the finer stage, the semantic information from earlier stages gradually permeates and enriches to the finer stage, rendering more intra-attention alone sufficient for enriching texture and local details. Notably, adaptive sampling is used to fine-tune the feature maps at finer stages. It maintains network efficiency but also leads to the loss of critical information in subsequent processes. Therefore, we present a dual-feature guided aggregation (DFGA), which embeds the coarse global feature into the finer cost volume construction, to guide our network to construct cost volume with global semantic awareness and local geometric details. This stratified design avoids imposing additional computational load on the network and ensures that every stage benefits from a mixture of local and global information. Also, our model maintains coarse-stage semantics and fine-granular details, which collectively enhances its stability and efficiency.

In addition, our AMT treats each pixel uniformly, resulting in that a depth map often appears overly smoothed. For this situation, we design a feature metric loss (FM Loss), which weakens discrepancies from feature matching across multiple views while considering geometric consistency and the stability of features post-transformations. Extensive experiments show that our network contributes to enhancing the overall reconstruction quality and achieves state-of-the-art performance on both DTU [1] dataset and Tanks and Temples [9] benchmark.

In summary, our method presents the following key contributions:

1) We introduce an adaptive matching-aware transformer (AMT) that effectively employs different attention combinations to capture intra-image context information and inter-image feature relationships at each stage of FPN.

2) We propose a dual-feature guided aggregation (DFGA) that aims to enhance the stability of cost volume construction by leveraging the feature information derived from different stages.

3) We design a novel feature metric loss (FM Loss) that penalizes errors in feature matching across different views to capture subtle details and minor variations within the reconstructed scene.

4) Our method outperforms existing methods, achieving state-of-the-art results on several benchmark datasets, including the DTU dataset and both intermediate and advanced sets of Tanks and Temples benchmark.

2 Related Works

2.1 Learning-Based MVS Methods

Over the past few decades, learning-based MVS has aimed to improve the accuracy and completeness of reconstructed scenes, which has been extensively researched. MVSNet [30], as a pioneering work, roughly divides learning-based MVS into four steps: image feature extraction using 2D CNN, variance-based cost fusion via homography warping, cost regularization through 3D U-Net, and depth regression. Among these steps, the use of 3D U-Net for processing cost volumes introduces significant computational and memory overhead to the entire network, prompting subsequent research to explore alternative methods. Recurrent MVSNet [26,31] leverage recurrent networks to regularize the cost volume along the depth dimension. Cascade MVSNet [3,7,28] employ a multi-stage strategy to refine the cost volume by narrowing hypothesis ranges, and this strategy is widely adopted by transformer-based methods.

2.2 Attention-Based Transformer for Feature Matching

Transformer architecture [21], initially renowned for its performance in natural language processing (NLP) tasks, has recently garnered considerable attention in the computer vision (CV) community [5,8,12], particularly in feature-matching tasks. Some works [16,17,20,23] have demonstrated the importance of the transformer in capturing long-range global context by using self-attention and cross-attention blocks for feature matching.

TransMVSNet [4] first introduces transformers into MVS, which leverages transformers to capture long-range context information within and across images at the coarsest stage. MVSTER [25] confines attention-based matching to the epipolar lines of the source views, effectively reducing matching redundancy, but it is sensitive to changes in the viewing angle and exhibits unstable matching results. Unlike these methods, we propose CT-MVSNet, which enables efficient and effective feature matching at multiple stages. Compared to TransMVSNet, our network extends long-range context aggregation and global feature interaction to higher-resolution reconstructions. Compared to MVSTER, our network is more robust to inaccurate camera calibrations and poses.

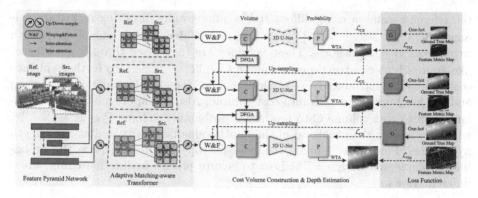

Fig. 1. Illustration of our CT-MVSNet. FPN first extracts multi-scale features and then introduces AMT, which employs different attention combinations to enhance features at each stage (Sect. 3.2). W&F is used to construct cost volume under depth hypotheses and the DFGA is added in finer stages to further refine cost volume construction (Sect. 3.3). The 3D U-Net is used to obtain a probability volume, and then *winner − take − all* (WTA) is used to get a depth map (Sect. 3.3). The CE Loss with one-hot labels is applied to probability volume and the FM Loss with feature metric is applied to the depth map (Sect. 3.4). Then we update the depth hypotheses by up-sampling to finer stages and finally estimate coarse-to-fine depth maps.

3 Methodology

This section introduces the main process of our network in detail. Firstly, we review the overall architecture of CT-MVSNet in Sect. 3.1, and then describe in turn the three main contributions: adaptive matching-aware transformer (AMT) in Sect. 3.2, dual-feature guided aggregation (DFGA) in Sect. 3.3 and feature metric loss (FM Loss) in Sect. 3.4.

3.1 Network Overview

Given a reference image $I_0 \in \mathbb{R}^{3 \times H \times W}$ and its neighboring $N-1$ source images $\{I_i\}_{i=1}^{N-1}$, as well as their corresponding camera intrinsics $\{K_i\}_{i=0}^{N-1}$ and extrinsics $\{T_i\}_{i=0}^{N-1}$, CT-MVSNet aims to estimate the reference depth map $D \in \mathbb{R}^{H' \times W'}$, which aligns I_i with I_0 in depth ranges $[d_{min}, d_{max}]$, to reconstruct a dense 3D point cloud, where H', W' denote the height and width at current stage.

The overall architecture of our network is shown in Fig. 1. Follow the coarse-to-fine manner, we first extract multi-scale feature maps $\left\{F_i \in \mathbb{R}^{C' \times H' \times W'}\right\}_{i=0}^{N-1}$ from FPN at resolutions of 1/4, 1/2, and full image scale. To expand the capability for long-range context aggregation and global feature interaction across scales, we introduce the AMT, which employs combinations of interleaved inter-attention and intra-attention blocks within and across multi-view images at each stage. Afterward, we warp 2D transformed features into the 3D frustum of the reference camera using differentiable homography, resulting in feature volumes $\left\{ \mathbf{V}_i \in \mathbb{R}^{C' \times M' \times H' \times W'} \right\}_{i=0}^{N-1}$, where C', M' denote the channels and depth hypotheses at current stage. These volumes are then fused through a variance-like approach to construct a cost volume $\mathbf{C} \in \mathbb{R}^{M' \times H' \times W'}$. To further enhance cost volume, we introduce DFGA, which embeds the global semantic information and the local texture details into cost volume. Also, we apply a 3D U-Net regularization to cost volume $\mathbf{C}$ to produce a probability volume $\mathbf{P} \in \mathbb{R}^{M' \times H' \times W'}$, which represents the likelihood of the depth value being situated at each depth hypotheses. Finally, we employ the cross-entropy loss to supervise the probability volume and design a FM Loss to ensure precise feature matching across different views.

3.2   Adaptive Matching-Aware Transformer

Considering the high memory and computational demands of transformers, most cascade MVS methods that leverage transformers are limited to extracting long-range context information at the coarsest stage and struggle to extend to finer stages. To solve this problem, we introduce AMT, which uses an ingenious method to dismantle the superimposed attention modules of previous methods and arrange and reorganize them to distribute them to appropriate stages. Next, we introduce the preliminaries of attention and then demonstrate how AMT can be seamlessly integrated at various scales to facilitate long-range aggregation and global interactions within and across views.

Intra-attention and Inter-attention. Common attention mechanisms [21] group feature maps $\{\mathbf{F}\}_{i=0}^{N-1}$ into queries $\mathbf{Q}$, keys $\mathbf{K}$ and values $\mathbf{V}$. $\mathbf{Q}$ retrieves relevant information from $\mathbf{V}$ according to the attention weights obtained from the dot product of $\mathbf{Q}$ and $\mathbf{K}$ corresponding to each $\mathbf{V}$. However, we follow the linear attention [8] in place of the common attention [21]. The linear attention effectively reduces the time complexity from $O(n^2)$ to $O(n)$ by leveraging the associative property of multiplication. The kernel function of linear attention is:

$$Linear\,Attention(\mathbf{Q}, \mathbf{K}, \mathbf{V}) = \Phi(\mathbf{Q})(\Phi(\mathbf{K}^\top)\mathbf{V}) \qquad (1)$$

where $\Phi(\cdot) = elu(\cdot) + 1$ and $elu(\cdot)$ denotes the activation function of the exponential linear unit.

Therefore, we apply linear attention to achieve intra-attention and inter-attention respectively. Figure 2(b) shows our attention architectures. For intra-attention, where $\mathbf{Q}$ and $(\mathbf{K}, \mathbf{V})$ originate from the same feature map, it focuses on capturing long-range context information of the given view and emphasizing crucial information for the current task. For inter-attention, where $\mathbf{Q}$ from the source feature maps and $(\mathbf{K}', \mathbf{V}')$ from the reference feature map, it enhances global feature interactions across views and enhance cross-relationships by amalgamating different perspectives. Notably, to prevent the reference feature $\mathbf{F}_0$ from being overly smoothed by inter-attention, this block just enhances the source features $\{\mathbf{F}\}_{i=1}^{N-1}$.

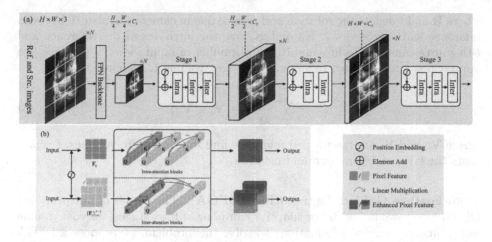

Fig. 2. Illustration of AMT. (a)Interleaved conbination of intra- and inter-attention at each stage. (b)Internal architecture of two attentions, each stage consist of interleaving-arranged intra-attention (w.r.t. **Q**, **K**, **V**) within images, and inter-attention (w.r.t. **Q**, **K**′, **V**′) across images.

AMT Architecture. The overall architecture of our proposed AMT is shown in Fig. 2(a). To adapt to the feature representation of the FPN at various stages, our AMT utilizes optimal interleaved combinations. These combinations consist of multiple intra-attention and inter-attention blocks tailored for feature maps of different scales. At stage 1, referred to as the coarsest stage, the receptive field for each pixel feature is relatively large, so a singular view cannot offer sufficient information for effective feature enhancement. Thus, we deploy more cross-attention blocks to refer to more different perspective images (i.e., one intra-attention and two inter-attentions). As we transition to the second and third stages, which are characterized by a finer granularity, the resolution noticeably increases. To ensure computational efficiency during AMT processing, we incorporate adaptive sampling, with sampling rates at $1/2$ and $1/4$ respectively. During the second phase, a combination of one intra-attention and inter-attention block serves as a bridge to the most refined phase. Then, at the finest stage, due to a smaller receptive field, the features exhibit rich texture details. Hence, we gravitate towards more intensive intra-attention blocks, involving two intra-attentions and one inter-attention.

3.3   Cost Volume Construction and Depth Estimation

Similar to most learning-based MVS methods [7,30], for each pixel **p** in the reference view, we utilize the differentiable homography to compute the corresponding pixel $\hat{\mathbf{p}}_i$ by warping the source feature $\mathbf{F}(\mathbf{p}_i)$ under the depth hypotheses d:

$$\hat{\mathbf{p}}_i = \mathbf{K}_i[\mathbf{R}(\mathbf{K}_0^{-1}\mathbf{p}d) + \mathbf{t}] \tag{2}$$

where $\mathbf{R}$ and $\mathbf{t}$ denote the rotation and translation in camera extrinsic $\mathbf{T}$ between reference and source views. $\mathbf{K}_0$ and $\mathbf{K}_i$ are the intrinsics of the reference and i-th source cameras. To fuse an arbitrary number of input views, a variance-like approach is often adopted to generate the cost volume $\mathbf{C}$, denoted as

$$\mathbf{C}^\ell = \sum_{i=1}^{N-1} \frac{1}{N-1}(\mathbf{V}_i - \mathbf{V}_0) \tag{3}$$

where $\mathbf{V}_0$ and $\mathbf{V}_i$ are the warped reference and source feature volumes, ℓ represents the layer of feature pyramid network.

Dual-Feature Guided Aggregation (DFGA). The details of our proposed DFGA are shown in Fig. 3. For adaptive sampling at finer stages, the cost volume loses some key feature information. To solve this problem, we propose a DFGA to enrich cost volumes. The method first uses a 3×3 2D convolution operation to extract the coarse-scale global semantic information of the previous stage and the local texture information in the initial cost volume of the current stage. Then, we fuse these two types of information and the initial cost volume through a skip connection strategy to obtain our cost volume. Also, we use a 1×1 2D convolution operation to further refine cost volume. The final cost volume that synthesizes global and local features is computed as follows:

$$\hat{\mathbf{C}}^\ell = DFGA[\mathbf{C}^\ell, \mathbf{C}_\uparrow^{\ell-1}] \tag{4}$$

$$DFGA[\,,\,] = \Sigma_{1\times1}\left[\Sigma_{3\times3}(\mathbf{C}_\uparrow^{\ell-1}), \Sigma_{3\times3}(\mathbf{C}^\ell), \mathbf{C}^\ell\right] \tag{5}$$

where $\mathbf{C}_\uparrow^{\ell-1}$ represents an up-sampled cost volume from the previous stage, and $\mathbf{C}^\ell$ is the cost volume at the current stage, $[\cdot,\cdot]$ is concatenation operation, $\Sigma_{1\times1}$ and $\Sigma_{3\times3}$ represent the convolution of two different receptive fields. The 3×3 receptive field excels at capturing spatially relevant local structures and contextual nuances, while the 1×1 receptive field bolsters feature representation,

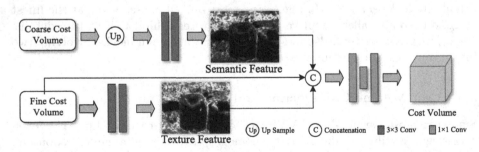

Fig. 3. Illustration of DFGA. Our DFGA is only applied in the second and third stage.

promotes cross-channel feature interaction, and curtails both parameters and computational complexity.

Depth Estimation. We employ a 3D U-Net to regularize the DFGA-enhanced cost volume into a probability volume $\mathbf{P}$. Applying $softmax$ and $winner-take-all$ (WTA) strategy to probability volume yields our final depth map:

$$\mathbf{D} = WTA \left(\underset{d \in \{d_m\}_{m=1}^M}{softmax} \left(\mathbf{P}^{(d)} \right) \right) \tag{6}$$

where d_m refers to the depth hypotheses of the m-th level.

3.4 Loss Functions

For MVS, depth estimation is often modeled as a probability distribution problem. Thus, pixel-wise classification methods [4,26] are more suitable than L1-based regression [7,30] in capturing accurate confidence maps. We use cross-entropy loss for pixel-wise supervision:

$$\mathcal{L}_{CE} = \sum_{\mathbf{p} \in \Psi} \sum_{d=d_1}^{d_M} -\mathbf{G}^{(d)}(\mathbf{p}) \log \left[\mathbf{P}^{(d)}(\mathbf{p}) \right] \tag{7}$$

where $\mathbf{G}^{(d)}(\mathbf{p})$ denote ground truth probability of depth hypotheses d at pixel $\mathbf{p}$. Ψ is the set of valid pixels with ground truth precision.

Moreover, we design a novel FM Loss to explore geometric and feature consistency between reference view and source views, which achieves more accurate depth estimation by supervising depth maps and penalizing errors in feature matching across different views. Using differentiable homography, we project pixel $\mathbf{p}$ from the reference image onto the i-th source image, resulting in $\mathbf{p}'_i$, and then warp back to the reference image, yielding $\mathbf{p}''$:

$$\begin{cases} \mathbf{p}'_i = \mathbf{T}_0^{-1} \mathbf{K}_0^{-1} \mathbf{D}_0(\mathbf{p}) \mathbf{p} \\ \mathbf{p}'' = (\mathbf{K}_i \mathbf{T}_i \mathbf{p}') / \mathbf{D}_i^{gt}(\mathbf{p}'_i) \end{cases} \tag{8}$$

where $\mathbf{D}_i^{gt}(\mathbf{p}'_i)$ denotes the ground truth depth value of $\mathbf{p}'_i$. FM Loss evaluates the consistency of each feature across views to ensure their stability and reliability under varying conditions. Our final FM Loss is:

$$\mathcal{L}_{FM} = (\Upsilon)(\left\| \mathbf{D}_0(\mathbf{p}) - \mathbf{D}_0^{gt}(\mathbf{p}) \right\|_2) \tag{9}$$

where $\mathbf{D}_0^{gt}(\mathbf{p})$ represents the depth value predicted by our network at pixel $\mathbf{p}$. To take into account in depth the differences between original features and their mappings after various transformations, we add feature metric weights Υ, which is denoted as:

$$\Upsilon = \underset{v \in [v_1, v_2]}{log} |v|, \quad v = \frac{\hat{\mathbf{F}}_0(\mathbf{p}'') - \hat{\mathbf{F}}_0(\mathbf{p})}{\mathbf{F}_0(\mathbf{p}'') - \mathbf{F}_0(\mathbf{p}) + \varepsilon} \tag{10}$$

where υ is feature bias, $\mathbf{F}_0(\mathbf{p}'')$ and $\mathbf{F}_0(\mathbf{p})$ denote the features and back-projected features before transformer, and $\hat{\mathbf{F}}_0(\mathbf{p}'')$ and $\hat{\mathbf{F}}_0(\mathbf{p})$ represent the features and back-projected features after transformer, ε ensures the denominator does not approach zero, $[\upsilon_1, \upsilon_2]$ defines the acceptable range for feature deviations.

In summary, the loss function consists of $\mathcal{L}_{CE}$ and $\mathcal{L}_{FM}$:

$$\mathcal{L} = \sum_{\ell=1}^{L} \lambda_1^\ell \mathcal{L}_{CE} + \lambda_2^\ell \mathcal{L}_{FM} \tag{11}$$

where $\lambda_1^* = 2$, $\lambda_2^* = 1.2$ among each stage in our experiments, and L is the number of stages.

4   Experiment

4.1   Datasets.

DTU [1] dataset is an indoor multi-view stereo dataset with 124 scenes with 49 views and 7 illumination conditions. It contains ground-truth point clouds captured under well-controlled laboratory conditions for evaluation. **Tanks and Temples(T&T)** [9] dataset is a large-scale dataset captured in realistic indoor and outdoor scenarios, which contains an intermediate subset of 8 scenes and an advanced subset of 6 scenes. **BlendedMVS** [32] dataset is a large-scale synthetic MVS training dataset containing cities, sculptures, and small objects, split into 106 training scans and 7 validation scans.

4.2   Implementation Details

We implemented CT-MVSNet with PyTorch, which is trained and evaluated on DTU dataset [1]. Also, BlendedMVS dataset [32] is used to fine-tune our model for evaluation on T&T dataset [9]. For the coarse-to-fine network, we use $L = 3$ layer pyramids, the layer number of depth hypotheses M is set to 48, 32, 8 for each level, and the depth sampling range is set to $425mm \sim 935mm$. We set υ_1 and υ_2 are 0.6 and 1.7 for acceptable ranges in feature deviations. For training on DTU, the number of input images is set to $N = 5$ with a resolution of 512×640. As for fine-tuning on BlendedMVS, we set $N = 5$ with 576×768 image resolutions. We used the Adam optimizer with a learning rate of 0.001 and trained the model for 16 epochs, reducing the learning rate by 0.5 after 6, 8, and 12 epochs. We trained the model on 8 NVIDIA GeForce RTX 2080Ti GPUs, with a batch size of 1. The training of CT-MVSNet typically takes around $18\,h$ and consumes approximately $11\,GB$ of memory on each GPU.

4.3   Benchmark Performance

DTU. We evaluated our proposed method on DTU evaluation set at 864×1152 resolution with $N = 5$. The visualized results are shown in Fig. 4, CT-MVSNet

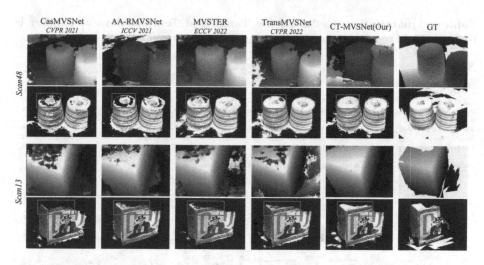

Fig. 4. Point clouds error comparison of state-of-the-art methods on DTU dateset. The first and third rows are estimated depth maps while others are point cloud reconstruction results.

Table 1. Quantitative comparison on the DTU dataset (lower is better). The lower is better for Accuracy (Acc.), Completeness (Comp.), and Overall. The best results are in **bold** and the second best results are in <u>underlined</u>.

Method	Years	Acc.(mm)	Comp.(mm)	Overall$\downarrow$(mm)
Gipuma [6]	ICCV 2015	**0.283**	0.873	0.578
COLMAP [18]	CVPR 2016	0.400	0.664	0.532
R-MVSNet [31]	CVPR 2019	0.385	0.459	0.422
CasMVSNet [7]	CVPR 2020	0.325	0.385	0.355
CVP-MVSNet [28]	CVPR 2020	<u>0.296</u>	0.406	0.351
AA-RMVSNet [26]	ICCV 2021	0.376	0.339	0.357
EPP-MVSNet [13]	CVPR 2021	0.413	0.296	0.355
Patchmatchnet [22]	CVPR 2021	0.427	0.277	0.352
RayMVSNet [27]	CVPR 2022	0.341	0.319	0.330
Effi-MVSNet [24]	CVPR 2022	0.321	0.313	0.317
MVS2D [29]	CVPR 2022	0.394	0.290	0.342
TransMVSNet [4]	CVPR 2022	0.321	0.289	<u>0.305</u>
UniMVSNet [15]	CVPR 2022	0.352	0.278	0.315
MVSTER [25]	ECCV 2022	0.350	<u>0.276</u>	0.313
CT-MVSNet (Ours)	MMM 2024	0.341	**0.264**	**0.302**

produces notably accurate depth estimations and comprehensive point clouds, particularly on low-texture and edge surfaces. Furthermore, the robustness of our method is further demonstrated by its performance on two scenes: *scan*13

Table 2. Quantitative results on the Tanks and Temples dataset (higher is better). **Bold** represents the best while <u>underlined</u> represents the second-best.

Method	Intermediate									Advanced						
	Mean↑	Family	Francis	Horse	L.H	M60	Panther	P.G	Train	Mean↑	Auditorium	Ballroom	Courtroom	Museum	Palace	Temple
COLMAP [18]	42.14	50.41	22.25	26.63	56.43	44.83	46.97	48.53	42.04	27.24	16.02	25.23	34.70	41.51	18.05	27.94
R-MVSNet [31]	50.55	73.01	54.46	43.42	43.88	46.80	46.69	50.87	45.25	29.55	19.49	31.45	29.99	42.31	22.94	31.10
CasMVSNet [7]	56.42	76.36	58.45	46.20	55.53	56.11	54.02	58.17	46.56	31.12	19.81	38.46	29.10	43.87	27.36	28.11
Patchmatchnet [22]	53.15	66.99	52.64	43.24	54.87	52.87	49.54	54.21	50.81	32.31	23.69	37.73	30.04	41.80	28.31	32.29
AA-RMVSNet [26]	61.51	77.77	59.53	51.53	**64.02**	**64.05**	59.47	<u>60.85</u>	55.50	33.53	20.96	40.15	32.05	46.01	29.28	32.71
Effi-MVSNet [24]	56.88	72.21	51.02	51.78	58.63	58.71	56.21	57.07	49.38	34.39	20.22	42.39	33.73	45.08	29.81	35.09
TransMVSNet [4]	<u>63.52</u>	<u>80.92</u>	65.83	<u>56.89</u>	62.54	63.06	<u>60.00</u>	60.20	<u>58.67</u>	37.00	24.84	<u>44.59</u>	34.77	46.49	<u>34.69</u>	36.62
GBi-Net [14]	61.42	79.77	**67.69**	51.81	61.25	60.37	55.87	60.67	53.89	<u>37.32</u>	**29.77**	42.12	**36.30**	<u>47.69</u>	31.11	36.93
MVSTER [25]	60.92	80.21	63.51	52.30	61.38	61.47	58.16	58.98	51.38	37.53	26.68	42.14	<u>35.65</u>	**49.37**	32.16	**39.19**
CT-MVSNet (Ours)	**64.28**	**81.20**	65.09	**56.95**	<u>62.60</u>	<u>63.07</u>	**64.83**	**61.82**	**58.68**	**38.03**	<u>28.37</u>	**44.61**	34.83	46.51	**34.69**	<u>39.15</u>

Fig. 5. Point clouds error comparison of state-of-the-art methods on the Tanks and Temples dataset. τ is the scene-relevant distance threshold determined officially and darker means larger errors.

and *scan*48. These scenes, which account for illumination changes and reflections, are considered the most challenging within the DTU evaluation set. The quantitative results are summarized in Table 1. Here, accuracy and completeness are reported using the official MATLAB code. The overall metric represents the average of accuracy and completeness measures. Our approach achieves the highest completeness and overall metric compared to other state-of-the-art methods.

Tanks and Temples. We further validate the generalization capability of our method on the T&T dataset. Figure 5 illustrates the error comparison of the reconstructed point clouds on the scene Pather of intermediate and Temple of advanced, and our method exhibits higher precision and recall, especially in sophisticated and low-texture surfaces. The quantitative results on both intermediate and advanced sets are reported in Table 2, the Mean is the average score of all scenes, CT-MVSNet achieves the highest results in many complex scenes. Besides, there is an official website for online evaluation of T&T benchmark.

4.4   Ablation Study

Table 3 shows the ablation results of our CT-MVSNet. Our baseline is CasMVS-Net [7], and all experiments were conducted with the same hyperparameters.

Effect of AMT. As shown in Table 4, we conducted different combinations of attention blocks. By adding a combination of attention blocks at each stage to make the network always focus on global information acquisition, as depicted in Fig. 4, we have achieved outstanding reconstruction results in texture holes. Also, the self-adaptive sampling strategy guarantees time and memory consumption.

Effect of DFGA. In Table 5, we conducted ablation study on semantic feature (SF) and texture feature (TF) to verify effectiveness. DFGA makes cost volume between stages to establish connections, thereby promoting the restoration and supplementation of information after sampling strategy and stage progress.

Effect of FM Loss. We unify the weight λ_1 of CE Loss to 2 and perform ablation study on the weight λ_2 of FM Loss in Table 6. Based on feature consistency, network focuses more on vital information, and the combination of AMT brings about feature enhancement and fully plays global feature perception.

Table 3. Ablation results with different components on DTU.

	Module Settings				Mean Distance		
	CE Loss	AMT	DFGA	FM Loss	Acc	Comp	Overall
(a)					0.360	0.338	0.349
(b)	✓				0.342	0.325	0.333
(c)	✓	✓			0.371	**0.262**	0.316
(d)	✓	✓	✓		0.346	0.269	0.308
(e)	✓	✓	✓	✓	**0.341**	0.264	**0.302**

Table 4. Ablation study of the number of optimization combinations on DTU.

	intra-	inter-	Acc	Comp	Overall	Mem.(GB)	Time(s)
(a)	4,0,0	4,0,0	0.337	0.293	0.315	3990	1.05
(b)	2,0,0	2,0,0	0.344	0.298	0.321	**3875**	**0.88**
(c)	1,2,0	2,1,0	0.356	0.274	0.309	4034	1.06
(d)	1,2,2	2,2,1	0.346	**0.262**	0.304	4276	1.20
(e)	1,1,2	2,1,1	**0.341**	0.264	**0.302**	4117	1.11

Table 5. Ablation study concerning the SF and TF in the construction process of cost volume.

	DFGA	Acc	Comp	Overall
(a)	SF	0.346	0.269	0.308
(b)	SF+TF	0.344	0.266	0.305
(c)	Ours	**0.341**	**0.264**	**0.302**

Table 6. Ablation study on the different weight parameters λ_1, λ_2 of loss function.

	λ_1	λ_2	Acc	Comp	Overall
(a)	2	0	0.346	0.269	0.308
(b)	2	2	0.356	0.265	0.310
(c)	2	1.5	0.342	0.266	0.304
(d)	2	1.2	**0.341**	**0.264**	**0.302**
(e)	2	1	0.345	0.274	0.309

5 Conclusion

In this paper, we propose a novel learning-based MVS network, named CT-MVSNet, designed for long-range context aggregation and global feature interaction at different scales. Specifically, we introduce an adaptive matching-aware transformer (AMT) that effectively leverages intra- and inter-attention blocks within and across images at each stage. In addition, we present a dual-feature guided aggregation (DFGA) to guide our network to construct cost volume with semantic awareness and geometric details. Meanwhile, we design a feature metric loss (FM Loss) to penalize errors in feature matching across different views. Extensive experiments prove that CT-MVSNet achieves superior performance in 3-D reconstruction. In the future, we intend to expand upon the merits of CT, positioning it to replace the role of FPN as the feature encoder in MVSNet.

Acknowledgements. This work is supported by the National Natural Science Foundation of China (NSFC) (No. 62101275).

References

1. Aanæs, H., Jensen, R.R., Vogiatzis, G., Tola, E., Dahl, A.B.: Large-scale data for multiple-view stereopsis. Int. J. Comput. Vis. **120**(2), 153–168 (2016)
2. Campbell, N.D.F., Vogiatzis, G., Hernández, C., Cipolla, R.: Using multiple hypotheses to improve depth-maps for multi-view stereo. In: Forsyth, D., Torr, P., Zisserman, A. (eds.) ECCV 2008. LNCS, vol. 5302, pp. 766–779. Springer, Heidelberg (2008). https://doi.org/10.1007/978-3-540-88682-2_58
3. Cheng, S., et al.: Deep stereo using adaptive thin volume representation with uncertainty awareness. In: Proceedings of the IEEE/CVF Conference on Computer Vision and Pattern Recognition, pp. 2524–2534 (2020)
4. Ding, Y., et al.: TransMVSNet: global context-aware multi-view stereo network with transformers. In: Proceedings of the IEEE/CVF Conference on Computer Vision and Pattern Recognition, pp. 8585–8594 (2022)
5. Dosovitskiy, A., et al.: An image is worth 16×16 words: transformers for image recognition at scale. arXiv preprint arXiv:2010.11929 (2020)
6. Galliani, S., Lasinger, K., Schindler, K.: Massively parallel multiview stereopsis by surface normal diffusion. In: Proceedings of the IEEE International Conference on Computer Vision, pp. 873–881 (2015)
7. Gu, X., Fan, Z., Zhu, S., Dai, Z., Tan, F., Tan, P.: Cascade cost volume for high-resolution multi-view stereo and stereo matching. In: Proceedings of the IEEE/CVF Conference on Computer Vision and Pattern Recognition, pp. 2495–2504 (2020)
8. Katharopoulos, A., Vyas, A., Pappas, N., Fleuret, F.: Transformers are RNNs: fast autoregressive transformers with linear attention. In: Proceedings of International Conference on Machine Learning, pp. 5156–5165 (2020)
9. Knapitsch, A., Park, J., Zhou, Q.Y., Koltun, V.: Tanks and temples: benchmarking large-scale scene reconstruction. ACM Trans. Graph. (ToG) **36**(4), 1–13 (2017)
10. Liao, J., et al.: WT-MVSNet: window-based transformers for multi-view stereo. Adv. Neural. Inf. Process. Syst. **35**, 8564–8576 (2022)
11. Lin, T.Y., Dollár, P., Girshick, R., He, K., Hariharan, B., Belongie, S.: Feature pyramid networks for object detection. In: Proceedings of the IEEE Conference on Computer Vision and Pattern Recognition, pp. 2117–2125 (2017)
12. Liu, Z., et al.: Swin transformer: hierarchical vision transformer using shifted windows. In: Proceedings of the IEEE/CVF International Conference on Computer Vision, pp. 10012–10022 (2021)

13. Ma, X., Gong, Y., Wang, Q., Huang, J., Chen, L., Yu, F.: EPP-MVSNet: epipolar-assembling based depth prediction for multi-view stereo. In: Proceedings of the IEEE/CVF International Conference on Computer Vision, pp. 5732–5740 (2021)

14. Mi, Z., Di, C., Xu, D.: Generalized binary search network for highly-efficient multi-view stereo. In: Proceedings of the IEEE/CVF Conference on Computer Vision and Pattern Recognition, pp. 12991–13000 (2022)

15. Peng, R., Wang, R., Wang, Z., Lai, Y., Wang, R.: Rethinking depth estimation for multi-view stereo: a unified representation. In: Proceedings of the IEEE/CVF Conference on Computer Vision and Pattern Recognition, pp. 8645–8654 (2022)

16. Ruan, C., Zhang, Z., Jiang, H., Dang, J., Wu, L., Zhang, H.: Vector approximate message passing with sparse Bayesian learning for gaussian mixture prior. China Communications (2023)

17. Sarlin, P.E., DeTone, D., Malisiewicz, T., Rabinovich, A.: SuperGlue: learning feature matching with graph neural networks. In: Proceedings of the IEEE/CVF Conference on Computer Vision and Pattern Recognition, pp. 4938–4947 (2020)

18. Schonberger, J.L., Frahm, J.M.: Structure-from-motion revisited. In: Proceedings of the IEEE Conference on Computer Vision and Pattern Recognition, pp. 4104–4113 (2016)

19. Stereopsis, R.M.: Accurate, dense, and robust multiview stereopsis. IEEE Trans. Pattern Anal. Mach. Intell. **32**(8) (2010)

20. Sun, J., Shen, Z., Wang, Y., Bao, H., Zhou, X.: LoFTR: detector-free local feature matching with transformers. In: Proceedings of the IEEE/CVF Conference on Computer Vision and Pattern Recognition, pp. 8922–8931 (2021)

21. Vaswani, A., et al.: Attention is all you need. In: Advances in Neural Information Processing Systems, vol. 30 (2017)

22. Wang, F., Galliani, S., Vogel, C., Speciale, P., Pollefeys, M.: PatchmatchNet: learned multi-view patchmatch stereo. In: Proceedings of the IEEE/CVF Conference on Computer Vision and Pattern Recognition, pp. 14194–14203 (2021)

23. Wang, Q., Zhang, J., Yang, K., Peng, K., Stiefelhagen, R.: MatchFormer: interleaving attention in transformers for feature matching. In: Proceedings of the Asian Conference on Computer Vision, pp. 2746–2762 (2022)

24. Wang, S., Li, B., Dai, Y.: Efficient multi-view stereo by iterative dynamic cost volume. In: Proceedings of the IEEE/CVF Conference on Computer Vision and Pattern Recognition, pp. 8655–8664 (2022)

25. Wang, X., et al.: MVSTER: epipolar transformer for efficient multi-view stereo. In: Proceedings of the European Conference on Computer Vision, pp. 573–591. Springer, Cham (2022). https://doi.org/10.1007/978-3-031-19821-2_33

26. Wei, Z., Zhu, Q., Min, C., Chen, Y., Wang, G.: AA-RMVSNet: adaptive aggregation recurrent multi-view stereo network. In: Proceedings of the IEEE/CVF International Conference on Computer Vision, pp. 6187–6196 (2021)

27. Xi, J., Shi, Y., Wang, Y., Guo, Y., Xu, K.: RayMVSNet: learning ray-based 1D implicit fields for accurate multi-view stereo. In: Proceedings of the IEEE/CVF Conference on Computer Vision and Pattern Recognition, pp. 8595–8605 (2022)

28. Yang, J., Mao, W., Alvarez, J.M., Liu, M.: Cost volume pyramid based depth inference for multi-view stereo. In: Proceedings of the IEEE/CVF Conference on Computer Vision and Pattern Recognition, pp. 4877–4886 (2020)

29. Yang, Z., Ren, Z., Shan, Q., Huang, Q.: MVS2D: efficient multi-view stereo via attention-driven 2d convolutions. In: Proceedings of the IEEE/CVF Conference on Computer Vision and Pattern Recognition, pp. 8574–8584 (2022)

30. Yao, Y., Luo, Z., Li, S., Fang, T., Quan, L.: MVSNet: depth inference for unstructured multi-view stereo. In: Proceedings of the European Conference on Computer Vision, pp. 767–783 (2018)
31. Yao, Y., Luo, Z., Li, S., Shen, T., Fang, T., Quan, L.: Recurrent MVSNet for high-resolution multi-view stereo depth inference. In: Proceedings of the IEEE/CVF Conference on Computer Vision and Pattern Recognition, pp. 5525–5534 (2019)
32. Yao, Y., et al.: BlendedMVS: a large-scale dataset for generalized multi-view stereo networks. In: Proceedings of the IEEE/CVF Conference on Computer Vision and Pattern Recognition, pp. 1790–1799 (2020)

A Coarse and Fine Grained Masking Approach for Video-Grounded Dialogue

Feifei Xu, Wang Zhou(✉), Tao Sun, Jiahao Lu, Ziheng Yu, and Guangzhen Li

Shanghai University of Electric Power, Shanghai, China
xufeifei@shiep.edu.cn,
{zhouwang,302664224,Jay-ho,zihengyu,liguangzhen}@mail.shiep.edu.cn

Abstract. The task of Video-Grounded Dialogue involves developing a multimodal chatbot capable of answering sequential questions from humans regarding video content, audio, captions and dialog history. Although existing approaches utilizing deep learning models have demonstrated impressive performance, their success often relies on fusion of multimodal information in the limited datasets rather than understanding the interactions and dependencies among individual modalities such as video-audio, caption, dialog history, question or answer. In this paper, we present CFM (Coarse and Fine Grained Masking), a novel approach based on the pre-training model GPT2, aiming at enhancing cross-modal understanding among individual modalities in video-grounded dialogue. CFM achieves this by employing distinct coarse-grained and fine-grained masking strategies to differentiate various inputs, including video-audio, caption, dialog history, question and answer. Furthermore, we improve GPT2 model to strengthen its ability to integrate video-audio with text information effectively by incorporating multimodal feedforward network. Through extensive experiments on the Audio Visual Scene-Aware Dialog (AVSD) datasets, our proposed approach demonstrates promising performance, highlighting the benefits of our method in effectively figuring out dependencies and interactions among individual modalities.

Keywords: Video-grounded dialogue · Coarse-grained and fine-grained · GPT2

1 Introduction

Multimodal machine learning has spurred the emergence of vision-language tasks, including visual dialog [1], video question-answering [2], and video-based dialog [3], which aim to integrate vision with text. Video-based dialog, in particular, simulates human-chatbot interactions in a multimodal environment, where the chatbot responds to questions based on a dynamic scene in the video-audio, and captions, as well as a dialog history comprising previous question-answer pairs. As shown in Fig. 1, when the chatbot is asked a follow-up question: "Anything else ?", to provide an accurate answer, the chatbot needs to answer the question based on the information above.

S. Rudinac et al. (Eds.): MMM 2024, LNCS 14555, pp. 409–422, 2024.
https://doi.org/10.1007/978-3-031-53308-2_30

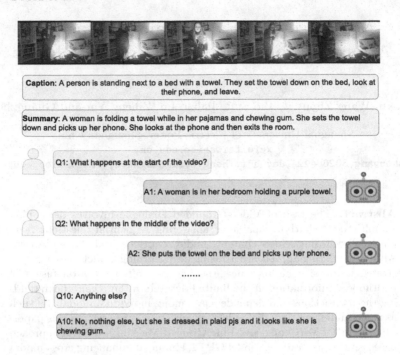

Fig. 1. An Example of video-grounded dialogue from the benchmark dataset of Audio-Visual Scene Aware Dialogue (AVSD) challenge. Q: question, A: answer.

Several state-of-the-art approaches have been proposed for video-grounded dialogue task. Hori et al. [4] incorporates multimodal attention to describe the video, which is integrated into an end-to-end dialogue system that improves the quality of generative video dialogues. Dat Tien Nguyen and Asri [5] propose a hierarchical recurrent encoder-decoder architecture that incorporates a FiLM-based approach to extract audio-visual features. This architecture is used for the purpose of encoding and generating responses, offering a unique and innovative approach. As the winner of the DSTC7@AVSD task, [6] employs the hierarchical attention to efficiently integrate text and image patterns. MTN [7] employs a Transformer decoder architecture to sequentially fuse the temporal features of different modalities. Kim et al. [8] proposes an approach known as structured common reference graph attention, which addresses the task of video-based dialogues dynamically replicating certain parts of a question to facilitate the decoding of answer sequences. Le et al. [9] proposes a model named VGNMN, consisting of dialogue and video understanding neural Modules.

In recent years, pre-trained models have demonstrated their strong ability to handle multimodal tasks, such as GPT2 [10], which is powerful to deal with vision-language task. Li et al. [11] tries to extend a pre-trained GPT2 model to learn joint video-audio and text features through multitask learning objectives. However, their method mainly makes use of attention mechanism to fuse video-audio and text representations, limiting the full exploitation of joint representations for inter modalities.

To address above problem, we propose CFM, a novel approach to reveal the interactions and dependences among individual modalities by incorporating an adaptive masking mechanism during the training phase. To this end, CFM employs coarse-grained and fine-grained masking strategies. In the former, we randomly select 50% of the samples for the coarse-grained masking process in each epoch. Each of these samples is decomposed into five individual modalities, including video-audio, caption, dialog history, question, and answer. During each training iteration, one of these five modalities in the selected samples is chosen at random for entire masking. In the latter, we adopt a fine-grained masking strategy for caption, dialog history, or question. Specifically, we mask some tokens and then recover them by minimizing the negative log-likelihood. In addition, we improve the GPT2 architecture. The output of the last layer of GPT2 is replaced by extending a multimodal feedforward network. The self-attention layer in GPT2 enables the interaction between multimodal features through the attention mechanism, while the multimodal feedforward network facilitates the fusion of information from different modalities. We conduct experiments with ablation analysis using the Audio Visual Scene-Aware Dialogues (AVSD) benchmark and the results show that our method can achieve promising performance in the video-grounded dialogue generation task.

The primary contributions of our work can be summarized as follows:

(1) We propose CFM, a novel approach that employs coarse-grained and fine-grained masking strategies, which aims to enhance the cross-modal understanding among individual modalities.
(2) We improve the GPT2 model by incorporating multimodal feedforward network, which aims to facilitate the fusion of information from different modalities.
(3) We conduct thorough experiments and ablation studies on the datasets DSTC7@AVSD [3] and DSTC8@AVSD [12]. The experimental results show the superior performance of our model. Various evaluating indicators indicate that CFM not only promotes cross-modal understanding by penetrating the relationship among modalities through masks, but also assists the fusion among different modalities.

2   Related Work

With the development of multimodal technologies, an increasing number of scholars pay attention to studying dialog systems in conjunction with video, audio or image. It is worth noting that video, which consists of multiple frames of images, inherently contains richer information than static images. Visual dialog [1] generates replies based on image content and dialog history, while video-based question answering focuses on solving single query originating from video content. Unlike the above studies, video-grounded dialogue is an emerging task that combines dialogue generation and video-audio understanding research. Early attempts employ the recurrent neural network to encode dialog history [5]. Then Attention mechanism is introduced to fuse multimodal features [4,6–8]

Along with the emergence of large models, for multimodal dialog, pre-trained models have been followed with much interest, especially the GPT2 [10] model, pretrained on an extensive corpus of English data. The GPT2 model employs a self-supervised learning approach and has demonstrated exceptional success in generating deep contextual linguistic representations. Due to its unsupervised nature, GPT2 can capture intricate nuances in language and generate diverse and coherent responses. The GPT2 model has been extensively discussed in previous works [13–15]. GPT2 model helps fast training so that experiments typically require only a few epochs to complete.

Building upon its success, many researchers and scholars have employed the GPT2 model for the AVSD task. Li et al. [11] puts forward a natural language generation pre-trained model for multimodal dialogue generation task. Hung Le et al. [16] proposes VGD+GPT2 by extending GPT2 model and formulates the video-grounded dialogue generation as a sequence-to-sequence task. However, these methods only rely on the multi-head attention mechanism in GPT2 to interact with multimodal information, which are not enough for cross-modal understanding. We believe that enhancing the cross-modal understanding at a fine-grained level can significantly facilitate a comprehensive cross-modal comprehension.

3 Approach

3.1 Task Formulation

In this section, we provide a formal definition of the video-grounded dialogue task. The task involves an input video-audio V and a dialog history $H_t = (C, (Q_1, R_1), \cdots, (Q_{n-1}, R_{n-1}))$, where C is the concatenation of the summary and the whole caption $C = \{c_1, c_2, \ldots, c_I\}$. The pair (Q_{n-1}, R_{n-1}) corresponds to the question-answer pair at round $n-1$. The goal is to generate a response $R_n = \{r_{n1}, r_{n2}, \ldots, r_{nm}\}$ given a follow-up question Q_n, taking into account the input features: V, C, $H_{1:n-1}$ and Q_n. The task can be formulated as follows:

$$P(R_n \mid V, C, H_{1:n-1}, Q_n; \theta) = \prod_{j=1}^{m} P\left(r_{nj} \mid V, C, H_{1:n-1}, Q_n, r_{n,<j}; \theta\right) \quad (1)$$

Here, θ denotes the trainable network parameter.

Video-Audio Features. The video features of the video are achieved based on the I3D model [3], which has been widely used in related studies. For the given video, it is split into T_i segments using a sliding window of length l video frames. Each segment $S_t = \{f_1, f_2, \ldots, f_I\}$ represents a sequence of frames. Pre-trained

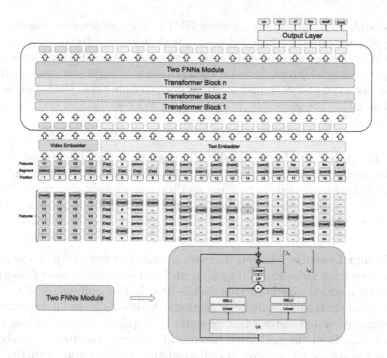

Fig. 2. Our universal multimodal transformer architecture, as illustrated in Fig. 2, extends the baseline model by incorporating TFM (Two FNNs Module).

I3D-rgb and I3D-flow are applied to extract V_{rgb} and V_{flow} respectively. Since the audio is synchronized with the video, we select the audio from the same segment and operate a pre-trained VGGish model to draw L_a-dimensional audio features as A_{vgg}. Finally, the video-audio features V_t are obtained by concatenating the I3D-flow features, I3D-rgb features, and VGGish features:

$$V_t = [V_{flow}, V_{rgb}, A_{vgg}], V_t \in R^{(2L_{I3D}+L_a) \times d}. \tag{2}$$

Textual Features. We improve GPT2 model and tokenize the input sentences into WordPieces [17] to generate a text embedding $E_t \in R^{L_t \times d}$. $\langle cap \rangle$ denotes the beginning of the caption. $\langle eos \rangle$ denotes the end of the final turn answer. $\langle user1 \rangle$ denotes the beginning of the question, and $\langle user2 \rangle$ denotes the beginning of the answer. The video features and text features are separately fed into fully connected layers and mapped to the same dimension:

$$E_{feat} = [FFN(V_t); FFN(E_t)], H_{fuse} \in R^{(2L_{I3D}+L_a+L_t) \times d}. \tag{3}$$

3.2 Model Overview

Fig. 2 illustrates the proposed universal multimodal transformer architecture. Our model incorporates the multimodal feedforward network composed of Two

FNNs Module (TFM) into the baseline GPT2. The architecture handles seven input types and aims to fuse video and text features effectively. Our model consists of a 12-layer decoder-only transformer based on GPT2 architecture. It can help multimodal interactions between object-level representations extracted from videos and word-level representations derived from text. The input features, denoted as E_{feat}, are gained by concatenating video-audio and text representations.

Video-Grounded Dialogue Task. Video-grounded Dialogue Task has two sub-tasks, including text-only task and video-text task. In the text-only task, only text features are used for training and testing, without employing TFM. In the video+text task, both text and video-audio features are used, and the model is trained with the TFM.

Transformer Decoder Model. The video-audio features and text features are passed through the video embedding layer and text embedding layer respectively. They are then added together to obtain the final feature embedding E_{feat}. Both of those layers are trainable linear projections. Additionally, we incorporate positional information and distinct feature types by acquiring the embedding vectors E_{pos} and E_{seg} from the positional token and segment token respectively. This allows us to capture the relative positions of the elements and explicitly differentiate different types of features. The input of the Transformer decoder blocks, denoted as $E^{(0)}$, is obtained by summing the feature, position, and segment embeddings. This combined input then passes through the LayerNorm layer as follows:

$$E^{(0)} = LN\left(E_{feat} + E_{pos} + E_{seg}\right). \tag{4}$$

$E^{(0)}$ is then processed through 12 layers of Transformer decoder blocks. Each layer, denoted by the index l, takes the output of the previous layer $E^{(11)}$ as inputs and performs the transformer operations. The feature embedding $E^{(0)}$ is converted into $E^{(11)}$ through 12 layers of Transformer decoder blocks as follows:

$$E^{(l)} = Transformer Decoder Block^{(l-1)}\left(E^{(l-1)}\right) \tag{5}$$

where l denotes the layer index of the Transformer decoder blocks.

Two FNNs Module Construction. To effectively fuse video-audio and text information, we employ the multimodal feedforward network. Each FFN layer incorporates Sub-LayerNorm [18] to improve training stability. Layer normalization is inserted before the input projection and output projection of each FFN layer. We employ the Sub-LayerNorm [18] because Sub-LayerNorm achieves better performance in comparison to Pre-LayerNorm [19]. Additionally, we enhance performance by replacing the activation function in the FFN with the GeGLU [20] activation function as follows:

$$R_s = GeGLU\left(E_{video-audio}^{(11)}\right) + GeGLU\left(E_{text}^{(11)}\right), \tag{6}$$

$$E^{(11)} = E^{(11)}_{video-audio} + E^{(11)}_{text}. \tag{7}$$

$E^{(11)}$ represents the sum of the video and audio embedding $E^{(11)}_{video-audio}$ and the text embedding $E^{(11)}_{text}$ obtained by applying Transformer to each respective modality.

The proposed architecture enables the output of the Transformer block to be disassembled into two branches. Each handles a specific task. The intermediate dimension of the feed-forward neural (FFN) layer is set to a value four times larger than the embedding dimension, following the practice of PaLM [21]. We introduce LayerScale [22] to enable dynamic adjustments of the output from each residual block. To achieve this, we introduce a multiplication operation between the output of the feed-forward neural (FFN) layer and a learnable diagonal matrix. This matrix, initialized to $1e-5$, plays a crucial role in fine-tuning the scaling factor before adding it to the residual. In our experiments, we observe that the inclusion of LayerScale has demonstrated notable advantages in terms of stabilizing the training process and enhancing overall performance. By incorporating LayerScale into our model, we achieve improved stability during the training phase and attain outstanding results in various evaluation metrics. The tokens R_s sent into the following LayerScale is represented as:

$$u = Drop\left(LN\left(R_s\right)\right) + E^{(11)}. \tag{8}$$

At last, the output probability for the j-th token $r_{(j)}$ is obtained by the output layer which consists of linear projection and the softmax function as follows:

$$P_\Theta\left(r_{(j)} \mid V, C, H, Q, r_{(<j)}\right) = softmax\left(W\left(u_{(j+\tau)}\right)\right), \tau = |V| + |C| + |H| + |Q|, \tag{9}$$

where W is a trainable matrix, and τ represents the sum of the lengths of V, C, H, and Q. The network parameter θ is optimized to minimize the cross-entropy loss between the predicted and reference output probabilities.

4   Cross-Modal Training Objectives

As shown in Fig. 2, for each input quintet $X = (V, C, H, Q, R)$, we construct five types of coarse-grained masked inputs, where V denotes the Video and audio, C denotes the Caption, H denotes the history dialog, Q denotes the question, R denotes response of the current question. To improve the model's cross-modal understanding among individual modalities, we use two cross-modal training losses: coarse-grained masked token loss and fine-grained masked token loss

$$L = L_{CMTL} + L_{GMTL}, \tag{10}$$

where L_{CMTL} is the coarse-grained masked token loss and L_{GMTL} is the fine-grained masked token loss.

Table 1. DSTC7@AVSD test results. Metrics are: BLEU-1 (B-1), BLEU-2 (B-2), BLEU-3 (B-3), BLEU-4 (B-4), METEOR (M), ROUGE-L (R), and CIDEr (CI). The best results are bold respectively.

AVSD@DSTC7							
Methods	B-1	B-2	B-3	B-4	M	R	CI
Input:text only							
Baseline [3]	0.629	0.485	0.383	0.309	0.215	0.487	0.746
CMU [6]	0.718	0.584	0.478	0.394	0.267	0.563	1.094
RLM [11]	0.747	0.627	0.527	0.445	0.287	0.594	1.261
CFM(our)	**0.765**	**0.648**	**0.547**	**0.464**	**0.299**	**0.605**	**1.312**
Input:text+video							
FiLM [5]	0.718	0.584	0.478	0.394	0.267	0.563	1.094
MTN [23]	0.731	0.597	0.494	0.410	0.274	0.569	1.129
BiST [24]	0.755	0.619	0.510	0.429	0.284	0.581	1.192
SCGA [8]	0.745	0.622	0.517	0.430	0.285	0.578	1.201
VGD+GPT2 [16]	0.749	0.620	0.520	0.436	0.282	0.582	1.194
JSTL [25]	0.727	0.593	0.488	0.405	0.273	0.566	1.118
RLM [11]	0.765	0.643	0.543	0.459	0.294	0.606	1.308
VGNMN [9]	-	-	-	0.429	0.278	0.578	1.188
CFM(Ours)	**0.768**	**0.647**	**0.546**	**0.461**	**0.302**	**0.610**	**1.319**

4.1   Coarse-Grained Masked Token Loss

We design six types of coarse-grained masked token loss. The first one is the coarse-grained masked video-audio (V^{cm}, C, H, Q, R). The second is the coarse-grained masked caption (V, C^{cm}, H, Q, R). The third is the coarse-grained masked history (V, C, H^{cm}, Q, R). The fourth is the coarse-grained masked question (V, C, H, Q^{cm}, R). The fifth is the coarse-grained masked answer (V, C, H, Q, R^{cm}), and final one is (V, C, H, Q, R). cm marks that the corresponding message is masked (i.e., $V^{cm} = \{[mask], [mask], \ldots, [mask]\}$). (V, C, H, Q, R) means that the original sample is preserved. The coarse-grained masked loss is defined as:

$$L_{CMTL}(\theta) = -E_{(V,C,H,Q,R)\ D} log P\left(r_{nj} | V, C, H_{1:n-1}, Q_n, r_{n,<j}\right), \qquad (11)$$

where θ is the trainable parameters and (V, C, H, Q) pairs are sampled from the whole training set D. The coarse-grained masked inputs include (V^{cm}, C, H, Q, R), (V, C^{cm}, H, Q, R), (V, C, H^{cm}, Q, R), (V, C, H, Q^{cm}, R) and (V, C, H, Q, R^{cm}), divide 50% of the training set and the remaining 50% (V, C, H, Q).

4.2 Fine-Grained Masked Token Loss

During each iteration, we employ a random masking strategy where we mask each input token with a probability of 15%. The masked token is then replaced with a special token $[mask]$. This approach helps introduce a form of token-level dropout and encourages the model to learn robust representations that can handle missing or incomplete information. The inputs are $X = (V, C^{gm}, H^{gm}, Q^{gm}, R)$, where gm denotes token masked randomly. The model is subsequently tasked with recovering the masked tokens by leveraging not only the surrounding tokens $w_{\backslash m}$ but also the video-audio V and answer R. This objective is achieved through the minimization of the negative log-likelihood :

$$L_{FMTL}(\theta) = -E_{(V,w,R) \ D} log P\left(w_m \mid w_{\backslash m}, V, R\right), \qquad (12)$$

where θ is the trainable parameters and w_m refers to the masked token.

Table 2. DSTC8@AVSD test results. Metrics are: BLEU-1 (B-1), BLEU-2 (B-2), BLEU-3 (B-3), BLEU-4 (B-4), METEOR (M), ROUGE-L (R), and CIDEr (CI). The best results are bold respectively.

AVSD@DSTC8							
Methods	B-1	B-2	B-3	B-4	M	R	CI
Input:text+video							
JMAN [26]	0.645	0.504	0.402	0.324	0.232	0.521	0.875
MDMN [27]	-	-	-	0.296	0.214	0.496	0.761
STSGR [28]	-	-	-	0.357	0.267	0.553	1.004
SCGA [8]	0.675	0.559	0.459	0.377	0.269	0.555	1.024
JSTL [25]	-	-	-	0.394	0.250	0.545	1.079
RLM [11]	0.746	0.626	0.528	**0.445**	0.286	0.598	**1.240**
CFM(Ours)	**0.752**	**0.631**	**0.530**	0.444	**0.291**	**0.600**	1.240

5 Experiments

5.1 Datasets

We employ Audio-Visual Scene-Aware Dialog (AVSD) dataset for evaluation. In this dataset, each dialog includes 10 pairs of question and answer about both static and dynamic scenes in a video collected from Charades human-activity dataset. For comparation, we use two different test splits, the Seventh Dialog System Technology Challenges (DSTC7) [3] and the Eighth Dialog System Technology Challenges (DSTC8) [12]. AVSD@DSTC7 contains 7659, 1787, and 1710 dialogues for training, validation and test, while AVSD@DSTC8 is only provided with 1710 dialogues for test.

5.2   Experimental Details

Metrics. We follow official objective metrics, including BLEU [29], METEOR [30], ROUGE-L [31], and CIDEr [32]. These metrics quantify the degree of word overlap between the generated responses and the reference responses, allowing us to assess the quality of the generated outputs.

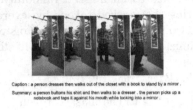

Fig. 3. A question-answering sample that decoding methods have difficulty in answering correctly.

Training Details. For the AVSD dataset, we fine-tune the pre-trained GPT2 model (12-layer, 768-hidden, 12-heads) provided by the Huggingface Transformers [33]. Our training process involves using a batch size of four and implementing the AdamW optimizer [34] with a learning rate of 6.25×10^{-5}. The models are trained for a total of seven epochs. Initially, we train the models for three epochs on a GPU with 24GB memory, utilizing both CMTL and GMTL losses with equal coefficients. This multi-task learning approach helps enhance the training process. Following the initial training phase, we continue training for an additional four epochs, focusing solely on the CMTL losses. This allows for further refinement and improvement of the models. During the answer sentence decoding stage, we employ beam search with a beam width of five, a maximum length constraint of 20, and a length penalty of 0.3. This decoding strategy helps generate more diverse and coherent answer sentences.

5.3   Results

Table 1 and Table 2 present the experimental results on the AVSD dataset, including the performance of our proposed CFM and baseline methods. The comparison contains official seven reference evaluations on AVSD@DSTC7 and AVSD@DSTC8.

As shown in Table 1 and Table 2, our CFM outperforms the existing approaches. This indicates our proposed method enables the model to learn cross-modal understanding, leading to the generation of high-quality responses and improved performance.

5.4   Ablation Study

To further evaluate the effectiveness of the proposed key components in CFM, we conduct an ablation study with several variants of the CFM model. Table 3 presents the ablation results of CFM with video on AVSD@DSTC7. The following observations can be made: (1) The full CFM model demonstrates a significant boost in performance. The coarse-grained masked token loss brings significant advantages. (2) Although there are some improvements through TFM or GMTL, the main progress is attributed to CMTL.

Table 3. Ablation study on model variants of CFM on the test dataset of AVSD@DSTC7 benchmark. "TFM" denotes the Two FNNs Module. "CMTL" denotes the coarse-grained masked token loss. "GMTL" denotes the fine-grained masked token loss.

Models	B-1	B-2	B-3	B-4	M	R	CI
Full CFM	0.768	0.647	0.546	0.461	0.302	0.610	1.319
w/o TFM	**0.768**	0.646	0.544	0.458	0.301	0.608	1.316
w/o CMTL	0.759	0.636	0.535	0.448	0.295	0.605	1.298
w/o GMTL	0.763	0.642	0.541	0.455	0.298	0.607	1.308

5.5   Decoding Methods Analysis

Table 4. Objective evaluation results compared with different decoding methods..

Decoding Methods	B-1	B-2	B-3	B-4	M	R	CI
Greedy Search	0.747	0.616	0.509	0.422	0.286	0.592	1.231
Nucleus Sampling	0.702	0.556	0.441	0.351	0.262	0.545	1.031
Beam Search	**0.768**	**0.647**	**0.546**	**0.461**	**0.302**	**0.610**	**1.319**

As shown in Table 4 and Fig. 3, we conduct a comparison of different decoding methods, including beam search, greedy generation, and nucleus sampling, along with case studies. Our findings point out that beam search outperforms the other two methods, as shown in Table 4. Additionally, in Fig. 3, when the question is "is he in the whole video?", the beam search method generates the response "yes he is there the whole video", which differs from the ground truth by only one word. On the other hand, the other two methods exhibit relatively poor performance, with the greedy search method performing a bit better than nucleus sampling. Above results highlight the effectiveness of beam search as a decoding method in generating high-quality responses, as it produces responses that closely match the ground truth. In contrast, both the greedy generation and nucleus sampling methods show limitations in capturing the desired response accurately.

6 Conclusion

We propose a coarse-grained and fine-grained masking approach for video-grounded dialogue task, named CFM. By leveraging coarse-grained and fine-grained masking, CFM captures the nuanced relationships between different input modalities, resulting in more accurate and contextually grounded responses. The inclusion of multimodal feedforward network further enhances the model's ability to fuse video-audio and text features, leading to prominent performance. Experiments show that our proposed approach improves the video dialog model's cross-model understanding and brings state-of-the-art results on the AVSD@DSTC7 and AVSD@DSTC8 dataset.

References

1. Das, A., et al.: Visual dialog. In: CVPR, pp. 326–335 (2017)
2. Lei, J., Yu, L., Bansal, M., Berg, T.L.: Tvqa: localized, compositional video question answering. arXiv preprint arXiv:1809.01696 (2018)
3. Alamri, H., et al.: Audio visual scene-aware dialog. In: Proceedings of the IEEE/CVF Conference on Computer Vision and Pattern Recognition, pp. 7558–7567 (2019)
4. Hori, C., et al.: End-to-end audio visual scene-aware dialog using multimodal attention-based video features. In: ICASSP 2019–2019 IEEE ICASSP, pp. 2352–2356. IEEE (2019)
5. Nguyen, D.T., Sharma, S., Schulz, H., Asri, L.E., et al.: From film to video: Multi-turn question answering with multi-modal context. arXiv preprint arXiv:1812.07023, 2018
6. Sanabria, R., Palaskar, S., Metze, F.: Cmu sinbad's submission for the dstc7 avsd challenge. In: DSTC7 at AAAI2019 workshop, vol. 6 (2019)
7. Le, H., Sahoo, D., Chen, N.F., Hoi, S.C., et al.: Multimodal transformer networks for end-to-end video-grounded dialogue systems. arXiv preprint arXiv:1907.01166, 2019
8. JKim, J., Yoon, S., Kim, D., Yoo, C.D.: Structured co-reference graph attention for video-grounded dialogue. In: Proceedings of the AAAI Conference on Artificial Intelligence, vol. 35, pp. 1789–1797 (2021)
9. Le, H., Chen, N., Hoi, S., et al.: Vgnmn: Video-grounded neural module networks for video-grounded dialogue systems. In: Proceedings of the 2022 Conference of the North American Chapter of the Association for Computational Linguistics: Human Language Technologies, pp. 3377–3393 (2022)
10. Radford, A., Jeffrey, W., Child, R., Luan, D., et al.: Language models are unsupervised multitask learners. OpenAI blog **1**(8), 9 (2019)
11. Li, Z., Li, Z., Zhang, J., et al.: Bridging text and video: a universal multimodal transformer for audio-visual scene-aware dialog. IEEE/ACM Trans. Audio, Speech, Lang. Process. **29**, 2476–2483 (2021)
12. Alamri, H., Hori, C., Marks, T.K., Batra, D., Parikh, D.: Audio visual scene-aware dialog (avsd) track for natural language generation in dstc7. In: DSTC7 at AAAI2019 Workshop, vol. 2 (2018)

13. Ji, Z., et al.: Survey of hallucination in natural language generation. ACM Comput. Surv. **55**(12), 1–38 (2023)
14. Biderman, S., et al.: Pythia: a suite for analyzing large language models across training and scaling. In: International Conference on Machine Learning, pp. 2397–2430. PMLR (2023)
15. Liu, P., Yuan, W., Jinlan, F., Jiang, Z., et al.: Pre-train, prompt, and predict: a systematic survey of prompting methods in natural language processing. ACM Comput. Surv. **55**(9), 1–35 (2023)
16. Le, H., Hoi, S.C.: Video-grounded dialogues with pretrained generation language models. arXiv preprint arXiv:2006.15319 (2020)
17. Wu, Y., et al.: Google's neural machine translation system: bridging the gap between human and machine translation. arXiv preprint arXiv:1609.08144 (2016)
18. Wang, H., et al.: Foundation transformers. arXiv preprint arXiv:2210.06423 (2022)
19. Brown, T., Mann, B., Ryder, N., Subbiah, M., et al.: Language models are few-shot learners. Adv. Neural. Inf. Process. Syst. **33**, 1877–1901 (2020)
20. Shazeer, N.: Glu variants improve transformer. arXiv preprint arXiv:2002.05202 (2020)
21. Chowdhery, A., Narang, S., et al.: Palm: scaling language modeling with pathways. J. Mach. Learn. Res. **24**(240), 1–113 (2023)
22. Touvron, H., Cord, M., Sablayrolles, A., Synnaeve, G., Jégou, H.: Going deeper with image transformers. In: Proceedings of the IEEE/CVF International Conference On Computer Vision, pp. 32–42 (2021)
23. Lee, H., Yoon, S., Dernoncourt, F., Kim, D.S., Bui, T., Jung, K.: Dstc8-avsd: multimodal semantic transformer network with retrieval style word generator. arXiv preprint arXiv:2004.08299 (2020)
24. Lee, H., Yoon, S., Dernoncourt, F., Kim, D.S., Bui, T., Jung, K.: Bist: Bi-directional spatio-temporal reasoning for video-grounded dialogues. arXiv preprint arXiv:2010.10095 (2020)
25. Shah, A., et al.: Audio-visual scene-aware dialog and reasoning using audio-visual transformers with joint student-teacher learning. In: ICASSP 2022–2022 IEEE ICASSP, pages 7732–7736. IEEE (2022)
26. Chu, Y.W., Lin, K.Y., Hsu, C.C., Ku, L.W., et al.: Multi-step joint-modality attention network for scene-aware dialogue system. arXiv preprint arXiv:2001.06206 (2020)
27. Xie, H., Iacobacci, I.: Audio visual scene-aware dialog system using dynamic memory networks. In: DSTC8 at AAAI2020 workshop (2020)
28. Geng, S., Gao, P., Marks, T., Hori, C., Cherian, A.: Spatio-temporal scene graph reasoning for audio visual scene-aware dialog at dstc8. In: DSTC8 at AAAI2020 workshop (2020)
29. Papineni, K., Roukos, S., Ward, T., Zhu, W.J.: Bleu: a method for automatic evaluation of machine translation. In: Proceedings of the 40th annual meeting of the ACL, pp. 311–318 (2002)
30. Denkowski, M., Lavie, A.: Meteor universal: Language specific translation evaluation for any target language. In: Proceedings of the Ninth Workshop on Statistical Machine Translation, pp. 376–380 (2014)
31. Lin, C.Y.: Rouge: A package for automatic evaluation of summaries. In: Text summarization branches out, pp. 74–81 (2004)
32. Vedantam, R., Lawrence Zitnick, C., Parikh, D.: Cider: consensus-based image description evaluation. In: Proceedings of the IEEE Conference on Computer Vision and Pattern Recognition, pp. 4566–4575 (2015)

33. Wolf, T., et al.: Transformers: state-of-the-art natural language processing. In: Proceedings of the 2020 Conference on Empirical Methods in Natural Language Processing: System Demonstrations, pp. 38–45 (2020)
34. Loshchilov, I., Hutter, F.: Decoupled weight decay regularization. arXiv preprint arXiv:1711.05101 (2017)

Deep Self-supervised Subspace Clustering with Triple Loss

Xiaotong Bu[1,2], Jiwen Dong[1,2], Mengjiao Zhang[1,2], Guang Feng[1,2], Xizhan Gao[1,2], and Sijie Niu[1,2(✉)]

[1] School of Information Science and Engineering, University of Jinan, Jinan 250022, China
sjniu@hotmail.com

[2] Shandong Provincial Key Laboratory of Network Based Intelligent Computing, Jinan, China

Abstract. Deep subspace clustering (DSC) methods are widely used in various fields such as motion segmentation, image segmentation, and text mining. It uses the deep neural network to map high-dimensional features into low-dimensional latent subspace to achieve effective division of data. Nevertheless, DSC simply tends to learn representations based on auto-encoder, which can't fully exploit the intrinsic structure of the data. In this paper, we design a novel approach called Deep self-supervised subspace clustering with triple data (DSSCT), which aims to uncover supervised information inherent in the data. Specifically, DSSCT leverages data augmentation and triple contrastive loss to obtain more effective low-dimensional representations that capture the similarity and difference among different samples. In addition, we introduce a dual self-expression matrix fusion strategy to further enhance the discriminant of the self-expression matrix used in DSSCT. To evaluate the performance of our proposed method, we conduct extensive experiments on several widely used public datasets and achieved excellent performance when compared with other state-of-art methods.

Keywords: Deep subspace clustering · Triple contrastive loss · Dual self-expression matrix

1 Introduction

Clustering is an unsupervised learning methodology, which separates samples into corresponding classes without labels' information. With the development of the times, more and more complex high-dimensional data have emerged. However, traditional clustering algorithms such as k-means [11] and spectral clustering [5,8] have shown adverse effects in those data. Subspace clustering [10] assume that different subsets of features may exhibit distinct clustering patterns. Subspace clustering methods can be broadly categorized into four main categories, i.e., Iteration-based methods [1], Algebra-based methods [9], Statistical-based methods [28], and Self-expression-based methods [25]. Among them, self-expression-based subspace clustering has attracted a lot of researchers'

S. Rudinac et al. (Eds.): MMM 2024, LNCS 14555, pp. 423–436, 2024.
https://doi.org/10.1007/978-3-031-53308-2_31

attention. Numerous traditional subspace clustering methods has focused on improving subspace clustering performance by introducing various constraints on the matrix [8,25], but they have weak performance in dealing with non-linear subspace data. With the development of deep neural networks (DNNs), a few researchers proposed deep subspace clustering methods that employ neural networks to learn a non-linear mapping relationship between the original data and its latent feature [7,16,18]. However, conventional subspace clustering methods often rely on a single feature extraction, which may not capture the crucial features and exploit the rich information contained in data [13] e.g., light, darkness, shape, position, etc. These limitations undoubtedly impede the efficiency of subspace clustering.

Self-supervised learning allows for leveraging large amounts of unlabeled data to learn useful representations or features from the data. In recent years, Contrastive learning [4,27] has gained significant attention and success in self-supervised domains. It is based on the principle of encouraging the model to learn a distinction between similar samples and dissimilar samples. It allows the model to reduce the redundant features in learning representations and learn more discriminative features [12].

Motivated by recent progress in deep subspace clustering and self-supervised learning, in this paper we introduce a novel end-to-end trainable framework called DSSCT (Deep Self-Supervised Subspace Clustering with Triple Data). Our proposed method overcomes the limitation of traditional subspace clustering and shows excellent performance on benchmark dataset. The framework of the model is shown in Fig. 2. In our proposed method, we use the triple contrast module and the dual self-expression module for training to learn better low-dimensional latent features and enhance the discriminant of self-expression matrix. The main contributions of this paper are as follows:

(1) We propose a novel end-to-end trainable framework called DSSCT (Deep Self-Supervised Subspace Clustering with Triple Data), which employs a triple contrastive module to capture significant differences and extract self-expression information in latent subspace.
(2) We assume that the positive sample pairs lie in the same subspace and keep difference with negative samples. Based on this assumption, we design a dual self-expression structure and fuse two self-expression matrices as the affinity matrix to improve the discriminant ability of the affinity matrix.
(3) We conduct extensive experiments on four benchmark datasets to verify the superiority of our DSSCT.

2 Related Work

2.1 Deep Subspace Clustering

Subspace clustering is an effective clustering technique, which assumes data points whose lie in the same subspace can well be represented by a low-dimensional linear subspace [8]. Traditional subspace clustering for dealing with linear subspace problems are based on spectral clustering [26]. They can be unified into two setps: a) building a affinity matrix; b) applying the matrix into spectral clustering. Ehsan et al. [25] proposed the Sparse Subspace Clustering (SSC) algorithm, which imposes an l_1 norm on self-expression matrix to make the feature in subspace as sparse as possible. But SSC is sensitive to the noise of data. To deal with this problem, René et al. [26] proposed the Low-Rank subspace clustering (LRR) method, which used kernel norm to optimize the self-expression matrix. However, some researchers [3,6,21,24,31] observed that traditional subspace clustering methods, both SSC and LRR, often exhibit limited performance when applied directly in the original subspace. Most classical methods can be formulated as:

$$\underset{\theta \in R^{N \times N}}{\arg \min} \|X - X\theta\|_F^2 + R(\theta), s.t.(\theta) = 0 \tag{1}$$

where $R(\theta)$ represent different norms. With the development of deep learning, more and more works leverage deep neural networks for data projection and clustering. They project the input data into a more suitable feature space by deep neural networks, and then applying clustering algorithms specifically designed for the projected representation [12]. In the past years, Deep subspace clustering network (DSC-NET) was proposed by Ji et al. [13], as shown in Fig. 1. DSC-NET inserted a fully connected linear as self-expression module between the encoder and decoder to generate the suitable subspace clustering coefficient. Inspired by low-rank representation, Kheirandishfard et al. [14] extended DSC-net and proposed Deep Low-Rank Subspace Clustering (DLRSC) to enable subspace clustering using the low-rank features that existed in original data separate samples. For a given data matrix $X \in R^{N \times D}$, the deep subspace clustering network can be described by the formulation:

$$l_\theta = \frac{1}{2} \|X - \bar{X}\|_F^2 + \varphi_1 \|\theta\|_p + \frac{\varphi_2}{2} \|Z - Z \cdot \theta\|_F^2, s.t.diag(\theta) = 0 \tag{2}$$

where Z is the latent features obtained after convolutional auto-encoder that its different classes of features belong to different subspaces.

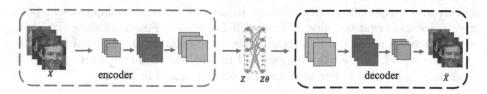

Fig. 1. The framework of Deep Subspace Clustering Network.

2.2 Self-supervised Deep Subspace Clustering

Self-supervised learning [12,15,17] can be broadly classified into two categories: generative and discriminative methods. Generative models take the original data as input to project it into the latent space and then perform pixel-level reconstruction. Since depth subspace clustering relies solely on the reconstruction of a single pixel of the auto-encoder to map the data into the latent space. Therefore, many researchers leverage self-supervised learning methods to acquire additional auxiliary information and optimize features in the latent subspace. By designing pretext tasks that do not require human annotations, these approaches aim to exploit the inherent structure and patterns within the data to learn meaningful representations. Inspired by self-supervised learning, Zhou et al. [31] made further advancements to the Deep Subspace Clustering (DSC) model by introducing an adversarial approach known as Discriminative Adversarial Subspace Clustering (DASC). Then Yu et al. [29] proposed DSC-DAG, which leverages the dual generative adversarial networks (GANs) for learning latent features of the input data through an adversarial training process. In addition, Zhang et al. [30] proposed $S^2ConvSCN$ to optimize the self-expression model using the auxiliary information contained in the spectral clustering fused into the subspace clustering. Inspired by contrastive learning, Peng et al. [3] proposed deep contrastive subspace clustering (DSCSC) network, which enhanced the performance of subspace clustering. Chen et al. [20] proposed DSCNSS, which used the supervised information provided by clustering pseudo-label to improve network performance.

3   Method

In this section, we provide a comprehensive overview and all the necessary details of our model.

3.1   Architecture of DSSCT

Different from previous work, DSSCT consists of three modules: a) Triple Autoencoder Module (TAM), which consists of a triplet generation module and stacked convolutional auto-encoder. TAM can reconstruct triple data and capture the latent feature; b) Triplet Contrastive Module (TCM) projects the latent feature into a triple contrastive subspace to capture the relationship between positive pairs and negative pairs; c) Dual Self-Expression Fusion Module (DSFM), which incorporates dual self-expression layers to enhance the discriminant ability of the model. In the following, we will provide detailed explanations of each module incorporated in our model and show our model in Fig. 2.

Triple Auto-encoder Module (TAM). We first introduce the basic module TAM in DSSCT. TAM contains the triple data generation module and the triple auto-encoder module. In the triplet data generation module, we can easily generate the triplet data through data augmentation. Specifically, for given data $X_{anc} = [x_1...x_i...x_n] \in R^{N \times D}$, we use random data augmentation strategy $\tau^i(.)$ for each x_i to get augment data $\hat{x}_i = \tau^i(x_i)$. 2020 Mahdi et al. [16] shows that selecting an appropriate data augmentation strategy plays an

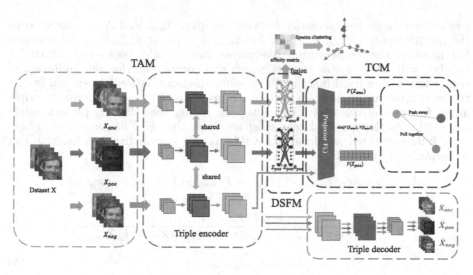

Fig. 2. The network architecture of the proposed DSSCT method, which consists of three modules divided by dotted line with different colors. The TAM is used to learn the latent embedded representation $Z_{anc}, Z_{pos}, Z_{neg}$, then DSFM combines Z_{anc}, Z_{pos} with θ, θ_{pos} to reconstruct the subspace latent representation. At the same time, TCM guides framework to learn the discriminative feature.

important role in improving subsequent task performance. Therefore, we follow the same augmentation strategy to construct our positive and negative pairs. In this work, six types of data augmentation methods are used i.e., Posterize, Sharpness, FlipLR, ShearX, TranslateY and Contrast. Then we can easily obtain positive sets $X_{pos} = \{\hat{x}_1...\hat{x}_i...\hat{x}_n\}$ and negative sets $X_{neg} = \left\{\hat{x}_1'...\hat{x}_i'...\hat{x}_n'\right\}$. Then triple data generation module combines the three sets to get triplet data $T = \{X_{anc}, X_{pos}, X_{neg}\}$. For a specific sample x_i, we use $\{x_i, \hat{x}_i\}$ as a positive pair and choose $\left\{x_i, \hat{x}_i, \hat{x}_i'\right\}$ as a group of triplet data, where x_i is anchor sample, $\hat{x}_i$ is a positive sample and $\hat{x}_i'$ is negative sample selected different from x_i and $\hat{x}_i$. Subsequently, the triple auto-encoder module, which consists of a few shared auto-encoders, maps the triplet data T into the latent subspace and gets the reconstructed data $\bar{X}_{anc}, \bar{X}_{pos}, \bar{X}_{neg}$ in the same triplet subspace. Finally, TAM network parameters are updated through the process of backpropagation with the following loss function:

$$l_{rec} = \frac{1}{2}\left(\left\|X_{anc} - \bar{X}_{anc}\right\|_F^2 + \left\|X_{pos} - \bar{X}_{pos}\right\|_F^2 + \left\|x_{neg} - \bar{X}_{neg}\right\|_F^2\right) \quad (3)$$

Triplet Contrastive Module (TCM). To preserve both local and overall structural information of the data, we propose a novel joint contrast loss function that combines the triplet loss and the temperature cross-entropy loss. This joint loss function encourages the model to learn an affinity matrix that captures more semantic information. Specifically, under the limitation of *margin*, triplet loss enforces a higher similarity between samples X_{anc} and X_{pos} compared to the negative pair $\{X_{anc}, X_{neg}\}$ i.e., $s(X_{anc}, X_{pos}) - s(X_{anc}, X_{neg}) > margin$. Furthermore, the temperature cross-entropy loss can also be seen as soft triplet loss but with an adaptive margin, which will compensate for the limitation of triplet loss with a fixed margin. To alleviate the information loss caused by the triplet contrastive module, we introduce a projector $F(\cdot)$ to transform the triple latent features into another latent subspace, rather than directly perform triple contrastive on the original latent subspace. Note that $F(\cdot)$ has consisted of two-layer fully connected nonlinear MLP. The triplet contrastive loss function can be formulated by Eq. 6:

$$l_{triple} = max(F(Z_{anc}) \cdot F(Z_{pos}) - F(Z_{anc}) \cdot F(Z_{neg}) + margin, 0) \tag{4}$$

$$l_{Ce} = \log(\frac{exp(F(Z_{anc}) \cdot F(Z_{pos})/\iota)}{exp(F(Z_{anc}) \cdot F(Z_{pos})/\iota) + exp(F(Z_{anc}) \cdot F(Z_{neg})/\iota)}) \tag{5}$$

$$l_{tripleCe} = \xi_1 \cdot l_{triplet} - \xi_2 \cdot l_{Ce} \tag{6}$$

where ι is temperature parameter that control the softness, margin is similarity parameter that restraint the triplet data, ξ_1, ξ_2 are trade-off parameters.

In TCM, the reason that we combine the triplet loss and the temperature cross-entropy loss can be summarized as follows: a) Triplet loss could preserve local structure information and maintain a stable self-expression metric. b) Temperature cross-entropy loss can be regarded as soft triplet loss, which will compensating for the limitation of the fixed margin. c) The joint loss function will reduce the influence of negative pairs and enhancing the performance of DSSCT.

Dual Self-expression Fusion Module (DSFM). The dual self-expression fusion module is designed to fuse dual self-expression matrices and get discriminant features. With the constraint in Eq. 6, our model will follow the subspace-preserving property in the latent space to construct the self-expression layer. Different from other works, we believe that the positive sample pairs not only lie in the same subspace, but also can express each other in the latent subspace i.e., $Z_{anc} = Z_{anc} \cdot \theta, Z_{pos} = Z_{pos} \cdot \theta_{pos}$. Therefore, we use a weighted fusion strategy to obtain the shared self-expression matrix $C = \zeta_1 \cdot \theta \oplus \zeta_2 \cdot \theta_{pos}$, where ζ_1 and ζ_2 are hyper-parameters, $\oplus$ denotes an addition with different weights. The loss function of the dual self-expression fusion module can be described by the following equation:

$$l_{self} = \frac{1}{2}(\alpha \|Z_{anc} - Z_{anc} \cdot \theta\|_F^2 + (1 - \alpha) \|Z_{pos} - Z_{pos} \cdot \theta_{pos}\|) + (\beta \|\theta\|_F$$
$$+ (1 - \beta) \|\theta_{pos}\|_F), s.t. diag(\theta) = 0, diag(\theta_{pos}) = 0 \tag{7}$$

where α and β are different trade-off parameters that apply different attention to the dual self-expression matrix.

3.2   Optimization and Training Strategy

Our framework can be regarded as a 'multi-task' model, so it is still difficult to train the million-parameters network model directly. Therefore, we divide the whole training strategy into two parts i.e., pre-training and fine-tuning phases to obtain faster convergence speed and better generalization ability. At first, we pre-train the auto-encoder using the triplet data with Eq. 2. Then we fine-tune the whole network with overall objective loss Eq. 8 and obtain the affinity matrix W through $W = \frac{|C| + |C|^T}{2}$.

$$l_{toal} = \lambda_1 l_{rec} + \lambda_2 l self + \lambda_3 l_{tripleCe} \tag{8}$$

where $\lambda_i, i \in (1, 3)$ is trade-off parameter of DSSCT, W is regarded as the input of spectral clustering algorithm to get clustering result. The optimization strategy of our model is presented in Algorithm 1.

Algorithm 1. Training procedure of our DSSCT-Net

Input: Data X, trade-off parameters $\lambda_1, \lambda_2, \lambda_3, \lambda_4, \alpha, \beta$ margin, temperature ι,
    maximum iteration T_{max}, and Test iteration T_{test}.
Pre-training: Pre-train auto-encoder via (5);run TAM module to initialize
    X_{pos} and X_{neg} ; initialize θ and θ_{pos} with $1.0e - 8$;
Fine-tuning:
  while $iter < T_{max}$ **do**
     update X_{pos} and X_{nbeg} by running TAM module;
     run the encoder to extract feature;
     update all the network parameters via minimizing l_{toal} with Adam solver;
     if $iter\%T_{test} = 0$ **then**
       run the DSFM module to get affinity matrix $C = \omega_1 \cdot \theta \oplus \omega_2 \cdot \theta_{pos}$.
       run the Spectral Clustering on C.
     end if
  end while
Output: label Y.

---

4   Experiments

In this section, we conducted a comprehensive evaluation of our proposed method on four public benchmark datasets. To assess its effectiveness, we compared our approach against state-of-art subspace clustering methods, and the details of experience will be shown in Sect. 4.3, 4.4 and 4.5.

4.1   Datasets

There are four public datasets included: COIL20 [19], ORL [22], COIL100 [19], Extended Yale B [10].

Coil20 dataset is a most common object dataset, which is consisted of 1440 grayscale images of 20 objects, each with the dimension of $32 \times 32 \times 1$.

ORL is a widely adopted face dataset in the field of feature extraction and image clustering. It comprises facial images collected from 40 distinct individuals. Each individual's face is represented by 10 images captured from different viewing angles, resulting in a total of 400 images on the dataset, we down-sample the dimension from 112×92 to 32×32.

Coil100 dataset comprises 1440 RGB images of 100 different objects. Following DSCN [13], we down-sample the dimension of those RGB images from 128×128 to 32×32 and use their grayscale version.

Extended Yale B is also a popular face dataset. It contains 38 individuals with 64 images, each recorded from different angles, for a total of 2432 face images. According to previous work, we down-sampled dimension to $48 \times 48 \times 1$. Some images of the dataset are shown in Fig. 3.

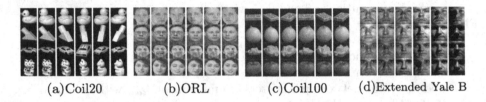

(a)Coil20   (b)ORL   (c)Coil100   (d)Extended Yale B

Fig. 3. Sample images of the dataset used in the paper

4.2   Evaluation Indicators

In our experience, we employed two widely adopted performance metrics i.e., Accuracy (ACC) [23] and Normalized Mutual Information (NMI) [2], to assess the effectiveness of our method. ACC is used to evaluate the clustering performance, and it can be described by the following formulation:

$$\psi(a,b) = \begin{cases} 1 & whilea \neq b \\ 0 & whilea = b \end{cases} \tag{9}$$

$$ACC = 1 - argmin\frac{\sum_{i=1}^{N} \psi(y_i, \hat{y}_i))}{N}, s.t.(y_i, \hat{y}_i)\epsilon K \tag{10}$$

where N is the number of images, x_i denotes a sample that input the model, $\hat{y}_i$ s the prediction label output by the model, y_i is the ground truth of x_i, $\psi(.)$

Table 1. DSSCT-net clustering performance on Coil20 and Coil100 datasets

Methods	Coil20		Coil100	
	ACC	NMI	ACC	NMI
SSC	0.8631	0.8892	0.5510	0.5841
AE+SSC	0.8711	0.8747	0.4607	0.4871
LRR	0.8118	0.8603	0.4018	0.4721
ENSC	0.8760	0.8952	0.5732	0.5924
EDSC	0.8371	0.8828	0.6187	0.6751
SSC-OMP	0.6410	0.7412	-	-
DSC-NET-l1	0.9314	0.9353	0.6638	0.6720
DSC-NET-l2	0.9368	0.9408	0.6904	0.7015
LRSC	0.7416	0.8452	0.4933	0.5810
DLRSC	0.9708	-	0.7186	-
DSSCN	0.9799	-	0.7253	-
$S^2 ConSCN$-l_1	0.9786	-	-	-
$S^2 ConSCN$-l_2	0.9767	-	-	-
DSCSC	0.9788	0.9742	-	-
DSCNSS-l_1	0.9606	-	0.6966	-
DSCNSS-l_2	0.9624	-	0.7142	-
DSSCT (ours)	**0.9833**	**0.9824**	**0.7428**	**0.7528**

represent discriminant function, K is the number of subspaces. NMI is another commonly used in clustering. It accounts for the clustering quality in terms of both cluster purity and cluster completeness, and it can be formulated by Eq. 11:

$$NMI(Y,\hat{Y}) = 2N \frac{\sum_{i=1}^{Y} \sum_{j=1}^{\hat{Y}} m_{ij} \log(\frac{N m_{ij}}{m_i m_j})}{\sum_{i=1}^{Y} \sum_{j=1}^{\hat{Y}} m_i m_j \log(\frac{m_i}{N}) \log(\frac{m_j}{N})} \quad (11)$$

where $Y, \hat{Y}$ are the collections of y_i and $\hat{Y}_i$, m_i is the number of i-th class and m_j is the prediction number of j-th class. m_{ij} is the number of predicted categories that do not match the true labels. Note that these metrics provide robust and objective measures for evaluating our model, and both have a range of [0, 1], where the value more closer to 1 indicates better performance.

4.3   Coil20 and Coil100 Clustering

To evaluate the effectiveness of our model, we performed experiments on two object datasets and compared with several classic and recent methods. According to the previous work, we down-sample images to 32×32 and use the same experimental setup as DSCN. Specifically, the auto-encoder consists of one layer of convolutional layers with 15 and 50 channels, and the kernel sizes are 3×3 and 5×5 respectively, and the network setting is shown in Table 4. For Coil20 (K = 20) dataset, the parameter of our model used as follows: $\lambda_1 = 10.0$,

$\lambda_2 = 75.0$, $\lambda_3 = 20.0$, $\lambda_4 = 10.0$, $\iota = 0.5$, margin $= 0.25$. The Coil100 dataset ($K = 100$) has a higher number of categories compared to COIL20, so we use the following trade-off parameters: $\lambda_1 = 10.0, \lambda_2 = 75.0$, $\lambda_3 = 20.0$, $\lambda_4 = 10.0$, $\iota = 0.5$, margin $= 0.2$. Table 1 summarizes the clustering results on two object datasets. It is able to see that the accuracy of our method is higher than DSC-net-l_2 by 4.58% and 4.18% on the two datasets, and both better than the current self-supervised methods. This demonstrates the excellence of our model on the item dataset.

Table 2. DSSCT-net clustering performance on ORL and Extended Yale B datasets

Methods	ORL		EYale B	
	ACC	NMI	ACC	NMI
SSC	0.7425	0.8459	0.7354	0.7796
AE+SSC	0.7450	0.8824	0.7475	0.7764
LRR	0.8110	0.8603	0.8499	0.8636
ENSC	0.7525	0.8540	0.7537	0.7915
EDSC	0.7038	0.7799	0.8814	0.8835
SSC-OMP	0.7100	0.7952	0.7372	0.7803
DSC-NET-l_1	0.8550	0.9032	0.9681	0.9687
DSC-NET-l_2	0.8600	0.9034	0.9733	0.9703
LRSC	0.7200	0.8156	0.7913	0.8264
DLRSC	-	-	0.9753	-
DSSCN	0.8850	-	0.9432	-
$S^2ConSCN$-l_1	0.8875	0.9214	0.9848	-
$S^2ConSCN$-l_2	0.8950	0.9238	0.9844	-
DSCSC	0.9075	0.9444	0.9864	0.9815
DSCNSS-l_1	0.8868	-	0.9792	-
DSCNSS-l_2	0.8921	-	0.9815	-
DSSCT(ours)	**0.9084**	**0.9519**	**0.9823**	**0.9756**

We also visualized the loss functions during alternation training. Figure 4 shows that the loss of each model gradually decrease until it become stable. With the optimization of other sub-loss, the affinity Loss demonstrates that affinity matrix θ strives to learn information as much as possible in the subspace.

4.4   ORL and Extend Yale B Clustering

Then we evaluate our method on the face image ORL and Extend Yale B. In the process of experience, the structure of the auto-encoder still consistent

Table 3. Accuracy clustering result (%) on Coil20 dataset with different combinations of modules

	Coil20		Coil100	
	ACC	NMI	ACC	NMI
TAM	0.9444	0.9448	0.6938	0.7086
TAM with TCM	0.9826	0.9819	0.7322	0.7455
TAM and TCM with DSFM	**0.9833**	**0.9824**	**0.7428**	**0.7528**

with previous work. Both encoder and decoder have three-layer convolutional structure with (5, 3, 3) and (10, 20, 30) channels, and the kernel sizes are 5×5 and 3 respectively. As shown in Table 2, our model achieves an accuracy of 90.84%, which is better than the DSC-net-l_2 method by 4.84% and higher than the compared methods. On the Extend Yale B dataset, we achieve an accuracy of 98.23%, which also has perfect performance. It is worth noting that we do not specifically train the pre-trained model, so there still have some potential to improve the results on the face dataset. Table 4 shows the details of the network on different real-word datasets.

Table 4. The network setting for different dataset

	encoder-1	encoder-2	encoder-3	self-expression	decoder-1	decoder-2	decoder-3
Coil20	3×3	-	-	1440×1440	-	-	3×3
Coil100	5×5	-	-	7200×7200	-	-	5×5
ORL	5×5	3×3	3×3	400×400	3×3	3×3	5×5
EYale B	5×5	3×3	3×3	2432×2432	3×3	3×3	5×5

4.5 Ablation Experiments

In order to assess the positive impact of different modules on DSSCT (Deep Self-Supervised Subspace Clustering with Triple Data), some ablation experiments are conducted. We performed ablation experiments on two object datasets and the results of the experiment are summarized in Table 3. In addition, we have visualized the block-diagonal structure matrix generated by our method and compared them with DSC-Net on Coil20, Coil100 and ORL datasets. Specifically, the affinity matrix is a mutual-expressed matrix that has high-dimensional data in a particular low-dimensional subspace. Therefore, the affinity matrix with subspace-preserving properties should have a distinct block diagonal structure and the more distinct block diagonal the better discriminant. As shown in Fig. 6, our model has obtained a more discriminative block diagonal matrix on different datasets.

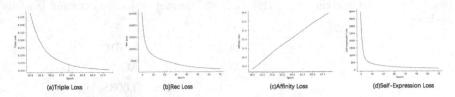

Fig. 4. The optimized trend of sub-loss (a) Triple Loss; (b) Rec Loss; (c) Affinity Loss (d) Self−Expression Loss

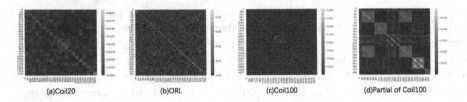

Fig. 5. The block-diagonal structure matrix of DSCN on different datasets

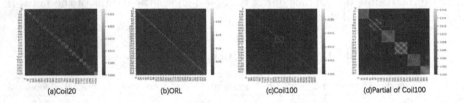

Fig. 6. The block-diagonal structure matrix of our method on different datasets

5   Conclusion

In this paper, we proposed a novel framework called DSSCT-net, which integrated the triplet contrastive loss into our model. We introduce a dual self-expression layer structure to enhance the discriminant of our model. The experimental result shows that our model gets excellent performance on many real-word datasets and is competitive compared with other state-of-art methods. In the future, we will continue improve the robustness and efficiency of our model.

Acknowledgements. This work is supported by the National Natural Science Foundation of China under Grant No. 62101213, No. 62103165, 62302191, the Natural Science Foundation of Shandong Province, China, under Grant No. ZR2020QF107, No. ZR2020MF137, ZR2023QF001, Development Program Project of Youth Innovation Team of Institutions of Higher Learning in Shandong Province

References

1. Bradley, P.S., Mangasarian, O.L.: K-plane clustering. J. Global Optim. **16**, 23–32 (2000)
2. Cai, D., He, X., Han, J.: Locally consistent concept factorization for document clustering. IEEE Trans. Knowl. Data Eng. **23**(6), 902–913 (2010)
3. Chen, C., Lu, H., Wei, H., Geng, X.: Deep subspace image clustering network with self-expression and self-supervision. Appl. Intell. **53**(4), 4859–4873 (2023)
4. Chen, T., Kornblith, S., Norouzi, M., Hinton, G.: A simple framework for contrastive learning of visual representations. In: International Conference on Machine Learning, pp. 1597–1607. PMLR (2020)
5. Chen, X., Zhexue Haung, J., Nie, F., Chen, R., Wu, Q.: A self-balanced min-cut algorithm for image clustering. In: Proceedings of the IEEE International Conference on Computer Vision, pp. 2061–2069 (2017)
6. Chen, Y., Xiao, X., Zhou, Y.: Multi-view subspace clustering via simultaneously learning the representation tensor and affinity matrix. Pattern Recogn. **106**, 107441 (2020)
7. Dang, Z., Deng, C., Yang, X., Huang, H.: Multi-scale fusion subspace clustering using similarity constraint. In: Proceedings of the IEEE/CVF Conference on Computer Vision and Pattern Recognition, pp. 6658–6667 (2020)
8. Elhamifar, E., Vidal, R.: Sparse subspace clustering: algorithm, theory, and applications. IEEE Trans. Pattern Anal. Mach. Intell. **35**(11), 2765–2781 (2013)
9. Gear, C.W.: Multibody grouping from motion images. Int. J. Comput. Vis. **29**, 133–150 (1998)
10. Georghiades, A.S., Belhumeur, P.N., Kriegman, D.J.: From few to many: illumination cone models for face recognition under variable lighting and pose. IEEE Trans. Pattern Anal. Mach. Intell. **23**(6), 643–660 (2001)
11. Hartigan, J.A., Wong, M.A.: Algorithm as 136: a k-means clustering algorithm. J. R. Stat. Soc. Ser. C (Appl. Stat.) **28**(1), 100–108 (1979)
12. Jaiswal, A., Babu, A.R., Zadeh, M.Z., Banerjee, D., Makedon, F.: A survey on contrastive self-supervised learning. Technologies **9**(1), 2 (2020)
13. Ji, P., Zhang, T., Li, H., Salzmann, M., Reid, I.: Deep subspace clustering networks. In: Advances in Neural Information Processing Systems, vol. 30 (2017)
14. Kheirandishfard, M., Zohrizadeh, F., Kamangar, F.: Deep low-rank subspace clustering. In: Proceedings of the IEEE/CVF Conference on Computer Vision and Pattern Recognition Workshops, pp. 864–865 (2020)
15. Le-Khac, P.H., Healy, G., Smeaton, A.F.: Contrastive representation learning: a framework and review. IEEE Access **8**, 193907–193934 (2020)
16. Li, C., Yang, C., Liu, B., Yuan, Y., Wang, G.: LRSC: learning representations for subspace clustering. In: Proceedings of the AAAI Conference on Artificial Intelligence, vol. 35, pp. 8340–8348 (2021)
17. Liu, Y., et al.: Graph self-supervised learning: a survey. IEEE Trans. Knowl. Data Eng. **35**(6), 5879–5900 (2022)
18. Lu, C.-Y., Min, H., Zhao, Z.-Q., Zhu, L., Huang, D.-S., Yan, S.: Robust and efficient subspace segmentation via least squares regression. In: Fitzgibbon, A., Lazebnik, S., Perona, P., Sato, Y., Schmid, C. (eds.) ECCV 2012, Part VII. LNCS, vol. 7578, pp. 347–360. Springer, Heidelberg (2012). https://doi.org/10.1007/978-3-642-33786-4_26
19. Nene, S.A., Nayar, S.K., Murase, H., et al.: Columbia object image library (COIL-20) (1996)

20. Peng, B., Zhu, W.: Deep structural contrastive subspace clustering. In: Asian Conference on Machine Learning, pp. 1145–1160. PMLR (2021)
21. Peng, X., Zhu, H., Feng, J., Shen, C., Zhang, H., Zhou, J.T.: Deep clustering with sample-assignment invariance prior. IEEE Trans. Neural Net. Learn. Syst. **31**(11), 4857–4868 (2019)
22. Samaria, F.S., Harter, A.C.: Parameterisation of a stochastic model for human face identification. In: Proceedings of 1994 IEEE Workshop on Applications of Computer Vision, pp. 138–142. IEEE (1994)
23. Shaham, U., Stanton, K., Li, H., Nadler, B., Basri, R., Kluger, Y.: SpectralNe: spectral clustering using deep neural networks. arXiv preprint arXiv:1801.01587 (2018)
24. Valanarasu, J.M.J., Patel, V.M.: Overcomplete deep subspace clustering networks. In: Proceedings of the IEEE/CVF Winter Conference on Applications of Computer Vision, pp. 746–755 (2021)
25. Vidal, E.E.R.: Sparse subspace clustering. In: 2009 IEEE Conference on Computer Vision and Pattern Recognition (CVPR), vol. 6, pp. 2790–2797 (2009)
26. Vidal, R., Favaro, P.: Low rank subspace clustering (LRSC). Pattern Recogn. Lett. **43**, 47–61 (2014)
27. Wu, Z., Xiong, Y., Yu, S.X., Lin, D.: Unsupervised feature learning via nonparametric instance discrimination. In: Proceedings of the IEEE Conference on Computer Vision and Pattern Recognition, pp. 3733–3742 (2018)
28. Yang, A.Y., Rao, S.R., Ma, Y.: Robust statistical estimation and segmentation of multiple subspaces. In: 2006 Conference on Computer Vision and Pattern Recognition Workshop, CVPRW 2006, pp. 99–99. IEEE (2006)
29. Yu, Z., Zhang, Z., Cao, W., Liu, C., Chen, C.P., Wong, H.S.: Gan-based enhanced deep subspace clustering networks. IEEE Trans. Knowl. Data Eng. **34**(7), 3267–3281 (2020)
30. Zhang, J., et al.: Self-supervised convolutional subspace clustering network. In: Proceedings of the IEEE/CVF Conference on Computer Vision and Pattern Recognition, pp. 5473–5482 (2019)
31. Zhou, P., Hou, Y., Feng, J.: Deep adversarial subspace clustering. In: Proceedings of the IEEE Conference on Computer Vision and Pattern Recognition, pp. 1596–1604 (2018)

LigCDnet:Remote Sensing Image Cloud Detection Based on Lightweight Framework

Baotong Su[1,2] and Wenguang Zheng[1,2(✉)]

[1] School of Computer Science and Engineering, Tianjin University of Technology, TianJin 300384, China
subaotong@stud.tjut.edu.cn, wenguangz@tjut.edu.cn
[2] Tianjin Key Laboratory of Intelligence Computing and Novel Software Technology, Tianjin University of Technology, TianJin 300384, China

Abstract. Cloud contamination is inevitable in remote sensing images, resulting in a large number of images that cannot be applied in various fields. Therefore, cloud detection is one of the important tasks in remote sensing image preprocessing, aimed at removing images obstructed by clouds. Most existing methods are mostly based on CNN and feature a complex network structure, requiring a significant amount of computational resources, making it challenging to deploy them in practical applications. To tackle this problem, we propose a lightweight cloud detection framework (LigCDnet) with a lightweight feature extraction module (LFEM), a channel attention module (CAM), and a lightweight feature pyramid module (LFPM). The LFEM serves as the backbone of the network to capture rich spatial and contextual information; the CAM adaptively adjusts the channel weights of the feature maps; and the LFPM extracts cloud features at multiple scales. The effectiveness of our approach is evaluated on two public datasets, GF-1 and LandSat8. Extensive experiments have demonstrated that the proposed LigCDnet achieves state-of-the-art detection accuracy while significantly reducing computational burden and having a smaller model size.

Keywords: Remote sensing images · Cloud detection · Lightweight Framework · GF-1 · LandSat8

1 Introduction

With the rapid development of remote sensing technology, optical remote sensing images have been extensively used in various fields such as agriculture engineering, geographical survey, military reconnaissance, natural disaster prediction, and environmental pollution monitoring [16]. However, cloud occlusion is an inevitable challenge in satellite imagery due to the extensive cloud cover that spans over 60% of the Earth's surface area [14]. The cloud cover obstructs the satellite sensor's ability to obtain a clear view of the Earth's surface, making

S. Rudinac et al. (Eds.): MMM 2024, LNCS 14555, pp. 437–450, 2024.
https://doi.org/10.1007/978-3-031-53308-2_32

many image analysis tasks difficult, such as remote sensing image classification and segmentation [29], image matching [8], etc. Therefore, it is necessary to quickly and accurately detect cloud cover in order to enhance the availability of remote sensing images.

Over the years, researchers have conducted in-depth studies on cloud detection algorithms in remote sensing imagery and have proposed numerous algorithms. These methods can be broadly categorized into two types: threshold-based methods and machine learning-based methods. Threshold-based methods rely on the physical characteristics of clouds and set appropriate thresholds based on these characteristics to classify pixels in an image into cloud and non-cloud categories. ISCCP [21] cloud mask algorithm utilized the fact that cloud and clear scenes differ in the amount of radiance variability they exhibit in space and time to detect clouds. Cihlar and Howarth [7] proposed a method that can identify clouds with different opacities as well as cloud shadows present in composite materials, effectively eliminating the impact of cloud contamination in AVHRR synthetic images on land. Huang et al. [12] used clear forest pixels as a reference to define cloud boundaries and separate clouds from clear surfaces in a spectral-temperature space. However, these methods lack a universal threshold and do not consider the structure and texture of clouds when dealing with complex scenes, resulting in low robustness. The principle of machine learning-based methods for cloud detection is to extract features from remote sensing images as input and then train a classification model by comparing these features with labeled samples. An and Shi [2] designed a scene learning-based cloud detection algorithm, this algorithm utilizes the color features, texture features, and structural features of the image. Li et al. [13] extracted brightness features, texture features, and average gray-level co-occurrence matrix features [10] from the image, they then used these features to train a support vector machine (SVM) [25] classifier. Shi et al. [23] proposed a ground-based cloud detection method using graph model built upon super-pixels [1] to integrate multiple sources of information. However, these methods extract shallow features from images through statistical means such as mean, maximum, minimum, variance, etc., which do not effectively comprehend the images, leading to a decrease in detection accuracy.

In recent years, convolutional neural networks (CNNs) have allowed the field of computer vision to grow rapidly with their powerful feature extraction capabilities. CNN-based approaches can improve the model's understanding of images by stacking convolutional layers and yield superior performance in target detection, image classification, and semantic segmentation [24,33]. Yang et al. [30] utilize thumbnail images to extract cloud masks, extracting multi-scale contextual information without losing resolution. Wu and Xu [28] present cross-supervised learning for cloud detection to address the issue of insufficient labeled cloudy images. However, most existing deep learning methods feature complex network structures and high computational resource requirements. In practical applications, the deployed devices often lack significant computational power and storage space. Therefore, this limits the applicability of these methods.

To solve the above problem, we design a lightweight cloud detection framework called LigCDnet, which achieves excellent performance with very few parameters. The contributions of this paper are briefly summarized as follows:

1. We propose a lightweight cloud detection framework based on the U-shaped architecture, called LigCDnet. It achieves state-of-the-art detection accuracy compared to existing cloud detection algorithms while having an extremely small number of parameters, only 2.39M.
2. We utilize a lightweight feature extraction module (LFEM) to capture spatial and contextual information and design a channel attention module (CAM) to adjust the impact of different channels on cloud detection performance. Additionally, we propose a lightweight feature pyramid module (LFPM) to extract cloud features at different scales.

2   Proposed Method

2.1   Overall Framework

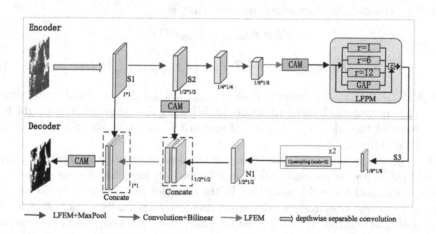

Fig. 1. Framework of the proposed LigCDnet.

We use a U-shaped encoder-decoder structure as the framework for our cloud detection network model, as shown in Fig. 1. High-level features contain rich semantic information, low-level features contain rich spatial information [32]. The abundant spatial information plays a crucial role in generating cloud masks. Therefore, in the encoder part, we performed three downsampling operations to preserve rich spatial information in the feature map. We utilize the lightweight feature extraction module (LFEM) that maximizes the extraction of cloud features while minimizing the increase in parameters. To enhance the feature maps channels that are favorable for cloud segmentation, designing a channel attention module (CAM) to adjust the channel weights. To better understand clouds of different scales, we propose the lightweight feature pyramid module (LFPM). In the decoder part, we gradually restore the resolution of the feature maps through upsampling and compensate for the spatial information lost during the encoding stage by connecting them with the feature maps in the encoder using skip connections.

Given a remote-sensing image I as input, the feature map $S1$ is first generated in the encoder by depthwise separable convolution operations. Depthwise separable convolution consists of depthwise convolution and pointwise convolution. that is

$$S1 = H_{conv_{dep}}\left(H_{conv_{poi}}(I)\right) \tag{1}$$

where $H_{conv_{dep}}(\cdot)$ represents depthwise convolution operation, and $H_{conv_{poi}}(\cdot)$ denotes pointwise convolution operation, $S1$ has the same size as the input image.

To reduce the computational complexity while extracting cloud information, we use a downsampling unit, that is

$$F_{downsampling} = MaxPool\left(H_{LFEM}(S)\right) \tag{2}$$

where $MaxPool(\cdot)$ is max pooling operation, and $H_{LFEM}(\cdot)$ denotes the lightweight feature extraction module. Then, the feature maps $S2, S3$ are generated by consecutive downsampling unit, that is

$$S2 = F_{downsampling}(S1) \tag{3}$$

$$S3 = H_{LFPM}\left(H_{CAM}\left(H_{LFEM}\left(F_{downsampling}^2(S2)\right)\right)\right) \tag{4}$$

where $F_{downsampling}^2(\cdot)$ means that $F_{downsampling}$ is executed two times, $H_{CAM}(\cdot)$ represents channel attention module, H_{LFPM} denotes lightweight feature pyramid module. $S2$ is $1/2 \times 1/2$ size of the input image, $S3$ is $1/8 \times 1/8$ size of the input image.

Due to the small spatial resolution of the feature map $S3$ generated in the encoder, it leads to problems such as information loss, insufficient contextual information, and blurred boundaries. In the decoder part, we restore the feature map to the same resolution as the input image by gradually upsampling it, the upsampling is limited to 2×. To reduce computational complexity, we employ a simple bilinear interpolation to directly upsample $S3$ twice, resulting in the generation of feature map $N1$, $N1$ is $1/2 \times 1/2$ size of the input image. And introduce the upsampling unit, that is

$$F_{upsampling} = bilinear\left(H_{conv_3}^2(I)\right) \tag{5}$$

where $bilinear(\cdot)$ denotes bilinear interpolation operation, $H_{conv_3}(\cdot)$ represents a convolution operation with a convolution kernel size of 3, and the predicted cloud detection result I_O can be described as

$$I_O = H_{CAM}\left(concat\left(S1, F_{upsampling}\left(concat(H_{CAM}(S2), N1)\right)\right)\right) \tag{6}$$

where $concate$ denotes the concatenate operation, I_O has the same size as the input image.

2.2   Lightweight Feature Extraction Module

In recent years, the main trend in improving the network's understanding of complex scenes has been the development of deeper and more complex networks. However, these networks require a significant amount of computational cost both during training and inference phases. To address this challenge, a plethora of lightweight network frameworks have been proposed. For instance, in MobileNet [22], depthwise separable convolution is employed, which consists of depthwise convolution and pointwise convolution. Depthwise convolution operates independently on each channel of the input feature map, pointwise convolution integrates information between different channels to enhance the network's representational capacity. In LEDNet [26], channel splitting and shuffle operations are applied to each residual block. Channel splitting divides the channels of the feature map into multiple groups, allowing the network to independently extract different types of features. Shuffle operations enable information exchange between different channel groups. Inspired by the MobileNet, we designed the lightweight feature extraction module shown in Fig. 2. The lightweight feature extraction module consists of two pointwise convolutions and one depthwise convolution. For a feature map with C channels, the first step is to increase its dimensionality to $2C$ by employing a pointwise convolution. The different channels of feature maps can be seen as the network's response to various characteristics of the data, enabling the network to understand the data from different perspectives. Subsequently, a depthwise convolution is utilized to capture spatial features of the cloud and extract local information for each channel. Lastly, another pointwise convolution is employed to reduce the dimensionality back to C while integrating features across the channels of the feature map.

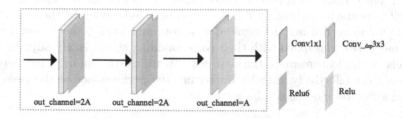

Fig. 2. Structure of LFEM.

For a 3×3 standard convolution, with an input feature map of $[H \times W \times C]$, output channel set to $2C$, and convolution layer depth set to 3, the number of parameters of the module is $3\times3\times C\times2C+3\times3\times2C\times2C+3\times3\times2C\times2C = 90C$. And the number of parameters of our lightweight feature extraction module is $1 \times 1 \times C \times 2C + 3 \times 3 \times 2C + 1 \times 1 \times 2C \times 2C = 24C$. With the same depth of convolution layers, the number of standard convolutional parameters is 3.74 times higher than ours, and the LFEM module greatly reduces the number of parameters while increasing the inference speed and computational efficiency.

The lightweight feature extraction module can be stated as

$$H_{\text{LFEM}} = H_{conv_{poi}}\Big(H_{\text{conv}_{\text{dep}}}\big(H_{conv_{poi}}(I)\big)\Big) \tag{7}$$

2.3 Channel Attention Module

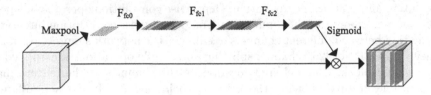

Fig. 3. Structure of CAM.

In computer vision tasks, which often rely on convolutional operations to extract features from images, different channels of the feature map play different important roles for the task in the process of network learning. Therefore, we designed a channel attention module and introduced a learnable weight vector to enable the network to automatically learn the importance of different channels in the task. It allows the network to adjust the weights of each channel, enhancing the dependency on important channels and reducing the dependency on unimportant channels, as shown in Fig. 3. First, we apply max-pooling operations to each channel of the feature map to generate initial channel weight vectors. Then, these vectors are fed into three layers of linear units to let the network learn the importance of different channels. Subsequently, the weight vectors are normalized using the sigmoid function, and finally, the weight vectors are element-wise multiplied with their corresponding channels. By adjusting the channels of the feature map, the network can utilize the information between channels more effectively, thereby improving the performance of the task. The channel attention module can be stated as

$$H_{\text{LFEM}} = F_{fc2}\Big(F_{fc1}\big(F_{fc0}(MaxPool(I))\big)\Big) \otimes I \tag{8}$$

where F_{fc} denotes Linear layer operation, $\otimes$ is element-wise multiplication operation.

2.4 Lightweight Feature Pyramid Module

Clouds have diverse morphologies, and accurately segmenting clouds of different sizes is a fundamental challenge for cloud detection algorithms. Capturing multi-scale cloud features and establishing contextual information can effectively

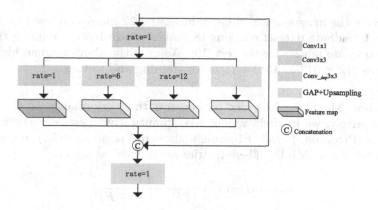

Fig. 4. Structure of LFPM.

enable the network to learn feature differences between cloud regions and backgrounds. Inspired by the ASPP [5] model, we propose a lightweight pyramid module, as shown in Fig. 4. The number of channels of the feature map is first adjusted by a pointwise convolution. Dilated convolution [31], also known as atrous convolution, can significantly expand the receptive field of convolutional neural networks. Combining multiple dilated convolutions with different sampling rates in parallel effectively captures multi-scale contextual information. Therefore, we use parallel dilated convolutions with dilation rates of 1, 6, and 12. To reduce computational complexity, the dilated convolutions are replaced with deepwise convolutions while keeping the dilation rates unchanged. Additionally, a global average pooling (GAP) layer is introduced to extract global contextual information. Subsequently, the features captured by the four parallel branches are concatenated along the channel dimension. To facilitate feature reuse and mitigate the gradient vanishing problem, short connections are introduced. The Lightweight Feature Pyramid Module can be stated as

$$H_{\mathrm{LFPM}} = concate(I, H_{conv_{poi}}(I), H_{conv_{dep-r6}}(I),$$
$$H_{conv_{dep-r12}}(I), bilinear(H_{GAP}(I))) \tag{9}$$

where $H_{conv_{dep-r}}(\cdot)$ is depthwise convolution with dilate rate r, H_{GAP} denotes global average pooling operation.

3 Experimental Results

3.1 Dataset and Experimental Setup

Dataset. We chose two widely used datasets, GF-1 remote sensing images and datasets of CloudSat8, to validate the effectiveness of our method, using only their visible channels. The GF-1 remote sensing images includes 108 GF-1 Wide Field of View (WFV) level-2A scenes and its reference cloud and cloud shadow masks. 86 of the images are used for training and 22 images are used for testing [15]. The CloudSat8 dataset contains 18 images of size 1000×1000 for training and 20 same-size images for the test [19]. We crop the above original high pixel image into $512 \times 512 \times 3$ sub-images for training and testing.

Evaluation Metrics. In order to measure the performance of the model comprehensively, we used six widely used quantitative metrics, including Jac-cardIndex, Precision, Recall, F1-score, and overall accuracy (OA), mean intersection over union (MIoU). These metrics are defined as follows:

$$JaccardIndex = \frac{TP}{(TP + FN + FP)} \tag{10}$$

$$Precision = \frac{TP}{(TP + FP)} \tag{11}$$

$$Recall = \frac{TP}{(TP + FN)} \tag{12}$$

$$F1 - score = 2 \times \frac{Precision \times Recall}{(Precision + Recall)} \tag{13}$$

$$OverallAccuracy = \frac{TP + TN}{(TP + TN + FP + FN)} \tag{14}$$

$$MIoU = \frac{1}{k} \sum_{i=1}^{k} \frac{n_{ii}}{\sum_{j=1}^{k} n_{ij} + \sum_{j=1}^{k} n_{ji} - n_{ii}} \tag{15}$$

where TP, TN, FP, and FN are the total number of true-positive, true-negative, false-positive, and false-negative pixels, respectively. The k represents the number of categories, n_{ii} represents the count of correctly predicted pixels, and n_{ij} represents the count of pixels where the true value is i and they were predicted as j.

Parameter Settings. Our model is implemented using the Pytorch framework [20], with the training step running on Ubuntu 22.04 and an RTX3090 GPU. Using the Stochastic Gradient Descent (SGD) [3] algorithm for optimization with an initial learning rate of 2×10^{-4}, decay strategy "poly" [4], batch size of 4, momentum of 0.9. All CNN-based methods are trained using the same configuration and settings without the need for pre-training.

3.2   Comparative Experiments

Comparative Methods. This paper compares a machine learning-based cloud detection method: SVM [11], and also compares six state-of-the-art deep learning-based cloud detection algorithms: FCN-8 [17], DeeplabV3+ [6], CDNetV1 [30], CDnetV2 [9], LWCDnet [18], Boundarynet [27]. Among these, LWCDnet is a lightweight cloud detection method.

Table 1. Quantitative comparisons with other cloud detection methods on the GF-1 test set. Cloud extraction accuracy (%)

Method	Jaccard index	precision	recall	F1-score	OA	MIoU
SVM	62.01	81.17	71.62	73.61	89.20	72.72
FCN-8	72.94	80.20	86.96	83.15	92.23	79.66
deeplabV3+	77.82	83.95	89.78	87.14	94.14	83.59
CDnetV1	80.77	86.45	91.34	89.22	94.65	85.85
CDnetV2	76.80	82.86	89.78	86.33	93.86	82.9
LWCDnet	75.57	80.99	87.69	83.81	93.03	82.61
Boundarynet	83.14	**90.68**	89.90	90.60	**95.88**	87.68
LigCDnet	**84.29**	90.42	**92.18**	**91.11**	95.88	**88.35**

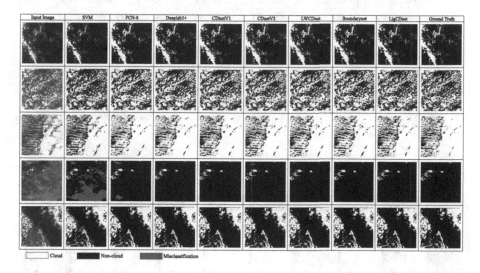

Fig. 5. Visual comparisons of different cloud detection methods on GF-1 dataset.

Cloud Detection Results on GF-1 Dataset: Table 1 reports the results of different cloud detection methods in the GF-1 dataset. From the results, our proposed LigCDnet outperforms most of them. Compared to the SVM machine learning method, deep learning methods have significant advantages in various metrics. FCN-8, in terms of Jaccard index, recall, and F1-score, shows an average improvement of 12% over SVM. Compared to cloud detection methods, the boundarynet achieves slightly higher precision, with the same score as ours on OA. However, in terms of Jaccard index and MIoU, our method is higher than it by 1.15% and 0.67%, respectively. Compared to the lightweight method LWCDnet, our proposed method outperforms LWCDnet by 7.3% and 5.74% in terms of F1-score and MIoU metrics, respectively. Fig. 5 shows a visual comparison of five typical examples of cloud segmentation methods in the GF-1 dataset, with a variety of cloud cover and backgrounds. For clarity, we use white to represent correctly labeled cloud pixels and black to represent non-cloud pixels. Red markings indicate misclassified pixels. From a visual perspective, SVM's performance is the poorest; it only extracts the physical features of the image and does not fully comprehend the context of the image. CDnetV2 tends to misclassify bright objects as clouds. Overall, our LigCDnet performs the best.

Table 2. Quantitative comparisons with other cloud detection methods on the Land-Sat8 test set. Cloud extraction accuracy (%)

Method	Jaccard index	precision	recall	F1-score	OA	MIoU
SVM	72.38	86.72	77.16	80.74	85.10	65.12
FCN-8	79.09	85.19	83.81	87.58	93.09	74.00
deeplabV3+	81.30	86.61	87.38	87.26	92.06	76.74
CDnetV1	84.83	90.72	86.52	87.74	94.61	81.91
CDnetV2	79.68	86.89	84.83	86.48	93.64	79.07
LWCDnet	82.09	85.48	86.78	88.42	93.44	76.20
Boundarynet	83.71	89.25	87.28	90.20	94.64	80.55
LigCDnet	**88.02**	**93.99**	**90.16**	**92.25**	**95.00**	**84.39**

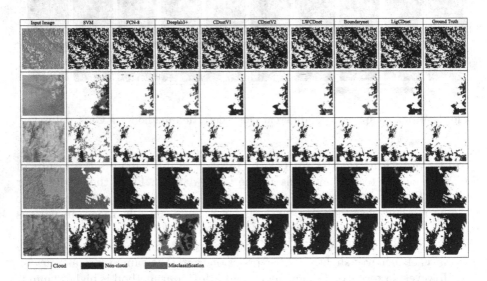

Fig. 6. Visual comparisons of different cloud detection methods on LandSat8 dataset.

Cloud Detection Results on LandSat8 Dataset: Table 2 reports the results of different cloud detection methods on the LandSat8 dataset. From the results, it can be observed that our proposed LigCDnet achieves better performance, especially in terms of Jaccard index and MIoU. While LigCDnet's OA is only 0.39% higher than CDnetV1, there is a significant improvement of 3.19% in MIoU. Compared to the lightweight network LWCDnet, our proposed network still demonstrates clear advantages, with a 3.38% higher recall and an 8.19% higher MIoU score. Figure 6 illustrates five examples from the LandSat 8 dataset, these examples encompass various backgrounds, such as situations where thin clouds and cloud ice coexist. From the visual results, it is evident that SVM performs poorly in handling scenarios where ice and snow coexist. DeeplabV3+ and LWCDnet also exhibit significant errors when dealing with scenes containing thin clouds. In contrast, our method demonstrates the best overall performance in handling all complex scenarios. It has fewer false positives (highlighted in red) compared to other methods.

Computational Complexity Analysis: In Table 3, we utilized floating point operations (FLOPs) and the number of trainable parameters to assess the computational complexity of these networks. Due to the results of the efficiency evaluation being directly proportional to the input image size, the FLOPs results were computed from input images sized at $224 \times 224 \times 3$. From the table, it can be observed that our proposed network has the fewest parameters. Although our proposed method has 7.69% higher GFLOPs compared to the lightweight model LWCDnet, we demonstrate significant advantages in both quantitative and qualitative analyses on various datasets. This is because we employ the Channel Attention Module (CAM) multiple times to adjust the weights of feature map channels, and we have designed a Lightweight Feature Pyramid Module (LFPM) to capture features of multi-scale clouds.

Table 3. Computational Complexity Analysis Based on CNN Method

Method	GFLOPs (224×224)	Params (M)
FCN-8	26.57	32.9
deeplabV3+	33.19	39.75
CDnetV1	59.94	47.50
CDnetV2	14.27	67.08
LWCDnet	**3.10**	2.55
Boundarynet	97.62	88.87
LigCDnet	10.79	**2.39**

3.3 Ablation Study

The LigCDnet proposed by us consists of three modules, namely the lightweight feature extraction module (LFEM), the channel attention module (CAM) and the lightweight feature pyramid module (LFPM). To investigate the performance of different components in the network, we conducted an ablation analysis on the GF-1 dataset. Table 4 provides detailed quantitative results.

From the results, LigCDnet demonstrates the best performance. Decreasing any of the blocks results in a certain degree of degradation in network performance. Removing CAM results in a deterioration of the metrics, indicating that CAM adjusts the weights of different channels in the feature maps, allowing channels favorable for the detection task to play a major role. Without LFPM, the metrics show a decrease, which suggests that LFPM can capture cloud features at different scales. Overall, these three modules play important roles in the cloud detection task.

Table 4. Ablation study on the GF-1 dataset by our LigCDnet with different modules

Method	Jaccard index	precision	recall	F1-score	OA	MIoU
LFEM	82.11	88.66	90.64	89.74	95.41	86.76
LFEM+CAM	82.85	88.31	91.88	90.23	95.70	87.40
LFEM+LFPM	83.15	89.44	91.54	90.07	95.76	87.66
LigCDnet	**84.29**	**90.42**	**92.18**	**91.11**	**95.88**	**88.35**

4 Conclusions

This article proposes a lightweight method (LigCDnet) for cloud detection. Compared with existing cloud detection models, LigCDnet achieves the best detection accuracy with a minimal number of parameters. In LigCDnet, we extensively extract multi-scale contextual features and further enhance segmentation accuracy by adjusting the channel weights of the feature maps. In the encoder, LFEM effectively extracts the semantic information of clouds, while CAM enhances feature map channels beneficial for the detection task and suppresses feature map channels that interfere with segmentation accuracy. Due to the diverse morphology of clouds, LFPM efficiently captures contextual features at different scales. In the decoder, the feature maps are gradually restored to the size of the input image through skip connections. Extensive experiments have been conducted on GF-1 and LandSat8 datasets, and the results show that LigCDnet can achieve excellent performance while reducing computational effort.

References

1. Achanta, R., Shaji, A., Smith, K., Lucchi, A., Fua, P., Süsstrunk, S.: Slic superpixels compared to state-of-the-art superpixel methods. IEEE Trans. Pattern Anal. Mach. Intell. **34**(11), 2274–2282 (2012)
2. An, Z., Shi, Z.: Scene learning for cloud detection on remote-sensing images. IEEE J. Sel. Top. Appl. Earth Observ. Remote Sens. **8**(8), 4206–4222 (2015)
3. Bottou, L.: Stochastic gradient descent tricks. In: Montavon, G., Orr, G.B., Müller, K.-R. (eds.) Neural Networks: Tricks of the Trade. LNCS, vol. 7700, pp. 421–436. Springer, Heidelberg (2012). https://doi.org/10.1007/978-3-642-35289-8_25
4. Chen, L.C., Papandreou, G., Kokkinos, I., Murphy, K., Yuille, A.L.: Deeplab: semantic image segmentation with deep convolutional nets, atrous convolution, and fully connected CRFs. IEEE Trans. Pattern Anal. Mach. Intell. **40**(4), 834–848 (2017)
5. Chen, L.C., Papandreou, G., Schroff, F., Adam, H.: Rethinking atrous convolution for semantic image segmentation. arXiv preprint arXiv:1706.05587 (2017)
6. Chen, L.-C., Zhu, Y., Papandreou, G., Schroff, F., Adam, H.: Encoder-decoder with atrous separable convolution for semantic image segmentation. In: Ferrari, V., Hebert, M., Sminchisescu, C., Weiss, Y. (eds.) ECCV 2018. LNCS, vol. 11211, pp. 833–851. Springer, Cham (2018). https://doi.org/10.1007/978-3-030-01234-2_49
7. Cihlar, J., Howarth, J.: Detection and removal of cloud contamination from AVHRR images. IEEE Trans. Geosci. Remote Sens. **32**(3), 583–589 (1994)
8. Guo, J.h., Yang, F., Tan, H., Wang, J.x., Liu, Z.h.: Image matching using structural similarity and geometric constraint approaches on remote sensing images. J. Appl. Remote Sens. **10**(4), 045007–045007 (2016)
9. Guo, J., Yang, J., Yue, H., Tan, H., Hou, C., Li, K.: Cdnetv2: CNN-based cloud detection for remote sensing imagery with cloud-snow coexistence. IEEE Trans. Geosci. Remote Sens. **59**(1), 700–713 (2020)
10. Hafizah, W.M., Supriyanto, E., Yunus, J.: Feature extraction of kidney ultrasound images based on intensity histogram and gray level co-occurrence matrix. In: 2012 Sixth Asia Modelling Symposium, pp. 115–120. IEEE (2012)
11. Hao, Q., Zheng, W., Xiao, Y.: Fusion information multi-view classification method for remote sensing cloud detection. Appl. Sci. **12**(14), 7295 (2022)
12. Huang, C., et al.: Automated masking of cloud and cloud shadow for forest change analysis using landsat images. Int. J. Remote Sens. **31**(20), 5449–5464 (2010)
13. Li, P., Dong, L., Xiao, H., Xu, M.: A cloud image detection method based on SVM vector machine. Neurocomputing **169**, 34–42 (2015)
14. Li, Y., Yu, R., Xu, Y., Zhang, X.: Spatial distribution and seasonal variation of cloud over china based on ISCCP data and surface observations. J. Meteorol. Soc. Jpn. Ser. II **82**(2), 761–773 (2004)
15. Li, Z., Shen, H., Li, H., Xia, G., Gamba, P., Zhang, L.: Multi-feature combined cloud and cloud shadow detection in gaofen-1 wide field of view imagery. Remote Sens. Environ. **191**, 342–358 (2017)
16. Long, J., Shi, Z., Tang, W., Zhang, C.: Single remote sensing image dehazing. IEEE Geosci. Remote Sens. Lett. **11**(1), 59–63 (2013)
17. Long, J., Shelhamer, E., Darrell, T.: Fully convolutional networks for semantic segmentation. In: Proceedings of the IEEE Conference on Computer Vision and Pattern Recognition, pp. 3431–3440 (2015)
18. Luo, C., et al.: LWCDnet: a lightweight network for efficient cloud detection in remote sensing images. IEEE Trans. Geosci. Remote Sens. **60**, 1–16 (2022)

19. Mohajerani, S., Saeedi, P.: Cloud-net: an end-to-end cloud detection algorithm for landsat 8 imagery. In: IGARSS 2019–2019 IEEE International Geoscience and Remote Sensing Symposium, pp. 1029–1032. IEEE (2019)

20. Paszke, A., et al.: Pytorch: an imperative style, high-performance deep learning library. In: Advances in Neural Information Processing Systems, vol. 32 (2019)

21. Rossow, W.B., Garder, L.C.: Cloud detection using satellite measurements of infrared and visible radiances for ISCCP. J. Clim. **6**(12), 2341–2369 (1993)

22. Sandler, M., Howard, A., Zhu, M., Zhmoginov, A., Chen, L.C.: Mobilenetv 2: inverted residuals and linear bottlenecks. In: Proceedings of the IEEE Conference on Computer Vision and Pattern Recognition, pp. 4510–4520 (2018)

23. Shi, C., Wang, Y., Wang, C., Xiao, B.: Ground-based cloud detection using graph model built upon superpixels. IEEE Geosci. Remote Sens. Lett. **14**(5), 719–723 (2017)

24. Sun, L., et al.: A cloud detection algorithm-generating method for remote sensing data at visible to short-wave infrared wavelengths. ISPRS J. Photogramm. Remote. Sens. **124**, 70–88 (2017)

25. Suthaharan, S., Suthaharan, S.: Support vector machine. Machine learning models and algorithms for big data classification: thinking with examples for effective learning, pp. 207–235 (2016)

26. Wang, Y., et al.: Lednet: a lightweight encoder-decoder network for real-time semantic segmentation. In: 2019 IEEE International Conference on Image Processing (ICIP), pp. 1860–1864. IEEE (2019)

27. Wu, K., Xu, Z., Lyu, X., Ren, P.: Cloud detection with boundary nets. ISPRS J. Photogramm. Remote. Sens. **186**, 218–231 (2022)

28. Wu, K., Xu, Z., Lyu, X., Ren, P.: Cross-supervised learning for cloud detection. GISci. Remote Sens. **60**(1), 2147298 (2023)

29. Yang, F., Guo, J., Tan, H., Wang, J.: Automated extraction of urban water bodies from zy-3 multi-spectral imagery. Water **9**(2), 144 (2017)

30. Yang, J., Guo, J., Yue, H., Liu, Z., Hu, H., Li, K.: CDnet: CNN-based cloud detection for remote sensing imagery. IEEE Trans. Geosci. Remote Sens. **57**(8), 6195–6211 (2019)

31. Yu, F., Koltun, V.: Multi-scale context aggregation by dilated convolutions. arXiv preprint arXiv:1511.07122 (2015)

32. Zhang, Z., Zhang, X., Peng, C., Xue, X., Sun, J.: ExFuse: enhancing feature fusion for semantic segmentation. In: Ferrari, V., Hebert, M., Sminchisescu, C., Weiss, Y. (eds.) ECCV 2018. LNCS, vol. 11214, pp. 273–288. Springer, Cham (2018). https://doi.org/10.1007/978-3-030-01249-6_17

33. Zhou, P., Han, J., Cheng, G., Zhang, B.: Learning compact and discriminative stacked autoencoder for hyperspectral image classification. IEEE Trans. Geosci. Remote Sens. **57**(7), 4823–4833 (2019)

Gait Recognition Based on Temporal Gait Information Enhancing

Qizhen Chen[1,2,3,4], Xin Chen[1,2,3,4], Xiaoling Deng[1,2,3,4(✉)],
and Yubin Lan[1,2,3,4]

[1] College of Electronic Engineering, College of Artificial Intelligence,
South China Agricultural University, Guangzhou 510642, China
icefloor@stu.scau.edu.cn, {chenxin,dengxl,ylan}@scau.edu.cn
[2] National Center for International Collaboration Research
on Precision Agricultural Aviation Pesticide Spraying Technology,
Guangzhou 510642, China
[3] Guangdong Laboratory for Lingnan Modern Agriculture,
Guangzhou 510642, China
[4] Guangdong Engineering Technology Research Center of Smart Agriculture,
Guangzhou 510642, China

Abstract. Gait recognition is a long range biometric technology that
identifies individuals by their walking patterns. Currently, gait recognition primarily extracts gait features using convolutional neural networks, which are based on either the global appearance or local human
body regions. However, the global feature methods are lack of long range
interactions in different local regions and lose temporal features by some
extent, and the local feature method segmenting gait silhouettes into
blocks limits the ability to characterize local feature weights. In this
paper, we propose a gait recognition method that enhances interactions
between local regions. To implement this method, we construct a new feature enhancement module, which is a global and local feature extractor
based on SENet (GLFES), to enhance the recognition of local features
using the attention mechanism. Extensive experiments based on our proposed method have been conducted on the public datasets CASIA-B and
OUMVLP to achieve state-of-the-art performances.

Keywords: Gait recognition · Squeeze and excitation · Convolutional
neural network · Human body alignment · Global and local features

1 Introduction

As a kind of unique biometric features, gait is long-range, non-contact and difficult to disguise. Moreover, gait samples can be obtained without subjects' cooperation. It has a very wide range of applications in security surveillance, criminal
investigation surveillance and the public domain (Fig. 1).

Existing global or local feature extraction methods [1,10,12,17,26] mainly
focus on extracting spatial features, while temporal features are limited by some
extent. Specifically, the max-pooling operation in the temporal dimension easily

© The Author(s), under exclusive license to Springer Nature Switzerland AG 2024
S. Rudinac et al. (Eds.): MMM 2024, LNCS 14555, pp. 451–463, 2024.
https://doi.org/10.1007/978-3-031-53308-2_33

Fig. 1. The gait silhouettes images are from individuals 33 and 53, taken every 4 frames. It can be observed that subtle changes need to be slowly detected through timing, as a single silhouette image alone cannot depict them. Therefore, the temporal features of gait are crucial, especially when individuals are wearing coats.

loses many gait temporal features. Moreover, existing methods mainly segment gait features into chunks and extract local features of different local regions separately, the interacting relationships between different human body parts have not been discovered enough, which thus limits gait recognition accuracies.

To address these issues, in this paper, we propose a gait recognition method that enhances the temporal gait features and interactions between local features. Specifically, we build a new feature enhancement module, called Global and Local Feature Extractor based on SeNet (GLFES), to enhance the interactions between local features by integrating attention mechanism. This method is realized by a squeeze and excitation Module (SEM) in the network, a module that is capable of enhancing inter-regional interactions. We can see the effect of SEM in Fig. 2. We show visualization maps under different views, including 0°, 54°, 90° and 126°.

At the same time, we design a Temporal Global and Local Aggregator (TGLA) to extract temporal global and local features in a principled way. The global timing feature extractor focuses on the timing features of the entire gait sequence, while the local timing feature extractor splits the gait sequence in the time dimension, focusing on the gait details of adjacent frames. This then gives the model better recognition abilities by merging both global and local features.

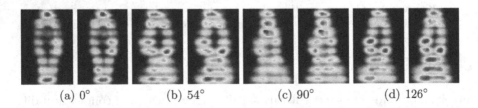

(a) 0°   (b) 54°   (c) 90°   (d) 126°

Fig. 2. The convolutional feature map visualization after SEM. The visualization on the left is without SEM module, and the visualization on the right is with SEM module.

Finally, we design a novel Temporal Feature Aggregator (TFA). The gait features have two dimensions, height and width, and the features in the higher dimension are more discriminative than the wider dimension. Therefore, by pooling the wide dimension and reducing the number of parameters of the gait features as input to the TFA, the recognition accuracy of the model can be improved. The highlighted contributions of our method are listed as follows:

(1) We propose a simple and lightweight application of convolution, called Temporal Global and Local Aggregator (TGLA), to facilitate refined learning of temporal features at the local level. The core idea of TGLA is to constraint the convolutional temporal perceptual wilderness and focus more on local timing information of gait-adjacent frames, then the local temporal feature are enhanced.
(2) We propose a novel local feature enhancement module (SEM) that maximises the usage of local features by interacting with local features in different regions.
(3) We propose a lightweight convolutional application, called Temporal Feature Aggregator (TFA), which is able to improve the comprehensive performance of the model.
(4) We conduct extensive experiments on the public datasets CASIA-B and OUMVLP. Experimental estimates show that the proposed method can achieve the state-of-the-art performance.

2 Related Work

Current gait recognition studies prove the importance of spatial feature extraction and modeling from time series [2–4,6,7,9,14,23,24,27,31,32]. In order to obtain more discriminative features from gait sequences, most of the existing models are based on CNNs, and they use 2D [2,32] or 3D [14,16,22–25] convolution for feature extraction along the spatial dimension with good success. The importance of different human body parts in gait recognition is different, and performing the same scanning operation for all gait sequences often overlooks this characteristic. To obtain more detailed information about the different human body parts, GaitSet [2,3], GaitPart [7], GLN [9], MT3D [15] tried to slice the output features into m blocks along the horizontal dimension, which can learn unique gait features of different body parts.

In addition, to better obtain discriminative gait features, many studies have integrated the entire gait sequence into one frame [14,29]. At the same time, there are many studies that extracted frame-level features from gait sequences by CNNs and applied a max-pooling operation on the temporal dimension [2,9], which easily limits the interrelationships and interactions between different gait frames.

In order to better obtain the relationships between consecutive gait frames, the original max-pooling operation is replaced by LSTM to integrate the gait features in the time series to generate the final gait features [5,13,18,32], and the whole pipeline retains the non-essential order constraint in the gait sequences. These methods are good at extracting spatial and temporal features from gait sequences, ignore the spatio-temporal dependencies between non-local features.

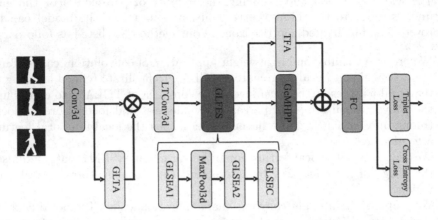

Fig. 3. The overview of the whole GaitSE framework. 'Conv3d' and 'LTConv3d' denote three-dimensional convolution. 'TGLA' represents the Temporal Global and Local Aggregator. 'GLFES' represents the Global and Local Feature Extrator based on SeNet. 'TFA' represents Temporal Feature Aggregator. 'GeMHPP' represents the Generalized-Mean Horizontal Pyramid pooling. 'FC' represents Fully Connected layer. 'Triplet Loss' and 'Cross Entropy Loss' represent two kinds of loss functions

3   Method

In this section, the pipeline of GaitSE is first described, then the Temporal Global and Local Aggregator (TGLA) is described, followed by the Global and Local Feature Extractor based on SeNet(GLFESN), ending with the Temporal Feature Aggregator (TFA) and implementation details. The overall framework is presented in Fig. 3.

3.1   Pipeline

To obtain more holistic gait features, we first extract shallow features from the original gait sequences by 3D CNNs. Next, the Temporal Global and Local Aggregator (TGLA) was designed to extract a combination of global and local temporal information. Later, the Local Temporal 3D Convolution (LTConv3d) is designed to replace the original Max-pooling operation to retain more spatio-temporal information to ensure more comprehensive temporal information. After that, the Global and Local Feature Extractor based on SeNet (GLFES) is designed to enhance global and local information. Then, we propose the Temporal Feature Aggregator (TFA) to integrate the global temporal information. Finally, the triplet loss and cross entropy loss are used as our loss functions to train the model.

3.2 Temporal Global and Local Aggregator

We propose a Temporal Global and Local Aggregator (TGLA). The TGLA module consists of two 3D CNNS, one for global temporal information and the other for local temporal information. Since global and local temporal information are considered at the same time, the gait features extracted by TGLA are comprehensive, as shown in Fig. 4.

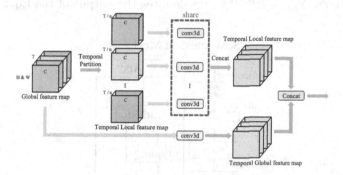

Fig. 4. Architectures of Temporal Global and Local Aggregator. 'H', 'W', 'C' and 'T' denote the height, width, number of channels and length of the gait sequence. 'n' denotes the number of cuts along the time dimension. 'Temporal Partition' denotes the segmentation of features along the time dimension and 'N' represents the number of segmented regions. 'Conv3d' is a 3D convolution operation, and 'share' denotes these 3D convolution shared parameters. 'Concat' indicates a concat operation in the temporal dimension

3.3 Global and Local Feature Extractor Based on SeNet

We propose a Global and Local Feature Extractor based on SeNet (GLFES). The first of this paragraph for the two forms of fusion methods, local features and global features, there are two classical methods, one is addition and the other is concat in this high dimension. The module based on additional fusion method we define as GLSEA, the module based on concat fusion method we define as GLSEC, the difference between the two lies in the final fusion method. The GLFES module consists of four layers 'GLSEA1-MaxPool3d-GLSEA2-GLSEC' as shown in Fig. 3.

The principle of TGLA has been shown above, and the difference between TGLA and GLSE lies in how the local feature map is partitioned and the Squeeze and Excitation Module (SEM). The segmentation of TGLA is along the time dimension, and GLSE is a horizontal segmentation. The SEM is shown in Fig. 5

Given $X_l \in \mathbb{R}^{H \times W \times T \times C_l}$ as the final local feature map. Therefore, the squeeze excitation module can be formulated as:

$$Y_f = F_{se}(Reshape(Max(X_l))) \tag{1}$$

$$Y_{end} = F_{scale}(Reshape(Y_f), x_l) \tag{2}$$

where $Max(.)$ in maximizes the width of the feature, and $Reshape(.)$ denotes merging the height and time dimensions of the feature. $F_{se}(.)$ indicates that 'FC-ReLU-FC-Sigmoid' operations are performed, and and the output dimension of the first FC is $\frac{C_l}{r}$, and the output dimension of the second FC is C_l. $Y_f \in \mathbb{R}^{H \times 1 \times T \times C_l}$ indicates the output of the Eq. 1. The $Reshape$ in Eq. 2 means to separate the dimensions. $F_{scale}(.,.)$ denotes width-wise multiplication between the feature map. $Y_{end} \in \mathbb{R}^{H \times W \times T \times C_l}$ denotes the output of the Eq. 2.

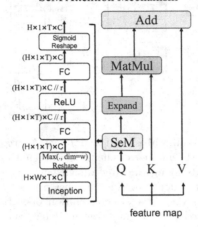

Fig. 5. The overview of SeM Attention Mechanism. The attention mechanism is widely used in Transformer.

3.4   Temporal Feature Aggregator

We propose an effective and low-consumption module, Temporal Feature Aggregator (TFA). The TFA module includes three layers, 'Conv2d_down-Conv2d_inter-Conv2d_up'.

Suppose the input to the module is $X_t \in \mathbb{R}^{C_{begin} \times T \times H \times W}$, where C_{begin} indicates the total number of channels of input, T represents the length of the input gait sequence and (H, W) denotes the height and width of each silhouette image. Therefore, the Temporal Feature Aggregator can be formulated as:

$$Y_{beg} = Max(X_t, dim = W) \tag{3}$$

$$Y_{mid} = F_r(F_b(F_u(F_{br}(F_i(F_{br}(F_d(Y_{beg})))))) + Y_{beg}) \tag{4}$$

$$Y_{end} = GMP(Y_{mid}) \tag{5}$$

where $Max(., dim = W)$ represents the maximization of the input along the wide(W) dimension, and $Y_{beg} \in \mathbb{R}^{C_{begin} \times T \times H \times 1}$ is the first output. $F_d(.)$, $F_i(.)$

and $F_u(.)$ denote three different 2d convolutions with output channel C_{down}, C_{inter} and C_{begin}. Their convolution kernel sizes are $(1,1)$, $(T_i,1)$ and $(1,1)$ respectively. $F_b(.)$ and $F_r(.)$ represent batch normalization and ReLU operations respectively, and $F_{br}(.)$ denotes that the input performs the batch normalization operation first and then the ReLU operation. $GMP(.)$ denotes global maximum pooling operation. $Y_{mid} \in \mathbb{R}^{C_{begin} \times T \times H \times 1}$ and $Y_{end} \in \mathbb{R}^{C_{begin} \times 1 \times 1 \times 1}$ denotes the output of the corresponding step.

3.5 Generalized-Mean Horizontal Pyramid Pooling

GeMHPP is an in-between method, determined by parameter learning. The GeMHPP module can be represented as:

$$Y_{GeMHPP} = (F_{Avg}(Y_{GeMin}.pow(p)))).pow(1/p) \qquad (6)$$

where Y_{GeMin} and Y_{GeMHPP} denotes the input and output of GeMHPP. F_{Avg} denotes the average pooling operation. $pow(.)$ denotes the power operation. p is a parameter to be learned, and a suitable parameter is derived by multiple training.

In order to better train the proposed model, we use both triplet loss [8] and cross entropy loss.

4 Experiments

In this section, we first compare the experimental results on CASIA-B dataset with the state-of-the-art methods, then perform ablation study to compare the influence of different modules. Then we compare the experimental results on OUMVLP dataset.

4.1 Datasets

(1) CASIA-B [28] is a multi-view large gait dataset. There are gait samples of 124 subjects in this dataset. The samples are collected from 11 views (0°, 18°, ..., 162°, 180°).

(2) OUMVLP [20] contains above 10,000 subjects. Each subject's samples are collected from 14 views (0°, 15°, 30°, 45°, 60°, 75°, 90°, 180°, 195°, 210°, 215°, 240°, 255°, and 270°).

(3) GREW [33] is currently recognized as the most extensive gait dataset in the wild according to available information. It consists of raw videos collected from 882 cameras positioned in a large public area, resulting in a substantial collection of nearly 3,500 h of video streams at a resolution of $1,080 \times 1,920$.

Table 1. Rank-1 accuracy (%) on CASIA-B dataset under all view angles, different settings and conditions

Gallery NM#1-4 Probe		Gallery view: 0° − 180°											Mean
		0°	18°	36°	54°	72°	90°	108°	126°	144°	162°	180°	
NM#5-6	ACL [30]	92	98.5	**100**	**98.9**	95.7	91.5	94.5	97.7	98.4	96.7	91.9	96
	GEINet [19]	40.2	38.9	42.9	45.6	51.2	42	53.5	57.6	57.8	51.8	47.7	48.1
	GaitSet [2]	90.8	97.9	99.4	96.9	93.6	91.7	95	97.8	98.9	96.8	85.8	95
	GaitPart [7]	94.1	98.6	99.3	98.5	94	92.3	95.9	98.4	99.2	97.8	90.4	96.2
	GLFE [14]	96	98.3	99	97.9	**96.9**	95.4	97	98.9	99.3	98.8	94	97.4
	GLN [9]	93.2	**99.3**	99.5	98.7	96.1	95.6	97.2	98.1	99.3	98.6	90.1	96.9
	Ours	**96.1**	98.5	99.1	97.9	96.4	**95.6**	**97.4**	**98.9**	**99.3**	99	**95.5**	97.6
BG#1-2	GEINet [19]	34.2	29.3	31.2	35.2	35.2	27.6	35.9	43.5	45	39	36.8	35.7
	GaitSet [2]	83.8	91.2	91.8	88.8	83.3	81	84.1	90	92.2	94.4	79.0	87.2
	GaitPart [7]	89.1	94.8	96.7	95.1	88.3	84.9	89	93.5	96.1	93.8	85.8	91.5
	GLFE [14]	92.6	96.6	96.8	95.5	93.5	89.3	92.2	96.5	**98.2**	96.9	91.5	94.5
	GLN [9]	91.1	**97.7**	**97.8**	95.2	92.5	**91.2**	**92.4**	96	97.5	95	88.1	94
	Ours	**92.7**	95.5	97	**95.6**	94.3	88.8	92.2	96.8	97.9	**97.2**	**92.2**	**94.6**
CL#1-2	GEINet [19]	19.9	20.3	22.5	23.5	26.7	21.3	27.4	28.2	24.2	22.5	21.6	23.45
	GaitSet [2]	61.4	75.4	80.7	77.3	72.1	70.1	71.5	73.5	73.5	68.4	50	70.4
	GaitPart [7]	70.7	85.5	86.9	83.3	77.1	72.5	76.9	82.2	83.8	80.2	66.5	78.7
	GLFE [14]	76.6	90	90.3	87.1	84.5	**79**	84.1	87	87.3	84.4	69.5	83.6
	GLN [9]	70.6	82.4	85.2	82.7	79.2	76.4	76.2	78.9	77.9	78.7	64.3	77.5
	Ours	**78.7**	**90.8**	**92.7**	**90.2**	**85.1**	78.9	**84.3**	**87.5**	**89.1**	**86.7**	**72.9**	**85.2**

4.2 Experiment Results on CASIA-B

To test the performance in cross-view scenarios, we compare GaitSE with the latest advanced methods. As shown in Table 1, GaitSE outperforms SOTA in most views. Specifically, GaitSE outperforms previous methods by at least 0.2%, 0.1% and 1.6% in three conditions (normal/bag/cloth). Most notably, in the most challenging CL condition, GaitSE achieves an accuracy of 85.2%, a 1.6% improvement compared to GaitGL [14], which validates the robustness of GaitSE in difficult scenarios.

From the results we see that the accuracy of our proposed network can be greatly improved under cl conditions, which proves that the potential of the model can be improved by SeM, thus improving the resistance of the network to interference. From the graphs, we can see that the accuracy improvement is much larger in the (0°, 180°) than in the other angles.

4.3 Experiment Results on OUMVLP

In this section, we evaluate the performance of our proposed model on a larger OUMVLP dataset. The experimental settings in this section follow the setup of GaitPart [7] and GaitGL [14]. In Table 2, we evaluate gait samples at 14 different

Table 2. The recognition accuracy (%) comparison on OUMVLP dataset under 14 probe views excluding identical views.

Method	Gallery view: 0° − 180°														
	0°	15°	30°	45°	60°	75°	90°	108°	195°	210°	225°	240°	255°	270°	Mean
GEINet [19]	23.2	38.1	48.0	51.8	47.5	48.1	43.8	27.3	37.9	46.8	49.9	45.9	45.7	41.0	42.5
GaitSet [2]	79.3	87.9	90.0	90.1	88.0	88.7	87.7	81.8	86.5	89.0	89.2	87.2	87.6	86.2	87.1
GaitPart [7]	82.6	88.9	90.8	91.0	89.7	89.9	89.5	85.2	88.1	90.0	90.1	89.0	89.1	88.2	88.7
GLN [9]	83.8	90.0	91.0	91.2	90.3	90.0	89.4	85.3	89.1	90.5	90.6	89.6	89.3	88.5	89.2
GLFE [14]	84.9	90.2	91.1	91.5	91.1	90.8	90.3	88.5	88.6	90.3	90.4	89.6	89.5	88.8	89.7
Ours	**86.6**	**90.8**	**91.4**	**91.7**	**91.5**	**91.1**	**90.7**	**89.8**	**89.3**	**90.6**	**90.7**	**90.2**	**90**		**89.5**	**90.3**

views. During the test, we used Seq#00 and Seq#01 as the probe and gallery sequence, respectively. From the results, it seems that our proposed method improves more in the larger scale dataset than in the smaller scale dataset.

4.4 Experiment Results on GREW

We have analyzed the effectiveness of the proposed method by comparing its performance with various gait recognition methods using the GREW dataset. The evaluated methods, namely GaitGraph [21], GaitSet [2], Gaitpart [7], GaitGL [14], and CSTL [11], have been thoroughly assessed, and their respective experimental outcomes are presented in Table 3. Our findings from this comparison reveal an important trend. It appears that the gait recognition methods, which demonstrate satisfactory results in controlled laboratory settings, exhibit a notable decline in performance when confronted with real-world scenarios and datasets.

Table 3. The recognition accuracy (%) comparison on GREW dataset

Method	R-1 %	R-5 %	R-10 %
GaitGraph [21]	6.25	16.23	5.18
GaitSet [2]	46.3	63.6	70.3
GaitPart [7]	44	60.7	67.3
GaitGL [14]	47.3	63.6	69.3
CSTL [11]	**50.6**	65.9	71.9
ours	49	**66.2**	**72.4**

4.5 Training Details

The alignment of the input silhouettes is referred to [20], and the resolution size of the final silhouette is 64×44. Adam as our optimizer sets the learning rate and momentum to 1e−4 and 0.9, respectively. The margin m in the Eq. ?? about triplet loss is set to 0.2. The length of the gait sequences T is set to 30. Four NVIDIA 3080TI GPUs are used as our computational resources to train our model.

4.6   Ablation Study

We design various pertinent ablation experiments to analyze the importance of different modules.

Analysis of Global and Local Feature Extractor Based on SeNet. In GLSE, the role of the SEM module is mainly to reorganize features. In GaitGL [14], only local features are fused together to form a new gait feature map ignoring the connections between regions, the SEM module helps to establish the interactions between regions (Table 4).

Table 4. The recognition accuracy (%) of different max-pooling strategies in SE module on CASIA-B dataset.

SEM			NM	BG	CL	MEAN
Height	Width	Temporal				
✓	✗	✗	97	94.1	83.6	91.6
✗	✓	✗	**97.6**	**94.6**	**85.2**	**92.5**
✓	✓	✗	97.3	94.1	83.7	91.7
✓	✓	✓	97.4	94.4	84.5	92.1

Analysis of Temporal Feature Aggregator. We set the TFA in different places and set different temporal feature convolution kernel sizes, and drew experimental conclusions in Table 5. In order to fully verify the best hyperparameters, we set two positions, which are after GLSEA2 and GLSEC. Meanwhile, we also set three convolution kernel sizes, i.e., (3, 1), (4, 1) and (5, 1). From the table we can see that the hyperparameters mentioned above have quite a strong influence on the model, especially under the CL condition.

Table 5. The recognition accuracy (%) of placing TFA in different positions on CASIA-B dataset.

TFA		NM	BG	CL	MEAN
Location	kernel size				
GLSEA2	(3, 1)	97.3	94.2	84	91.8
GLSEA2	(4, 1)	97.3	94.6	84.1	92
GLSEA2	(5, 1)	97.1	94	83.8	91.6
GLSEC	(3, 1)	97.4	94.5	84.2	92
GLSEC	(5, 1)	97.2	94.3	84.1	91.9
GLSEC	(4, 1)	**97.6**	**94.6**	**85.2**	**92.5**

5   Conclusions

In this paper, we propose a novel gait recognition framework that is capable of enhancing temporal global and local gait information, which can better generate interactions among the local regions and thus improve the robustness of the gait recognition task. First, we propose to partition the features into multiple local regions along the temporal dimension to extract discriminative features separately, i.e., Temporal Global and Local Aggregator. Second, we propose to introduce SEM into the local features in order to better utilize the local features, which enhances the interaction between regions. Our experiments on public datasets including both CASIA-B and OUMVLP demonstrate the superiority of our proposed framework.

Acknowledgment. This work was supported by National Natural Science Foundation of China (Grant Nos. 61906074, 32371984), Guangdong Basic and Applied Basic Research Foundation (Grant No. 2019A1515011276), Guangzhou Basic and Applied Basic Research Foundation (Grant No. 2023A04J1669), Key-Area Research and Development Program of Guangdong Province (Grant No. 2019B020214003, 2023B0202090001), China Agriculture Research System (CARS-15-23).

References

1. Ben, X., Gong, C., Zhang, P., Yan, R., Wu, Q., Meng, W.: Coupled bilinear discriminant projection for cross-view gait recognition. IEEE Trans. Circuits Syst. Video Technol. **30**(3), 734–747 (2019)
2. Chao, H., He, Y., Zhang, J., Feng, J.: Gaitset: regarding gait as a set for cross-view gait recognition. In: Proceedings of the AAAI Conference on Artificial Intelligence, vol. 33, pp. 8126–8133 (2019)
3. Chao, H., Wang, K., He, Y., Zhang, J., Feng, J.: Gaitset: cross-view gait recognition through utilizing gait as a deep set. IEEE Trans. Pattern Anal. Mach. Intell. (2021)
4. Chen, X., Luo, X., Weng, J., Luo, W., Li, H., Tian, Q.: Multi-view gait image generation for cross-view gait recognition. IEEE Trans. Image Process. **30**, 3041–3055 (2021)
5. Chen, X., Weng, J., Lu, W., Xu, J.: Multi-gait recognition based on attribute discovery. IEEE Trans. Pattern Anal. Mach. Intell. **40**(7), 1697–1710 (2018)
6. Chen, X., Xu, J.: Uncooperative gait recognition: re-ranking based on sparse coding and multi-view hypergraph learning. Pattern Recogn. **53**, 116–129 (2016)
7. Fan, C., et al.: GaitPart: temporal part-based model for gait recognition. In: Proceedings of the IEEE/CVF Conference on Computer Vision and Pattern Recognition, pp. 14225–14233 (2020)
8. Hermans, A., Beyer, L., Leibe, B.: In defense of the triplet loss for person re-identification. arXiv preprint arXiv:1703.07737 (2017)
9. Hou, S., Cao, C., Liu, X., Huang, Y.: Gait lateral network: learning discriminative and compact representations for gait recognition. In: Vedaldi, A., Bischof, H., Brox, T., Frahm, J.-M. (eds.) ECCV 2020. LNCS, vol. 12354, pp. 382–398. Springer, Cham (2020). https://doi.org/10.1007/978-3-030-58545-7_22
10. Huang, T., Ben, X., Gong, C., Zhang, B., Yan, R., Wu, Q.: Enhanced spatial-temporal salience for cross-view gait recognition. IEEE Trans. Circuits Syst. Video Technol. **32**(10), 6967–6980 (2022)

11. Huang, X., et al.: Context-sensitive temporal feature learning for gait recognition. In: Proceedings of the IEEE/CVF International Conference on Computer Vision, pp. 12909–12918 (2021)
12. Li, G., Guo, L., Zhang, R., Qian, J., Gao, S.: Transgait: multimodal-based gait recognition with set transformer. Appl. Intell. 53(2), 1535–1547 (2023)
13. Li, X., Makihara, Y., Xu, C., Yagi, Y., Yu, S., Ren, M.: End-to-end model-based gait recognition. In: Proceedings of the Asian Conference on Computer Vision (2020)
14. Lin, B., Zhang, S., Yu, X.: Gait recognition via effective global-local feature representation and local temporal aggregation. In: IEEE International Conference on Computer Vision (ICCV), pp. 14648–14656 (2021)
15. Lin, B., Zhang, S., Yu, X., Chu, Z., Zhang, H.: Learning effective representations from global and local features for cross-view gait recognition. arXiv preprint arXiv:2011.01461, 1(2) (2020)
16. Liu, W., Zhang, C., Ma, H., Li, S.: Learning efficient spatial-temporal gait features with deep learning for human identification. Neuroinformatics 16(3), 457–471 (2018)
17. Qin, H., Chen, Z., Guo, Q., Wu, Q.J., Lu, M.: RPnet: gait recognition with relationships between each body-parts. IEEE Trans. Circuits Syst. Video Technol. 32(5), 2990–3000 (2021)
18. Sepas-Moghaddam, A., Etemad, A.: View-invariant gait recognition with attentive recurrent learning of partial representations. IEEE Trans. Biometrics Behav. Identity Sci. 3(1), 124–137 (2020)
19. Shiraga, K., Makihara, Y., Muramatsu, D., Echigo, T., Yagi, Y.: Geinet: view-invariant gait recognition using a convolutional neural network. In: International Conference on Biometrics (ICB) (2016)
20. Takemura, N., Makihara, Y., Muramatsu, D., Echigo, T., Yagi, Y.: Multi-view large population gait dataset and its performance evaluation for cross-view gait recognition. IPSJ Trans. Comput. Vision Appl. 10(1), 4 (2018)
21. Teepe, T., Khan, A., Gilg, J., Herzog, F., Hörmann, S., Rigoll, G.: Gaitgraph: graph convolutional network for skeleton-based gait recognition. In: 2021 IEEE International Conference on Image Processing (ICIP), pp. 2314–2318. IEEE (2021)
22. Thapar, D., Nigam, A., Aggarwal, D., Agarwal, P.: VGR-net: a view invariant gait recognition network. In: 2018 IEEE 4th International Conference on Identity, Security, and Behavior Analysis (ISBA), pp. 1–8. IEEE (2018)
23. Wolf, T., Babaee, M., Rigoll, G.: Multi-view gait recognition using 3D convolutional neural networks. In: 2016 IEEE International Conference on Image Processing (ICIP), pp. 4165–4169. IEEE (2016)
24. Wu, Z., Huang, Y., Wang, L., Wang, X., Tan, T.: A comprehensive study on cross-view gait based human identification with deep CNNs. IEEE Trans. Pattern Anal. Mach. Intell. 39(2), 209–226 (2016)
25. Xing, W., Li, Y., Zhang, S.: View-invariant gait recognition method by three-dimensional convolutional neural network. J. Electron. Imaging 27(1), 013010 (2018)
26. Xu, C., Makihara, Y., Li, X., Yagi, Y., Lu, J.: Cross-view gait recognition using pairwise spatial transformer networks. IEEE Trans. Circuits Syst. Video Technol. 31(1), 260–274 (2020)
27. Xu, Z., Lu, W., Zhang, Q., Yeung, Y., Chen, X.: Gait recognition based on capsule network. J. Vis. Commun. Image Represent. 59, 159–167 (2019)

28. Yu, S., Tan, D., Tan, T.: A framework for evaluating the effect of view angle, clothing and carrying condition on gait recognition. In: 18th International Conference on Pattern Recognition (2006)
29. Zhang, K., Luo, W., Ma, L., Liu, W., Li, H.: Learning joint gait representation via quintuplet loss minimization. In: Proceedings of the IEEE/CVF Conference on Computer Vision and Pattern Recognition, pp. 4700–4709 (2019)
30. Zhang, Y., Huang, Y., Yu, S., Wang, L.: Cross-view gait recognition by discriminative feature learning. IEEE Trans. Image Process. **29**, 1001–1015 (2020)
31. Zhang, Z., Tran, L., Liu, F., Liu, X.: On learning disentangled representations for gait recognition. IEEE Trans. Pattern Anal. Mach. Intell. (2020)
32. Zhang, Z., et al.: Gait recognition via disentangled representation learning. In: Proceedings of the IEEE/CVF Conference on Computer Vision and Pattern Recognition, pp. 4710–4719 (2019)
33. Zhu, Z., et al.: Gait recognition in the wild: a benchmark. In: Proceedings of the IEEE/CVF International Conference on Computer Vision, pp. 14789–14799 (2021)

Learning Complementary Instance Representation with Parallel Adaptive Graph-Based Network for Action Detection

Yanyan Jiao, Wenzhu Yang[✉], and Wenjie Xing

School of Cyber Security and Computer, Hebei University,
Baoding 071002, China
wenzhuyang@hbu.edu.cn

Abstract. Temporal action detection (TAD) aims to find action boundaries in untrimmed videos. Video sequences contain multiple actions with various durations, which brings about great challenges to accurate boundary location. Many methods are usually ineffective in multi-scale issues in complex scenes. Besides, there is a crucial fact that local information is vital for clear boundaries. However, the traditional convolution receptive field is limited in local methods and lacks critical frame-level attention. To deal with these problems, we propose the **P**arallel **A**daptive **G**raph-Based **N**etwork (PAGN), which constructs a multi-branch parallel subnetwork that retains multiple video resolutions and enables information interaction between different levels. This results in feature outputs that can represent precise location information and rich semantic information simultaneously, making it more efficient and adaptable to changes in action scale. Additionally, we also propose a novel Complementary Graph Module (CGM) that assigns differential attention to different numbers of neighbors at the current timestamp. Extensive experimental validations are conducted on the challenging datasets ActivityNet-1.3 and THUMOS-14, respectively, and PAGN all consistently exhibit significantly better performance.

Keywords: Temporal action detection · Parallel adaptive graph-based network · Complementary graph module

1 Introduction

In recent years, the popularity of online videos has led to widespread attention to action understanding in untrimmed videos. TAD has been widely applied in various tasks, such as video comprehension [8,20] and monitoring [17].

Although many effective TAD models have achieved remarkable results, detecting actions with varying durations remains challenging. In real-world videos, actions usually last from seconds to minutes. As shown in Fig. 1, for the action type 'clean and jerk', the longest action is 15 times larger than the shortest in the same video, and a large discrepancy in duration across different

S. Rudinac et al. (Eds.): MMM 2024, LNCS 14555, pp. 464–478, 2024.
https://doi.org/10.1007/978-3-031-53308-2_34

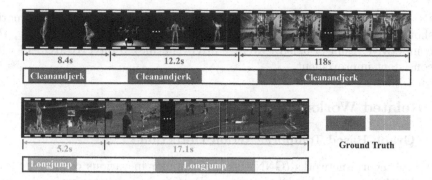

Fig. 1. Different action durations in videos of THUMOS-14.

videos. This challenge brings about two problems. On the one hand, although multi-scale approaches [9] have been explored in TAD, for each candidate action proposal, corresponding temporal regions with different resolutions can generate valuable contextual features. On the other hand, Many of the past methods exploited local [10, 27] or global [2, 18] information. However, short actions with fewer frames and clearer boundaries require more positional information, while long actions with similar backgrounds need higher-level semantics to capture continuous features. These contradictions motivate us to devise a multi-resolution parallel network that emphasizes intrinsic connections within action instances. We exploit the complementarity between high-level and low-level feature maps to retain accurate position information while suppressing background interference.

Therefore in this paper, we propose a novel Parallel Adaptive Graph-Based Network (PAGN) to balance conflicts between long and short actions. We construct a multi-resolution parallel network through temporal convolution, which preserves the original resolution of the video while capturing global contexts at multiple resolutions. Specifically, the architecture enhances the intrinsic connections within actions at different scales by focusing on the close interaction between feature sequences at different levels. This enables better adaptation to targets of different scales and more accurate start and end times localization. Additionally, in order to distinguish the importance of features at different stages, we introduce a well-designed Complementary Graph Module (CGM) in the parallel structure. CGM dynamically selects neighbors and adjusts the receptive field for different-level features. This allows for accurate boundary detection with fewer neighbors in high-resolution stages and enriched contextual information with more neighbor features in low-resolution stages.

To sum up, our main contributions are as follows:

· In order to adapt to actions of different scales, we design a multi-resolution parallel network to enhance the intrinsic connection between actions, and effectively integrate rich location information and semantic information.
· In response to the feature requirements at different stages, we design CGM to adjust the receptive field and assign corresponding weights according to the importance of the features.

· Extensive experiments on THUMOS-14 and ActivityNet-1.3 benchmarks show that our proposed method achieves favorable performance, *e.g.*, the competitive result of 52.98% average mAP is achieved on THUMOS-14 with consistent improvement.

2   Related Works

2.1   Graph-Based Temporal Action Detection

As Graph neural networks (GNN) become popular in various computer vision fields [4,24,29], researchers find their application in temporal action localization [1,27]. G-TAD [27] utilizes the temporal context of video actions, treating each clip as a node in the graph and connecting relationships between pairs of segments as edges, but also attempts to capture semantic context through edge convolution in graph learning methods. BC-GNN [1] models proposal boundaries and content as nodes and edges, in a graph neural network to improve localization ability. G-TACL [24] believes that instances of actions within the same category can help each other. It captures relationships between proposals by constructing graphs using three different adjacency matrices.

2.2   Multi-scale Solution

One of the main difficulties in temporal action detection is effectively generating proposals at different scales. The feature pyramid network (FPN) [12] is one of the most used methods, which mainly works through encoders and decoders to generate multi-scale features. It has also been applied in temporal action detection in recent years [13,18]. HRNet [23] is also a method for constructing multi-scale features. HRNet handles multi-scale features by preserving information from multiple resolutions of images. After HRNet, some methods have been proposed to improve the structure further [22].

3   Methodology

3.1   Feature Extraction and Feature Pre-processing

We encode the raw video sequence into a set of features by a two-stream network [14]. Given an untrimmed video V that contains L frames, we process video frames with regular frame intervals σ to $T = \lceil L/\sigma \rceil$ video segments to reduce the cost. The feature vector of the video is represented as $F_A \in R^{T \times C}$ containing T snippets of C-dimension. Moreover, the position information P is embedded into the initial input video features to form the embedded features $F'_A = F_A + P$.

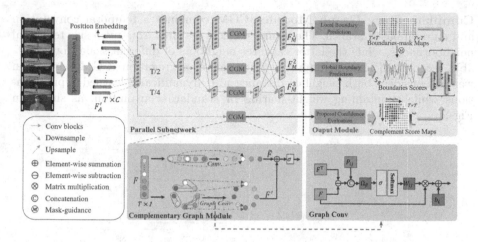

Fig. 2. Illustration of the architecture of PAGN.

3.2 Parallel Adaptive Graph-Based Network

Our novel framework employs a parallel network to generate features with diverse temporal resolutions. This facilitates effective feature extraction for both longer and shorter actions. Additionally, we introduce a graph-based module to enhance local modeling, emphasizing the significance of features at different timestamps and improving location accuracy.

Parallel Subnetwork. The parallel subnetwork in PAGN maintains the representation of the original scale throughout the pipeline. It takes the full-length video sequence as input and incorporates long and short sequences to form high-to-low resolutions in parallel, as shown in Fig. 2. Through repeated multi-scale fusion, it can obtain rich information from actions of different lengths, leading to improved feature fusion. The resulting subnet branch can be denoted as:

$$F_M = \left\{ F_M^1, F_M^2, \ldots, F_M^R \right\}, \tag{1}$$

where F_M^r represents the feature map of the rth branch of F_M, and R is the total number of subnetwork branches. Shorter duration actions are easier to locate accurately using lower-level feature maps with small receptive fields. However, actions with complex backgrounds may have error boundaries that are earlier or later, requiring higher-level features for more detailed associations. Therefore, we use low-level feature maps for local prediction and high-level semantics for global prediction before the final boundary prediction, expressed as:

$$S_l = \Theta_1 \left(F_M^1, F_M^2, \ldots, F_M^{r_1} \right), S_g = \Theta_2 \left(F_M^{r_2}, F_M^{r_2+1}, \ldots, F_M^R \right) \tag{2}$$

Among them, S_l and S_g represent local and global probabilities, $\Theta_1 (\cdot)$ is stacked 1D convolution with a kernel of size 3, and $\Theta_2 (\cdot)$ is a global module in our previous work.

Complementary Graph Module. CGM combines a temporal convolution and a graph-based branch. The former collects features from fixed-size temporal neighborhoods, while the latter captures neighbors of different scales for adaptable receptive fields and improved information aggregation. CGM's attention weight considers both neighbor distance and dynamically learned representations while maintaining weight-sharing in standard convolution, as shown in Fig. 2.

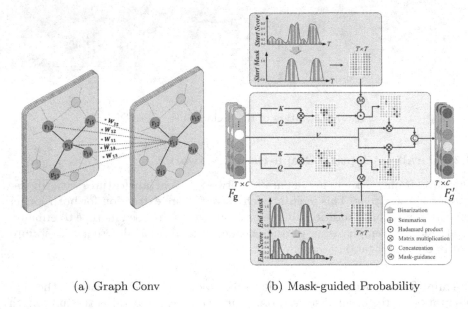

(a) Graph Conv   (b) Mask-guided Probability

Fig. 3. (a) Describes graph attention convolution in CGM. (b) is a detailed guidance process with boundary masks.

After the video feature sequence is given, we first add the sinusoidal position encoding to assign a unique encoding vector about each timestamp in the video sequence to capture the position information: $F = F_{in} + P$, where $F_{in} \in R^{L \times C}$ is the input feature sequence, $P \in R^{L \times C}$ is a position encoding vector, then the elements in the vector can be expressed as:

$$P_{a,2b} = \sin\left(a/10000^{2b/C}\right), P_{a,2b+1} = \cos\left(a/10000^{2b/C}\right), \tag{3}$$

the above formulation respectively represent the elements on column $2b$ and column $2b + 1$ of row a in the vector.

We construct a graph $G = \{V, E\}$ from the input video sequence. $V = \{v_m\}^M$ represents each video clip as a point in the graph, $\mathcal{E} = \{\mathcal{E}_t\}_{t=1}^M$ represents the edges pointing to each node, and M is the number of vertices. The set of neighbors of vertex i is denoted as $\mathcal{N}(i) = \{j : (i, j) \in \mathcal{E}\} \cup i$. Each node has K edges, expressed as $\mathcal{E}_t = (v_{tk}, v_t)$. The determination of these edges is based on the cosine similarity between nodes. The input vertex feature is denoted as $F = \{f_l\}_{l=1}^{L} \in R^{L \times C}$.

We introduce a graph attention function $G : F \rightarrow F'$ to summarize the features of all relevant nodes for a given timestamp node f_t^i. This function maps the input features to a new vertex feature set $F' = \{f_l'\}_{l=1}^{L} \in R^{L \times C}$. We utilize a shared attention mechanism β to focus on the differences between neighbors and the current node, assigning appropriate attention weights to different feature channels, as shown in Fig. 3(a). The relative position index of neighbors is also considered to assist in learning associations among them. This enables the CGM to dynamically adapt to the relationship between the current timestamp and its surrounding neighbors.

The graph attention weights is described as $\widetilde{W}_{ij} = \beta(P_{ij}, \Delta f_{ij})$, $j \in \mathcal{N}(i)$, where $\widetilde{W}_{ij} = \{\widetilde{W}_{ij,k}\}_{k=1}^{K} \in R^{K \times C}$ indicates the attentional weight vector timestamp j to i: $\Delta f_{ij} = \Omega(f_j) - \Omega(f_i)$. $\Omega(\cdot)$ is a multi-layer perceptron (MLP). P_{ij} is the positional index of j relative to i, which helps to span relations of the neighbor timestamps. While the difference between pairs of nodes can guide assigning more attention to similar scene frames. The sharing attention mechanism β is implemented using a MLP : $\beta(P_{ij}, \Delta f_{ij}) = \Omega_\beta([P_{ij} \|_C \Delta f_{ij}])$, where $\|_C$ is the concatenation operation along the channel dimension, Ω_β indicates the applied multilayer perception. More details are shown in Fig. 2.

Therefore, the final output of the graph-based convolution can be formulated as follows: $f' = \sum_{j \in \mathcal{N}(i)} \phi(f_j, W_{ij}) + b_i$, ϕ is calculated by the shared attention W_{ij} and the neighbor feature f_j, this operation produces the elementwise production of two vectors and add bias b_i, and finds out where to handle the size-varying neighbors across different points. The attention weights W_{ij} are normalized across all the neighbors of vertex i.

$$W_{ij} = \frac{\exp(\widetilde{W}_{ij,k})}{\sum_{n \in \mathcal{N}(i)} \exp(\widetilde{W}_{in,k})} \tag{4}$$

Finally, we fuse the CGM results with the temporal branch features by summing the element-wise features. The output features are obtained after applying the activation function: $F'' = \sigma(F' + \widetilde{F})$, where $\widetilde{F}$ represents the feature output of the temporal branch, σ is the activation function.

3.3 Output Module

After obtaining the multi-scale features of the video sequence, we send it to the subsequent output module to obtain the boundary of the action and the integrity score of the action required in the reasoning stage. Finally, it generates proposals and ranks them.

Temporal Boundary Prediction. We adopt the two-branch prediction boundary method, which involves sending features processed by PAGN into the local and global modules based on the characteristics of long and short actions.

The local prediction module consists of two 1D convolutions and activation functions, while the global prediction module utilizes a self-attention mechanism. This allows us to obtain the start and end probabilities for each video clip, denoted as $S_{s,g}, S_{e,g}, S_{s,l}, S_{e,l} \in R^{T \times 1}$.

However, Local methods capture frame location information, while global modeling lacks focus on foreground action representation and may introduce background noise. To address this, we use the foreground mask from local probabilities to guide global attention and learn semantic similarity.

As shown in Fig. 3(b), we first evaluate the sequence of local probabilities $S_{s,l}, S_{e,l}$ for each action segment and then implement the binarization procedure, the mask sequence is illustrated as follows:

$$\tilde{S}_i = \begin{cases} 0, S_i < \gamma \cdot \rho(S) \\ 1, S_i > \gamma \cdot \rho(S), \end{cases}, i = 1, 2, \ldots, T \tag{5}$$

where γ is the binary probability threshold, we set it to 0.5 by default. The function $\rho(\cdot)$ selects the maximum value in the probability sequence. Moreover, by repeating the mask sequence T times in the temporal dimension, $M_s, M_e \in R^{T \times T}$, represents the mask of the predicted start and end probabilities. Therefore, the correlation between any two segments and the trend of the start and end of the action can be expressed. Finally, the corresponding mask is multiplied with the similarity map $SM \in R^{T \times T}$ of Q and K formed from the input integration feature F_g in the attention to obtain the enhanced attention map $SM' \in R^{T \times T}$, thus draw attention near action boundaries, formulated as:

$$SM'_v = SM_v \odot M_v, v = \{s, e\}, \tag{6}$$

where $\odot$ denotes the Hadamard product.

Completeness Regression. In order to improve detection performance, we make inferences on proposal relationships and predict completeness scores in addition to generating rough proposals. Based on the boundary-matching mechanism, we define a shared sampling matrix. The network then evaluates the sampled proposal-level features to obtain classification and regression scores.

4   Experiments

4.1   Datasets and Experimental Settings

Datasets. Experimental validation is conducted on two datasets: THUMOS-14 [6] and ActivityNet-1.3 [5]. The THUMOS-14 dataset consists of 413 untrimmed videos with 20 sports action categories. Its validation and test sets consist of 200 and 213 videos, with an average of 15 action instances per video. The validation set is used for network training, and the test set is used for evaluation. ActivityNet-1.3 dataset contains 19993 untrimmed videos with 200 labeled action types; each video has an average of 1.5 action instances. The dataset is divided into training, test, and validation sets in a 2:1:1 ratio.

Table 1. Comparison of PAGN with other SOTA methods on THUMOS-14 in terms of mAP (%).

Method	$mAP_{0.3}$ (%)	$mAP_{0.4}$ (%)	$mAP_{0.5}$ (%)	$mAP_{0.6}$ (%)	$mAP_{0.7}$ (%)
TURN [3]	46.3	35.3	24.5	14.1	6.3
BSN [11]	53.5	45.0	36.9	28.4	20.0
BMN [10]	56.0	47.4	38.8	29.7	20.5
G-TAD [27]	54.5	47.6	40.2	30.8	23.4
P-GCN [28]	60.1	54.3	45.5	33.5	19.8
TCANet [16]	60.6	53.2	44.6	36.8	26.7
ABN [21]	59.9	54.0	46.1	37.0	25.6
BC-GNN [1]	57.1	49.1	40.4	31.2	23.1
RTD-Net [19]	58.5	53.1	45.1	36.4	25.0
TadTR [15]	67.1	61.1	52.0	39.9	26.2
SoLa [7]	59.1	53.1	44.3	33.6	23.2
PAGN (Ours)	**68.7**	**62.9**	**54.5**	**45.2**	**33.6**

Implementation Details. We utilize the pre-trained TSN model [14] to extract video features. For ActivityNet-1.3, we set the sampling time interval σ to 16, the constant value of each feature sequence to $T = 100$, and the maximum duration D to 100. For THUMOS-14, we use two different model contexts with σ values of 8 and 5. The sliding window length is set to $T = 128$ and $D = 64$. In the training process, we optimize using Adam with a batch size of 16 and 8 epochs. The initial learning rate for the first seven epochs is 10^{-3}, which then decays to 10^{-4} for subsequent epochs.

4.2   Comparison with SOTA Methods

The evaluation results on the test sets of THUMOS-14 and ActivityNet-1.3 are shown in Table 1 and Table 2, respectively. We take video classification scores by classification model UntrimmedNet [25] for THUMOS-14 and [26] for ActivityNet-1.3. The best record in the column is highlighted in bold.

Our PAGN outperforms other SOTA methods on THUMOS-14, as shown in Table 1. Our method achieves superior performance across different tIoU thresholds. Compared with methods such as the competitive TCANet and BC-GNN that do not perform multi-scale processing actions, PAGN improves $mAP_{0.5}$ by 9.9% and 14.1%, respectively, and increases $mAP_{0.7}$ by 6.9% and 10.5%, respectively, reaching 33.6%. These results demonstrate that PAGN enables more accurate temporal action detection boundaries with its multi-scale aggregation of action information and graph attention.

PAGN achieves competitive performance on ActivityNet-1.3 compared to others like TadTR and RTD-Net, as shown in Table 2. THUMOS-14 videos are more complex, with over 15 action instances, while ActivityNet-1.3 videos have

Table 2. Comparison between PAGN with SOTA methods on ActivityNet-1.3. The results are measured by mAP (%) at different tIoU thresholds and average mAP (%).

Method	$mAP_{0.5}$ (%)	$mAP_{0.75}$ (%)	$mAP_{0.95}$ (%)	Avg. (%)
SSN [30]	39.12	23.48	5.49	23.98
BSN [11]	46.45	29.96	8.02	30.03
BMN [10]	50.07	34.78	8.29	33.85
G-TAD [27]	50.36	35.02	9.02	34.09
P-GCN [28]	48.26	33.16	3.27	31.11
TCANet [16]	51.91	34.92	7.46	34.43
BC-GNN [1]	50.56	34.75	9.37	34.26
RTD-Net [19]	47.21	30.68	8.61	30.83
ContextLoc [31]	51.24	31.40	2.83	30.59
TadTR [15]	41.40	28.85	7.86	28.21
PAGN (Ours)	**53.45**	**34.60**	**10.13**	**34.91**

Table 3. The mAP results of our proposed method PAGN on THUMOS-14 dataset (%) with different numbers of layers at various tIoU thresholds.

Layers	$mAP_{0.3}$ (%)	$mAP_{0.5}$ (%)	$mAP_{0.7}$ (%)	Avg. (%)
2	65.61	52.88	32.15	50.90
3	**68.65**	**54.50**	**33.61**	**52.98**
4	66.49	53.79	32.95	51.70

an average of 1.5 actions. The data shows that PAGN excels in detecting videos with multiple actions. ContextLoc improves detection performance by global modeling through video-level representation. This is because, when there is only one action instance in a video or across multiple clips (e.g., ActivityNet-1.3), the global context provides consistent prior knowledge to better distinguish individual action frames from background frames. However, when there are multiple action instances in a video (e.g., THUMOS-14), the discriminative information provided by the global context is weakened due to the different categories of action instances. Therefore, we propose an adaptive graph attention module to model actions of various categories with different lengths and strengthen proposal-level representations.

4.3   Ablation Study

In this section, we further investigate the performance of each component and suitable settings to understand PAGN better. All experiments are validated on THUMOS-14. The other output video feature sequences are all from the TSN method as the backbone network. The best record in the column is highlighted in bold.

Table 4. The detection result on THUMOS-14 (%), using different fusion methods on the parallel branch.

Fusion	$mAP_{0.3}$ (%)	$mAP_{0.5}$ (%)	$mAP_{0.7}$ (%)	Avg. (%)
Add	**68.65**	**54.50**	**33.61**	**52.98**
Concat	65.91	53.01	31.85	50.74

Impact of the Number of Layers. We build our parallel network using convolutions with stride 3 for multi-resolution processing of actions. We evaluate multi-scale importance by different layer numbers, as shown in Table 3. The result when the number of layers is three is better than two, which shows that if there is no multi-level feature interaction, it will affect the detection performance. This can be explained by lower-level features' inability to model the relationship between action instances reasonably. When there are four layers of features, When there are four layers of features, the Avg. is 51.70%, which is 1.28% lower than the three layers. which shows that overly rich feature interactions may interfere with the adjacent timestamps.

Table 5. The effect of different neighbors about local and global in CGM on THUMOS-14 in terms of mAP(%).

Count	L1(3,5)L3(3,5)	L1(6,12)L3(6,12)	L1(3,5)L3(6,12)
$mAP_{0.3}$(%)	66.12	66.56	**68.65**
$mAP_{0.5}$(%)	53.82	53.39	**54.50**
$mAP_{0.7}$(%)	32.38	32.70	**33.61**

Effectiveness of Different Components in PAGN. Parallel branches and CGM are the core components of PAGN. We evaluate several variants by eliminating parallel branches and GCNs with #parameters in millions (M) and mAP. The baseline represents the version without parallel branches and CGM. As shown in Table 4, joint use of all components can improve detection performance. As for aggregating neighbor information, the results show that adaptive graph attention is beneficial for this task, and removing position embedding or offset will degrade the performance.

Impact of Feature Fusion in Parallel Network. In the parallel branch for multi-scale action interaction, we must fuse information between high-level and low-level feature maps and enter the next stage of different scale integration. We evaluate two feature fusion methods for experiments: add and concatenate. The experimental results are shown in Table 5. We can observe that the effect of

Table 6. Performance evaluation on different components of our model at various tIoU thresholds on THUMOS-14 dataset (%). PE is position embedding, and OS is offset of neighbors.

Variants of model	mAP$_{0.3}$ (%)	mAP$_{0.5}$ (%)	mAP$_{0.7}$ (%)	#Params (M)
Baseline	63.39	50.09	31.09	18.21
Baseline+parallel network	64.98	52.28	31.72	17.34
Baseline+CGM wo/PE	66.60	54.06	32.57	–
Baseline+CGM wo/OS	65.26	52.84	32.38	–
Baseline+CGM	**68.65**	**54.50**	**33.61**	16.74

Table 7. The mAP results of our proposed method PAGN on THUMOS-14 (%) with integrating multi-level features at various tIoU thresholds.

Integration	mAP$_{0.3}$ (%)	mAP$_{0.5}$ (%)	mAP$_{0.7}$ (%)	Avg. (%)
$(F_M^1)/(F_M^2, F_M^3)$	66.20	53.13	32.17	51.15
$(F_M^1, F_M^2)/(F_M^2, F_M^3)$	65.85	53.06	32.01	50.98
$(F_M^1)/(F_M^1, F_M^2, F_M^3)$	**68.65**	**54.50**	**33.61**	**52.98**
$(F_M^1, F_M^2)/(F_M^1, F_M^2, F_M^3)$	65.78	52.79	32.83	51.17

add is better than that of concatenate. The concatenate operation improves the overall information density by increasing the dimension but does not enrich the semantic information in each channel. The add operation makes each channel contain more features, paying more attention to important temporal features.

Effects of the Different Neighbor Count of CGM. The performance of two graph module selections with different neighbors in CGM is shown in Table 6. $L1$ and $L3$ represent the first and third layers' features. We can conclude that when the feature keeps a small receptive field, it is beneficial to perform small-scale local modeling and help locate shorter actions. For richer advanced features, more neighbor features can result in deeper semantic information required for longer actions to capture more accurate detection results.

Analysis of the Input Feature Before Prediction. We validate input features for local and global predictions through experiments with different representation methods. We present the characteristics of each layer (F_M^i) in the parallel network in Table 7. (,) denotes feature mixing and addition after upsampling or downsampling without changing the number of channels. The notation before and after / represents the input feature for the local and global branches. Results show that using only the first branch F_M^1 for local prediction generally outperforms using multiple branches. This highlights the importance of clear position information in achieving high-recall detection. While deep global information emphasizes information integrity, which necessitates location and rich semantic information to supplement action boundary details.

4.4 Qualitative Results

To validate the effectiveness of temporal boundary location and action classification, we visualize qualitative results in Fig. 4. Compared with ContextLoc, PAGN predicts a more precise temporal boundary and classification for these examples. In addition, we also show the confidence map visualization results of the proposals in a specific video, as shown in Fig. 5(b). The area with the highest score corresponds to the ground truth.

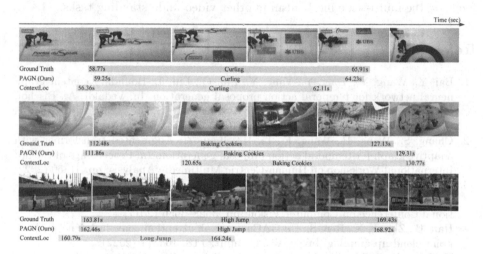

Fig. 4. Qualitative results on an example from the ActivityNet-1.3 validation set (top) and an example from the THUMOS-14 test set (middle) show that our PAGN locates the temporal boundaries more accurately than ContextLoc. And PAGN correctly classifies action instances (bottom) which ContextLoc fails on.

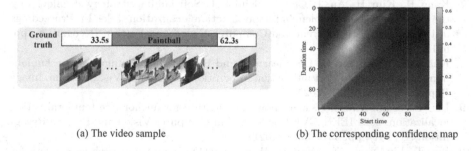

(a) The video sample   (b) The corresponding confidence map

Fig. 5. The confidence map of the video sample from the ActivityNet-1.3 dataset.

5   Conclusion

In this paper, we propose a novel network PAGN for temporal action detection in untrimmed videos. We design a parallel network that can complement and interact with feature maps of different levels. The purpose is to balance the conflicts between long and short instances in the video while retaining precise location information and enriching semantic information. Combining local and global components, it provides reliable boundary probability for action detection. In addition, a joint convolution module based on graph attention is proposed to endow a dynamic mechanism for local modeling. Extensive experiments and studies indicate that our PAGN achieves superior performance in temporal action detection on THUMOS-14 and ActivityNet-1.3 datasets. The future work will focus on the multi-scale mechanism in other video understanding tasks.

References

1. Bai, Y., Wang, Y., Tong, Y., Yang, Y., Liu, Q., Liu, J.: Boundary content graph neural network for temporal action proposal generation. In: Vedaldi, A., Bischof, H., Brox, T., Frahm, J.-M. (eds.) ECCV 2020. LNCS, vol. 12373, pp. 121–137. Springer, Cham (2020). https://doi.org/10.1007/978-3-030-58604-1_8
2. Chang, S., Wang, P., Wang, F., Li, H., Shou, Z.: Augmented transformer with adaptive graph for temporal action proposal generation. In: Proceedings of the 3rd International Workshop on Human-Centric Multimedia Analysis, pp. 41–50 (2022)
3. Gao, J., Yang, Z., Chen, K., Sun, C., Nevatia, R.: Turn tap: temporal unit regression network for temporal action proposals. In: Proceedings of the IEEE International Conference on Computer Vision, pp. 3628–3636 (2017)
4. Han, B., Zhang, X., Ren, S.: Pu-GACnet: graph attention convolution network for point cloud upsampling. Image Vis. Comput. **118**, 104371 (2022)
5. Heilbron, F.C., Escorcia, V., Ghanem, B., Niebles, J.C.: Activitynet: A large-scale video benchmark for human activity understanding. In: 2015 IEEE Conference on Computer Vision and Pattern Recognition (CVPR), pp. 961–970. IEEE (2015)
6. Idrees, H., et al.: The thumos challenge on action recognition for videos "in the wild." Comput. Vis. Image Underst. **155**, 1–23 (2017)
7. Kang, H., Kim, H., An, J., Cho, M., Kim, S.J.: Soft-landing strategy for alleviating the task discrepancy problem in temporal action localization tasks. In: Proceedings of the IEEE/CVF Conference on Computer Vision and Pattern Recognition, pp. 6514–6523 (2023)
8. Khaire, P.A., Kumar, P.: Deep learning and RGB-D based human action, human-human and human-object interaction recognition: a survey. J. Vis. Commun. Image Represent. **86**, 103531 (2022)
9. Lin, C., et al.: Learning salient boundary feature for anchor-free temporal action localization. 2021 IEEE/CVF Conference on Computer Vision and Pattern Recognition (CVPR), pp. 3319–3328 (2021)
10. Lin, T., Liu, X., Li, X., Ding, E., Wen, S.: BMN: boundary-matching network for temporal action proposal generation. In: Proceedings of the IEEE/CVF International Conference on Computer Vision, pp. 3889–3898 (2019)

11. Lin, T., Zhao, X., Su, H., Wang, C., Yang, M.: BSN: boundary sensitive network for temporal action proposal generation. In: Ferrari, V., Hebert, M., Sminchisescu, C., Weiss, Y. (eds.) ECCV 2018. LNCS, vol. 11208, pp. 3–21. Springer, Cham (2018). https://doi.org/10.1007/978-3-030-01225-0_1

12. Lin, T.Y., Dollár, P., Girshick, R., He, K., Hariharan, B., Belongie, S.: Feature pyramid networks for object detection. In: Proceedings of the IEEE Conference on Computer Vision and Pattern Recognition, pp. 2117–2125 (2017)

13. Liu, Q., Wang, Z.: Progressive boundary refinement network for temporal action detection. In: Proceedings of the AAAI Conference on Artificial Intelligence, vol. 34, pp. 11612–11619 (2020)

14. Liu, S., Zhao, X., Su, H., Hu, Z.: TSI: temporal scale invariant network for action proposal generation. In: Ishikawa, H., Liu, C.-L., Pajdla, T., Shi, J. (eds.) ACCV 2020. LNCS, vol. 12626, pp. 530–546. Springer, Cham (2021). https://doi.org/10.1007/978-3-030-69541-5_32

15. Liu, X., et al.: End-to-end temporal action detection with transformer. IEEE Trans. Image Process. **31**, 5427–5441 (2022)

16. Qing, Z., et al.: Temporal context aggregation network for temporal action proposal refinement. In: Proceedings of the IEEE/CVF Conference on Computer Vision and Pattern Recognition, pp. 485–494 (2021)

17. Ratre, A., Pankajakshan, V.: Deep imbalanced data learning approach for video anomaly detection. In: 2022 National Conference on Communications (NCC), pp. 391–396 (2022)

18. Shang, J., Wei, P., Li, H., Zheng, N.: Multi-scale interaction transformer for temporal action proposal generation. Image Vis. Comput. **129**, 104589 (2023)

19. Tan, J., Tang, J., Wang, L., Wu, G.: Relaxed transformer decoders for direct action proposal generation. In: 2021 IEEE/CVF International Conference on Computer Vision (ICCV), pp. 13506–13515 (2021)

20. Vahdani, E., Tian, Y.: Deep learning-based action detection in untrimmed videos: a survey. IEEE Trans. Pattern Anal. Mach. Intell. **45**, 4302–4320 (2021)

21. Vo, K., Yamazaki, K., Truong, S., Tran, M.T., Sugimoto, A., Le, N.: ABN: agent-aware boundary networks for temporal action proposal generation. IEEE Access **9**, 126431–126445 (2021)

22. Wang, J., Long, X., Chen, G., Wu, Z., Chen, Z., Ding, E.: U-HRnet: delving into improving semantic representation of high resolution network for dense prediction. arXiv preprint arXiv:2210.07140 (2022)

23. Wang, J., et al.: Deep high-resolution representation learning for visual recognition. IEEE Trans. Pattern Anal. Mach. Intell. **43**(10), 3349–3364 (2020)

24. Wang, L., Zhai, C., Zhang, Q., Tang, W., Zheng, N., Hua, G.: Graph-based temporal action co-localization from an untrimmed video. Neurocomputing **434**, 211–223 (2021)

25. Wang, L., Xiong, Y., Lin, D., Van Gool, L.: Untrimmednets for weakly supervised action recognition and detection. In: Proceedings of the IEEE Conference on Computer Vision and Pattern Recognition, pp. 4325–4334 (2017)

26. Xiong, Y., et al.: Cuhk & ethz & siat submission to activitynet challenge 2016. arXiv preprint arXiv:1608.00797 (2016)

27. Xu, M., Zhao, C., Rojas, D.S., Thabet, A.K., Ghanem, B.: G-tad: sub-graph localization for temporal action detection. In: 2020 IEEE/CVF Conference on Computer Vision and Pattern Recognition (CVPR), pp. 10153–10162 (2020)

28. Zeng, R., et al.: Graph convolutional networks for temporal action localization. In: 2019 IEEE/CVF International Conference on Computer Vision (ICCV), pp. 7093–7102 (2019)

29. Zhang, W., et al.: Graph attention multi-layer perceptron. In: Proceedings of the 28th ACM SIGKDD Conference on Knowledge Discovery and Data Mining (2022)
30. Zhao, Y., Xiong, Y., Wang, L., Wu, Z., Tang, X., Lin, D.: Temporal action detection with structured segment networks. Int. J. Comput. Vision **128**(1), 74–96 (2020)
31. Zhu, Z., Tang, W., Wang, L., Zheng, N., Hua, G.: Enriching local and global contexts for temporal action localization. In: Proceedings of the IEEE/CVF International Conference on Computer Vision, pp. 13516–13525 (2021)

CESegNet:Context-Enhancement Semantic Segmentation Network Based on Transformer

Xu Chen[1,2] and Zhibin Zhang[1,2(✉)]

[1] School of Computer Science, Inner Mongolia University, Hohhot 010021, China
cschenxu@mail.imu.edu.cn, cszhibin@imu.edu.cn
[2] Key Laboratory of Wireless Networks and Mobile Computing, School of Computer Science, Inner Mongolia University, Hohhot 010021, China

Abstract. CNN-based methods have achieved success in semantic segmentation. However, research on improving network robustness in this domain has been limited. Similarly, transformer and its variants have recently shown state-of-the-art results in many vision tasks, from image classification to dense prediction, because transformer has a global receptive field, but transformer-based methods have much higher computational complexity compared to CNN-based methods. To address these problems, we introduce a context-enhancement network based on the transformer. Firstly, we enhance global contextual information through a hierarchical simplified transformer encoder. Then, we design two different Context-Enhancement Modules (CEM) to enrich contextual features further. Finally, we propose a Contextual Fusion Decoder (CFD) to fuse multi-scale contextual information. Extensive experiments demonstrate that our method achieves significant performance and robustness compared to previous counterparts. Our best model, CESegNet-Large, achieves 82.21% mIoU and 48.77% mIoU on the Cityscapes and the ADE20K validation sets, and it also demonstrates excellent zero-shot robustness on the Cityscapes-N dataset.

Keywords: Semantic segmentation · Transformers · Contextual enhancement

1 Introduction

Semantic segmentation is a fundamental task in computer vision that assigns a class label to each pixel in an image. This enables applications in fields like autonomous driving, assisted healthcare, etc. Unlike image classification, which performs classification at the image level, semantic segmentation requires pixel-level classification. In 2015, Trevor Darrell [15] from UC Berkeley proposed Fully Convolutional Networks (FCN), pioneering the application of convolutional neural networks for semantic segmentation.

Since then, many methods based on FCNs have been improved, such as U-Net [17], which is divided into two parts: downsampling and upsampling.

S. Rudinac et al. (Eds.): MMM 2024, LNCS 14555, pp. 479–493, 2024.
https://doi.org/10.1007/978-3-031-53308-2_35

Downsampling is used for encoding, and upsampling is used for restoring the encoding to its original size to finish pixel-level predictions. Then SegNet [1] was proposed, a semantic segmentation network based on the modification of VGG-16 [18], which, for the first time, divided the network structure into encoder and decoder. This structure has been widely adopted and proven successful, such as Deeplab series [2,3], Mask R-CNN [10], PSPNet [24], ConvNeXt [14], etc. However, the robustness of convolution-based networks is not good for corrupted images.

Recently, Transformers have been introduced to computer vision and achieved promising results. Dosovitskiy et al. [8] proposed a Vision Transformer (ViT) for image classification, which splits an image into patches and applies patch embedding, then feeds it into a standard transformer architecture. ViT achieved excellent results on the image classification benchmark ImageNet [7]. In semantic segmentation, Zheng et al. [25] first adopted ViT as the backbone and proposed SETR, which obtained strong performance but with high computational cost. Due to the higher complexity of transformer compared to convolutions, subsequent researchers have proposed new networks such as Swin-Transformer [13], PVT [19], SegFormer [21], CPVT [5], etc. in order to solve this problem. These methods mainly rely on shortening the sequence length to solve the computationally intensive problem [8].

However, unlike convolutions, transformers hold inherent global receptive fields for image feature extraction. When reducing computational cost, the receptive field of transformers will also cause a loss. Most researchers focus on improving the feature extraction capability of the encoder. However, they ignore the fact that the decoder is equally important for semantic parsing. In contrast, the convolution-based networks as decoders do not need to provide extensive receptive fields, but can help to curb effectively parameter growth. For dense prediction tasks such as semantic segmentation, the contextual information allows model to understand the image content better. Thereby, this paper presents a transformer-based semantic segmentation network that incorporates the contextual information of images, where two different CEMs are designed to fully use contextual information and a transformer-based encoder to obtain global receptive fields. Meanwhile, the growth of parameters can also be effectively controlled by the convolution-based the Contextual Fusion Decoder(CFD). Our method demonstrates strong performance on public datasets. The main contributions are as follows:

1. We propose a semantic segmentation network based on a hierarchical simplified transformer and design an efficient CFD.
2. Two different CEMs are designed to counteract the loss of contextual information due to the simplified transformer.
3. Our method demonstrates good robustness in imitating real-world corrupted image datasets.

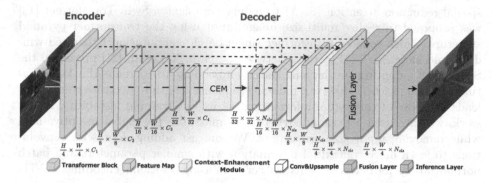

Fig. 1. The framework of the transformer-based semantic segmentation networks proposed in this paper, in which the encoder consists of a transformer-based multilevel network and a Context-Enhancement Module (CEM); the decoder consists of continuous upsampling layers and a feature fusion layer.

2 Related Work

The feature extraction module has been modified as the backbone for semantic segmentation in many image classification methods. Researchers designed corresponding networks for pixel-level prediction in segmentation. For instance, FCN [15] replaces fully connected layers with convolutions for pixel-level prediction. Later improvements like UNet [17] and FastFCN [20] were built on based FCN, as backbones like ResNet50 etc. The encoder-decoder architecture was first introduced for computer vision in SegNet [1] where the encoder is used to extract features, and the decoder to decode the features for prediction. Since then, the encoder-decoder structure has been extensively explored by researchers.

Transformer Backbone. In 2020, Dosovitskiy et al. [8] first proposed a Vision Transformer (ViT) for image classification, pioneering the application of transformers in computer vision, demonstrating the potential of transformers for dense prediction tasks. This sparked a broad interest in applying transformers across various downstream tasks in computer vision. For instance, Carion et al. [25] proposed DETR, enabling a transformer for object detection for the first time. Later, in 2021, Zheng et al. [25] introduced SETR, adapting transformers as the backbone for semantic segmentation. However, these transformers have had the following drawbacks for vision tasks:

– The computational load is large, and the computational complexity is positive correlation to the sequence length.
– The input image size is limited.
– Insufficient local features are produced after patch embedding.

To alleviate these problems, Wang et al. [19] proposed the Pyramid Vision Transformer (PVT), which first introduced a multi-stage pyramid structure into transformers, removing the constraint due to single size during training, and utilized spatial-reduction attention (SRA) to reduce complexity. Swin Transformer [13] was proposed to realize multi-size image input using the transformer pyramid structure, which reduces transformer single batch computation by shifted windows and increases local features between patches. Xie et al. [21] presented the SegFormer, adopting overlapping patch embedding to increase the local feature between patches. However, reducing sequence length to decrease complexity leads to a loss of contextual information, as well as training instability. In our work, we have used a pyramid structure to maximize the global receptive field while removing input size constraints. We suggest adding depthwise convolutions to learn more contextual information with smaller parameters and batch normalizations for better stability, rather than using positional encodings in ViT.

Global Context Prior. In real-world scenes, objects tend to have correlations with surrounding entities and environments to form global context priors like positional, scale, or occurrence probability information. Zhao et al. [24] proposed the Pyramid Scene Parsing Network (PSPNet), capturing homogeneous context via spatial pyramid pooling. Fu et al. [9] proposed the Dual Attention Network (DAnet), which learns channel attention and spatial attention based on an attention-based approach to aggregate heterogeneous contextual information selectively. Yu et al. [23] proposed the Context Prior Network (CPnet), which improves context information by adding context branches through deep separable convolution. In our work, we have designed two context enhancement modules to counteract the contextual information loss caused by the simplified transformer, and we also designed a decoder with multi-level feature fusion to decode the semantic information.

3   Method

This section introduces CESegNet, as shown in Fig. 1. It is an efficient and accurate semantic segmentation network that can fully extract contextual information with good robustness. In this section, we first present the overall architecture of the model. Then, we describe in detail the Context Enhancement Modules(CEM) and decoder tailored for the backbone network and segmentation tasks.

3.1   Overall Architecture

As shown in Fig. 1, our backbone network is inspired by the hierarchical feature extraction architectures of popular CNN models [11,14,15,18]. We adopt a hierarchical feature extraction architecture where the transformer block is taken as the core. Firstly, given an input image with the size of $H \times W \times 3$ pixels, in order to adapt to the requirement of dense prediction for semantic segmentation, it is divided into patches with the size of 4×4 pixels. The patches are fed into Transformer blocks and reshaped to produce feature maps for the next stage. There are 4 stages in total. As shown in Fig. 1, the feature maps F_i have resolutions of

$\{\frac{1}{4}, \frac{1}{8}, \frac{1}{16}, \frac{1}{32}\}$ from the input image, where $i \in \{1, 2, 3, 4\}$. Especially in stage-4, the feature map F_4 goes through the CEM and is input to the decoder, where it undergoes 3 successive upsampling layers operations. The decoder fuses multi-level features F_i from the encoder outputs to obtain $\hat{F}_i$ of size $\frac{H}{2^{i+1}} \times \frac{W}{2^{i+1}} \times N_{cls}$, where $i \in \{1, 2, 3, 4\}$, and N_{cls} is the number of classes. Then 4 layers of features $\hat{F}_i$ are fed into the fusion layer to get the feature map of size $\frac{H}{4} \times \frac{W}{4} \times N_{cls}$, and finally fed into the Inference Layer for inference to get the result.

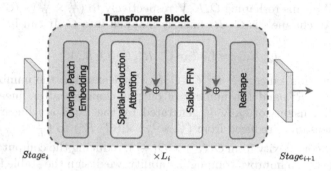

Fig. 2. Illustration of a transformer block, in which the Batch Normalization (BN) added to the Stable FNN is to make the model training more stable. L_i is the number of transformer blocks between neighboring stages.

3.2   Transformer Encoder

Since the image cannot be directly input to the transformer, we need to apply patch embedding to the input image before feature extraction. Also, the sequence output from transformers cannot be directly passed as input to the next stage. Therefore, researchers usually add multilayer perceptron (MLP) [8] after transformer and then reshape the sequence into multi-level feature maps. The transform block designed in this paper is illustrated in Fig. 2.

Transformer Block. To be specific, an image is taken as input. To maintain local context continuity, referring to the approach in [21], we adopt an overlapping patch embedding block consisting of a 2D convolution followed by a batch normalization (BN). Before entering the Transformer, the size of the input feature map is reduced by half. In visual multi-head attention, each head of Q, K, V has the same dimension of $N \times C$, where C is the number of channels and the sequence length $N = H \times W$ for each image. It can be formulated as:

$$Attention\,(Q, K, V) = Softmax\left(\frac{QK^T}{\sqrt{d_{head}}}\right)V \qquad (1)$$

In Eq. 1, d_{head} is the dimension of the feature vector in each attention head, and T denotes the matrix transpose. The computational complexity of the transformer is $O(N^2)$, growing exponentially. Thereby, we introduce the method from [19] to downscale the sequence length to decrease its computational complexity. The query Q, K, V is processed in this paper as follows:

$$\hat{Q}, \hat{K}, \hat{V} = \text{Reshape}\left(\frac{H}{R} \times \frac{W}{R}, C \cdot R^2\right)(Q, K, V) \qquad (2)$$

The dimensions of Q, K, V are $N \times C$, where $N = H \times W$. Reshape($\frac{H}{R} \times \frac{W}{R}, C \cdot R^2$)($Q, K, V$) means reshaping Q, K, V respectively to $(\frac{H}{R} \times \frac{W}{R}) \times (C \cdot R^2)$. Then the number of channels is $C \cdot R^2$ linearly mapped into C. It can be formulated as:

$$Q, K, V = \text{Linear}(C \cdot R^2, C)(\hat{Q}, \hat{K}, \hat{V}) \qquad (3)$$

where the function Linear($C \cdot R^2, C$)($\hat{Q}, \hat{K}, \hat{V}$) means that the number of input channels is $C \cdot R^2$, and the number of output channels after linear mapping is C. The dimensions of $\hat{Q}, \hat{K}, \hat{V}$ generated by the Eq. 3 are linearly mapped and the dimension is reduced from $(\frac{H}{R} \times \frac{W}{R}) \times (C \cdot R^2)$ to $(\frac{H}{R} \times \frac{W}{R}) \times C$. This decreases the complexity from $O(N^2)$ to $O(\frac{N^2}{R^2})$, reducing the computational cost of the attention. To improve training instability, we design the stable feedforward neural network (Stable FFN) after attention. Different from MixFFN [21], a BN is added after the activation function to improve model stability, formulated as:

$$x_{out} = \text{Conv}_{1\times 1}\left(\text{GELU}\left(\text{Conv}_{3\times 3}\left(\text{BN}\left(\text{Conv}_{1\times 1}\left(x_{in}\right)\right)\right)\right)\right) + x_{in} \qquad (4)$$

In Eq. 4, x_{in} is the output of attention, $\text{Conv}_{k\times k}$ is a convolution with kernel size k, and GELU is the activation function. The depthwise convolution reduces computation, meanwhile, allowing the network to learn local information [14]. It was validated in subsequent experiments.

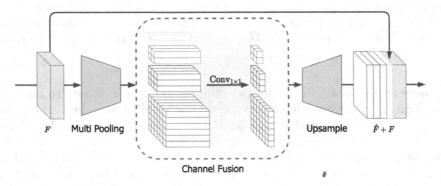

Fig. 3. Illustration of the Pyramid Pooling Module (PPM) in this paper.

Context-Enhancement Module. In deep neural networks, the receptive field size can roughly indicate the degree to which the network utilizes contextual information. For CEM, we design two structures to enhance contextual information respectively, one is a pyramid pooling based context information enhancement method, and the other is a row convolution and column convolution using asymmetric kernel to enhance the pipe context information.

First, we adopt the same configuration as the Pyramid Pooling Module (PPM) in [24], which has been empirically shown to provide an effective global context prior. As shown in Fig. 3, the input feature map F is subjected to multi-scale global average pooling with a pooling kernel size of (1, 2, 3, 6), and then up-sampled to obtain $\hat{F}$, and finally the context-enhanced features are obtained by concatenating $\hat{F}$ and F.

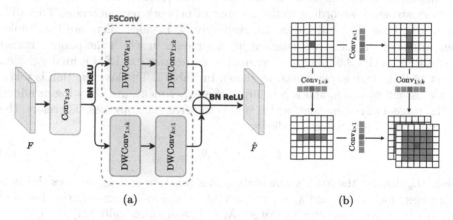

(a)   (b)

Fig. 4. Illustration of the Fully Separable Convolution Module (FSM). (a) DWConv stands for deep convolution, FSConv stands for fully separable convolution, and k is the filter size. (b) Illustration of the receptive field of the FSM.

We also designed another fully separable convolution enhancement module (FSM) to control the parameter volume growth further. We propose a solution that uses asymmetric kernels in row and column convolutions to enlarge the receptive field in each direction separately, as shown in Fig. 4. The row convolution kernel has a larger height, and the column convolution kernel has a larger width. In this way, the horizontal and vertical context information can be enhanced as much as possible. Finally, the row context information and column context information are fused to further extract the context information of targets in the image.

3.3   Contextual Fusion Decoder

We designed the Contextual Fusion Decoder (CFD) inspired by the concept of Feature Pyramid Network (FPN) [12]. To enhance semantic information, the decoder in CESegNet receives outputs from all Transformer blocks in the encoder and progressively fuses them, as shown in Fig. 1. The decoder upsamples feature maps using bilinear interpolation and receives a corresponding encoder output of the same spatial size at each level. The output channels of the decoder are N_{cls}. After going through 3 upsampling layers, we obtain 4-level feature maps F_i with the sizes of $\frac{H}{2^{i+1}} \times \frac{W}{2^{i+1}} \times N_{cls}$, where $i \in \{1, 2, 3, 4\}$. The multi-scale feature maps are upsampled to $\frac{H}{4} \times \frac{W}{4} \times N_{cls}$ and fed into the Fusion Layer for aggregation. The final feature F with the size of $\frac{H}{4} \times \frac{W}{4} \times N_{cls}$.

3.4   Model Instantiation

In this work, the configuration of the overlapping patch embedding layer follows the design strategy suggested in previous works [19,21]. We set up 4 sizes of network structure according to the number of network parameters as Tiny (T), Small (S), Base (B), and Large (L), respectively. Besides, there are two implementations of the CEM we designed in this paper, in which the pooling kernel size of PPM is (1,2,3,6), and the asymmetric kernel size k of FSM is finally chosen to set to 9 through experiments, as shown in Table 2. The loss function is taken as a cross-entropy loss, and according to the experiment, auxiliary loss is added to the stage-4 of the encoder, which is also the cross-entropy loss. Therefore, the loss function L is denoted as follows:

$$L = \lambda_s L_s + \lambda_a L_a \tag{5}$$

where L_s denotes the main segmentation loss function and L_a denotes the auxiliary loss function. λ_s and λ_a are the weights of the main segmentation loss and the auxiliary loss, respectively. We set $\lambda_s = 1$, and empirically [24] set $\lambda_a = 0.4$ in the experiments.

4   Experiments

4.1   Datasets

We use two publicly available datasets: Cityscapes [6] and ADE20K [26]. Cityscapes is a driving scene dataset for segmentation, containing 5000 finely annotated high-resolution images with 19 classes. ADE20K is a scene parsing dataset covering 150 fine-grained semantic concepts with 20210 images. Additionally, we have built the Cityscapes Noise dataset (Cityscapes-N) based on Cityscapes according to [16], which is only used to test the zero-shot robustness of the network model with a total of 15 common image noises added by us in this paper.

4.2   Implementation Details

We conduct training on servers with 1 NVIDIA A100 GPU. During the training, we applied the data enhancement to ADE20K and Cityscapes with the random resizing by 0.5–2.0 ratios, the random horizontal flipping, and the random cropping to 512×512 and 1024×1024, respectively. We trained the models using the AdamW optimizer for 160k iterations on ADE20K and Cityscapes, respectively. For the ablation study, we trained the model with only 40,000 iterations due to the limitation of the video memory, but the iteration amount was enough in the experiment. We used a batch with the size of 4 for ADE20K and 8 for Cityscapes. The initial learning rate is set to 0.00006 by default using the dynamic adjustment with a factor of 1.0. We mainly used the mean Intersection over Union (mIoU) to evaluate the semantic segmentation performance of models, and the mean Accuracy (mAcc) and the average Accuracy (aAcc) as auxiliary metrics in the ablation experiments.

Table 1. Ablation studies related to CEPPM, CEFSM and CFD. CEPPM, CEFSM by using context enhancement module PPM and FSM respectively

Backbone	Decoder	MParams↓	Cityscapes	
			GFLOPs↓	mIoU↑
MiT-b0	MLP	**3.72**	44	76.34
MiT-b0+CEPPM	MLP	6.93	47	76.46
MiT-b0+CEFSM	MLP	5.51	46	**76.64**
MiT-b0	CFD	**7.52**	232	78.00
MiT-b0+CEPPM	CFD	10.74	236	78.07
MiT-b0+CEFSM	CFD	9.31	234	**78.39**

4.3   Ablation Studies

We conducted the ablation experiments on the dataset Cityscapes for both CEM and CFD, as shown in Table 1. The table shows that CEM with PPM and FSM designed in this paper demonstrates the effectiveness, regardless of whether the decoder is CFD or MLP. When the decoder is MLP, the two modules are improved with the increasements of 0.12% and 0.30% of mIoU, respectively. When the decoder is set to CFD, they are improved with the increasements of 0.07% and 0.39% of mIoU, respectively. Table 1 verifies the effectiveness of CFD under the same backbone and also shows that the addition of CFD in the three different models is improved with the increasements of 1.66%, 1.61%, and 1.75% of mIoU, respectively. Therefore, it is validated that CEM and CFD can better extract contextual information in this paper.

For the ablation experiments of the FSM, the asymmetric filter of size k we set has 6 values respectively, and the results are shown in Table 2. The table shows that the mIoU reaches 73.81% when $k = 7$ and 9, but the aAcc and the mAcc are higher when $k = 9$, compared with $k = 7$. Therefore, we set $k = 9$. The introduced auxiliary loss helps to optimize the learning process without

Table 2. Ablation experimental results when the filter size is set to different k values.

k	Cityscapes		
	aAcc↑	mIoU↑	mAcc↑
none FSM	95.20	71.89	80.88
$k = 3$	95.29	71.88	80.42
$k = 5$	95.35	72.44	82.17
$k = 7$	95.44	**73.81**	82.45
$k = 9$	**95.51**	**73.81**	**82.82**
$k = 11$	95.46	73.8	83.02
$k = 15$	95.30	72.58	81.22

Table 3. Ablation experimental results when using auxiliary loss in the stage-i $(i = 2, 3, 4)$.

stage-i	Cityscapes		
	aAcc↑	mIoU ↑	mAcc ↑
none AL	95.37	77.01	84.28
stage-2	95.99	77.17	84.95
stage-3	96.08	77.07	84.55
stage-4	**96.17**	**78.47**	**85.75**
stage-2, 3, 4	96.10	77.44	84.88

affecting the main branch. We try adding the auxiliary loss to stage-2, 3, and 4, with results shown in Table 3. The effect enhancement is more obvious when we add the auxiliary head to stage-4, the mIoU has improved by 1.46% compared with no auxiliary head, and the aAcc and the mAcc are also better than the baseline model.

4.4   Comparison to State of the Art Methods

We compare our method with the state-of-the-art methods on the Cityscapes and ADE20K datasets. As shown in Table 4, TS as the abbreviation for Transformer Structure, shows that it is contained in the corresponding models. The

Table 4. Comparison results of the state-of-the-art methods and ours on ADE20K and Cityscapes, where TS is the abbreviation for transformer structure.

Method	Backbone	TS	MParams↓	Cityscapes		ADE20K	
				GFLOPs↓	mIoU↑	GFLOPs↓	mIoU↑
SegFormer-b3 [21]	MiT-b3	√	45.20	239	81.94	**43**	46.87
Mask2Former [4]	Swin-T	√	47.00	232	81.41	74	47.70
SETR Naïve [25]	ViT-L	√	305.67	2068	78.10	363	48.06
SETR PUP [25]	ViT-L	√	318.31	2286	79.21	426	48.24
PSPNet [24]	ResNet-101		68.10	1026	79.74	257	44.39
DeepLabv3+ [3]	ResNet-101		62.70	2032	80.01	255	44.08
PIDNet [22]	PIDNet-L		**37.31**	**138**	80.92	-	-
Ours-L	CEMiT-L	√	52.41	431	**82.21**	90	**48.77**

evaluation indicators of MParams and GFLOPs of SETRs (SETR PUT and SETR Naïve) methods are much more than that of our method, meanwhile, the improvement of our method over SETR reaches 4.11% mIoU, 3% mIoU respectively on Cityscapes dataset. Although the indicator of Mparams of our method is slightly higher than those of SegFormer and Mask2Former method methods, the indicator of mIoU ours reaches 82.21% and 48.77%, which is higher than theirs on Cityscapes and ADE20K dataset. Compared with the PSPNet and PIDNet methods, with the close indicators of the MParams, our method has only a relatively small increase in the GFLOPs but improves the indicator of mIoU by 2.47%, 1.29% respectively on Cityscapes dataset. On the ADE20k dataset, our method achieves 48.77% mIoU, ahead of the other methods in the experiments. Our method achieves the best semantic segmentation performance, proving its effectiveness and generality.

We compared our approach with SegFormer (including SegFormer-b0 and

Table 5. Comparison results of ours and SegFormer on ADE20K and Cityscapes.

Method	Backbone	MParams↓	Cityscapes		ADE20K	
			GFLOPs↓	mIoU↑	GFLOPs↓	mIoU↑
SegFormer-b0 [21]	MiT-b0	**3.72**	**44**	76.54	9	36.30
Ours-Tiny	CEMiT-Tiny	10.74	236	**78.54**	56	**38.76**
SegFormer-b3 [21]	MiT-b3	**45.20**	**239**	81.94	**43**	46.87
Ours-Large	CEMiT-Large	52.41	431	**82.21**	90	**48.77**

SegFormer-b3) in detail, as shown in Table 5. From the evaluation indices of MParams and GFLOPs for our methods (Ours-Tiny and Ours-Large), although the index value in each comparison group is slightly larger than that of the corresponding SegForme-b0 or SegFormer-b3, these reveal that their network backbone structure configurations are close overall in the respective comparison groups. From this point of view, it is reasonable to believe that our method exceeds SegFormer due to the improvements in the model accuracy of mIoU on both public datasets after we embedded the CEM and CFD contextual information augmentation into the transformer backbone. Our method has a much lower parameter growth rate compared to SegFormer. Meanwhile, our method outperforms SegFormer in both public datasets, with mIoU improvements of 2%, 2.46%, 0.65%, and 1.9% respectively. Our models achieve better segmentation quality.

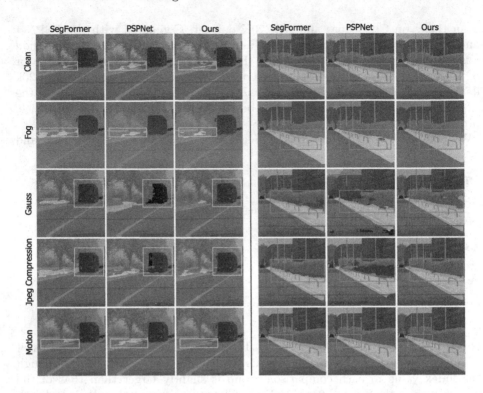

Fig. 5. Zero-shot robustness comparison results on Cityscapes. Our method has a more stable performance in noise image sets

4.5   Robustness Studies

We demonstrate the excellent robustness of our method by evaluating indicator mIoU on the public Cityscapes dataset in the experiment, in which we construct the Cityscapes-N dataset by referring to the method in [16] to imitate real-world noise images. The dataset consists of 15 typical image noises, as shown in Table 6. From the results shown in Table 4, we compared our method with SegFormer-b3 (with the transformer structure) and the methods (without the transformer structure) including PSPNet, PIDNet, and DeepLabv3+, respectively, on the Cityscapes-N dataset designed in this paper. The experimental results are shown in Table 6. We find that our method achieves the leading mIoU values on all noisy image sets except for IM noisy image set, compared with SegFormer-b3; our method also achieves larger mIoU values than those of those methods (PSPNet, PIDNet, and DeepLabv3+) on all of the noisy image sets except for PI digital image set. These are sufficient to show a good robustness of our method. Figure 5 shows the qualitative results. Through enlarging the local images details marked by red, yellow, and blue rectangles, it is obvious that our method can segment accurately in relatively complex environments compared with SegFormer and PSPNet. This reveals our method has more stable and accurate segmentation effects on the corrupted images than the other methods in the experiment, and our method's excellent zero-shot robustness greatly improves its practicability and adaptability in the real world.

Table 6. Demonstration of robustness results on the Cityscapes-N dataset where the abbreviations for gauss, shot, impulse, defocus, glass, motion, zoom, snow, frost, fog, bright, contrast, elastic, pixel, and jpeg for various noises added by referring to [16]. The 'SF' and 'DLv3+' refer to Segformer and DeepLabv3+. The evaluation indicator is mIoU in the table.

Method	Clean	Noise			Blur				Weather				Digital			
		GA	SH	IM	DE	GL	MO	ZO	SN	FR	FO	BR	CO	EL	PI	JP
SF-b3	81.9	63.4	67.9	**62.6**	77.9	73.1	78.7	36.1	59.3	**69.5**	79.9	81.9	**81.6**	81.2	77.4	68.0
PSPNet	79.7	10.4	17.8	16.9	76.8	69.1	76.4	27.8	31.7	55.1	77.2	79.4	76.8	79.0	79.3	37.1
DLv3+	80.1	26.4	36.4	17.9	77.0	72.1	77.1	29.2	34.8	57.6	76.9	79.8	78.5	79.3	79.6	48.6
PIDNet	80.8	12.7	22.4	14.0	78.1	70.5	77.8	29.1	25.8	59.6	77.8	80.5	78.4	80.1	**79.8**	54.4
Ours-L	**82.2**	**65.3**	**70.1**	62.1	**78.4**	**75.4**	**79.1**	**38.3**	**60.5**	69.5	**80.2**	**82.1**	81.6	**81.6**	79.0	**70.1**

5   Conclusions

In this work, we propose an efficient semantic segmentation network with excellent robustness. We designed a network based on a hierarchical simplified transformer by combining CEM encoder and CFD. The corresponding experiments validate that our method outperforms state-of-the-art methods on two public datasets. It is demonstrated that our method has held effective enhancing effects on the contextual information, and the good robustness, meanwhile, effectively controlling the growth of the network parameters. Besides, our method exhibits strong zero-shot robustness on the Cityscapes-N dataset, which will significantly enhance the model's utility in the real world.

Acknowledgements. This work is supported by the Major Special Project of the Inner Mongolia Autonomous Region (No. 2021SZD0043).

References

1. Badrinarayanan, V., Kendall, A., Cipolla, R.: SegNet: a deep convolutional encoder-decoder architecture for image segmentation. IEEE Trans. Pattern Anal. Mach. Intell. **39**(12), 2481–2495 (2017). https://doi.org/10.1109/TPAMI.2016. 2644615
2. Chen, L.C., Papandreou, G., Schroff, F., Adam, H.: Rethinking atrous convolution for semantic image segmentation. arXiv preprint arXiv:1706.05587 (2017)
3. Chen, L.-C., Zhu, Y., Papandreou, G., Schroff, F., Adam, H.: Encoder-decoder with atrous separable convolution for semantic image segmentation. In: Ferrari, V., Hebert, M., Sminchisescu, C., Weiss, Y. (eds.) ECCV 2018. LNCS, vol. 11211, pp. 833–851. Springer, Cham (2018). https://doi.org/10.1007/978-3-030-01234-2_49

4. Cheng, B., Misra, I., Schwing, A.G., Kirillov, A., Girdhar, R.: Masked-attention mask transformer for universal image segmentation. In: Proceedings of the IEEE/CVF Conference on Computer Vision and Pattern Recognition, pp. 1290–1299 (2022). https://doi.org/10.1109/cvpr52688.2022.00135

5. Chu, X., et al.: Conditional positional encodings for vision transformers. arXiv preprint arXiv:2102.10882 (2021)

6. Cordts, M., et al.: The cityscapes dataset for semantic urban scene understanding. In: Proceedings of the IEEE Conference on Computer Vision and Pattern Recognition, pp. 3213–3223 (2016)

7. Deng, J., Dong, W., Socher, R., Li, L.J., Li, K., Fei-Fei, L.: ImageNet: a large-scale hierarchical image database. In: 2009 IEEE Conference on Computer Vision and Pattern Recognition, pp. 248–255. IEEE (2009)

8. Dosovitskiy, A., et al.: An image is worth 16×16 words: transformers for image recognition at scale. arXiv preprint arXiv:2010.11929 (2020)

9. Fu, J., et al.: Dual attention network for scene segmentation. In: Proceedings of the IEEE/CVF Conference on Computer Vision and Pattern Recognition, pp. 3146–3154 (2019). https://doi.org/10.1109/cvpr.2019.00326

10. He, K., Gkioxari, G., Dollár, P., Girshick, R.: Mask R-CNN. In: Proceedings of the IEEE International Conference on Computer Vision, pp. 2961–2969 (2017)

11. He, K., Zhang, X., Ren, S., Sun, J.: Deep residual learning for image recognition. In: Proceedings of the IEEE Conference on Computer Vision and Pattern Recognition, pp. 770–778 (2016)

12. Lin, T.Y., Dollár, P., Girshick, R., He, K., Hariharan, B., Belongie, S.: Feature pyramid networks for object detection. In: Proceedings of the IEEE Conference on Computer Vision and Pattern Recognition, pp. 2117–2125 (2017). https://doi.org/10.1109/cvpr.2017.106

13. Liu, Z., et al.: Swin transformer: hierarchical vision transformer using shifted windows. arXiv preprint arXiv:2103.14030 (2021). https://doi.org/10.1109/iccv48922.2021.00986

14. Liu, Z., Mao, H., Wu, C.Y., Feichtenhofer, C., Darrell, T., Xie, S.: A convnet for the 2020s. In: Proceedings of the IEEE/CVF Conference on Computer Vision and Pattern Recognition, pp. 11976–11986 (2022). https://doi.org/10.1109/cvpr52688.2022.01167

15. Long, J., Shelhamer, E., Darrell, T.: Fully convolutional networks for semantic segmentation. In: Proceedings of the IEEE Conference on Computer Vision and Pattern Recognition, pp. 3431–3440 (2015). https://doi.org/10.1109/cvpr.2015.7298965

16. Michaelis, C., et al.: Benchmarking robustness in object detection: autonomous driving when winter is coming. arXiv preprint arXiv:1907.07484 (2019)

17. Ronneberger, O., Fischer, P., Brox, T.: U-Net: convolutional networks for biomedical image segmentation. In: Navab, N., Hornegger, J., Wells, W.M., Frangi, A.F. (eds.) MICCAI 2015. LNCS, vol. 9351, pp. 234–241. Springer, Cham (2015). https://doi.org/10.1007/978-3-319-24574-4_28

18. Simonyan, K., Zisserman, A.: Very deep convolutional networks for large-scale image recognition. arXiv preprint arXiv:1409.1556 (2014)

19. Wang, W., et al.: Pyramid vision transformer: a versatile backbone for dense prediction without convolutions. In: Proceedings of the IEEE/CVF International Conference on Computer Vision, pp. 568–578 (2021). https://doi.org/10.1109/iccv48922.2021.00061

20. Wu, H., Zhang, J., Huang, K., Liang, K., Yu, Y.: FastFCN: rethinking dilated convolution in the backbone for semantic segmentation. arXiv preprint arXiv:1903.11816 (2019)

21. Xie, E., Wang, W., Yu, Z., Anandkumar, A., Alvarez, J.M., Luo, P.: SegFormer: simple and efficient design for semantic segmentation with transformers. Adv. Neural. Inf. Process. Syst. **34**, 12077–12090 (2021)

22. Xu, J., Xiong, Z., Bhattacharyya, S.P.: PIDNet: a real-time semantic segmentation network inspired by PID controllers. In: Proceedings of the IEEE/CVF Conference on Computer Vision and Pattern Recognition, pp. 19529–19539 (2023). https://doi.org/10.1109/cvpr52729.2023.01871

23. Yu, C., Wang, J., Gao, C., Yu, G., Shen, C., Sang, N.: Context prior for scene segmentation. In: Proceedings of the IEEE/CVF Conference on Computer Vision and Pattern Recognition, pp. 12416–12425 (2020). https://doi.org/10.1109/cvpr42600.2020.01243

24. Zhao, H., Shi, J., Qi, X., Wang, X., Jia, J.: Pyramid scene parsing network. In: Proceedings of the IEEE Conference on Computer Vision and Pattern Recognition, pp. 2881–2890 (2017). https://doi.org/10.1109/cvpr.2017.660

25. Zheng, S., et al.: Rethinking semantic segmentation from a sequence-to-sequence perspective with transformers. In: Proceedings of the IEEE/CVF Conference on Computer Vision and Pattern Recognition, pp. 6881–6890 (2021). https://doi.org/10.1109/cvpr46437.2021.00681

26. Zhou, B., Zhao, H., Puig, X., Fidler, S., Barriuso, A., Torralba, A.: Scene parsing through ADE20K dataset. In: Proceedings of the IEEE Conference on Computer Vision and Pattern Recognition, pp. 633–641 (2017)

MoCap-Video Data Retrieval with Deep Cross-Modal Learning

Lu Zhang[1,2], Jingliang Peng[1,2], and Na Lv[1,2(✉)]

[1] Shandong Provincial Key Laboratory of Network Based Intelligent Computing,
University of Jinan, Jinan, China
[2] School of Information Science and Engineering, University of Jinan, Jinan, China
ise_lvn@ujn.edu.cn

Abstract. Cross-modal retrieval between video and motion capture (MoCap) data facilitates efficient reuse of human motion data in either skeletal or video format. For this purpose, we propose a deep cross-modal learning model for cross-modal retrieval between MoCap data and video data. First, we use a graph convolution-based network and a 3D convolution-based network to extract features from MoCap data and video data, respectively. In addition, we propose to use a pre-defined common subspace to maximize the inter-class variation and minimize the intra-class variation. Furthermore, we employ a similarity matrix to achieve the alignment between these two modalities and exploit their underlying correlations. For the purpose of experimental evaluation, due to the small amount of video data corresponding to the MoCap data in the public HDM05 dataset, we recorded a video dataset corresponding to the HDM05 motion capture dataset and performed cross-modal retrieval on it. The experimental results proved the effectiveness of the proposed scheme.

Keywords: Motion capture data · Video data · Cross modal retrieval · Common subspace · Similarity matrix

1 Introduction

The 3D motion capture (MoCap) technology has been widely utilized in diverse domains, including animation production, game development and sports training. Realistic simulation of human motion is required in these domains, and motion capture has emerged as a dominant approach for this purpose.

Acquiring MoCap data requires expensive equipment and specialized expertise. In addition, the capture process is usually time-consuming. Therefore, it is essential to effectively reuse the existing MoCap data instead of capturing all motions from the scratch whenever they are needed. To effectively enable the reuse of motion data, it is desirable to efficiently search from a MoCap repository in a user-friendly manner.

In recent years, numerous methods have been proposed for MoCap data retrieval, which can be broadly categorized into two main approaches: text

© The Author(s), under exclusive license to Springer Nature Switzerland AG 2024
S. Rudinac et al. (Eds.): MMM 2024, LNCS 14555, pp. 494–506, 2024.
https://doi.org/10.1007/978-3-031-53308-2_36

label-based methods and content-based methods. The text label-based method requires significant manual effort to annotate all the motion sequences in the database. This process consumes a considerable amount of time and human resources, resulting in inefficiency. In addition, text labels may not provide sufficient descriptive information. During the retrieval process, accurately describing a motion segment with a certain level of complexity by only text labels can be challenging. This highlights the need for individuals to engage in extensive research on content-based MoCap data retrieval methods. So many algorithms have been proposed for content-based MoCap data retrieval. Researchers have used a variety of query modes, including MoCap clips [17,24], hand-drawn sketches [23], puppet actions [20], and Kinect skeleton actions [6]. while these approaches have shown promising results, they also come with inherent drawbacks. For example, the quality and style of hand-drawn sketches can have a significant impact on retrieval performance. The construction and execution of a puppet device can be challenging and complex. In addition, Kinect-based approaches often impose limitations on the users, restricting them to a limited range of distances and orientations to achieve optimal results.

Among the various modalities for representing human motion, video is probably the most widely used one. Video is easy to acquire, and the availability of video data is higher than that of MoCap data. Therefore, it is convenient to search a MoCap repository with a video query. Since a video sequence (either captured from the real world or rendered from a sequence of 3D models) visualizes human motion in a more concrete form than a MoCap sequence, searching for similar motions in a video repository with a MoCap query can help in the conception of film, animation *etc.* In addition, motion capture data contains only the motion information of the joints, when it is used to retrieve video data, it can make the retrieval results more focused on the motion itself, regardless of the background, illumination, *etc.* Meanwhile, it enables a deeper investigation of human body movement and posture, allowing for a comprehensive understanding of the context and applications of these movements.

Observing that cross-modal retrieval between MoCap data and video data is highly demanded, we are motivated to solve this problem in this work. And our method is primarily applied in single-player sports scenarios. For this purpose, we propose a deep learning based model in this work and our main contributions are summarized as follows.

- The proposed model is the first to be published that systematically addresses cross-modal retrieval between human motion video and MoCap data.
- An original and complete model based on deep learning is designed. Graph convolution and 3D convolution are used for feature extraction, and a predefined common subspace and a similarity matrix are used for feature embedding and cross-modal alignment.
- A video dataset corresponding to the HDM05 MoCap dataset has been created and will be published for research purposes.

2 Related Work

Cross-modal retrieval is mainly divided into three steps: the first step is feature extraction of multi-modal data, the second step focuses on cross-modal correlation modeling to learn the common representation of various modes of data, and the third step is to achieve cross-modal retrieval through appropriate search results sorting and indexing of search results. Specifically, queries are mapped to a common representation space, and cosine similarity is used to evaluate the similarity of samples, thereby ranking the retrieval results.

2.1 Feature Extraction of Multi-modal Data

Efficient representation of MoCap data is critical in tasks involving the retrieval of cross-modal data. Recently, there has been a surge in the use of deep learning models, such as Convolutional Neural Networks (CNNs) [2,13], Recurrent Neural Networks (RNNs) [10,26], and Graph Convolutional Networks (GCNs), to study motion characteristics. GCNs, due to its adeptness in representing the natural connection of the skeleton, has become a popular choice for skeleton-based action classification tasks. HD-GCN [9], an advanced GCN model, is exceptional in its capacity to partition each joint node into numerous groups, allowing the identification of meaningful nearby and far-reaching edges. These edges are then used to create an HD-Graph that encompasses the same semantic space as a human skeleton. This method allows for a more complete characterization and understanding of the underlying structure of skeletal data.

Optical flow is a powerful method for representing the motion of scenes or objects. Simonyan *et al.* [22] proposed two-stream networks consisting of a spatial stream and a temporal stream to optimize its effectiveness. The final prediction is derived by calculating the average of the scores obtained from both streams. However, the pre-computation of the optical flow requires considerable computational resources, which leads to the use of 3DCNNs as efficient processing units that efficiently model temporal information. Xie *et al.* [25] designed an S3D architecture that achieved superior performance. With the proposal of Transformer, Video Transformer Network [19] and Video Swin Transformer [15] have attained impressive recognition accuracy. However, their computational complexity appears to be high.

2.2 Cross-Modal Correlation Modeling

Methods for cross-modal correlation modeling can be divided into two main categories: retrieval methods based on binary values and retrieval methods based on real values.

Binary-value-based methods [3,12,14], typically represent the modalities using binary codes. These methods aim to learn compact binary representations for each modality and establish cross-modal similarity measure based on these binary codes. However, the structure of the original data will be destroyed in the process of binary transformation of the real-valued features of the data, resulting in a loss of accuracy.

On the other hand, real-value-based retrieval methods [1,4,16], rely on learning real-valued representations for different modalities. These methods leverage deep learning models to extract rich and discriminative features from the input data. Real-value-based retrieval methods often achieve superior performance compared to binary-value methods, as they can capture more fine-grained similarities between modalities.

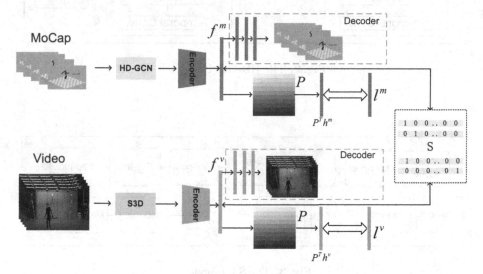

Fig. 1. The flow chart of our proposed approach. Our retrieval framework mainly consists of auto-encoders, similarity matrix **S** and matrix **P**. The MoCap features and video features are extracted by HD-GCN [9] model and S3D model [25], respectively. The extracted features from each part are individually fed into their respective encoders and then projected into a common feature space defined by the matrix **P**. Then the data of different modalities can be compared and sorted in the common space.

3   Method

In this section, we present the details of our MoCap-Video data cross-modal retrieval (MoViCR) method, including the model formulation and the learning algorithm. From Fig. 1, we can see that each modality consists of three parts: an encoder, a decoder, and a supervised label projection. Furthermore, the correspondence between MoCap data and video data is established by computing the similarity matrix **S** between them.

The goal of deep cross-modal learning is to obtain transformations that can efficiently project data from different modalities into a predefined common subspace. Data from different modalities, when projected into this common subspace, can be made to maximize the inter-class distance and minimize the intra-class distance to better distinguish and represent multi-modal data from different classes.

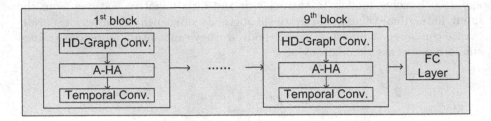

Fig. 2. The HD-GCN model

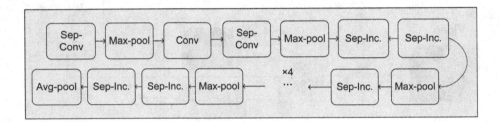

Fig. 3. The S3D model

3.1 Problem Definition

In our approach, we denote the MoCap modality as X and the video modality as Y. Let $X = \{x_i\}_{i=1}^{N_x}$ and $Y = \{y_j\}_{j=1}^{N_y}$ denote the input MoCap and video, respectively. Furthermore, let $(x_i, y_i)_{i=1}^{N_s}$ be the matched MoCap-video pairs, where (x_l, y_l) is a MoCap-video pair with the same category labels. The MoCap features are extracted using the HD-GCN model [9], while the video features are extracted using the S3D model [25]. The structures of the two models are shown in Fig. 2 and 3, respectively.

Cross-modal learning involves learning modality-specific transform functions for different modalities into a common subspace, which can then be directly compared for similarity between them. In this paper, we use $f_x(x_i, \theta_i) \in R^d$ and $f_y(y_j, \theta_j) \in R^d$, where d is the dimensionality of the features in the common subspace, to denote the transform functions of MoCap and video data, respectively, where θ_i and θ_j are the parameters to be learned.

3.2 Deep Auto-encoder

We use deep auto-encoders for each modality to learn the transform function. The primary goal of the auto-encoder is to learn a concise representation of the input data by compressing it into a lower-dimensional space and then reconstructing the original data. To improve the deep auto-encoder module, we introduce a supervised loss function at the representation layer. As proposed in [8], training an auto-encoder to reconstruct the input allows us to extract underlying features that can significantly improve prediction accuracy through supervision.

Denoting the encoder by $h^m = f_x(x_i, \theta_i)$ and $h^v = f_y(y_j, \theta_j)$, and the decoder by $\widehat{x} = g_x(h_m, \Phi_i)$ and $\widehat{y} = g_y(h_v, \Phi_j)$, we formulate the objective function of the deep auto-encoder as follows:

$$L_{res} = L_{res_x} + L_{res_y} = \sum_{i=1}^{N_x} \sum_{j=1}^{N_y} (\alpha_1 \|\widehat{x}_i - x_i\|_2 + \alpha_2 \|\widehat{y}_j - y_j\|_2), \qquad (1)$$

where $\widehat{x}_i$ and $\widehat{y}_i$ are the outputs of the MoCap decoder and the video decoder, respectively, x_i and y_i are the inputs of the MoCap encoder and the video encoder, respectively, and N_x and N_y are the total number of samples for the MoCap and video, respectively, α_1 and α_2 are hyperparameters.

3.3 Predefined Common Subspace

The key to our proposed MoViCR is that it predefines a common subspace by a fixed matrix $\mathbf{P}$. Through the matrix $\mathbf{P}$, the representations of the samples are projected into the label space. The additional supervised loss is derived from the label information to maximize the discrimination within the predefined common subspace, and is used to guide the learning of the networks.

Thus, we formulate the objective function of the predefined common space as follows:

$$L_{\mathbf{P}} = L_{\mathbf{P}_x} + L_{\mathbf{P}_y} = \sum_{i=1}^{N_x} \sum_{j=1}^{N_y} (\beta_1 \|P^T h_i^m - l_i^m\|_2 + \beta_2 \|P^T h_j^v - l_j^v\|_2), \qquad (2)$$

where l^m and l^v are the semantic label vectors for their respective modalities, h^m and h^v have the same meanings as in Formula (1), and β_1 and β_2 are hyperparameters.

Suppose the matrix $\mathbf{P}$ has r rows and c columns, where r is the number of encoder output units and c is the number of semantic categories. This ensures compatibility and alignment between the corresponding encoder outputs and the semantic categories. The matrix $\mathbf{P}$ is specifically configured as an orthogonal matrix with orthonormal columns. This configuration ensures that the one-dimensional subspaces corresponding to different categories are orthogonal to each other, thereby enhancing the discriminative nature of the predefined common subspace.

3.4 Similarity Matrix

As mentioned above, let $f_x(x_i, \theta_i) \in R^d$ denote the learned feature for MoCap data x_i, which corresponds to the output of the HD-GCN for MoCap modality. Similarly, $f_y(y_j, \theta_j) \in R^d$ denotes the learned video feature for video data y_j, which corresponds to the output of the S3D for video modality. Here, θ_i and θ_j are the network parameters for MoCap and video modality, respectively.

We provide a cross-modal similarity matrix $\mathbf{S}$. If MoCap data x_i and video data y_j have the same class label, then $\mathbf{S}_{ij}=1$; conversely, if MoCap data x_i and video data y_j have different class labels, then $\mathbf{S}_{ij}=0$. The objective function of this module is defined as follows:

$$L_{\mathbf{S}} = -\sum_{i=1}^{N_x}\sum_{j=1}^{N_y}\left(S_{ij}\Theta_{ij} - \log\left(1 + e^{\Theta_{ij}}\right)\right), \tag{3}$$

where $\Theta_{ij} = \frac{1}{2}\left(h_i^m\right)^T\left(h_j^\nu\right)$. This formula above is the negative log likelihood of the cross-modal similarities with the likelihood function defined as follows:

$$p_{ij}\left(\mathbf{S}_{ij}|h_i^m h_j^\nu\right) = \begin{cases} \sigma\left(\Theta_{ij}\right) & \mathbf{S}_{ij} = 1, \\ 1 - \sigma\left(\Theta_{ij}\right) & \mathbf{S}_{ij} = 0. \end{cases} \tag{4}$$

where $\sigma\left(\Theta_{ij}\right) = \dfrac{1}{1 + e^{-\Theta_{ij}}}$.

It is easy to see that minimizing this negative log-likelihood in Formula (3), which is equivalent to maximizing the likelihood in Formula (4). It will make the similarity (inner product) between h_i^m and h_j^ν be large when $\mathbf{S}_{ij} = 1$ and be small when $\mathbf{S}_{ij} = 0$. Therefore, optimizing this term in (3) can preserve the cross-modal similarity in $\mathbf{S}$ with the MoCap feature representation h^m and the video feature representation h^ν.

3.5 Loss Function

Combining the reconstruction loss L_{res}, the supervised loss $L_\mathbf{P}$, and the similarity loss $L_\mathbf{S}$, the final objective function L is obtained as,

$$L = L_{res} + L_\mathbf{P} + L_\mathbf{S} \tag{5}$$

3.6 Implementation Details

The entire alternating learning algorithm for MoViCR is briefly outlined in Algorithm 1. In the test phase, given a query in one modality (a MoCap query or a video query), the goal of the MoCap-Video cross-modal data retrieval task is to find the most similar sample from the database in another modality. The decoder is ignored and the outputs of the encoder are the common representations of the samples. The cosine distance between these representations is used as the similarity metric for cross-modal retrieval.

Algorithm 1. The learning algorithm for MoViCR

Input: MoCap set $\mathbf{X}$, Video set $\mathbf{Y}$, cross-modal similarity matrix $\mathbf{S}$, and the matrix $\mathbf{P}$.

Output: Parameters θ_i and θ_j of the deep neural networks

1: Initialize neural network parameters θ_i and θ_j, mini-batch size $n_b = 128$, and iteration number $t_x = \lceil n_b/N_x \rceil, t_y = \lceil n_b/N_y \rceil$.

2: **repeat**

3:    **for** $iter = 1, 2, ..., t_x$ **do**

      Randomly sample N_x points from $\mathbf{X}$ to construct a mini-batch.

      Compute the representations h_i^m and $\widehat{x}_i$ for the samples in the mini-batch by forward-propagation.

      Calculate the loss for the MoCap modal neural network using the Loss Function.

      Update the parameter θ_i using back propagation.

4:    **end for**

5:    **for** $iter = 1, 2, ..., t_y$ **do**

      Randomly sample N_y points from $\mathbf{Y}$ to construct a mini-batch.

      Compute the representations h_j^v and $\widehat{x}_j$ for the samples in the mini-batch by forward-propagation.

      Calculate the loss for the video modal neural network with the Loss Function.

      Update the parameter θ_j using back propagation.

6:    **end for**

7: **until** a fixed number of iterations

---

4 Experiment

The proposed MoViCR model is trained on NVIDIA GeForce RTX 3090 GPU using PyTorch. We divided the data into training set, validation set, and test set in the ratio of 8:1:1. On the first two datasets, we trained and validated our proposed MoViCR. For training, we used the ADAM [7] optimizer with a learning rate of 10^{-4} and set the maximum number of epochs to 200. Then, the retrieval performance of the test set is evaluated.

4.1 Dataset and Features

To evaluate the proposed method, experiments were conducted on the HDM05 dataset [18]. The HDM05 dataset consists of 130 categories of MoCap data recorded by five individuals. However, the semantics of some motion categories are similar, such as walking 3 steps and walking 1 step. To eliminate the possible effect of repeated actions, we finally grouped them into 13 categories. Due to the limited amount of video data corresponding to MoCap in the HDM05 dataset, we provide an additional recorded video dataset as a supplement, including action categories such as sneak, squat, stand up, walk.

For the MoCap data, we normalized the skeleton and the length of the motion sequence. Then we fed the motion trajectory information of 31 joints into the HD-GCN model to extract the 256-dimensional outputs as the MoCap features. For the video data, we performed center clipping, normalization, and finally extracted 7,168-dimensional feature vectors by S3D network as the video features.

4.2 Evaluation Metric

To evaluate the performance of the method, we perform the cross-modal retrieval tasks as querying the data of one modality to retrieve the same semantic data of another modality, such as retrieving video by MoCap query (MoCap → Video) and retrieving MoCap by video query (Video → MoCap). The commonly used mean average precision (mAP) and precision-recall curve (P-R curve) are used as the performance evaluation. To compute mAP, we first evaluate the average precision (AP) of a set of R retrieved results:

$$AP = \frac{1}{R} \sum_{i=1}^{n} p(i) \, \delta(i), \tag{6}$$

where R is the number of relevant results in the retrieved set, and $p(i)$ denotes the precision of the top i retrieved results, $\delta(i) = 1$ if the ith retrieved result is relevant (where 'relevant' means belonging to the class of the query) and $\delta(i) = 0$ otherwise.

Table 1. The ablation experiment results. Best results are denoted in boldface.

Method	Video → MoCap	MoCap → Video	Average
MoViCR (without L_{res})	0.676	0.581	0.629
MoViCR (without L_P)	0.697	0.605	0.651
MoViCR (without L_S)	0.776	0.736	0.756
MoViCR	**0.860**	**0.781**	**0.821**

4.3 Ablation Study

To explore the effectiveness of different components in our model, we conducted the ablation experiment. As shown in Table 1, we compare the performance of our model with the following three ablated versions: first, training the MoViCR model without the reconstruction loss, L_{res}; second, training the MoViCR model without the predefined common subspaces; third, evaluating the model by removing the similarity matrix from the architecture. These ablation results show that each part of our model is essential for cross-modal retrieval of MoCap and video data.

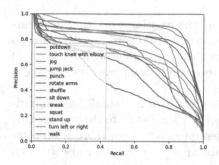

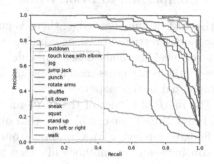

Fig. 4. P-R curves of our method for MoCap-Video data retrieval for all the action classes.

Fig. 5. P-R curves of our method for Video-MoCap data retrieval for all the action classes.

Table 2. Performance comparison in terms of mAP scores. Best results are denoted in boldface.

Method	Video → MoCap	MoCap → Video	Average
NDVP [11]	0.568	-	
Jiang et al.'s method [5]	0.727	-	
ours (HDM05)	**0.860**	**0.781**	**0.821**
ours (NTU-RGB+D)	0.813	0.745	0.779

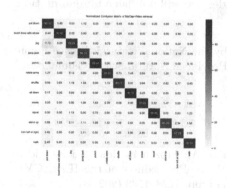

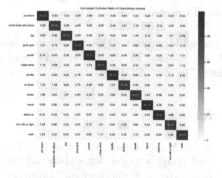

Fig. 6. The confusion matrix of MoCap-Video data retrieval.

Fig. 7. The confusion matrix of Video-MoCap data retrieval.

4.4 Comparison to Prior Work

To the best of our knowledge, there is currently no existing work that performs cross-modal retrieval between MoCap and video data. There are two MoCap data retrieval methods based on video data [5,11]. Considering that both of these methods are cross-modal MoCap data retrieval, a comparative experiment is conducted in this paper.

The mAP statistics of the three methods are shown in Table 2 for comparison. From Table 2, we can observe that our method is superior to the two existing methods in retrieving MoCap data from video data. Although there is no comparison between the methods of video to MoCap retrieval, it can also be seen that the effect of our method is significant. And our MoViCR method achieves a retrieval time under 50 milliseconds. To guarantee scalability, we conducted experiments on the NTU-RGB+D [21] partial video dataset for cross-modal retrieval, with the results displayed in Table 2.

For this experiment, we obtain P-R curves over all the action classes of the MoCap-Video data retrieval and the Video-MoCap data retrieval in Fig. 4 and Fig. 5. The observed variations in the P-R curves can be attributed to the different levels of difficulty in distinguishing and retrieving specific action classes. We also obtained the confusion matrix for the top 10% of retrieved results, as shown in Fig. 6 and Fig. 7.

5 Conclusion

We have presented a novel deep cross-modal learning approach for cross-modal retrieval of MoCap data and video data. After extracting the respective modal features, we use a deep cross-modal learning framework to facilitate the alignment of the two modal features. We are the first to systematically address cross-modal retrieval between MoCap data and video data, and our proposed method yields significant and promising results. Besides, we will publish a human motion video dataset for research uses.

Acknowledgement. This work is supported by the National Natural Science Foundation of China (No. 61802144) and Shandong Provincial Natural Science Foundation, China (No. ZR2022MF294).

References

1. Bain, M., Nagrani, A., Varol, G., Zisserman, A.: Frozen in time: a joint video and image encoder for end-to-end retrieval. In: Proceedings of the IEEE/CVF International Conference on Computer Vision, pp. 1728–1738 (2021)
2. Caetano, C., Sena, J., Brémond, F., Dos Santos, J.A., Schwartz, W.R.: SkeleMotion: a new representation of skeleton joint sequences based on motion information for 3D action recognition. In: 2019 16th IEEE International Conference on Advanced Video and Signal Based Surveillance (AVSS), pp. 1–8. IEEE (2019)

3. Gu, W., Gu, X., Gu, J., Li, B., Xiong, Z., Wang, W.: Adversary guided asymmetric hashing for cross-modal retrieval. In: Proceedings of the 2019 on International Conference on Multimedia Retrieval, pp. 159–167 (2019)
4. He, K., Fan, H., Wu, Y., Xie, S., Girshick, R.: Momentum contrast for unsupervised visual representation learning. In: Proceedings of the IEEE/CVF Conference on Computer Vision and Pattern Recognition, pp. 9729–9738 (2020)
5. Jiang, Z., Li, Z., Li, W., Li, X., Peng, J.: Generic video-based motion capture data retrieval. In: 2019 Asia-Pacific Signal and Information Processing Association Annual Summit and Conference (APSIPA ASC), pp. 1950–1957. IEEE (2019)
6. Kapadia, M., Chiang, I., Thomas, T., Badler, N.I., Kider, J.T., Jr.: Efficient motion retrieval in large motion databases. In: Proceedings of the ACM SIGGRAPH Symposium on Interactive 3D Graphics and Games, pp. 19–28 (2013)
7. Kingma, D.P., Ba, J.: Adam: a method for stochastic optimization. arXiv preprint arXiv:1412.6980 (2014)
8. Le, L., Patterson, A., White, M.: Supervised autoencoders: improving generalization performance with unsupervised regularizers. In: Advances in Neural Information Processing Systems, vol. 31 (2018)
9. Lee, J., Lee, M., Lee, D., Lee, S.: Hierarchically decomposed graph convolutional networks for skeleton-based action recognition. arXiv preprint arXiv:2208.10741 (2022)
10. Li, L., Zheng, W., Zhang, Z., Huang, Y., Wang, L.: Skeleton-based relational modeling for action recognition. arXiv preprint arXiv:1805.02556 (2018)
11. Li, W., Huang, Y., Kuo, C.C.J., Peng, J., et al.: Video-based human motion capture data retrieval via normalized motion energy image subspace projections. In: 2017 IEEE International Conference on Multimedia & Expo Workshops (ICMEW), pp. 243–248. IEEE (2017)
12. Li, X., Hu, D., Nie, F.: Deep binary reconstruction for cross-modal hashing. In: Proceedings of the 25th ACM International Conference on Multimedia, pp. 1398–1406 (2017)
13. Li, Y., Xia, R., Liu, X., Huang, Q.: Learning shape-motion representations from geometric algebra spatio-temporal model for skeleton-based action recognition. In: 2019 IEEE International Conference on Multimedia and Expo (ICME), pp. 1066–1071. IEEE (2019)
14. Li, Z., Guo, C., Feng, Z., Hwang, J.N., Jin, Y., Zhang, Y.: Image-text retrieval with binary and continuous label supervision. arXiv preprint arXiv:2210.11319 (2022)
15. Liu, Z., et al.: Video swin transformer. In: Proceedings of the IEEE/CVF Conference on Computer Vision and Pattern Recognition, pp. 3202–3211 (2022)
16. Luo, H., et al.: CLIP4Clip: an empirical study of clip for end to end video clip retrieval and captioning. Neurocomputing **508**, 293–304 (2022)
17. Lv, N., Jiang, Z., Huang, Y., Meng, X., Meenakshisundaram, G., Peng, J.: Generic content-based retrieval of marker-based motion capture data. IEEE Trans. Vis. Comput. Graph. **24**(6), 1969–1982 (2017)
18. Müller, M., Röder, T., Clausen, M., Eberhardt, B., Krüger, B., Weber, A.: Mocap database HDM05. Institut für Informatik II, Universität Bonn **2**(7) (2007)
19. Neimark, D., Bar, O., Zohar, M., Asselmann, D.: Video transformer network. In: Proceedings of the IEEE/CVF International Conference on Computer Vision, pp. 3163–3172 (2021)
20. Numaguchi, N., Nakazawa, A., Shiratori, T., Hodgins, J.K.: A puppet interface for retrieval of motion capture data. In: Proceedings of the 2011 ACM SIGGRAPH/Eurographics Symposium on Computer Animation, pp. 157–166 (2011)

21. Shahroudy, A., Liu, J., Ng, T.T., Wang, G.: NTU RGB+D: a large scale dataset for 3d human activity analysis. In: 2016 IEEE Conference on Computer Vision and Pattern Recognition (CVPR), pp. 1010–1019. IEEE Computer Society (2016)
22. Simonyan, K., Zisserman, A.: Two-stream convolutional networks for action recognition in videos. In: Advances in Neural Information Processing Systems, vol. 27 (2014)
23. Xiao, J., Tang, Z., Feng, Y., Xiao, Z.: Sketch-based human motion retrieval via selected 2D geometric posture descriptor. Sig. Process. **113**, 1–8 (2015)
24. Xiao, Q., Siqi, L.: Motion retrieval based on dynamic Bayesian network and canonical time warping. Soft. Comput. **21**, 267–280 (2017)
25. Xie, S., Sun, C., Huang, J., Tu, Z., Murphy, K.: Rethinking spatiotemporal feature learning: speed-accuracy trade-offs in video classification. In: Ferrari, V., Hebert, M., Sminchisescu, C., Weiss, Y. (eds.) ECCV 2018. LNCS, vol. 11219, pp. 318–335. Springer, Cham (2018). https://doi.org/10.1007/978-3-030-01267-0_19
26. Zhang, P., Xue, J., Lan, C., Zeng, W., Gao, Z., Zheng, N.: Adding attentiveness to the neurons in recurrent neural networks. In: Ferrari, V., Hebert, M., Sminchisescu, C., Weiss, Y. (eds.) ECCV 2018. LNCS, vol. 11213, pp. 136–152. Springer, Cham (2018). https://doi.org/10.1007/978-3-030-01240-3_9

LRATNet: Local-Relationship-Aware Transformer Network for Table Structure Recognition

Guangjie Yang[1], Dajian Zhong[2], Yu-jie Xiong[3], and Hongjian Zhan[1,2]([✉])

[1] Shanghai Key Laboratory of Multidimensional Information Processing, East China Normal University, Shanghai, China
51215904041@stu.ecnu.edu.cn, ecnuhjzhan@foxmail.com
[2] College of Information Engineering, Shanghai Maritime University, Shanghai, China
djzhong@shmtu.edu.cn
[3] Shanghai University of Engineering Science, Shanghai, China
xiong@sues.edu.cn

Abstract. Table structure recognition is a challenging task due to complex background and various styles of tables. Existing methods address this challenge by exploring adjacency relationship prediction, image-to-text generation, logical position prediction, etc. However, these methods either adopt Graph Convolutional Network (GCN) structures, which mainly focus on the local context information, or Multi-Head Attention (MHA) structures, which mainly focus on the global context information. Both of them ignore the correlation between local and global features. In this paper, we propose a Local-Relationship-Aware Transformer Network (LRATNet) for table structure recognition. LRATNet constructs a robust correlation between local and global information using the LRAT module. The LRAT model has been adapted into three distinct variants: Row-LRAT, Col-LRAT, and Spa-LRAT. These variants are designed to emphasize specific aspects of information: row information, column information, and spatial information, respectively. This is achieved through the exploration of different adjacency relationships. This improves the performance of logical location prediction. Additionally, we have developed a new loss function called Lstage, which is designed to improve accuracy in predicting logical positions. Experimental results demonstrate that our method outperforms existing approaches on three public datasets.

Keywords: Table structure recognition · GCN · Transformer · Adjacency encoding

1 Introduction

Table structure recognition (TSR) is an important task in the field of computer vision, aiming to automatically understanding and processing the structure and content of tables in images or documents. This facilitates various applications, including information extraction, document automation, database population,

S. Rudinac et al. (Eds.): MMM 2024, LNCS 14555, pp. 507–520, 2024.
https://doi.org/10.1007/978-3-031-53308-2_37

question answering and dialogue systems. Although many relevant works have
been proposed and they have achieved impressive progress in addressing the TSR
problem. It remains a challenging task due to its complex table styles, such as
merged cells, spanning rows and columns, dense cell layouts, various table layouts, table images distortion and etc. Some methods [1, 2, 11–13, 16, 17] involve
predicting relationships between table cells, specifically determining whether
adjacent cells are in the same row or column. However, these approaches do
not provide direct solutions for determining the logical positions of individual
cells within tables. As a result, they often necessitate the use of complex post-
processing or graph optimization algorithms to obtain the desired structural
information. Some methods, as discussed in [3, 7, 10, 25] involve modeling the
TSR problem as a sequence prediction task. In this task, the table structure is
predicted using corresponding LaTeX or HTML tags, which are then evaluated
using Tree-Edit-Distance-based Similarity (TEDS) [25]. Other methods [20–22]
adopt a more direct approach by predicting the logical positions of individual
table cells. They aim to predict the starting row, starting column, ending row,
and ending column of each cell, allowing for table reconstruction without the
need for intricate post-processing, as depicted in Fig. 1. This approach offers a
more intuitive way for machines to understand table structure. For instance,
TGRNet [22] utilizes Graph Convolutional Networks (GCN) [9] and ordered
node classification to predict the logical positions of rows and columns separately.
GCNs are primarily designed for aggregating local context information. Although
TGRNet incorporates global edge distance information as graph weights, the performance of GCNs in aggregating global context information remains limited.
In contrast, LORE [20] employs Multi-Head Attention (MHA) blocks [19] to

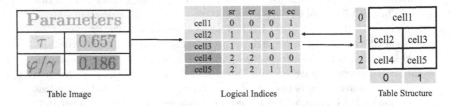

| | Table Image | | Logical Indices | | | Table Structure |

Fig. 1. Overview of logical location prediction paradigm. Here, sr, er, sc, ec refer to
the starting row, ending row, starting column and ending column respectively.

directly aggregate context information for regression-based predictions. Attention is renowned for its ability to capture global context information and have
achieved impressive results in LORE. However, it is worth noting that LORE
places relatively less emphasis on the role of local context information.

In response to the existing problems for logical position prediction, we propose a novel approach called LRATNet, which is based on GCN and MHA architectures. LRATNet aims to solve the TSR challenge by aggregating both local
and global context information. In the traditional MHA structure, the query (q),
key (k), and value (v) are derived from a single feature and individually processed
through three linear layers before entering MHA for information aggregation.

However, MHA excels at extracting global context information but falls short in addressing local context information. To address this problem, we propose a novel module named LRAT, which sends the key (k) and value (v) through a GCN for local context information aggregation. As a result, when the query (q) interact with the key (k) during searches, they now consider not only the global context information but also pay attention to the local context information. Inspired by [23], we have also introduced a novel strategy for the adjacency encoding and incorporated a convolutional layer within the LRAT structure to further emphasize the significance of local context information. According to the different ways of aggregating information, we also proposed three other modules, Row-LRAT, Col-LRAT, Spa-LRAT. They are employed to aggregate the context information of rows, columns and spaces respectively. Furthermore, an LRAT module is employed to aggregate these three types of context information. The above four modules together constitute a LRAT blcok. In addition, we have also designed a new loss function called the Lstage loss, which is primarily used to emphasize the accuracy of predicting all the correct logical indices for each cell, rather than solely focusing on the accuracy of a single logical index.

We summarize our contributions as follows:

- We proposed a TSR model for predicting the logical position of the table, named LRATNet. Multiple modules, such as Row-LRAT, Col-LRAT, Spa-LRAT, and LRAT, are employed to aggregate context information from various aspects.
- We proposed a LRAT module based on GCN and Transformer to pay attention to both global and local context information.
- We designed a loss function named Lstage to focus on predicting the accuracy of four logical indices for each cell, rather than solely focusing on the single logic index.

2   Related Work

Early works to TSR heavily relied on predefined rules and hand-crafted features [5,8,18]. However, rule-based methods often struggle to provide robust results for diverse table structures. In recent years, deep learning-based approaches have emerged to tackle TSR, demonstrating promising outcomes. These methods can be categorized into three main groups: adjacency relation prediction methods, image-to-text generation methods, and logical location prediction methods.

2.1   Adjacency Relation Prediction

The prediction of adjacency relationships aims to determine the positional relationship between two candidate cells in a table, specifically whether two cells are in the same row, the same column, or if they represent the same cell. Methods [1, 2, 11, 16, 21] classify relationships between cells by treating cells as nodes and the connections between them as edges, and then a graph structure can be constructed.

FLAG-NET [13], NCGM [12], and TabStruct-Net [17] employ DGCNN [15] to model the positional relationships between adjacent cells. These methods utilize precise network structures or multimodal information to enhance relationship prediction, achieving impressive results. However, they primarily focus on predicting relationships between local cells. To capture the overall structural information of the entire table, complex post-processing or graph optimization algorithms are often required. Additionally, in specific scenarios, predicting only adjacent relationships may not suffice to infer the complete structure of the entire table.

2.2   Image-to-Text Generation

Image-to-text generation methods treat table structure prediction as a sequence prediction task. In this framework, the sequence represents the structural information of the table, often encoded in formats such as HTML or LaTeX. Sequence decoders are employed to generate textual label tags that describe the table's structure. Recent research within this paradigm predominantly adopts an end-to-end methods. Deng et al. [3] utilize an attention mechanism with an LSTM [6] structure to generate LaTeX labels that define the table's structure. EDD [25] and TableBank [10] employ an encoder-decoder architecture to produce corresponding sequence labels in an end-to-end manner. In addition, VAST [7] has introduced a visual-alignment loss, which maximizes the utilization of local information from table cells. However, it's important to note that these methods often encounter challenges related to learning noisy markup sequences. This can lead to difficulties in achieving training convergence and result in highly time-consuming sequential decoding processes.

2.3   Logical Location Prediction

Logical location prediction involves determining the starting and ending rows and columns of each cell in a table, providing vital structural information. TGR-Net [22] was the pioneer in this field to our knowledge, utilizing a GCN [9] structure to integrate visual and positional cell data and employing ordered node classification for separate row and column predictions. While effective for simpler tables, its performance falls short for more complex structures. Conversely, LORE [20] employs cascaded MHA [19] blocks for direct cell position regression. However, its exclusive reliance on the MHA structure for global context aggregation overlooks the importance of local context information.

3   Methodology

3.1   Overall Architecture

The overall architecture of our proposed LRATNet model is illustrated in Fig. 2. It primarily consists three main components: a CNN-based visual feature extractor, three LRAT blocks, and a Regression module. Firstly, we use a feature

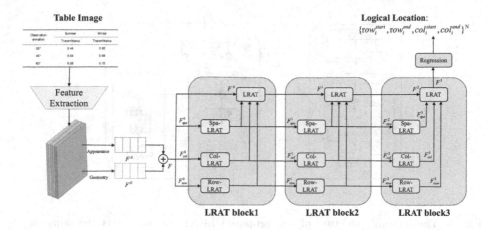

Fig. 2. The architecture of our proposed LRATNet.

extractor to acquire the visual appearance features F^A and geometric features F^G of the table cells. Then, these two sets of features are fused and fed into multiple stacked LRAT blocks. Finally, the aggregated context information is fed into the Regression Module, which predicts the logical positions of each table cell, including the starting row, ending row, starting column, and ending column.

3.2 Feature Extraction

Inspired by [14, 26], for a given input table image, we employ a key point segmentation network to extract features from the table image. Based on the ground truth for each cell, we model the visual features of each cell as the sum of its four corner points, denoted as F^{cr}, and the center point, denoted as F^{ct}. $F^{cr} = \{f^1_{cr}, f^2_{cr}, ..., f^N_{cr}\} \subset R^{N \times d}$, where N and d represent the number of each table cells and the embedding dimension, respectively. $F^{ct} = \{f^1_{ct}, f^2_{ct}, ..., f^N_{ct}\} \subset R^{N \times d}$. Here, f^i_{cr} corresponds to the features on the feature map of the bounding box corners, f^i_{ct} corresponds to the features on the feature map of the center point, and i denotes the i-th cell. Therefore, the visual embedding for the table image is denoted as:

$$F^A = F_{ct} + F_{cr}. \tag{1}$$

The geometric embedding, denoted as $F^G = \{f^1_b, f^2_b, ..., f^N_b\}$, is formed by concatenating the corner points and the center point of each cell. These embeddings are obtained using a Fully-Connected layer for embedding. Subsequently, we obtain the final feature representation $F = F^A + F^G = \{f^1, f^2, ...f^N\} \subset R^{N \times d}$.

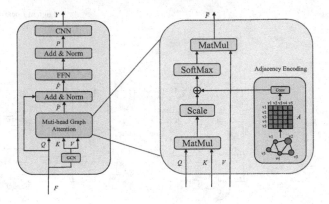

Fig. 3. The internal structure of our proposed LRAT module. A is the adjacency matrix.

3.3   LRAT Module

In this section, we will provide a detailed overview of the LRAT module. As shown in Fig. 3. The LRAT module mainly consists of convolutional layers, linear layers, a GCN layer, and multi-head graph attention components. For the input F, and Q, K and V are first calculated when the feature F is fed to the LRAT module. Next, we use F_Q, F_K and F_V to respectively represent the features calculated for Q, K and V. For F_Q, $F_Q = F$. Regarding K and V, to improve aggregation of local context information, and to facilitate enhanced focus on both local and global context information when querying K with Q later on, we employ GCN [9] to aggregate local context information. Which are denoted as:

$$F_V = F_K = GCN(F, A). \tag{2}$$

Secondly, we utilize Multi-Head Graph Attention (MGHA) to aggregate global context information. Following the MGHA operation, we obtain $\tilde{F} = MGHA(F_Q, F_K, F_V)$. Similar to the approach presented in [19], we apply residual connections and layer normalization to the LRAT module. After obtaining P, we introduce an additional CNN layer to further aggregate adjacent information at the end.

$$\hat{F} = Add\&Norm(\tilde{F}, F_Q), \tag{3}$$

$$P = Add\&Norm(FFN(\hat{F}), \hat{F}), \tag{4}$$

$$Y = CNN(P). \tag{5}$$

For the MGHA operation, drawing inspiration from prior work [23], we start by passing Q and K through a MatMul layer followed by a Scale layer. Additionally, we introduce a 1×1 convolution operation on the adjacency matrix to encode it into the same dimension as K and V. Subsequently, we sum these components together and apply a SoftMax function. This process further enhances the emphasis on local context information. Finally, $\tilde{F}$ can be obtained through a MatMul layer.

3.4   LRAT Block

In Sect. 3.3, we provided a detailed overview of the LRAT module. In this section, we introduce our proposed LRAT block. The LRAT block consists of four modules: Row-LRAT, Col-LRAT, Spa-LRAT, and LRAT. The structures of these four modules are the same as the LRAT module introduced in Sect. 3.3. However, we distinguish them here primarily based on how they aggregate context information. For Row-LRAT and Col-LRAT, their primary focus is on aggregating context information along rows and columns, respectively. To achieve this, when constructing the adjacency matrix for GCN input, we use Row-LRAT to exclusively capture row adjacency relationships in an adjacency matrix, while employing Col-LRAT to exclusively address column adjacency relationships in an adjacency matrix. This approach follows a similar methodology as described in [1]. In our modeling, each cell is considered a node, and we represent its edges as e_{ij} (where i is not equal to j, and $1 <= i, j <= N$). Consequently, when constructing edges, we calculate edges between each node and all other nodes, utilizing the Euclidean distance as the corresponding weight for each edge. The specific weight calculation is as follows:

$$A_{i,j}^{row} = exp\{-(\frac{b_i^y - b_j^y}{H} \times \beta)^2\}, \tag{6}$$

$$A_{i,j}^{col} = exp\{-(\frac{b_i^x - b_j^x}{W} \times \beta)^2\}, \tag{7}$$

Here, $A_{i,j}^{row}$ represents the row relationship, $A_{i,j}^{col}$ represents the column relationship, b_i^x and b_i^y correspond to the x and y coordinates of each cell's center point, and H and W denote the width and height of the image, respectively. β is a hyperparameter. In this process, we focus solely on one-dimensional location relationships within rows or columns, rather than spatial relationships. In contrast to the previous two modules, Spa-LRAT primarily aggregates information based on adjacency relationships along the row and column dimensions, which mainly focuses on spatial location relationships. The structure of Spa-LRAT is identical to that of LRAT, and the final information aggregation is performed by LRAT. The overall feature representation is defined as $F^c = F_{row}^c + F_{col}^c + F_{spa}^c$. Here, F_{row}^c, F_{col}^c, and F_{spa}^c represent the features after aggregation by Row-LRAT, Col-LRAT, and Spa-LRAT at a specific layer denoted by c.

3.5   Regression Module

After aggregating local and global context information through three LRAT blocks, we obtain the final context information. Subsequently, we employ a Multi-Layer Perceptron (MLP) for the regression task, predicting the starting row, ending row, starting column, and ending column for each cell. This process produces the ultimate results, denoted as $L = \{l^1, l^2, ..., l^N\}$, where $l^i = \{row_i^{start}, row_i^{end}, col_i^{start}, col_i^{end}\}$.

3.6 Loss Function

Our loss function is computed as $L_{all} = aL_{log} + bL_{stage}$, where a and b are hyperparameters. It comprises two main components: L_{log} and L_{stage}. L_{log} quantifies the difference between the predictions directly generated by the regression module and the corresponding ground truth values. Regarding L_{stage}, we only consider the logical position of a cell as correct when all four logical indices are accurately predicted. Even if three out of four indices are correct, it is still considered an incorrect prediction. To address this, we have devised a multi-stage loss function called L_{stage}. Specifically, l_c^i is calculated as the number of correctly predicted logical indices based on l^i for each cell. $\tilde{l}^i$ represents the penalty value we have designed based on different l_c^i values. $\hat{l}^i$ represents the ground truth. The specific calculation formula is as follows:

$$L_{log} = \frac{1}{N} \sum_{i=1}^{N} ||l^i - \hat{l}^i||_1, \tag{8}$$

$$\tilde{l}^i = \begin{cases} 2.5 , & l_c^i = 0 \\ 2 , & l_c^i = 1 \\ 1.5 , & l_c^i = 2 \\ 1 , & l_c^i = 3 \\ 0 , & l_c^i = 4 \end{cases} \tag{9}$$

$$L_{stage} = \frac{1}{N} \sum_{i=1}^{N} \tilde{l}^i. \tag{10}$$

4 Experiments

In this section, we evaluate the performance of our proposed LRATNet on three datasets and compare it with existing methods for predicting table logical positions. Additionally, we conduct comprehensive ablation experiments to validate the effectiveness of LRATNet.

4.1 Datasets

We evaluate the performance of LRATNet on three datasets related to table structure recognition, namely WTW [14], TableGraph-24K [22], and ICDAR2013 [4]. Due to the limited size of the ICDAR2013 dataset, we adopted a fine-tuning approach following initial training on the TableGraph-24K dataset.

WTW [14]: This dataset comprises wired tables found in natural scenes and spans across multiple categories, making it a significant challenge. It includes 10,037 training images and 3,600 testing images.

TableGraph-24K [22]: This dataset is a subset of TableGraph-350K, consisting of 20,000 training images and 2,000 testing images. It contains both wired and wireless textual tables.

ICDAR2013 [4]: This dataset consists of 156 PDF format tables from EU/US government websites. However, since it lacks a dedicated testing set, we have adopted the data processing approach used in TGRNet [22] to ensure a fair comparison. We randomly divided the dataset into two halves, with one half serving as the training set and the other as the testing set.

4.2 Implementation Details

All experiments were conducted using an RTX 3090 GPU with CUDA 11.3 and PyTorch 1.11. The labels for the ICDAR2013 and TableGraph-24K (TG24K) datasets were transformed following the TGRNet repository, whereas the labels for the WTW dataset were transformed based on the LORE repository. For TG24K and ICDAR13, we proportionally scaled the images to 768×768 and trained the proposed network for 200 epochs. The results for ICDAR2013 were obtained by fine-tuning the model pre-trained on TG24K. As for the WTW dataset, we proportionally scaled the images to 1024×1024 and trained the model for 100 epochs with a batch size of 6. We employed the DLA-34 [24] as the backbone with a dimension d set to 256. Additionally, we incorporated 3 layers of LRAT blocks. The initial learning rate was set to 0.0001 and was reduced by 10% at the 70th and 90th epochs. Hyperparameters were configured as follows: $a = 2$, $b = 1$, and $\beta = 3$.

4.3 Results on Benchmarks

We compared our proposed LRATNet with existing methods for table logical location prediction, including ReS2TIM [21], TGRNet [22], and LORE [20]. While ReS2TIM is similar to ours in that it directly uses the ground truth of the table cell as input, both TGRNet and LORE are end-to-end methods, which may not achieve complete accuracy in table cell detection. To ensure a fair comparison, we made slight modifications to their code. Instead of using the partially accurate detected table cells for logical position prediction, we provided the ground truth of the table cell information before the phase of logical position prediction. The experimental results are summarized in Table 1.

Table 1. Comparison results with existing methods for logical position prediction. The symbol '*' denotes methods that have been reproduced.

Method	ICDAR13	TG24K	WTW
ReS2TIM [21]	17.4	-	-
TGRNet* [22]	33.4	91.4	-
LORE* [20]	75.2	96.5	85.0
Ours	79.5	98.0	89.6

Our proposed LRATNet pays attention to both local and global context information. Experimental results demonstrate that our approach outperforms

ReS2TIM, TGRNet, and LORE on three datasets, highlighting the effectiveness of our model in considering both local and global context information. The ReS2TIM primarily relies on a cell relationship network, but its ability to capture both local and global context information is limited, resulting in poor performance. The TGRNet employs a GCN and uses global edge connections as weights, which address both local and global context information to some extent. However, when compared to a Transformer-based architecture [20], the capacity of GCN for aggregating context information is limited. Consequently, TGRNet performs better than ReS2TIM but falls short of LORE. LORE adopts a Transformer-based structure and achieves commendable results. However, it overlooks the significance of local context information.

The relatively high accuracy of the three methods for TG24K can be attributed to the fact that this dataset primarily consists of scientific literature, features relatively simple structures, and benefits from ample training data. However, the results of ICDAR2013, fine-tuned from models pretrained on the TG24K, did not match the high performance of the TG24K. We speculate that this discrepancy may arise from the limited size of the training dataset for the ICDAR2013, hindering the model of ability to effectively capture its structural characteristics. For the WTW dataset, known for its intricate and diverse structures, it presented a significant challenge. However, our model achieved an impressive accuracy rate of 89.6%, highlighting its robust performance.

Considering that our method relies on the ground truth of table cells, it might not always be possible to achieve perfectly accurate table cell detection results in real situations. To validate the effectiveness of our approach, given that the current accuracy of most table cell detection methods exceeds 80%, we randomly reduced the number of cells by 10% and 20% for training our model. The results are shown in Table 2. From the experimental results, it can be observed that our proposed LRATNet maintains strong robustness even on incomplete table structures.

Table 2. Robustness analysis results on the WTW dataset.

Method	100% cells	90% cells	80% cells
LRATNet	89.6	87.0	85.3

4.4   Ablation Study

In this section, we have performed a comprehensive set of ablation experiments on the WTW dataset to evaluate the effectiveness of our proposed LRATNet. The specific experiments are outlined as follows:

Analysis of the Effects of Our Proposed Row-LRAT, Col-LRAT, Spa-LRAT, and LRAT. Our experimental results are presented in Table 3. From

Table 3. Evaluation the significance of Row-LRAT, Col-LRAT, Spa-LRAT, and LRAT modules. Acc represents the accuracy of predicting the logical position of each cell.

Exp	Row-LRAT	Col-LRAT	Spa-LRAT	LRAT	Acc
1			✓		88.4
2	✓	✓			87.7
3	✓	✓	✓		88.5
4	✓	✓		✓	89.1
5	✓	✓	✓	✓	89.6

the Exp. 1–2 and Exp. 4, it can be observed that our proposed LRAT and Spa-LRAT have played a crucial role in improving the performance. This can be attributed to their ability to focus on both local and global context information. From Exp. 1 and Exp. 2, it is evident that Row-LRAT and Col-LRAT do not perform as well as the single Spa-LRAT. This is because they separately aggregate context information from rows and columns, lacking spatial context information. However, from Exp. 2 and Exp. 4, we observe that their performance improves significantly after aggregation through LRAT. Therefore, it is evident that each module we proposed plays a crucial role in enhancing the performance.

Analysis of the Internal Component Functions in LRAT. The experimental results are presented in Table 4. In Exp. 1, only a 3-layer MHA block was used, while in Exp. 2–5, GCN, Adjacency Encoding, CNN, and Lstage loss were progressively incorporated. The experimental results indicate that the introduction of GCN and Adjacency Encoding leads to noticeable performance improvements, further emphasizing the importance of considering both global and local context information. The influence of the Lstage loss on performance is moderate. Our analysis indicates that the primary contributions to improvements come from the L1 loss, while the Lstage loss plays a supplementary role in enhancing the accuracy of the model in predicting logical positions.

Table 4. Effectiveness of the LRAT in the proposed method.

Exp	GCN	Adjacency Encoding	CNN	Lstage	Acc
1					84.1
2	✓				88.0
3	✓	✓			89.0
4	✓	✓	✓		89.3
5	✓	✓	✓	✓	89.6

Analysis of Weight Parameter Sharing for CNN and GCN Within Each LRAT Module. The experimental results are detailed in Table 5. The experimental findings demonstrate an accuracy improvement of 0.2% points whether using CNN or GCN. Furthermore, this approach results in a reduction in the total number of model parameters. Our analysis indicates that this improvement can be attributed to the roles of CNN and GCN, both of which primarily focus on aggregating local information. Given that different layers of the Multi-Head Graph Attention (MGHA) module already perform information aggregation at various depths, there is no need to further increase the depth of CNN and GCN layers. A single layer suffices for effective local information aggregation. Consequently, sharing these parameters within each LRAT module proves to be a more effective strategy.

Table 5. Results of weight parameter sharing for CNN or GCN within each type of LRAT module.

Exp	GCN	GCN (shared)	CNN	CNN (shared)	Acc
1		✓	✓		89.4
2	✓			✓	89.4
3		✓		✓	89.6

Analysis of LRATNet with Different Number of LRAT Blocks. We conducted LRATNet experiments with two-layer, three-layer, and four-layer LRAT blocks. The experimental outcomes are presented in Table 6. The results indicate that the three-layer LRAT block exhibits a significant improvement compared to the two-layer experimental results, while the four-layer LRAT block does not yield a substantial improvement in performance. Taking into consideration both model complexity and performance, we ultimately opted for a three-layer LRAT block structure.

Table 6. The results of LRATNet with varying number of LRAT blocks.

Exp	two-layers	three-layer	four-layer	Acc
1	✓			88.9
2		✓		89.6
3			✓	89.9

5   Conclusion

In this paper, We proposed LRATNet, a TSR model designed for predicting table logical positions. LRATNet incorporates four modules: Row-LRAT, Col-LRAT, Spa-LRAT, and LRAT, each individually aggregating information from rows, columns, spaces, and the overall table. This approach maximizes the utilization of both local and global context information for modeling. LRATNet also utilizes a multi-stage loss function to emphasize the accuracy of the four logical indices for each cell. Experimental results demonstrate that our approach outperforms existing methods on three public datasets.

References

1. Chi, Z., Huang, H., Xu, H.D., Yu, H., Yin, W., Mao, X.L.: Complicated table structure recognition. arXiv preprint arXiv:1908.04729 (2019)
2. Clinchant, S., Déjean, H., Meunier, J.L., Lang, E.M., Kleber, F.: Comparing machine learning approaches for table recognition in historical register books. In: IAPR International Workshop on Document Analysis Systems (DAS), pp. 133–138 (2018)
3. Deng, Y., Rosenberg, D., Mann, G.: Challenges in end-to-end neural scientific table recognition. In: International Conference on Document Analysis and Recognition (ICDAR), pp. 894–901 (2019)
4. Göbel, M., Hassan, T., Oro, E., Orsi, G.: ICDAR 2013 table competition. In: International Conference on Document Analysis and Recognition (ICDAR), pp. 1449–1453 (2013)
5. Hirayama, Y.: A method for table structure analysis using DP matching. In: International Conference on Document Analysis and Recognition (ICDAR), pp. 583–586 (1995)
6. Hochreiter, S., Schmidhuber, J.: Long short-term memory. Neural Comput. 9(8), 1735–1780 (1997)
7. Huang, Y., et al.: Improving table structure recognition with visual-alignment sequential coordinate modeling. In: International Conference on Computer Vision and Pattern Recognition (CVPR), pp. 11134–11143 (2023)
8. Kieninger, T., Dengel, A.: The t-recs table recognition and analysis system. In: Lee, S.-W., Nakano, Y. (eds.) DAS 1998. LNCS, vol. 1655, pp. 255–270. Springer, Heidelberg (1999). https://doi.org/10.1007/3-540-48172-9_21
9. Kipf, T.N., Welling, M.: Semi-supervised classification with graph convolutional networks. arXiv preprint arXiv:1609.02907 (2016)
10. Li, M., Cui, L., Huang, S., Wei, F., Zhou, M., Li, Z.: TableBank: table benchmark for image-based table detection and recognition. In: Proceedings of the Twelfth Language Resources and Evaluation Conference, pp. 1918–1925 (2020)
11. Li, Y., Huang, Z., Yan, J., Zhou, Y., Ye, F., Liu, X.: GFTE: graph-based financial table extraction. In: Del Bimbo, A., Cucchiara, R., Sclaroff, S., Farinella, G.M., Mei, T., Bertini, M., Escalante, H.J., Vezzani, R. (eds.) ICPR 2021, Part II. LNCS, vol. 12662, pp. 644–658. Springer, Cham (2021). https://doi.org/10.1007/978-3-030-68790-8_50
12. Liu, H., Li, X., Liu, B., Jiang, D., Liu, Y., Ren, B.: Neural collaborative graph machines for table structure recognition. In: International Conference on Computer Vision and Pattern Recognition (CVPR), pp. 4533–4542 (2022)
13. Liu, H., et al.: Show, read and reason: table structure recognition with flexible context aggregator. In: ACM International Conference on Multimedia (ACM MM), pp. 1084–1092 (2021)
14. Long, R., et al.: Parsing table structures in the wild. In: International Conference on Computer Vision (ICCV), pp. 944–952 (2021)

15. Qasim, S.R., Kieseler, J., Iiyama, Y., Pierini, M.: Learning representations of irregular particle-detector geometry with distance-weighted graph networks. Eur. Phys. J. C **79**(7), 1–11 (2019)
16. Qasim, S.R., Mahmood, H., Shafait, F.: Rethinking table recognition using graph neural networks. In: International Conference on Document Analysis and Recognition (ICDAR), pp. 142–147 (2019)
17. Raja, S., Mondal, A., Jawahar, C.V.: Table structure recognition using top-down and bottom-up cues. In: Vedaldi, A., Bischof, H., Brox, T., Frahm, J.-M. (eds.) ECCV 2020. LNCS, vol. 12373, pp. 70–86. Springer, Cham (2020). https://doi.org/10.1007/978-3-030-58604-1_5
18. Tupaj, S., Shi, Z., Chang, C.H., Alam, H.: Extracting tabular information from text files. EECS Department, Tufts University, Medford, USA (1996)
19. Vaswani, A., et al.: Attention is all you need. In: Advances in Neural Information Processing Systems, vol. 30 (2017)
20. Xing, H., et al.: LORE: logical location regression network for table structure recognition. In: Association for the Advancement of Artificial Intelligence Conference (AAAI), pp. 2992–3000 (2023)
21. Xue, W., Li, Q., Tao, D.: ReS2TIM: reconstruct syntactic structures from table images. In: International Conference on Document Analysis and Recognition (ICDAR), pp. 749–755 (2019)
22. Xue, W., Yu, B., Wang, W., Tao, D., Li, Q.: TGRNet: a table graph reconstruction network for table structure recognition. In: International Conference on Computer Vision (ICCV), pp. 1295–1304 (2021)
23. Ying, C., et al.: Do transformers really perform bad for graph representation? arXiv preprint arXiv:2106.05234 (2019)
24. Yu, F., Wang, D., Shelhamer, E., Darrell, T.: Deep layer aggregation. In: International Conference on Computer Vision and Pattern Recognition (CVPR), pp. 2403–2412 (2018)
25. Zhong, X., ShafieiBavani, E., Jimeno Yepes, A.: Image-based table recognition: data, model, and evaluation. In: Vedaldi, A., Bischof, H., Brox, T., Frahm, J.-M. (eds.) ECCV 2020. LNCS, vol. 12366, pp. 564–580. Springer, Cham (2020). https://doi.org/10.1007/978-3-030-58589-1_34
26. Zhou, X., Wang, D., Krähenbühl, P.: Objects as points. arXiv preprint arXiv:1904.07850 (2019)

Author Index

Printed in the United States
by Baker & Taylor Publisher Services